Medicinal Mushrooms

Dinesh Chandra Agrawal
Muralikrishnan Dhanasekaran
Editors

Medicinal Mushrooms

Recent Progress in Research and Development

Springer

Editors
Dinesh Chandra Agrawal
Department of Applied Chemistry
Chaoyang University of Technology
Taichung, Taiwan

Muralikrishnan Dhanasekaran
Department of Drug Discovery
and Development, Harrison School
of Pharmacy
Auburn University
Auburn, AL, USA

ISBN 978-981-13-6381-8 ISBN 978-981-13-6382-5 (eBook)
https://doi.org/10.1007/978-981-13-6382-5

This Springer imprint is published by the registered company Springer Nature Singapore Pte Ltd.
The registered company address is: 152 Beach Road, #21-01/04 Gateway East, Singapore 189721, Singapore

Editors dedicate this book to their beloved families:
Manju, Somya, Neha, and Mihir (Family Agrawal)
and
Madhu, Rishi (Family Dhanasekaran)

Preface

The present book is the continuation of Professor Agrawal's previous two Springer books: *Medicinal Plants – Recent Advances in Research and Development* (Springer link: http://www.springer.com/in/book/9789811010842) and *Medicinal Plants and Fungi – Recent Advances in Research and Development* (Springer link: https://www.springer.com/gp/book/9789811059773).

In this volume, chapters (mostly review articles) on medicinal mushrooms have been included considering their importance in the human health. Medicinal mushrooms are now gaining worldwide attention because of its pharmacologically bioactive compounds which have demonstrated potent and unique clinical properties. Scientific studies carried out during the last decade have validated evidence of their efficacy in a wide range of diseases. Extracts and bioactive compounds obtained from different mushrooms have been used medicinally as anticancer, immunomodulator, antibacterial, antiviral, anti-inflammatory, anti-atherosclerotic, neuroprotectant, cardioprotectant, antioxidant, and anti-hypoglycemic agents and in stem cell-based therapies. There are ongoing research efforts on various aspects of medicinal mushrooms in different parts of the world. The editors wish to bring their recent research and development works into light in the form of this book.

The book contains chapters, mostly review articles, contributed by eminent researchers working with different disciplines of medicinal mushrooms in different countries across the globe. This book not only extends our knowledge about medicinal mushrooms and confirms the great potential of mushrooms for the development of new drugs but hopefully also inspires the readers to get involved in medicinal mushroom research.

The editors hope that this compendium of review articles will be very useful as a reference book for advanced students, researchers, academics, business houses, and all individuals concerned with medicinal mushrooms.

Taichung, Taiwan — Dinesh Chandra Agrawal
Auburn, AL, USA — Muralikrishnan Dhanasekaran
24 August 2018

Acknowledgments

The editors thank all the invited authors to this book for preparing their valuable manuscripts. Without their contributions, this book would not have been possible.

The coeditor, Dr. Dhanasekaran, wishes to place on record special appreciation and thanks to Professor Agrawal for initiating the book proposal; handling the entire correspondence with Springer and authors; dealing with editing, reviewing, and revision process of manuscripts; and managing them from start to finish. Without his untiring efforts, this book would not have become a reality.

Editor Professor Agrawal thanks Professor Tao-Ming Cheng, President of the Chaoyang University of Technology (CYUT); Professor Wen-Goang Yang, Vice-President, CYUT; Professor Sung-Chi Hsu, Dean, R&D Office, and Assistant Vice-President, CYUT; Professor Chia-Chi Cheng, Dean, College of Science and Engineering, CYUT, and Professor Hsin-Sheng Tsay, Emeritus Chair Professor, CYUT, Taichung, Taiwan, for their constant support and encouragement during the progress of the book. Editor Professor Muralikrishnan thanks Professor Timothy Moore, Head of the Department, Harrison School of Pharmacy; Professor Randall Clark, Harrison School of Pharmacy; and all his beloved students (Ms. Fujihashi, Mr. Majrashi, Mr. Almaghrabi, Dr. Sindhu, Dr. Manoj) for their relentless care and inspiration.

The editors sincerely thank the entire Springer Nature Singapore Pte. Ltd., the team concerned with the publication of this book, and place on record a special appreciation to Ms. Aakanksha Tyagi, Editor, Life Sciences, Springer, for her encouragement and prompt support during the publication of this book.

Editors express profound gratitude toward "God, the Infinite Being" for providing strength to accomplish the arduous task of handling this book.

Contents

About the Editors and Contributors

Editors

Dinesh Chandra Agrawal graduated in 1976 from Aligarh Muslim University (national university) and obtained his Ph.D. in 1982. Professor Agrawal has more than 37 years of research experience in plant biotechnology of diverse species including medicinal plants and medicinal mushrooms. After serving for more than 31 years, in 2013, he superannuated as a chief scientist and professor of biological sciences at the CSIR-National Chemical Laboratory (NCL), Pune, the top-ranking institute in chemical sciences under the umbrella of the Council of Scientific and Industrial Research (CSIR), Ministry of Science and Technology, Government of India. Currently, he is working as a professor in the Department of Applied Chemistry, Chaoyang University of Technology (CYUT), Taiwan. While in CSIR-NCL, Prof. Agrawal worked as a coordinator and project leader of several research projects funded by the Government of India. He has more than 175 publications including 4 books to his credit on the different aspects of plant biotechnology including medicinal plants and medicinal mushrooms. More than 35 M.Tech./M.Sc. and 7 Ph.D. students have completed their thesis work under his guidance. Professor Agrawal has been bestowed several prestigious awards and fellowships such as the Alexander von Humboldt Fellowship (Germany), DBT Overseas Associateship (USA), British Council Scholar (UK), European Research Fellow (UK), and INSA Visiting Scientist. During these fellowships, he had opportunities to work in the USA, Germany, and the UK. Also, he had a research collaboration with UMR Vigne et Vins, INRA,

Centre de Recherché Colmar, France. For more than 10 years, he has been a member of the executive committee of the Humboldt Academy, Pune Chapter, and held the position of treasurer. Professor Agrawal has reviewed a large number of research papers for several SCI journals on plant biotechnology and served as a member of the editorial board of *Medicinal and Aromatic Plants Abstracts*, NISCAIR, Government of India. Presently, he is on the editorial board of the *International Journal of Applied Sciences and Engineering* (Scopus), serving as associate editor in chief of the journal.

Muralikrishnan Dhanasekaran completed his Bachelor of Pharmacy from Annamalai University and Master of Pharmacy from Jadavpur University, West Bengal, India. He obtained his Ph.D. from the Indian Institute of Chemical Biology, Kolkata, India. Following which, he attained his postdoctoral training from renowned scientists Dr. Manuchair Ebadi (Professor at the University of North Dakota, Grand Forks, ND) and Dr. Bala Manyam (Scott & White Clinic/Texas A&M, Temple, TX). Dr. Dhanasekaran joined Auburn University in the year 2005 and is currently working as a full professor at Harrison School of Pharmacy, Auburn University, USA. Dr. Dhanasekaran's area of research and interest focuses on neuropharmacology, toxicology, dietary, and natural products. Dr. Dhanasekaran successfully completed the New Investigator Research Grant from Alzheimer's Association, several Auburn University grants, and several research projects from a pharmaceutical company. He has graduated 6 students (as a mentor) and currently has 3 graduate and 30 undergraduate students in his lab. Dr. Dhanasekaran has received several teaching awards from Auburn University, for teaching Pharm.D. and graduate students. He has published more than 100 scientific abstracts, 70 peer-reviewed publications, and several book chapters. With regard to professional service, he is the current chair of the "Faculty Grievance Committee" and "Teaching Effective Committee." Dr. Dhanasekaran also serves as a reviewer and member of the editorial board in several scientific journals.

Contributors

Daniel A. Abugri Department of Chemistry and Department of Biology, Laboratory of Ethnomedicine, Parasitology and Drug Discovery, Laboratory of General Mycology and Medical Mycology, Tuskegee University, Tuskegee, AL, USA

Dinesh Chandra Agrawal Department of Applied Chemistry, Chaoyang University of Technology, Taichung, Taiwan

Mohammad Feraz Ahsan Inter University Centre for Biomedical Research & Super Speciality Hospital, Kottayam, Kerala, India

Abdul Khader Akbar Department of Botany, C. Abdul Hakeem College, Vellore, Tamil Nadu, India

Mohammed Almaghrabi Department of Pharmaceutical Chemistry, College of Pharmacy, Taibah University, Almadinah Almunawwarah, Kingdom of Saudi Arabia

Joseph A. Ayariga Department of Microbiology, Alabama State University, Montgomery, AL, USA

Susanna M. Badalyan Laboratory of Fungal Biology and Biotechnology, Institute of Pharmacy, Department of Biomedicine, Yerevan State University, Yerevan, Armenia

Anush Barkhudaryan Department of Cardiology, Clinic of General and Invasive Cardiology, University Clinical Hospital No.1, Yerevan State Medical University, Yerevan, Armenia

Manal Buabeid Department of Drug Discovery and Development, Harrison School of Pharmacy, Auburn University, Auburn, AL, USA

College of Pharmacy and Health Science, Ajman University, Ajman, UAE

Goutam Chandra Inter University Centre for Biomedical Research & Super Speciality Hospital, Mahatma Gandhi University at Thalappady Campus, Kottayam, Kerala, India

Hsiao Hsiang Chao Department of Biochemistry and Molecular Biology, New York Medical College, Valhalla, NY, USA

Malvi Choudhary School of Biotechnology, University of Jammu, Jammu, Jammu and Kashmir, India

Yang Chu Department of Biochemistry and Molecular Biology, New York Medical College, Valhalla, NY, USA

Department of Pharmacy, The First Affiliated Hospital of China Medical University, Shenyang, China

Jasmina Ćilerdžić Faculty of Biology, University of Belgrade, Belgrade, Serbia

Randall Clark Department of Drug Discovery and Development, Harrison School of Pharmacy, Auburn University, Auburn, AL, USA

Jack Deruiter Department of Drug Discovery and Development, Harrison School of Pharmacy, Auburn University, Auburn, AL, USA

Muralikrishnan Dhanasekaran Department of Drug Discovery and Development, Harrison School of Pharmacy, Auburn University, Auburn, AL, USA

Manoj K. Dhar School of Biotechnology, University of Jammu, Jammu, Jammu and Kashmir, India

Barbara B. Doonan Department of Biochemistry and Molecular Biology, New York Medical College, Valhalla, NY, USA

Maali Fadan Department of Drug Discovery and Development, Harrison School of Pharmacy, Auburn University, Auburn, AL, USA

Eddie Fahoury Department of Drug Discovery and Development, Harrison School of Pharmacy, Auburn University, Auburn, AL, USA

Ayaka Fujihashi Department of Drug Discovery and Development, Harrison School of Pharmacy, Auburn University, Auburn, AL, USA

Shin-Yee Fung Medicinal Mushroom Research Group (MMRG), Department of Molecular Medicine, University of Malaya, Kuala Lumpur, Malaysia

Centre for Natural Products Research and Drug Discovery (CENAR), University of Malaya, Kuala Lumpur, Malaysia

University of Malaya Centre for Proteomics Research (UMCPR), University of Malaya, Kuala Lumpur, Malaysia

Maria Letizia Gargano Department of Earth and Marine Sciences, University of Palermo, Palermo, Italy

Manoj Govindarajulu Department of Drug Discovery and Development, Harrison School of Pharmacy, Auburn University, Auburn, AL, USA

Suruchi Gupta School of Biotechnology, University of Jammu, Jammu, Jammu and Kashmir, India

Tze-Chen Hsieh Department of Biochemistry and Molecular Biology, New York Medical College, Valhalla, NY, USA

Chih-Yang Huang Medical Research Center for Exosome and Mitochondria Related Diseases, China Medical University and Hospital, Taichung, Taiwan

Department of Biological Science and Technology, Asia University, Taichung, Taiwan

Graduate Institute of Basic Medical Science, School of Chinese Medicine, China Medical University and Hospital, Taichung, Taiwan

Ruckmani Kandasamy Department of Pharmaceutical Technology, University College of Engineering, Bharathidasan Institute of Technology Campus, Anna University, Tiruchirappalli, Tamil Nadu, India

Palaniyandi Karuppaiya Institute of Nutrition, College of Biopharmaceutical and Food Sciences, China Medical University, Taichung, Taiwan

Sanjana Kaul School of Biotechnology, University of Jammu, Jammu, Jammu and Kashmir, India

Wooseok Lee Department of Drug Discovery and Development, Harrison School of Pharmacy, Auburn University, Auburn, AL, USA

Mohammed Majrashi Department of Drug Discovery and Development, Harrison School of Pharmacy, Auburn University, Auburn, AL, USA

Department of Pharmacology, Faculty of Medicine, University of Jeddah, Jeddah, Kingdom of Saudi Arabia

Vidya Manju Inter University Centre for Biomedical Research & Super Speciality Hospital, Kottayam, Kerala, India

K. P. Mohanakumar Inter University Centre for Biomedical Research & Super Speciality Hospital, Mahatma Gandhi University at Thalappady Campus, Kottayam, Kerala, India

Timothy Moore Department of Drug Discovery and Development, Harrison School of Pharmacy, Auburn University, Auburn, AL, USA

Frank Mrema Department of Agriculture and Applied Sciences, Alcorn State University, Lorman, MS, USA

Vanisree Mulabagal Department of Civil Engineering, Auburn University, Auburn, AL, USA

Saraswathy Nachimuthu Department of Biotechnology, Kumaraguru College of Technology, Coimbatore, Tamil Nadu, India

Rishi M. Nadar Department of Drug Discovery and Development, Harrison School of Pharmacy, Auburn University, Auburn, AL, USA

Seema Patel Bioinformatics and Medical Informatics Research Center, San Diego State University, San Diego, CA, USA

Ramalingam Ponnusamy Department of Biotechnology, Kumaraguru College of Technology, Coimbatore, Tamil Nadu, India

Prabin Pradeep Inter University Centre for Biomedical Research & Super Speciality Hospital, Kottayam, Kerala, India

Sindhu Ramesh Department of Drug Discovery and Development, Harrison School of Pharmacy, Auburn University, Auburn, AL, USA

Sylvie Rapior Laboratoire de Botanique, Phytochimie et Mycologie, Faculté de Pharmacie, CEFE CNRS – Université de Montpellier – Université Paul-Valéry Montpellier – EPHE – IRD, Montpellier Cedex 5, France

V. R. Remya Inter University Centre for Biomedical Research & Super Speciality Hospital, Mahatma Gandhi University at Thalappady Campus, Kottayam, Kerala, India

Paola Saporita Department of Agricultural, Food and Forest Sciences, University of Palermo, Palermo, Italy

Tamilselvi Shanmugam Medical Research Center for Exosome and Mitochondria Related Diseases, China Medical University and Hospital, Taichung, Taiwan

Graduate Institute of Basic Medical Science, School of Chinese Medicine, China Medical University and Hospital, Taichung, Taiwan

Marthandam Asokan Shibu Medical Research Center for Exosome and Mitochondria Related Diseases, China Medical University and Hospital, Taichung, Taiwan

Graduate Institute of Basic Medical Science, School of Chinese Medicine, China Medical University and Hospital, Taichung, Taiwan

Mirjana Stajić Faculty of Biology, University of Belgrade, Belgrade, Serbia

Chon-Seng Tan Ligno Research Foundation, Balakong Jaya, Selangor, Malaysia

Sivasudha Thilagar Department of Environmental Biotechnology, School of Environmental Sciences, Bharathidasan University, Tiruchirappalli, Tamil Nadu, India

Boniface J. Tiimob Department of Chemistry and Department of Biology, Laboratory of Ethnomedicine, Parasitology and Drug Discovery, Laboratory of General Mycology and Medical Mycology, Tuskegee University, Tuskegee, AL, USA

Giuseppe Venturella Department of Agricultural, Food and Forest Sciences, University of Palermo, Palermo, Italy

Jelena Vukojević Faculty of Biology, University of Belgrade, Belgrade, Serbia

William H. Witola Department of Pathobiology, College of Veterinary Medicine, University of Illinois, Urbana-Champaign, IL, USA

Joseph M. Wu Department of Biochemistry and Molecular Biology, New York Medical College, Valhalla, NY, USA

Clement G. Yedjou Natural Chemotherapeutics Research Laboratory, RCMI Center for Environmental Health, Jackson State University, Jackson, MS, USA

Kenneth Anchang Yongabi Phytobiotechnology Research Foundation, Bamenda, Cameroon

Ebonyi State University, Abakaliki, Nigeria

Imo State University, Owerri, Nigeria

Recent Progress in Research on the Pharmacological Potential of Mushrooms and Prospects for Their Clinical Application

1

Susanna M. Badalyan, Anush Barkhudaryan, and Sylvie Rapior

Contents

S. M. Badalyan (✉)
Laboratory of Fungal Biology and Biotechnology, Institute of Pharmacy, Department of Biomedicine, Yerevan State University, Yerevan, Armenia
e-mail: s.badalyan@ysu.am

A. Barkhudaryan
Department of Cardiology, Clinic of General and Invasive Cardiology, University Clinical Hospital No 1, Yerevan State Medical University, Yerevan, Armenia

S. Rapior
Laboratoire de Botanique, Phytochimie et Mycologie, Faculté de Pharmacie, CEFE CNRS – Université de Montpellier – Université Paul-Valéry Montpellier – EPHE – IRD, Montpellier Cedex 5, France
e-mail: sylvie.rapior@umontpellier.fr

D. C. Agrawal, M. Dhanasekaran (eds.), *Medicinal Mushrooms*,
https://doi.org/10.1007/978-981-13-6382-5_1

Abstract

Fungi are considered one of the most diverse, ecologically significant, and economically important organisms on Earth. The edible and medicinal mushrooms have long been known by humans and were used by ancient civilizations not only as valuable food but also as medicines. Mushrooms are producers of high- and low-molecular-weight bioactive compounds (alkaloids, lectins, lipids, peptidoglycans, phenolics, polyketides, polysaccharides, proteins, polysaccharide-protein/peptides, ribosomal and non-ribosomal peptides, steroids, terpenoids, etc.) possessing more than 130 different therapeutic effects (analgesic, antibacterial, antifungal, anti-inflammatory, antioxidant, antiplatelet, antiviral, cytotoxic, hepatoprotective, hypocholesterolemic, hypoglycemic, hypotensive, immunomodulatory, immunosuppressive, mitogenic/regenerative, etc.).

The early record of *Materia Medica* shows evidence of using mushrooms for treatment of different diseases. Mushrooms were widely used in the traditional medicine of many countries around the world and became great resources for modern clinical and pharmacological research. However, the medicinal and biotechnological potential of mushrooms has not been fully investigated. This review discusses recent advances in research on the pharmacological potential of mushrooms and perspectives for their clinical application.

Keywords

Bioactive compounds · Clinical application · Ethno-mycopharmacology · Medicinal mushrooms · Pharmacological potential

Abbreviations

ACE	Angiotensin-converting enzyme
AIDS	Acquired immune deficiency syndrome
BDNF	Brain-derived neurotrophic factor
CL	Cultural liquid
COX-1	Cyclooxygenase-1
COX-2	Cyclooxygenase-2
CSF	Colony-stimulating factor
CVD	Cardiovascular diseases
DENV-2	Dengue virus type 2
DS	Dietary supplement
EPS	Exopolysaccharides
FIP	Fungal immunomodulatory protein
GLPS	*G. lucidum* polysaccharide

GSK-3	Glycogen synthase kinase 3
HIV	Human immunodeficiency virus
HK2	Hexokinase 2
HMG-CoA	5-Hydroxy-3-methylglutaryl-coenzyme A
HPV-1	Human papillomavirus 1
HSV-2	Herpes simplex virus 2
IC_{50}	The half maximal inhibitory concentration
IFN	Interferon
IL	Interleukin
iNOS	Inducible NO synthase
LPS	Lipopolysaccharide
MDCK	Madin-Darby canine kidney cells
MIC	Minimal inhibitory concentrations
MMDD	Medicinal mushroom-derived drug
MS	Mycosterol
NGF	Nerve growth factor
NO	Nitric oxide
NSAID	Nonsteroidal anti-inflammatory drug
OLTT	Oxygenated lanostane-type triterpenoid
PAMP	Pathogen-associated molecular pattern
PPAR	Peroxisome proliferator-activated receptor
PRR	Pattern recognition receptor
PI3K/Akt	Phosphatidylinositol-3-kinase and protein kinase B
PSK	Polysaccharide K
PSP	Polysaccharide-protein
PSPC	Polysaccharide-protein complex
PTP1B	Protein tyrosine phosphatase 1B
QoL	Quality of life
STAT3	Signal transducer and activator of transcription 3
TCM	Traditional Chinese medicine
TNBC	Triple-negative breast cancer
TNF	Tumor necrosis factor
TNFα	Tumor necrosis factor alfa
VDM	Vitamin D-enriched mushroom

1.1 Introduction

Mushrooms have widely been appreciated all over the world for their nutritional and medicinal properties (Moore and Chiu 2001; Chang and Miles 2008; Barros et al. 2008; Wasser 2010, 2011; Bandara et al. 2015, 2017; Valverde et al. 2015; de Mattos-Shipley et al. 2016; Chang and Wasser 2017; Gupta et al. 2018). The early civilizations of Greek, Egyptian, Roman, Japanese, and Mexican people prized mushrooms for their therapeutic value (Hobbs 1995; Guzmán 2015). Mushrooms have been used in traditional Chinese medicine (TCM) for more than 3000 years for prevention and treatment of different diseases. There are many sources of ethno-mycopharmacological information on wild mushrooms in Eastern and Western countries, but their medicinal and biotechnological potential has not been fully exploited yet (Boa 2004; Chang and Miles 2008; Lindequist 2011; Grienke et al. 2014; Money 2016; Karun and Sridhar 2017).

The ethno-mycological and ethnographic evidence has documented that mushrooms, particularly bracket species (e.g., *Coriolopsis gallica*, *Daedalea quercina*, *Daldinia concentrica*, *Fomes fomentarius*, *Ganoderma adspersum*, *Lenzites warnieri*, *Piptoporus betulinus*, and *Skeletocutis nivea*), could be used by humans not only as food and medicine but also as tinder for fire-lighting (Roussel et al. 2002; Boa 2004; Uzunov and Stoyneva-Gärtner 2015; Berihuete-Azorín et al. 2018).

Mushrooms or macrofungi are widely distributed worldwide. They are taxonomically placed in two phyla, the *Basidiomycota* (class *Agaricomycetes*) and *Ascomycota* (class *Pezizomycetes*) in the subkingdom Dikarya. From overall numbers of fungal species estimated between 0.5-(1.5)-(5.1) million, about 140,000–160,000 species are macrofungi from which around 10% (14,000–16,000) have been identified (Hawksworth 1991, 2012; Kirk et al. 2008; Blackwell 2011; Hibbet and Taylor 2013; Tedersoo et al. 2014; Dai et al. 2015; Peay et al. 2016). From about 7000 known mushroom species, possessing various degrees of edibility, more than 3000 species from 231 genera are considered prime edible mushrooms. About 3% of the known species (at least 170, currently around 500) are poisonous, while around 700 species from 2000 known safe mushroom species have medicinal properties (Boa 2004; Barceloux 2008; Chang and Wasser 2017).

Edible/medicinal wild and cultivated mushroom species produce a broad spectrum of high- and low-molecular-weight bioactive compounds (alkaloids, lectins, lipids, peptidoglycans, phenolics, polyketides, polysaccharides, polysaccharide-proteins/peptides, proteins, ribosomal and non-ribosomal peptides, steroids, terpenoids, etc.) which possess more than 130 therapeutic effects (analgesic, antibacterial, antifungal, anti-inflammatory, antioxidant, antiviral, cytotoxic, hepatoprotective, hypocholesterolemic, hypoglycemic, hypotensive, immunomodulatory, immunosuppressive, mitogenic/regenerative, etc.) (Mizuno et al. 1995; Wasser and Weis 1999; Hawksworth 2001; Wasser 2002, 2010, 2011; Zjawiony 2004; Lindequist et al. 2005; Poucheret et al. 2006; Cerigini et al. 2007; Park et al. 2007; Stanikunaite et al. 2007; Lee et al. 2008; Saltarelli et al. 2009, 2015; Baggio et al. 2010; Ferreira et al. 2010, Xu et al. 2011; Badalyan 2012; Chang and Wasser 2012; De Silva et al. 2012a, b, 2013; Patel et al. 2012; Wang et al. 2012; Grienke et al. 2014; Park 2014; Schueffler and Anke 2014; Bandara et al. 2015, 2017; Duru and Çayan 2015; Stadler and Hoffmeister 2015; Fu et al. 2016; Kosanić et al. 2016; Venkatachalapathi and Paulsamy 2016; Chaiyasut and Sivamaruthi 2017; Chen et al. 2017; Kolundčzić et al. 2017; Kües and Badalyan 2017; Phan et al. 2017a; Sánchez 2017a, b; Shen et al. 2017; Surup et al. 2018).

The medicinal properties and therapeutic potential of mushrooms were revealed in species from different ecological groups, such as xylotrophs (e.g., *F. fomentarius*, *Fomitopsis pinicola*, *Ganoderma lucidum*, *Grifola frondosa*, *Lentinula edodes*, *Pleurotus ostreatus*, *Trametes versicolor*, *Schizophyllum commune*) and mycorrhiza-forming mushrooms (e.g., *Boletus edulis*, *Cantharellus cibarius*, *Tuber borchii*) (Chang 1996; Wasser 2010) (Figs. 1.1 and 1.2). In Asian countries, wild and cultivated mushrooms (fresh or dried) and mushroom-based bio-products are used to prevent and treat different diseases (Wasser 2010; Bandara et al. 2015), while the usage of medicinal mushrooms in Western societies is limited to

Fig. 1.1 Wild-growing fruiting bodies of edible medicinal mushrooms tested in clinical trials: (**a**) *Agaricus subrufescens* (Photo Courtesy of Guinberteau J), (**b**) *Flammulina velutipes* (Photo Courtesy of Moingeon JM), (**c**) *Grifola frondosa* (Photo Courtesy of Verstraeten P), (**d**) *Hericium erinaceus* (Photo Courtesy of Moingeon JM), (**e**) *Lentinula edodes*. (Photo Courtesy of Fourré G), (**f**) *Pleurotus ostreatus*. (Photo Courtesy of Maurice JP)

Fig. 1.2 Wild-growing fruiting bodies of inedible medicinal mushrooms tested in clinical trials: (**a**) *Ganoderma lucidum* (Photo Courtesy of Moingeon JM), (**b**) *Ophiocordyceps sinensis* (Photo Courtesy of Rioult JP), (**c**) *S. commune* (Photo Courtesy of Angelini C), (**d**) *T. versicolor*. (Photo Courtesy of Perrone L)

"nutraceuticals" (healthy dietary food) and "nutriceuticals" (functional food or DSs) (Wasser 2002, 2011; Cheung 2008; De Silva et al. 2012a, b; Chen et al. 2017). The cultivation and commercialization of natural mushroom resources will create an opportunity to study their nutraceutical and pharmacological potential for developing health-enhancing mycofood and myco-pharmaceuticals (Badalyan and Zambonelli 2019).

Mushrooms are also widely used in the production of natural cosmetic products ("cosmeceuticals" and "nutricosmetics") (Wisitrassameewong et al. 2012; Bandara et al. 2015; Taofiq et al. 2016a, c, 2017a, b; Wu et al. 2016; Badalyan and Zambonelli 2019).

The pharmacological potential of mushrooms has not been fully investigated yet. Further clinical trials are needed to substantiate the pharmacological properties or side effects of mushroom consumption before they could be recommended as health-enhancing dietary food or myco-pharmaceuticals (Money 2016; Taofiq et al. 2017a; Wasser 2017; Mustonen et al. 2018; Zmitrovich et al. 2019).

We refer our review to recent advances in the study of medicinal properties and pharmacological potential of mushrooms and prospects for their clinical application.

1.2 Biological and Genetic Resources of Medicinal Mushrooms and Their Application

The study and conservation of genetic resources and biodiversity of edible and medicinal agaricoid, polyporoid, and other groups of mushrooms, as valuable biological resources possessing high exploratory potential, including the production of nutraceuticals, nutriceuticals, pharmaceuticals, and cosmeceuticals, are carried out in different countries (Lindequist 2011; Badalyan et al. 2012, 2015; Al-Fatimi et al. 2013; Allen and Lendemer 2015; Bhattacharjee et al. 2015; Degreef et al. 2016; Badalyan and Gharibyan 2017a; Gargano et al. 2017; Badalyan and Zambonelli 2019; Bhatt et al. 2018; Gargano 2018).

Prehistoric artifacts dating back to over 5000 years ago describe the tradition of using agaricomycetous polypore fungi (order *Polyporales*) for various applications, including food and medicinal usage. Several polypores (*F. fomentarius*, *F. pinicola*, *Laetiporus sulphureus*, *Laricifomes officinalis*, and *P. betulinus*) were used in the traditional medicine of central European countries for the treatment of bladder diseases, cancer, dysmenorrhoea, hemorrhoids, pyretic diseases, and rheumatism (Lindequist et al. 2005; Lindequist 2011). Modern chemical studies have reported the presence of several primary and secondary bioactive metabolites in crude extracts of bracket fungi and a wide spectrum of their medicinal activities (anti-inflammatory, antimicrobial, cytotoxic, and others) (Suay et al. 2000; Roussel et al. 2002; Anke and Antelo 2011; Grienke et al. 2014).

Nowadays, the biological resources of edible and medicinal mushrooms are used as *dietary food* (world mushroom production was 33 million tons in 2015), *dietary supplements* (DSs) (the market for mushroom-based products is rapidly expanding and comprises more than the US $20 billion/year), *biocontrol agents* (bactericides, fungicides, herbicides, insecticides, and nematocides), *cosmeceuticals* (activator of epidermal growth factor, stimulator of collagen activity, etc.), and *mushroom-derived pharmaceuticals* or *myco-pharmaceuticals* (Wasser et al. 2000; Lindequist 2013; Kumar 2015; Gargano et al. 2017; Glamočlija and Soković 2017; Carocho et al. 2018).

1.3 Mushroom-Derived Bioactive Compounds

Fungi, including mushrooms, are considered active producers of different primary and secondary bioactive metabolites (alkaloids, fatty acids, lectins, nucleic acids, nucleosides, peptides, phenolics, polyketides, polysaccharides, proteins, statins, steroids, terpenoids, etc.) which are responsible for their pharmacological properties (antibacterial, antifungal, anti-inflammatory, antinociceptive, antioxidant, antiproliferative, antitumor, antiviral, hypocholesterolemic, hypoglycemic, hypotensive, immunomodulatory, etc.) (Badalian et al. 1996, 1997a, b, 1999, 2001; Badalian and Serrano 1999; Badalyan 2001, 2015, 2016; Badalyan and Hambardzumyan 2001;

Fang et al. 2006; Zhang et al. 2007; Wasser 2010, 2014; Lindequist 2011, 2013; Chang and Wasser 2012; De Silva et al. 2013; Khatua et al. 2013, 2017; Grienke et al. 2014; Paterson and Lima 2014; Stojković et al. 2014a, b, 2017; Duru and Çayan 2015; Valverde et al. 2015; Friedman 2016; Tao et al. 2016; Hapuarachchi et al. 2017; Khadhri et al. 2017; Masri et al. 2017; Nielsen and Nielsen 2017; Sánchez 2017a, b; Yang et al. 2017). A wide spectrum of biological activities of mushroom-derived biomolecules could be used to develop health-enhancing mycopharmaceuticals for human and animal use, such as the immunomodulatory β-glucan lentinan from *L. edodes* (Novak and Vetvicka 2008); anti-quorum-sensing agents from *Agaricus brasiliensis* (syn. *Agaricus subrufescens*) (Kerrigan 2005), as potential antibiotics (Soković et al. 2014); the antimalarial alkaloid 4-hydroxymethylquinoline from *T. versicolor* (Liu 2005); pain-suppressive enkephalinase inhibitors from *P. betulinus* (Rathee et al. 2012); and nephroprotective polysaccharides, phenolics, and flavonoids from *Pleurotus tuber-regium* (Okolo et al. 2018). The medicinal species *A. subrufescens*, *Ganoderma* spp., *G. frondosa*, *Hericium erinaceus*, *L. edodes*, *Phellinus linteus*, *P. ostreatus*, and *Polyporus umbellatus* were also reported as producers of different groups of bioactive compounds and have been recommended for a variety of therapeutic applications (Donatini 2011; Thongbai et al. 2015; Hapuarachchi et al. 2017; Thongklang et al. 2017). Among the above-mentioned species, *Ganoderma* mushrooms produce the highest diversity of bioactive compounds with different pharmacological activities (Mizuno et al. 1995; Paterson 2006; Saltarelli et al. 2009, 2015; Welti et al. 2010; Yang et al. 2013; Ma et al. 2014b, 2015; Peng et al. 2015, 2016; Klupp et al. 2015; Wang et al. 2015b, 2017a; Hapuarachchi et al. 2016a, b; Klaus et al. 2016; Meneses et al. 2016; Xu et al. 2016a; Pu et al. 2017; Stojković et al. 2017; Suárez-Arroyo et al. 2017; Subramaniam et al. 2017; Taofiq et al. 2017a; Rubel et al. 2018; Zeng et al. 2018). Alkaloids, fatty acids, nucleosides, polysaccharides, proteins, sterols, triterpenoids, and other compounds were isolated and identified from *Ganoderma* spp. They are responsible for antiaging, antibacterial, anticancer, antidiabetic, antifungal, antihypertensive, anti-inflammatory, antioxidant, antiviral, hepatoprotective, hypoglycemic, immunomodulatory, neuroprotective, wound healing, and other medicinal effects (Paterson 2006; Dai et al. 2009; De Silva et al. 2012a, b, 2013; Cheng et al. 2013; Baby et al. 2015; Bishop et al. 2015; Liu et al. 2015a; Hennicke et al. 2016; Chang et al. 2017; Chen et al. 2017; Sánchez 2017a, b; Wang et al. 2017a).

Anticancer, antimicrobial, antioxidant, antiviral, hypolipidemic, immunomodulatory, and estrogen-like activities were observed in edible oyster mushroom *Pleurotus eryngii* due to the production of diterpenoids, as eryngiolide A, hemolysins, polysaccharides, pentacyclic triterpenoids, ubiquinone-9, and other pharmacologically active biomolecules (Shibata et al. 2010; Ma et al. 2014a; Fu et al. 2016; Zhang et al. 2016; Yen et al. 2018). Recent studies of agaricomycetous species *Suillus bellinii* and *P. eryngii* demonstrated that the mycelium of *S. bellinii* possessed a higher content of ergosterol and phenolic compounds with strong antioxidant effects, while the mycelium of *P. eryngii* showed anti-inflammatory and cytotoxic effects (Souilem et al. 2017).

Genome sequencing, comparative genomics, and phylogenetic analysis of medicinal polypore mushroom *Lignosus rhinocerotis* revealed sesquiterpenoid biosynthesis genes. Moreover, the genome of *L. rhinocerotis* encodes for 1,3-β- and 1,6-β-glucans, as well as for laccase, lectin, and other fungal immunomodulatory proteins (FIP) (Yap et al. 2014, 2015).

Nowadays, pharmaceutical companies consider the medicinal mushrooms a rich source of innovative biomedical molecules extracted not only from fruiting bodies but also from cultivated mycelial biomass and cultural broth. Moreover, the mycelia and the cultural broth might be considered potential sources of bioactive compounds, due to their shorter incubation time and affordable culture conditions (e.g., requiring less space, low probability of contamination, and higher production of biomass) (Zhang et al. 2016; Bandara et al. 2017; Souilem et al. 2017). A recent review on 40 mushroom-derived pharmaceuticals available in the markets of Australia and New Zealand has demonstrated that fungal-based natural products have a major role in the future of medicine (Beekman and Barrow 2014).

1.3.1 Polysaccharides

Modern phytochemical and pharmacological studies have shown that polysaccharides are one of the major bioactive compounds in mushrooms (Wasser and Weis 1999; Zhang et al. 2007). Macrofungal β-glucans are mainly represented by β-1,3 and β-1,6 glycosidic bonds and used in the treatment of cancer because of their immunomodulatory and antitumor effects (Aleem 2013; Yoon et al. 2013; Khan et al. 2014; Piotrowski et al. 2015; Meng et al. 2016; Wang et al. 2017b). Moreover, fungal polysaccharides have less toxic side effects, unlike many existing chemotherapeutic drugs, and can be used to develop alternative medicines for supportive healthcare for the treatment of cancer (Ramberg et al. 2010). Mushroom-derived β-glucans also exhibit significant antioxidant, antiviral, and other bioactivities (Wasser and Didukh 2005; Khan et al. 2014; Kozarski et al. 2014; He et al. 2017a, b).

About 80–85% of all the existing medicinal mushroom products are (1–3)-, (1–6)-β-D-glucans, such as lentinan extracted from *L. edodes*, polysaccharide K (PSK) or Krestin derived from fruiting bodies of *T. versicolor*, schizophyllan from *S. commune*, a polysaccharide-protein complex (PSPC) or polysaccharide peptide (PSP) from *Tricholoma laboyense*, pleuran from *P. ostreatus*, as well as polysaccharides from fruiting bodies and mycelium of *H. erinaceus* (Zhu et al. 2015). The polysaccharides extracted from *Auricularia auricula-judae* possess anti-inflammatory, antioxidant, and cardioprotective effects, whereas exopolysaccharides from *P. eryngii* exhibit higher antitumor activity on human hepatoma cells (Jing et al. 2013) and suppress the proliferation of HepG-2 cells (Ma et al. 2014a).

Although there are many reports on the bioactivity and structure of fungal glucans, studies on the quantitative assessment of these compounds are scarce. The overall β-glucan content in the wild-growing mushrooms *A. auricula-judae*, *Boletus pinophilus*, *Craterellus cornucopioides*, *Gyroporus cyanescens*, *Hydnum repandum*, *Suillus granulatus*, *S. variegatus*, and *Tricholomopsis rutilans* in comparison with edible cultivated mushroom *Agaricus bisporus* has recently been analyzed (Mirończuk-Chodakowska et al. 2017). The highest percentage of 1,3- and 1,6-β-D-glucan content in relation to the total β-glucan content was detected in *G. cyanescens* (54%), *S. granulatus* (49.8%), *A. auricula-judae* (47.9%), and *S. variegatus* (40.6%). It was shown that the average β-D-glucan content was higher in wild-growing mushrooms compared to cultivated specimens. The studied species are recommended as a source of dietary food (Mirończuk-Chodakowska et al. 2017).

The functional properties of polysaccharides obtained from medicinal mushroom *Ganoderma neo-japonicum*, as well as the presence of total phenolics, protein, and sugar in crude extracts, confirm that β-glucan-inhibiting carbohydrate-hydrolyzing enzymes may be used in the treatment of diabetes mellitus (Subramaniam et al. 2017). Biological and chemical characteristics of *Ganoderma* polysaccharides; their antioxidant, antitumor, and antimicrobial activities; as well as the structure-bioactivity relationship were recently discussed by Ferreira et al. (2015). Although *Ganoderma* polysaccharides are suggested as a healthy dietary food supplement, particularly for cancer patients, the authors considered that further clinical trials are required.

1.3.2 Terpenoids, Steroids, and Sterols

Many medicinal mushrooms contain different bioactive triterpenes, steroids, and sterols with antibacterial, antimitotic, antiviral, cytotoxic, immunomodulatory, and apoptosis-inducing effects (Baby et al. 2015; Castellano and Torrens 2015; Duru and Çayan 2015; Hadda et al. 2015a; Zhao et al. 2015, 2016; Corrěa et al. 2017; Wang et al. 2017b; Kovács et al. 2018; Morales et al. 2018). Currently, 431 secondary metabolites, over 380 terpenoids, and 30 steroids (lanostane triterpenes as ganoderic and lucidenic acids, meroterpenoids, pentacyclic triterpenes, prenyl hydroquinone sesquiterpenoids, steroids), as well as alcohols, aldehydes, alkaloids, esters, glycosides, ketones, and lactones with significant bioactivities, were isolated from 22 *Ganoderma* species (*G. amboinense*, *G. annulare*, *G. applanatum*, *G. australe*, *G. boninense*, *G. capense*, *G. carnosum*, *G. cochlear*, *G. colossum*, *G. concinna*, *G. fornicatum*, *G. hainanense*, *G. lipsiense*, *G. mastoporum*, *G. neo-japonicum*, *G. orbiforme*, *G. pfeifferi*, *G. resinaceum*, *G. sinense*, *G. theaecolum*, *G. tropicum*, and *G. tsugae*) (Baby et al. 2015). The structure-activity relationship of 71 triterpenoids and steroids isolated from *Ganoderma* species was revealed (Castellano and Torrens 2015). The chemopreventive agents for cancer, a new lanostanoid tsugaric acid F, and a novel palmitamide with antioxidant and weak cytotoxic activities against PC3 cells were isolated and characterized from the fruiting bodies of *G. tsugae* (Lin et al. 2016). Twelve new highly oxygenated lanostane nor-triterpenoids and 9 known ganoderic acids with moderate inhibitory effects against α-glucosidase were isolated from the fruiting body of *G. lucidum* (Zhao et al. 2015), while 6 new lanostane-type triterpenoids, leucocontextins, and 12 compounds with toxicity against K562, SMMC-7721, and MCF-7 cells were identified from the fruiting bodies of *Ganoderma leucocontextum* (Zhao et al. 2016). Fourteen lanostane triterpenoids, including nine *Ganoderma* acids and five *Ganoderma* alcohols, were isolated from the fruiting bodies of *Ganoderma hainanense*. Considering that *G. hainanense*, similar to *G. lucidum* and *G. sinense*, contains lanostane triterpenoids, it could also possess a broad spectrum of activities, particularly against HL-60, SMMC-7721, A-549, and MCF-7 cells (Peng et al. 2015). Three new nortriterpenes, ganoboninketals, and highly complex polycyclic systems with antiplasmodial activity against *Plasmodium falciparum*, as well as with weak cytotoxicity against A549 and HeLa cells, and no inhibitory activity against the lipopolysaccharide (LPS)-induced macrophages were detected in other *Ganoderma* species – *G. boninense* (Ma et al. 2014b). Sixteen new lanostane triterpenes (ganoleucoins A-P) together with ten known triterpenes were originally isolated from the cultivated fruiting bodies of *G. leucocontextum* belonging to *G. lucidum* complex (Zhou et al. 2015). These compounds showed inhibitory effects on HMG-CoA reductase and α-glucosidase, as well as cytotoxicity against the K562 and PC-3 cell lines (Wang et al. 2015b). The triterpene lactones (colossolactone G), seven new triterpene lactones (ganodermalactones A–G), five known triterpene lactones (colossolactone B, colossolactone E, colossolactone I, colossolactone IV, and schisanlactone B), and ergosterol have been isolated from mycelial biomass of *Ganoderma* strain KM01 (Lakornwong et al. 2014). Among these biomolecules, several compounds exhibited antimalarial activity against *P. falciparum*.

Fifteen undescribed and five known lanostane-type C31 triterpenoid derivatives (palustrisoic and polyporenic acids) were isolated from the aqueous-ethanolic extract of the fruiting bodies of cultivated *Fomitopsis palustris*. Strong cytotoxicity of polyporenic acid B against HCT116, A549, and HepG2 cell lines and weak cytotoxicity of palustrisolides A, C, and G were revealed (Zhao et al. 2018). Eight undescribed lanostane triterpenoids, pardinols A–H, and one previously reported lanostane triterpenoid saponaceol B were isolated from the fruiting bodies of agaricomycetous species *Tricholoma pardinum* (Zhang et al. 2018). Pardinols B and pardinols E–H exhibited certain inhibitory effects of nitric oxide (NO) production with IC_{50} value ranging from 5.3 to 14.7 mM, as well as cytotoxicity against human cancer cell lines.

It is known that a high rate of aerobic glycolysis which occurs in malignant tumors is one of the most fundamental metabolic alternatives during tumor development and progression (Patra et al. 2013; Ros and Schulze 2013; Tan and Miyamoto 2015). The enzyme hexokinase 2 (HK2) plays a pivotal role in the glycolytic pathway of cancer cells, promotes tumor progression in animal models, and provides a new target for cancer therapy. A new steroid isolated from *Ganoderma sinense* possesses a high binding affinity to HK2 with significant binding free energy. It was identified as an HK2 inhibitor and can be considered as a potential drug targeting the HK2 for cancer therapy (Bao et al. 2018). Highly oxygenated lanostane-type triterpenoids (OLTT) obtained from the fruiting bodies of *Ganoderma gibbosum*, their bioactivity (Pu et al. 2017), and chemotaxonomic significance (Welti et al. 2015) have recently been investigated.

Nine new sesquiterpenoids, clitocybulol derivatives, clitocybulols G–O, and three known sesquiterpenoids, clitocybulols C–E, with moderate inhibitory effects against protein tyrosine phosphatase 1B (PTP1B) were isolated from the solid culture of edible fungus *Pleurotus cystidiosus* (Tao et al. 2016). The piptolinic acid A isolated from a methanolic extract of the fruiting bodies of *P. betulinus* showed cytotoxic activity against human promyelocytic leukemia cell line HL-60 and human acute monocytic leukemia cell line THP-1 (Tohtahon et al. 2017). Fungal bioactive sesquiterpenoid eremophilanes possessing antibacterial, anti-inflammatory, anti-obesity, antiviral, and cytotoxic effects were reported in several mushrooms, particularly belonging to genera *Xylaria* (Yuyama et al. 2017a). Alliacane sesquiterpenoids were isolated from submerged cultures of *Inonotus* sp. (Isaka et al. 2017). The metabolites and multicomponent pharmacokinetics of ergostane and lanostane triterpenoids were identified in the mushroom *Antrodia cinnamomea* (Qiao et al. 2015). Further chemical screening of mushrooms allows discovering new pharmacologically promising terpenoids, steroids, and sterols with therapeutic effects.

1.3.3 Phenolics and Other Compounds

Phenols are a diverse group of biocompounds which include a large number of subclasses, such as flavonoids, phenolic acids, quinones, tocopherols, tannins, etc. Fungal phenolic derivatives are primarily known for their anticarcinogenic, anti-inflammatory, antioxidant, and antimutagenic effects (Palacios et al. 2011; Kozarski et al. 2015; Islam et al. 2016). It has been found that mushroom phenolics are outstanding antioxidants which lack mutagenic properties (Khatua et al. 2013).

The extraction of total phenolics and flavonoids from wild and cultivated edible, medicinal mushrooms by different solvents was reported (Abugri and McElhenney 2013). Phenolic compounds, flavonoids, ascorbic acid, β-carotene, and lycopene were detected in a methanol-soluble extract from fruiting bodies of *G. frondosa* and *Volvariella volvacea* (Acharya et al. 2015, 2016). Five phenolic acids (vanillic acid, *m*-hydroxybenzoic acid, *o*-hydroxybenzene acetic acid, 3-hydroxy-5-methyl benzoic acid, and *p*-hydroxybenzoic acid) with analgesic effects were isolated from *S. commune* (Yao et al. 2016).

The analysis of phenolic constituents, as well as antimicrobial and anti-radical activities of edible mushrooms growing in Poland, have recently been reported (Nowacka et al. 2014). Bioactivities and phenolic contents of 25 wild edible mushrooms from Northeastern Thailand were analyzed for their antioxidant properties, proteins, sugars, β-glucan, and phenolic profiles (Butkhup et al. 2018). The strongest scavenging activity (83.07 and 86.6%) and reductive power (9.79 and 8.42 g Fe^{2+}/kg) were revealed in *Termitomyces clypeatus* and *V. volvacea*, respectively. Both species were identified as sources of healthy compounds (β-glucans and flavonoids) which could be used to mitigate diseases involving free radicals.

The antioxidant capacity and total phenolic contents of wild *T. versicolor* and *T. gibbosa* were evaluated in Romania (Pop et al. 2018). Twenty-eight compounds were identified as coumarins, flavanols, flavones, flavonols, isoflavonoid derivatives, and phenolic acids. The methanolic extract revealed the highest antioxidant activity, while the highest total polyphenol and flavonoid contents were detected in a water extract of mushrooms. It was suggested that *Trametes* species could be considered a source of bioactive phenolics.

Recent studies of antioxidant activity of cultivated (*A. bisporus*, *P. ostreatus*) and wild-growing (*Agaricus campestris*, *B. edulis*, *C. cibarius*, *Macrolepiota procera*, *P. ostreatus*, *Russula alutacea*, and *R. vesca*) edible mushrooms identified them as sources of bioactive phenolics and flavonoids (Buruleanu et al. 2018).

Fungal alkaloids (N-containing heterocyclic compounds) are known for their toxicological relevance, e.g., ergot alkaloids in *Claviceps purpurea* and psilocybin in *Psilocybe* species (Liu 2005; Streith 2011; Tylš et al. 2014). Eight new alkaloids isolated from the medicinal mushroom *H. erinaceus* with inhibitory effects against protein tyrosine phosphatase-1B (PTP1B), α-glucosidase and moderate cytotoxicity against K562 cells were described (Wang et al. 2015a).

The nucleic acid constituents (purine and pyrimidine nucleobases, uridine, guanosine, adenosine, and cytidine nucleosides) play an important role in the regulation of various physiological processes in the human body. The nucleic acid constituents and other bioactive compounds (alkaloids, polysaccharides, and terpenoids) were recently identified in several edible and nonedible mushrooms (Phan et al. 2017b, 2018).

1.4 Pharmacological Activity of Mushrooms

The scientific research and case studies from a traditional medicine show that mushrooms possess promising pharmacological potential. Bioactive compounds and extracts from medicinal mushrooms showed mainly anti-allergic, antibacterial, antidepressant, antifungal, anti-inflammatory, antioxidant, antiviral, cardioprotective, hepatoprotective, neuroprotective, cytotoxic, hypotensive, and immunomodulatory activities (Badalyan 2003a, b, 2004a, b, 2012; Lindequist et al. 2005; Fan et al. 2006; Wasser 2010, 2014; Chang and Wasser 2012; Roupas et al. 2012; Stachowiak and Reguła 2012; De Silva et al. 2013; Ivanova et al. 2014; Paterson and Lima 2014; Muszynska et al. 2015; Xu et al. 2016a, b; Sharma et al. 2017; Özcan and Ertan 2018). Therefore, mushrooms can be considered as prospective sources of new myco-pharmaceuticals or natural product-derived drugs.

Several mushrooms (*A. brasiliensis*, *Agrocybe cylindracea*, *A. auricula-judae*, *Coprinus comatus*, *G. applanatum*, *G. lucidum*, *G. frondosa*, *Gymnopus confluens*, *H. erinaceus*, *Inonotus obliquus*, *Ophiocordyceps* (syn. *Cordyceps*) *sinensis*, and *Tremella fuciformis*) are used in the production of functional foods for the prevention and treatment of diabetes mellitus (Perera and Li 2011). Triterpenes isolated from the medicinal mushroom *Wolfiporia cocos* also reduce blood glucose levels in mice (Sato et al. 2002).

One of the most valuable medicinal fungi in China is ascomycetous mushroom *O. sinensis* known for its invigorating effects in strengthening the body and restoring energy. The fungus parasitizes larvae of moths and converts them into sclerotia from which grows the fruiting body of the fungus (Fig. 1.2b). A considerable effort has been devoted to the study of host insects related to the fungus. The information on the management of insect resources and sustainable use of *O. sinensis* has been reported (Wang and Yao 2011). The extract derived from mycelia and ascomata of *Cordyceps sinensis* and *C. militaris* possess anticancer effects by modulating the immune system and inducing cell apoptosis (Khan et al. 2010; Xu et al. 2016b). The availability of laboratory cultures of these entomopathogenic fungi will further expand their pharmacological usage (Wang et al. 2012).

Another well-known medicinal mushroom *A. subrufescens* contains a number of bioactive compounds with immunomodulatory, anticancer, and other pharmacological effects (Ahn et al. 2004; Kim et al. 2005; Ellertsen and Hetland 2009; Johnson et al. 2009; Bouike et al. 2011; Lima et al. 2011; Wang et al. 2013; Da Silva de Souza et al. 2017) (Fig. 1.1a). We recommend following Kerrigan (2005, 2016), Ludwig (2007), Cappelli (2011), Parra (2013), and Thongklang et al. (2014, 2017) for taxonomy and synonymy of *A. subrufescens*. The main synonyms of *A. subrufescens* are *Agaricus blazei* Murrill sensu Heinemann, *A. rufotegulis* Nauta, *A. brasiliensis* Wasser, M. Didukh, Amazonas and Stamets, *A. albopersistens* Zuccher, and *A. bambusae* Beeli var. *bambusae*. Previously, *A. subrufescens* has not only been misidentified as *A. blazei* but also to *A. sylvaticus* Schaeff. (Souza Dias et al. 2004). In other respects, Kerrigan (2005) hybridized between each other strains isolated from North, South America, Brazil, France, and Thailand (Thongklang et al. 2014, 2016).

The compounds extracted from edible medicinal *Pleurotus* mushrooms showed efficacy for treatment of several chronic disease (Gunde-Cimmerman 1999) due to their antibacterial, anti-inflammatory, anti-mitogenic, antitumor, antioxidant, antiviral, hypoglycemic, immunomodulatory, cardioprotective, and other pharmacological activities (Jose et al. 2002; Filipic et al. 2002; Hossain et al. 2003; Hu et al. 2006; Badalyan et al. 2008b; Baggio et al. 2010; Papaspyridi et al. 2011; Patel et al. 2012; Schillaci et al. 2013; Jayasuriya et al. 2015; Khatun et al. 2015; Fu et al. 2016; Venturella et al. 2016; Zhang et al. 2016; Acharya et al. 2017; Adebayo et al. 2018; Baskaran et al. 2017; Debnath et al. 2017; Ebrahimi et al. 2017; Masri et al. 2017; Owaid et al. 2017; Abidin et al. 2018; Finimundy et al. 2018a, b).

Bioactive compounds extracted from inedible bracket fungi also possess pharmacological potential and could be used to develop healthcare products with different formulations (Barros et al. 2008; Dai et al. 2009; Reis et al. 2014; Heleno et al. 2015a, b). The bioactive compounds isolated from *T. versicolor* exhibited health-beneficiary effects, as inhibitors of aflatoxins (Scarpari et al. 2016), while compounds (amino acids, aromatic acids, flavones, polysaccharides, triterpenes, etc.) isolated from *Ph. linteus* showed anticancer, hypoglycemic, anti-inflammatory, antioxidant, and immunoregulatory activities (Chen et al. 2016). The induction of ovulation by the extract of edible Maitake mushroom (*G. frondosa*) in patients with polycystic ovary syndrome (Chen et al. 2010) and medicinal properties of *F. fomentarius* and *F. pinicola* have been reported (Badalyan and Shahbazyan 2015).

A comparative chemical analysis, acute toxicity, and pharmacological activity of fruiting body extracts of several agaricoid mushrooms, particularly of nonedible

Nematoloma species, poisonous species *Cortinarius armillatus* and *T. pardinum*, as well as of edible species *Flammulina velutipes* and *Paxillus involutus*, were performed (Badalyan et al. 1994, 1995, 1996, 1997a, b, 2001). The antioxidant activity (Badalyan 2003b), cytokine induction, and immunomodulatory effect of fruiting body extract of Enokitake mushroom *F. velutipes* (Badalyan and Hambardzumyan 2001), antiprotozoal activity, and mitogenic/regenerative effects of mycelia of shiitake mushroom *L. edodes* (Badalyan 2004b) and other culinary-medicinal *Agaricomycetes* species (Badalyan 2003a) have also been reported. The latest advancements in the study of the pharmacological potential of mushrooms are further discussed in this chapter.

1.4.1 Antimicrobial (Antifungal and Antibacterial) Activity

Agaricoid and polypore mushrooms are known as active producers of antimicrobial (antibacterial and antifungal) compounds. Many of these compounds, such as velutin and flammulin from *F. velutipes*; applanoxidic acid, ganodermadiol, ganomycin, and ganoderiol from *G. lucidum*; lentinan from *L. edodes*; schizophyllan from *S. commune*; and others, have been shown to possess antifungal and antibacterial properties (Wasser and Weis 1999; Suay et al. 2000; Badalyan 2004a, b, 2008a; Badalyan et al. 2002, 2004; Lindequist et al. 2005; Dyakova et al. 2011; Alves et al. 2012; Khatun et al. 2012; Mohanarji et al. 2012; Nowacka et al. 2014; Ferreira et al. 2015; Heleno et al. 2015b; Jaszek et al. 2015; Bedlovičová et al. 2016; Canli et al. 2016; Ćilerdžić et al. 2016; Kosanić et al. 2016; Falade et al. 2017; Gargano et al. 2017; Kolundčzić et al. 2017; Morris et al. 2017; Shen et al. 2017; Waithaka et al. 2017; Chepkirui et al. 2018; Özcan and Ertan 2018).

In vitro screening of methanolic extracts from fruiting bodies of 28 Yemeni basidiomycetous mushrooms revealed antibacterial and antifungal effects against Gram-positive (*Bacillus subtilis, Micrococcus flavus, Staphylococcus aureus*) and Gram-negative (*Escherichia coli, Pseudomonas aeruginosa*) bacteria, as well as fungal pathogens, including dermatophytes, molds, and yeasts (*Aspergillus fumigatus, Candida albicans, C. krusei, Microsporum gypseum, Mucor* sp., and *Trichophyton mentagrophytes*) (Al-Fatimi et al. 2013). The highest antibacterial activity was observed in *Agaricus bernardii, Agrocybe pediades, Chlorophyllum molybdites, Coriolopsis polyzona, Ganoderma xylonoides, Pycnoporus sanguineus, Trametes cingulate*, and *T. lactinea* species, while the extracts from *C. molybdites, G. xylonoides*, and *P. sanguineus* possessed significant antifungal activity.

Antifungal and antibacterial activities against *B. subtilis, Ceratocistys pilifera, E. coli, E. faecalis, Fusarium oxysporum, Penicillium notatum, P. aeruginosa, Rhizoctonia solani*, and *S. aureus* were evaluated in 26 compounds isolated from cultural broth of *Collybia butyracea, Entoloma nubigenum, Inocybe geophylla, Mycena hialinotricha, Psathyrella* species, and *Stropharia semiglobata* mushrooms collected from sub-Antarctic forests in Southern Chile (Reinoso et al. 2013). Based on the results, tested mushrooms have been recommended for further studies as sources of antimicrobial compounds.

Biocontrol properties and the potential role of basidiomycetous mushrooms as antiphytobacterial, antiphytofungal, antiphytoviral, larvicidal, and nematicidal agents were recently described by Sivanandhan et al. (2017). The screening of antiphytopathogenic effects in 20 polypore mushrooms against phytopathogens (e.g., *Bipolaris sorokiniana*, *Fusarium culmorum*, *F. oxysporum*, *Gaeumannomyces graminis* var. *tritici* (syn. *Gaeumannomyces tritici*), *Ophiostoma ulmi*, *Pestalotiopsis funerea*, *Rhizoctonia cerealis*) and their antagonists (*Gliocladium roseum*, *Trichoderma asperellum*, *T. harzianum*, *T. pseudokoningii*, and *T. viride*) revealed the highest activity in white rot species *Climacodon septentrionale*, *F. fomentarius*, *Ganoderma resinaceum*, and *Lentinus tigrinus* (Badalyan and Gharibyan 2017b).

The methanolic and aqueous extracts from sclerotia of the wild-growing mushroom *L. rhinocerotis* were tested against 15 pathogenic bacteria from genera *Bacillus*, *Corynebacterium*, *Escherichia*, *Klebsiella*, *Micrococcus*, *Pseudomonas*, *Salmonella*, *Staphylococcus*, and *Streptococcus* and 4 fungi from genera *Candida* and *Mucor* (Mohanarji et al. 2012). The extracts (30 mg/mL) showed a significant inhibition against all test organisms, except for *Streptococcus pyogenes* and *Serratia marcescens*. The chemical analysis of extracts revealed the presence of alkaloids, proteins, gums, mucilage and flavonoids.

The antimicrobial activity of extracts obtained from fruiting bodies of *A. bisporus* and *T. gibbosa*, collected in Kenya, against several Gram-positive and Gram-negative bacteria, as well as phytopathogenic and keratinophilic fungi, was reported (Waithaka et al. 2017). The concentration-dependent bacteriostatic and bacteriocidic effects against *Helicobacter pylori* strain were revealed in ethanolic extracts from *A. bisporus*, *A. brasiliensis*, *Agrocybe aegerita*, *C. comatus*, *C. militaris*, *F. velutipes*, *G. lucidum*, *G. frondosa*, *H. erinaceus*, *Hypsizygus marmoreus*, *L. edodes*, *Ph. igniarius*, *P. eryngii*, and *P. ostreatus* (Shang et al. 2013). However, further research is needed to identify the role of active compounds of tested mushrooms in the treatment of *H. pylori*-associated gastrointestinal disorders.

Previously seven terpenoids named microporenic acids (A-G) isolated from the cultures of polypore *Microporus* species showed antibacterial activity against biofilm formation by Gram-positive *S. aureus* bacteria, as well as antifungal activity against *Candida tenuis* and *C. albicans* (Chepkirui et al. 2018). The extracts obtained from wild mushrooms *Fistulina hepatica*, *Lepista nuda*, *Leucopaxillus giganteus*, *Mycena rosea*, and *Russula delica* were originally reported as antimicrobials against in vitro biofilm formation by multiresistant clinical isolates of Gram-positive bacteria (*Acinetobacter baumannii*, *E. coli*, *Proteus mirabilis*, and *P. aeruginosa*) to solve multidrug resistance problems in public healthcare (Alves et al. 2014). The antibacterial and anti-biofilm compounds were originally isolated from *Hypoxylon fragiforme* mushroom (Yuyama et al. 2017b).

In vitro antibacterial, antifungal, and antioxidant properties of mycelial and fruiting body extracts obtained from several agaricomycetous fungi, such as *G. applanatum* (Klaus et al. 2016), *Pleurotus* spp. (Owaid et al. 2017), *Pycnoporus sanguineus* (Jaszek et al. 2015), *Rigidoporus microporus* (Falade et al. 2017), *Xylaria hypoxylon* (Canli et al. 2016), and other species, were recently reported (Dyakova et al. 2011; Ren et al. 2014; Ferreira-Silva et al. 2017; Gaylan et al. 2018).

The aqueous extracts of edible desert truffles *Picoa juniperi*, *Terfezia claveryi*, and *Tirmania pinoyi* (phylum *Ascomycota*) showed in vitro antibacterial activity against the Gram-positive human pathogenic reference strain *S. aureus* ATCC 29213 and the Gram-negative *P. aeruginosa* strain ATCC 15442. The acid-soluble protein extracts of *T. pinoyi* and *T. claveryi* showed minimum inhibitory concentrations (MIC) of 50 μg/mL against the tested pathogens (Schillaci et al. 2017).

Further studies directed to the mechanisms behind antimicrobial activities of mushroom-derived compounds are needed.

1.4.2 Immunomodulatory and Anticancer Activities

The immunomodulatory activity is the most prominent pharmacological property of medicinal mushrooms (Badalian 2000; Borchers et al. 2008; Novak and Vetvicka 2008; Ferreira et al. 2010; Tangen et al. 2015; Ahmad et al. 2016; Del Buono et al. 2016; Xu et al. 2016a; Diling et al. 2017; Wasser 2017; Rubel et al. 2018; Wang et al. 2018).

Mushroom derived bioactive molecules, particularly β-glucans schizophyllan, lentinan, and grifolan, proteins, glycoproteins, and lipopolysaccharides (LPS), have been identified as immunomodulators and are widely used in the treatment of several types of cancer (Keong et al. 2007; Wasser 2017). They can prevent oncogenesis by possessing direct inhibitory effects on tumor metastasis and exhibit antitumor effects by inducing immune response in the host (Wasser and Weis 1999; Wasser 2002, 2017; Brown and Gordon 2003; Wasser and Didukh 2005; Oba et al. 2007; Novak and Vetvicka 2008; Chan et al. 2009; Volman et al. 2010; Adotey et al. 2011; Aleem 2013; Mizuno and Nishitani 2013; Yoon et al. 2013; Guggenheim et al. 2014; Ivanova et al. 2014; Khan et al. 2014; Kozarski et al. 2014; Ma et al. 2014a; Dai et al. 2015; Fritz et al. 2015; Del Buono et al. 2016; Tsai et al. 2016; Gargano et al. 2017; Glamočlija and Soković 2017; Hapuarachchi et al. 2017; He et al. 2017a; Wang et al. 2017a; Pandya et al. 2018).

Mushroom-derived glucans belong to PAMPs (pathogen-associated molecular patterns) and after per os administration are recognized by PRR (pattern recognition receptors) on the surface of dendritic cells and macrophages in the gastrointestinal tract (Muta 2006; Batbayar et al. 2012; Wasser 2017; Patin et al. 2018). After recognition by PRR, the glucan molecules are internalized into and fragmented within the cells. The fragments are taken up by the lymph and transported to other parts of the immune system. They bind to specific CR3 receptors of the complement system on the surface of immune cells (neutrophil granulocytes and NK cells) and activate them which is followed by secretion of cytokines, such as tumor necrosis factors alfa (TNF-α), interferons (IFN-γ), and interleukins (IL-6, IL-8, IL-12), and leads to activation of cytotoxic T lymphocytes, T helper cells, and B cells. Increased phagocytosis, production of NO, formation of antibodies occur, and innate and adaptive immunity are activated which react against invading microorganisms and abnormal cells, such as tumor cells (Wasser 2002, 2017; Brown and Gordon 2003; Chen and Seviour 2007; Novak and Vetvicka 2008; Chan et al. 2009; Barsanti et al. 2011; Batbayar et al. 2012; Ren et al. 2012; Giavasis 2014; Guggenheim et al. 2014).

It is known that immunomodulatory effects depend on biological conditions of the organism. The immunomodulatory activity of pharmaceuticals also depends on the dosage, mode, timing of administration, and on pharmaceutical formulation (Wasser 2002, 2017).

The immune stimulation by fungal β-glucans can be used in the adjunct therapy of cancer patients together with surgery, chemotherapy, and radiation treatment (Oba et al. 2007; Ramberg et al. 2010; Zong et al. 2012; Guggenheim et al. 2014; Twardowski et al. 2015) (see part 6).

Apart from immunomodulation, β-glucans possess also direct tumoricidal effects by inhibiting the expression of aromatase, an enzyme responsible for the conversion of androgens to estrogens which is often upregulated in breast cancer patients (Adams et al. 2008). Furthermore, fungal β-glucans may modify cell cycle-regulating genes, arrest the cell cycle, and induce apoptosis (Jiang and Sliva 2010). Maitake's (*G. frondosa*) β-glucan stimulates differentiation of haematopoietic progenitor cells, production of granulocyte colony-stimulating factor, and the recovery of peripheral blood leukocytes after bone marrow injury in phase II study involving patients with a preleukemic myelodysplastic syndrome (Wesa et al. 2015). These compounds change adhesion molecules on the surface of cancer cells; inhibit migration, invasion, and adhesion of cells; and possibly affect the development of metastases (Masuda et al. 2008; Jiang and Sliva 2010). A recent study investigated the effect of Maitake D-fraction to attenuate the aggressive activity of triple-negative breast cancer (TNBC) cells which revealed that the proteoglucan fraction induces MDA-MB-231 cell apoptosis, reduces their metastatic potential, and decreases invasive capacity (Alonso et al. 2018).

The polysaccharide fucogalactan isolated from the aqueous extract of giant mushroom *Macrocybe titans* showed anticancer activity against murine melanoma cells B16-F10 by reducing their migration in vitro, however did not alter the viability, proliferative capacity, and morphology of cells (Da Silva Milhorini et al. 2018). The 14,942 Da polysaccharide, isolated from *Collybia radicata* with potent immunomodulatory activity, may be considered a novel immunomodulator to be used in medical and food industries (Wang et al. 2018).

The therapeutic effects of β-glucans against oncological diseases are supported by low-molecular-weight compounds of mushrooms, such as triterpenes, isolated from *G. lucidum*, and other *Ganoderma* species, *A. brasiliensis*, *I. obliquus*, and *W. cocos*. These compounds exhibit immunomodulatory, cytotoxic, and apoptosis-inducing effects by inhibiting cell cycle and phosphorylation of Erk1/2, increasing the level of p53 and *Bax*, upregulating NF-κB and AP-1, and reducing the activity of topoisomerase II (Ríos et al. 2012; Rubel et al. 2018).

Methanolic extracts of 29 different wild edible mushrooms were examined for antioxidant, antiproliferative, cytotoxic, and pro-apoptotic activities toward a lung adenocarcinoma cell line A549 (Vasdekis et al. 2018). Certain species exhibited a high antioxidant activity which was related to their high total phenolic content, whereas *C. cibarius*, *Cantharellus cinereus*, *Craterellus cornucopioides*, and *Hydnum repandum* (order *Cantharellales*) showed high cytotoxicity and induced apoptosis in A549 cells. As an active ingredient, piceatannol with antiproliferative activity was originally identified in studied mushrooms. A novel antitumor glucan extracted from the fruiting bodies of *Coriolus* (syn. *Trametes*) *versicolor* was purified and characterized (Awadasseid et al. 2017). The effects of *Inonotus hispidus* on human immunocompetent cells (Gründemann et al. 2016), anticancer and anti-inflammatory effects of *G. lucidum* on melanoma cells (Barbieri et al. 2017), antiproliferative effects of extract obtained from *P. ostreatus* on acute leukemia cell lines (Ebrahimi et al. 2017), RNA fraction from *B. edulis* against myelogenous leukemia cells (Lemieszek et al. (2017), as well as novel acid polysaccharides (Zhang et al. 2017) and anticancer recombinant latcripin 11 from *L. edodes* (Gao et al. 2018) were recently reported. A ganoderic acid with antiproliferative activities from neotropical *Ganoderma* mushrooms (Welti et al. 2010) and a recombinant fungal immunomodulatory protein (FIP) from *Ganoderma atrum* inducing apoptosis in breast cancer cells (Xu et al. 2016a) were characterized.

Seven steroids based on ergostane, as well as lanostane and ceramide derivatives, were originally isolated and identified from the methanolic extract of agaricomycetous mushroom *Scleroderma bovista* (Kovács et al. 2018). The lanostane derivatives and ergosterol peroxide 3-glucoside revealed significant antiproliferative properties on one or more tested human cancer cell lines (HeLa, A2780, MDA-MB-231, and MCF-7).

The antitumor effects of a new pyrrole compound By-1 obtained from the submerged culture of medicinal mushroom *Taiwanofungus camphoratus* have recently been revealed against various cancer cells (Yang et al. 2018). It has been shown that combinations of pro-apoptotic and anti-autophagy agents from *T. camphoratus* could be an effective new treatment strategy for non-small cell lung adenocarcinoma.

Further studies of the signaling pathways modulating cancer development and progression will promote the transfer of bioactive mushroom molecules to clinically effective therapeutic agents (Aras et al. 2018).

1.4.3 Hypocholesterolemic, Hypoglycemic, and Anti-obesity Effects

Hyperglycemia, hyperlipidemia, insulin resistance, obesity, and arterial hypertension are important symptoms of metabolic syndrome, preceding signs of type 2 diabetes and risk factors of cardiovascular diseases (CVD) (Francia et al. 1999; Gunde-Cimmerman 1999; Lo and Wasser 2011; De Silva et al. 2012b; Martel et al. 2017; Vitak et al. 2017; Morales et al. 2018). The existing drugs, such as insulin, statins, and angiotensin-converting enzyme (ACE) inhibitors used for the treatment of metabolic syndrome, have limited therapeutic efficacy and several side effects. There are other medications (e.g., HMG-CoA reductase, aldose reductase, and α-glucosidase) which are also used for the treatment of metabolic syndrome. Nevertheless, considerable effort has been made in the pharmaceutical industry to develop new preparations to improve glucose and lipid metabolism without significant side effects.

Mushrooms and mushroom-rich nutrition are regarded as healthy dietary food supplements to prevent and treat such widespread pathological conditions (Guillamón et al. 2010). The results of in vitro animal assays and several human studies suggest that medicinal mushrooms *A. bisporus* (Yamac et al. 2010), *A. brasiliensis* (Kim et al. 2005), *G. lucidum* (Ma et al. 2015), *G. frondosa* (Hong et al. 2007), *H. erinaceus* (Thongbai et al. 2015), *Ph. linteus* (Yamac et al. 2016), and *Pleurotus* species (Jayasuriya et al. 2015; Anjana and Savita 2017) can normalize blood glucose and lipid levels. Several edible and nonedible mushrooms, such as *L. edodes* (Poucheret et al. 2006; De Silva et al. 2013; Yang et al. 2013) and *Hypholoma fasciculare* (Badalian and Serrano 1999), have revealed beneficial effects on lowering the levels of blood glucose, lipids, and arterial pressure in experimental animals. Moreover, the eritadenine extracted from *L. edodes* has been identified as an anti-atherogenic compound which not only improves lipid metabolism but also inhibits the activity of ACE in vitro (Afrin et al. 2016).

Several new bioactive compounds, including lanostane triterpenoids (e.g., lucidenic acids, ganolucinins, new natural product ganomycin, etc.) isolated from fruiting bodies of *G. lucidum*, were suggested as promising bioactive agents for the treatment of metabolic syndrome (Chen et al. 2017). The obtained results provide evidence for the usage of *G. lucidum* extract as a myco-medicine and food supplement to control hyperglycemia and hyperlipidemia.

Evaluation of antidiabetic properties of agaricomycetous culinary-medicinal mushrooms *Calocybe indica*, *P. ostreatus*, and *V. volvacea* in mice revealed that *C. indica* could be identified as a natural source of antidiabetic compounds (Singh et al. 2017). The aqueous extract of bracket fungi *Inocutis levis* improves insulin resistance and glucose tolerance in high sucrose-fed Wistar rats (Ehsanifard et al. 2017). The hypoglycemic effect of *Pleurotus giganteus* is manifested by enhancing adipocyte differentiation and glucose uptake via activation of peroxisome proliferator-activated receptor (PPAR) and glucose transporters 1 and 4 in 3T3-L1 cells (Paravamsivam et al. 2016).

In vitro anti-atherogenic effects of extracts derived from *P. pulmonarius* based on inhibition of ACE and HMG-CoA reductase and the protective effects of extracts on the endothelial membrane against oxidative stress were recently evaluated (Abidin et al. 2018). The results of this study could be applied in nutriceutical and pharmaceutical industries.

Literature data and advancements on the treatment of type 1 diabetes mellitus using medicinal mushrooms were analyzed by Vitak et al. (2017). However, data concerning molecular mechanisms of action of hypocholesterolemic and hypoglycemic compounds obtained from edible mushrooms are still rare (Gil-Ramirez et al. 2017).

1.4.4 Antioxidant and Anti-inflammatory Effects

The maintenance of balance between free radical production and antioxidant defense is an essential condition for normal functioning of biological organisms.

Free radicals may damage cellular lipids, proteins, and DNA, affecting their normal function and leading to various diseases. Natural products with antioxidant activity may help the endogenous defense system, and from these perspective antioxidants present in diet are considered possible protective agents reducing oxidative damage. In this regard, mushrooms, including edible and medicinal species, are regarded as a natural source of antioxidants (Saltarelli et al. 2009, 2015; Palacios et al. 2011; Chang and Wasser 2012; Khatua et al. 2013; Ferreira et al. 2015; Jaworska et al. 2014; Heleno et al. 2015a; Khatun et al. 2015; Ćilerdžić et al. 2016; Bandara et al. 2017; Debnath et al. 2017; Souilem et al. 2017; Carocho et al. 2018; Choi et al. 2018; Da Silva de Souza et al. 2018; Khaskheli et al. 2018; Morel et al. 2018; Anwar et al. 2018). Mushrooms are also described as natural sources of anti-inflammatory agents (Jose et al. 2002; Ruthes et al. 2013; Lee et al. 2014a; Taofiq et al. 2016a, c; Souilem et al. 2017).

The antioxidant and anti-inflammatory compounds identified in mushrooms include fatty acids, phenolic acids, polysaccharides, steroids, terpenes, tocopherols, and other biomolecules. Among them, phenolics, polysaccharides, and terpenoids are considered the determinants of antioxidant and anti-inflammatory effects. However, further clinical studies are needed out to confirm the antioxidant and anti-inflammatory potential of these compounds (Taofiq et al. 2016a).

The antioxidant properties of wild mushrooms have been extensively investigated by Ferreira et al. (2009). Many antioxidant compounds, such as phenolics, tocopherols, ascorbic acid, and carotenoids, have been identified, and the mechanisms of action of antioxidant properties were discussed. It was suggested that wild or cultivated edible mushrooms could be used as dietary healthy food products, taking advantage of synergistic effects of all present bioactive compounds (polyphenols, polysaccharides, vitamins, carotenoids, and minerals) (Kozarski et al. 2014, 2015).

The study on fermentation broth of three *Ganoderma* species exhibiting a large number of phenol derivatives and flavonoids showed that antioxidant activity of *G. lucidum* was higher than the activity of *G. applanatum* and *G. carnosum* (Ćilerdžić et al. 2016). Moreover, *G. lucidum* strains were the most effective antibacterial and antifungal agents. The cultural broth of tested *Ganoderma* species was suggested as a potent natural antioxidant and antimicrobial agent.

Antioxidant potential of an aqueous extract from *A. brasiliensis* was evaluated in adjuvant-induced arthritic rats (Da Silva de Souza et al. 2018). The antioxidant defenses, diminished by arthritis, were improved by treatment with *A. brasiliensis* extract. Aqueous preparations of *A. brasiliensis* can be recommended as potential auxiliaries in the treatment of patients with rheumatoid arthritis due to their capacity of reducing oxidative stress.

A potent antioxidant effect was detected in methanolic extracts from mycelia of ten *Pleurotus* species, including *P. citrinopileatus* and *P. sajor-caju*. The inclusion of these mushrooms in diet may help to prevent diseases caused by oxidative damage (Debnath et al. 2017). Screening of standardized hydroalcoholic extracts of four *Pleurotus* species (*P. levis*, *P. ostreatus*, *P. pulmonarius*, and *P. tuber-regium*) revealed more remarkable antioxidant and antibacterial properties in *P. levis* and *P.*

tuber-regium compared to other studied species that are commercially cultivated (Adebayo et al. 2018). It was shown that *Pleurotus* mushrooms *P. citrinopileatus*, *P. florida*, and *P. pulmonarius* are comparable to best antioxidants, which is attributed to their catalase, phenolics, and peroxidase contents (Khatun et al. 2015).

Antioxidant and antiproliferative effects of cyclohexane, chloroform, ethanol, and water extracts of 28 wild *Boletales* mushrooms (*Caloboletus calopus*, *Gyroporus castaneus*, *Omphalotus olearius*, *Rubroboletus lupinus*, *R. pulchrotinctus*, *R. satanas*, and *Suillus luteus*) from France were recently evaluated (Morel et al. 2018). Among the tested mushroom extracts, 11 presented antiproliferative activities against HCT116 cells, while the highest antioxidant effects were revealed for chloroform, ethanol, or aqueous extracts, depending on mushroom species. Further studies are necessary to identify bioactive compounds and valorize mushrooms either for edible species, as healthy foods, or for inedible mushrooms, as natural ingredients in the pharmaceutical and food industries (Morel et al. 2018).

A strong antioxidant effect was described in methanolic extracts derived from *C. molybdites*, *G. xylonoides*, *Hexagonia velutina*, *P. sanguineus*, *Trametes lactinea*, and *T. cingulate* mushrooms (Al-Fatimi et al. 2013). The extract of the medicinal mushroom *A. auricula-judae* was reported as an antioxidant promoting procollagen biosynthesis in HaCaT cells (Choi et al. 2018). The antioxidant potential and antiproliferative activity were evaluated in edible and medicinal agaricoid (*A. subrufescens*, *B. edulis*, *F. velutipes*, *Ganoderma capense*, *H. erinaceus*, and *Pleurotus djamor*) (Badalyan 2003a, b; Llarena-Hernández et al. 2015; Panthong et al. 2016; Peng et al. 2016; Acharya et al. 2017; Novakovič et al. 2017; Sánchez 2017b; Carocho et al. 2018), as well as in culinary-prepared *S. luteus* (Jaworska et al. 2014) mushrooms.

Fungal polysaccharides play a vital role as dietary free radical scavengers in the prevention of oxidative damage in living organisms (De Silva et al. 2013). The study of chemical characteristics and antioxidant properties of water-soluble and crude polysaccharides obtained from *A. auricula*, *L. edodes*, and *Poria* (syn. *Wolfiporia*) *cocos* revealed that they are composed of β-glycoside linkages (Ke and Chen 2016; Khaskheli et al. 2018). The main compositions of monosaccharides were D-galactose, D-glucose, L-rhamnose, arabinose, and D-mannose for *A. auricular*a and *L. edodes*, while *P. cocos* consisted of D-mannose and D-galactose. The exopolysaccharide (EPS) isolated from agaricomycetous *R. microporus* as a natural antioxidant is recommended for application in food and medicines (Jia et al. 2018).

Antinociceptive and anti-inflammatory effects, as well as structure/activity correlation of (1–3)- and (1–6)-β-D-glucans from *Lactarius rufus* (Ruthes et al. 2013) and induction of pro-inflammatory cytokine production by glucan lentinan from *L. edodes* (Ahn et al. 2017), were reported. The anti-inflammatory potential and molecular mechanisms underlying the protective effect of ethanolic extract of culinary-medicinal mushroom *P. giganteus* against LPS and combination of LPS and hydrogen peroxide-induced inflammation on RAW 264.7 macrophages were revealed (Baskaran et al. 2017).

Data concerning the anti-inflammatory effects of aqueous, ethanolic, and methanolic extracts obtained from fruiting bodies and sclerotia of tiger milk mushroom *L.*

rhinocerotis was recently published (Lee et al. 2014a; Baskaran 2015; Nallathamby et al. 2016, 2018). The ethanolic extract revealed a significant decrease of NO production, unlike the aqueous extract. Moreover, the ethanolic extract was able to stimulate signal transducer and activator of transcription 3 (STAT3) pathways by reducing expression of inducible NO synthase (iNOS) and cyclooxygenase-2 (COX-2) and increasing the expression of interleukin 10 (Baskaran 2015; Nallathamby et al. 2018). The ethyl acetate fraction of the ethanolic extract of *L. rhinocerotis* significantly reduced the NO production in microglial BV2 cells (Nallathamby et al. 2016). The major compounds of ethyl acetate fraction were linoleic and oleic fatty acids and ethyl linoleate. The treatment with linoleic acid significantly decreased the expression of iNOS and COX-2 by 1.2-fold compared to control. Studies of LPS-induced brain microglial BV2 cells pre-treated with hot aqueous extract (500 μg/mL), n-butanol fraction of hot aqueous extract (250 μg/mL), and ethyl acetate fraction of hot aqueous extract (250 μg/mL) of *L. rhinocerotis* showed maximal inhibition of NO production by 88.95, 86.50, and 85.93%, respectively (Seow et al. 2017).

Further studies are needed to elucidate the antioxidant and anti-inflammatory potential of wild and cultivated mushroom resources for their further usage as healthy food supplements and myco-pharmaceuticals.

1.4.5 Cardioprotective Effect

Cardiovascular diseases (CVD) (e.g., heart attack or stroke) affect the heart and circulatory system and are considered the leading cause of death worldwide. The main risk factors for CVD include elevated arterial pressure, high levels of blood glucose and cholesterol. Hypocholesterolemic and hypoglycemic properties of different medicinal mushrooms allow using them as natural, healthy food to prevent the development of diseases and improve cardiovascular health (see parts 4.3, 5, and 6).

Mushrooms, particularly *G. frondosa*, *L. edodes*, and *P. ostreatus*, are almost ideal for low-calorie diets to prevent CVD due to a high content of fiber, proteins, and microelements (Khatun et al. 2007, 2012). Recent experimental and clinical data has demonstrated that many bioactive compounds, i.e., terpenoids, peptides, isoflavones such as biochanin A and formononetin, lanosterone derivative as fomiroid A, and lovastatin extracted from *Boletus aestivalis*, *Clitocybe nuda*, *G. lucidum*, *G. frondosa*, *H. marmoreus*, *L. edodes*, and *Pleurotus* species, can regulate the levels of low, high-density lipoproteins and homocysteine, total cholesterol, and fasting triglycerides and prevent the development of arterial hypertension, oxidative stress, diabetes, and other CVD (Shibu et al. 2017).

Thrombosis can also cause CVD due to fibrin aggregation in the blood (Previtali et al. 2011). Fibrinolysis is the process of dissolving the fibrin in blood clots by proteolytic enzymes. Fibrin clot formation and fibrinolysis are balanced processes in biological systems. However, accumulation of fibrin clots in blood vessels causes thrombosis and leads to myocardial infarction or other CVD.

The tissue-type, the urokinase-type, and the bacterial plasminogen activators, nattokinase and streptokinase, are well-known fibrinolytic agents with a wide range of clinical applications. However, uncontrolled use of these drugs results in side effects (internal hemorrhage, allergic reactions, limitation in specificity toward fibrin, etc.) and is costly (Blann et al. 2002). Therefore, the search for natural sources of fibrinolytic agents is topical.

Mushrooms belonging to different taxonomic and ecological groups are considered active producers of proteolytic (fibrinolytic, thrombolytic, caseinolytic, etc.) enzymes (Denisova 1982, 2010; Denisova et al. 1989; Badalyan et al. 2008b, 2016; Sharjahan et al. 2017). The proteases with fibrinolytic activity were originally isolated from *F. velutipes* (Morozova et al. 1982). Later, other basidiomycetous mushrooms were screened for fibrinolytic and thrombolytic effects (Denisova 1982, 2010; Denisova et al. 1989; Kudryavtseva et al. 2008; Lu and Chen 2012; Badalyan et al. 2016). A number of systematic studies have discovered fibrinolytic proteases from fruiting bodies and mycelia of medicinal mushrooms, such as *Armillariella mellea* (Kim and Kim 1999), *Auricularia polytricha* (Ali et al. 2014, 2017), coprini species (Badalyan et al. 2008b), *C. militaris* (Kim et al. 2006), *F. pinicola* (Badalyan et al. 2016), *Ganoderma* and *Fomes* species (Choi and Sa 2000; Kumaran et al. 2011; Hadda et al. 2015a, b), *H. erinaceus* (Choi et al. 2013), *Perenniporia fraxinea* (syn. *Fomitella fraxinea*) (Kim et al. 2008; Lee et al. 2002, 2006), *Pleurotus ferulae* (Choi et al. 2017), and *P. ostreatus* (Choi and Shin 1998). Among other screened species (*A. bisporus*, *H. erinaceus*, *L. edodes*, *P. cystidiosus*, *P. floridanus*, *P. pulmonarius*, and *P. salmoneostramineus*), the highest proteolytic activity (up to 54.28 U/mg) exhibited a crude extract of *L. edodes*. The molecular weight of the fibrinolytic enzyme was 50 kDa (Ali et al. 2017). The antiplatelet activity of aqueous extract of *L. rhinocerotis* was revealed using fresh human blood samples (Teo 2014). The molecular weight of the partially purified protease-like enzyme was 50–55 kDa. Fibrinolytic enzymes were also isolated from sclerotia of *L. rhinocerotis* as well (Ahmad et al. 2014).

The cultural liquid (CL) samples of eight polypore species *F. fomentarius*, *F. fraxinea*, *F. pinicola*, *L. sulphureus*, *T. gibbosa*, *T. hirsute*, *T. ochracea*, and *T. versicolor* have recently been screened for thrombolytic activity on human thrombus samples (Badalyan et al. 2016). The highest activity was detected in CL of *F. fraxinea* (up to 100%), *F. pinicola* (up to 85%), *F. fomentarius* (up to 83%), and *L. sulphureus* (up to 69%) strains, whereas the activity was weaker (20–55%) in *Trametes* species.

The screening of fibrinolytic activity of CL samples of two agaricoid (*F. velutipes* and *P. ostreatus*) and two polypore (*F. pinicola* and *G. lucidum*) mushrooms revealed the highest activity in *F. pinicola* (95%), followed by *G. lucidum* (55%), *P. ostreatus* (54.0), and *F. velutipes* (51%) (Badalyan et al. unpublished data). Milk-coagulating and fibrinolytic activities in *F. pinicola* and coprini species have also been detected (Badalyan et al. 2008b, 2016). It can be concluded that medicinal mushrooms may be considered alternative natural sources to develop novel pharmaceuticals with thrombolytic/fibrinolytic effects.

1.4.6 Neuroprotective and Antidepressant Effects

Inflammation, oxidative stress, mitochondrial dysfunction, and axonal transport deficits play a role in development many neurodegenerative disorders. Different neuroprotective strategies aim to limit these neurotoxic processes to delay disease progression. In spite of unsuccessful clinical trials, new therapies are emerging to ameliorating the process of traumatic and ischemic damage and to delay the progression of neurodegenerative processes.

Neurotrophins, such as nerve growth factor (NGF) and brain-derived neurotrophic factor (BDNF), are of great interest for survival, maintenance, and regeneration of specific neuronal populations in the adult brain. Several medicinal mushrooms, such as *Antrodia camphorata*, *G. lucidum*, and *H. erinaceus*, possess neurotrophic effects and could be considered as useful therapeutic agents in the management and/or treatment of depression and Alzheimer's, Huntington's, and Parkinson's diseases (Chen et al. 2006; Donatini 2011; De Silva et al. 2013; Lee et al. 2014b; Liu et al. 2015b; Muszynska et al. 2015; Phan et al. 2015, 2017a; Carhart-Harris et al. 2017; Wong et al. 2017; Rupcic et al. 2018).

The role of edible and medicinal mushrooms *A. bisporus*, *A. brasiliensis*, *C. militaris*, *G. lucidum*, *G. frondosa*, *H. erinaceus*, *L. edodes*, *L. rhinocerotis*, *O. sinensis*, *P. giganteus*, *T. versicolor*, *Termitomyces albuminosus*, and *T. fuciformis* in the treatment of neurodegenerative diseases and study of molecular mechanisms of neuroprotective and cognitive effects were recently reported (Wong et al. 2017; Tsuk et al. 2018). It was revealed that extracts and compounds obtained from mushrooms possess neuroprotective, and neuroregenerative effects for in vitro and in vivo assessments by targeting multiple physiological mechanisms and actions.

The neuroprotective effect has mainly been attributed to the discovery of various terpenoids that stimulate the production of NGF or BDNF. *Hericium* species are among the most praised edible medicinal mushrooms known as producers of bioactive compounds used for treatment of neurodegenerative diseases (Mori et al. 2009; Nagano et al. 2010; Donatini 2011; Lee et al. 2014b; Liu et al. 2015b; Samberkar et al. 2015; Thongbai et al. 2015; Brandalise et al. 2017; Chiu et al. 2018). Experimental studies have shown that polysaccharides and terpenoids (hericenones and erinacines) extracted from fruiting bodies and cultured mycelia of *H. erinaceus*, stimulate the synthesis of NGF, promote the growth and differentiation of neurons, and protect the cells against oxidative stress (Thongbai et al. 2015). The neuroprotective effects of *H. erinaceus* bio-products from homogenized, fresh fruiting bodies are manifested by attenuating the Aβ25–35-triggered damage of PC12 cells by significantly increasing cell viability and decreasing the release of lactate dehydrogenase (Liu et al. 2015b).

Hericenones and erinacines were also isolated from other *Hericium* species. NGF- and BDNF-induced corallocins and hericerin were detected in *H. coralloides* (Wittstein et al. 2016) and novel cyathane erinacines Z1 and Z2 with neurotrophin-inducing effects in mycelial cultures of *H. erinaceus* and *H. flagellum* (syn. *H. alpestre*) (Rupcic et al. 2018). Erinacine A-enriched mycelium of *H. erinaceus* produced antidepressant-like effects by modulating BDNF/PI3K/Akt/GSK-3 signaling pathway in mice (Chiu et al. 2018).

The stimulation of neurite outgrowth by aqueous and ethanol extracts from sclerotia of *L. rhinocerotis* in rat pheochromocytoma adherent (PC-12Adh) cells at 20 μg/mL was revealed (Eik et al. 2012). Furthermore, the combination of 20 μg/mL aqueous extract and 30 ng/mL of NGF enhanced the neurite outgrowth by 42.1% compared to either aqueous extract (24.4%) or NGF (24.6%) alone. The neurite outgrowth activity was also revealed by using an aqueous extract of *L. rhinocerotis* mycelium and in combination with other natural products, such as *Ginkgo biloba* and curcumin (John et al. 2013).

The neurite outgrowth of aqueous extracts of sclerotium and mycelium of *L. rhinocerotis* in mouse neuroblastoma (Neuro-2a) cells was detected (Phan et al. 2013). At 20 μg/mL, the extract of sclerotium resulted in 38.1% of neurite-bearing cells, which was approximately twice the number of NGF-treated neurite-bearing cells. However, the aqueous extract of mycelium did not cause a significant increase in neurite outgrowth when compared to NGF treatment. An increase of neuritogenic activity of medicinal mushrooms in rat pheochromocytoma cells (Seow et al. 2013) and neuritogenic activity in PC-12 cell lines treated with *L. rhinocerotis* extracts (Seow et al. 2015) was also reported. The maximal neuritogenic activity in PC12 at 25 μg/mL of aqueous extract was 20.99% followed by ethanol extract (17.4%) and crude polysaccharides (16.4%). The hot aqueous extract (25 μg/mL) stimulated neuritogenesis similar to NGF (50 μg/mL). However, all extracts promoted neuritogenesis without stimulating the release of NGF by PC12 cells (Eik et al. 2012; Seow et al. 2015).

The aqueous extract obtained from sclerotia of *L. rhinocerotis* was also able to stimulate neurite outgrowth in dissociated cells of the brain, spinal cord, and retina of chick embryo (Samberkar et al. 2015). The 50 μg/mL extract induced maximal neurite outgrowth in the brain and spinal cord (20.8 and 24.7%, respectively), while this induction in retinal cells (20.8%) was achieved at 25 μg/mL.

1.4.7 Antiviral and Anti-allergic Effects

The prevention and treatment of viral infections remain a serious problem in modern medicine. Therefore, the search for effective antiviral medications, including those of natural origin, is in demand. Mushrooms are considered the producers of different bioactive compounds with antiviral effects; however, despite considerable work to reveal various features of mushrooms, the studies of antiviral effects are inadequate (Adotey et al. 2011; Santoyo et al. 2012; Teplyakova et al. 2012; Krupodorova et al. 2014; Teplyakova and Kosogova 2016; Doğan et al. 2018).

The antiviral effect of protein fractions isolated from mycelium of *L. rhinocerotis* against human papillomavirus (HPV) was revealed (Abdullah et al. 2013). The antiviral activity against dengue virus type-2 (DENV-2) strain was detected in the hot aqueous extract of *L. rhinocerotis* at IC_{50} of 520 μg/mL (Ellan et al. 2013). The extract also inhibited viral RNA synthesis by 99.7% but did not have a significant effect on DENV-2 virusicidal activity and viral attachment. The anti-dengue activity correlated with carbohydrate content in the extract (2.0 mg/mL) suggesting that antiviral activity is attributed to polysaccharide content in the sclerotium.

The antiviral activities of extracts and polysaccharide fractions obtained from edible agaricoid mushrooms *B. edulis*, *L. edodes*, and *P. ostreatus* were detected against herpes simplex virus type 1 (HSV-1) (Santoyo et al. 2012). Aqueous extracts from mycelia of 11 polyporoid mushrooms showed antiviral activity against influenza A virus of birds A/chicken/Kurgan/05/2005 (H_5N_1) and humans A/Aichi/2/68 (H_3N_2) (Teplyakova et al. 2012). Among the tested species, *Daedaleopsis confragosa*, *Datronia mollis*, *Ischnoderma benzoinum*, *L. officinalis*, *Lenzites betulina*, *T. gibbosa*, and *T. versicolor* were suggested as perspective fungi to develop novel antiviral myco-medicines. It has been shown that polysaccharides, glycoproteins, melanins, nucleoside proteins, and terpenoids from several *Agaricomycetes* fungi exhibit antiviral effects against hepatitis, herpes, human immunodeficiency virus (HIV), influenza, West Nile viruses, as well as orthopox viruses, including the variola virus (Teplyakova and Kosogova 2016). Preparations from these mushrooms were also suggested for prevention of cancers with viral etiology.

The screening of antiviral activities of mycelia of ten basidiomycetous mushrooms against influenza A (serotype H_1N_1) and herpes simplex virus type 2 (HSV-2) showed that tested species inhibited the reproduction of influenza virus strain A/FM/1/47 (H_1N_1) in MDCK cells (Krupodorova et al. 2014). The species *Auriporia aurea*, *F. fomentarius*, *P. ostreatus*, and *T. versicolor* were also effective against HSV-2 strain BH in RK-13 cells, with similar levels of inhibition, for influenza virus. This is the original report of anti-influenza effects of *F. velutipes*, *Lyophyllum shimeji*, and *P. eryngii*. The anti-influenza and anti-herpes activities were also observed in *A. aurea*. The strain 353 of *T. versicolor* was found to be promising for the pharmaceutical industry as an anti-influenza and anti-herpes agent possessing low toxicity.

In a recent study, in vitro antiviral properties of *F. fomentarius*, *L. sulphureus*, *Morchella conica*, *M. esculenta*, *Ph. igniarius*, *P. ostreatus*, *Porodaedalea pini*, *Pyrofomes demidoffii*, *Terfezia boudieri*, and *Tricholoma anatolicum* were evaluated (Doğan et al. 2018). The results have demonstrated that the aqueous extract of *F. fomentarius*, *Ph. igniarius*, and *P. pini* showed considerable anti-HSV-1 activity, while methanolic and aqueous extracts of other species did not reveal any effects.

Recently the development of a mushroom-derived oral subunit vaccine from *Pleurotus* species has been suggested (Pérez-Martínez et al. 2015). However, antiviral bioactive compounds of fungal origin and mechanisms of their action remain subjects for further research.

Presently, allergies are considered an increasing problem worldwide. New strategies and a search for natural resources of anti-allergic compounds for preventive therapy are urgently needed. The anti-allergic potential of mushrooms has not been fully investigated yet. Using in vitro and animal assays, it has been shown that the influence of β-glucans on balance between Th1 and Th2 immune cells can result in an anti-allergic effect which has been described in *A. subrufescens*, *Armillaria ostoyae*, *F. velutipes*, *G. lucidum*, *G. tsugae*, *I. obliquus*, *Ph. linteus*, *P. ostreatus*, *P. pulmonarius*, *Tricholoma populinum*, and other *Agaricomycetes* fungi. However, the bioactive compounds responsible for this effect have not been identified yet (Ellertsen and Hetland 2009; Bouike et al. 2011; Jesenak et al. 2014; Merdivan and Lindequist 2017).

The sesquiterpenes originally isolated from dichloromethane extracts of mycelium and fruiting bodies of honey mushroom *A. ostoyae* exhibited a significant degranulation-inhibiting effect on RBL-2H3 cells at non-cytotoxic concentrations (Merdivan et al. 2017). The purified compounds melleolides H and J significantly inhibited degranulation. Therefore *A. ostoyae* could be recommended as a natural preparation to support the treatment of allergies.

1.5 Nutritional and Dietary Values of Medicinal Mushrooms

Mushrooms are regarded as gourmet food with high nutritional and dietary values, as well as healthy DSs and myco-pharmaceuticals (Hobbs 2005; Khatun et al. 2012; Glamočlija et al. 2015; Kumar 2015; Sękara et al. 2015; Badalyan et al. 2016; Wu et al. 2016; Atila et al. 2017; Biswas et al. 2017; Gargano et al. 2017; Glamočlija and Soković 2017; Reis et al. 2017; Badalyan and Zambonelli 2019; Phan et al. 2018). Therefore, there is a significant potential to develop mushrooms as nutraceuticals and functional food for human wellness and their bioactive molecules for the production of drugs (Dutta 2013; Degreef et al. 2016; Süfer et al. 2016; Landi et al. 2017). The best implementation of mushroom products and myco-pharmaceuticals is to maintain a good quality of life (QoL), especially in immune-deficient and immune-suppressed patients; cancer patients receiving chemotherapy or radiotherapy; chronic blood-borne viral infections of hepatitis B, C, and D; different types of anemia; human acquired immunodeficiency syndrome (AIDS); herpes simplex virus (HSV); chronic gastritis and gastric ulcers caused by *H. pylori*; etc. (Cucuianu et al. 2004; El Dine et al. 2009; Wasser 2010, 2014, 2017; Adotey et al. 2011; Chang and Wasser 2012; Santoyo et al. 2012; De Silva et al. 2013; Thongbai et al. 2015).

The nutritional value and health benefits of mushrooms are determined by their chemical composition. Many reports on chemical composition and nutritional value of culinary-medicinal mushrooms belonging to different taxonomic and ecological groups were published (Badalian et al. 1997a; Badalyan and Rapior 1999; Hobbs 2005; Nunes et al. 2012; Palazzolo et al. 2012; Kalač 2013, 2016; Badalyan 2015, 2016; Da Silva et al. 2015; Adejumo et al. 2015; Teklit 2015; Thongbai et al. 2015; Badalyan 2016; Lalotra et al. 2016; Bandara et al. 2017; Kostic et al. 2017; Taofiq et al. 2017a). Mushrooms contain essential minerals, trace elements, vitamins, high levels of dietary fiber (chitin), proteins, unsaturated fatty acids, and nearly free of cholesterol and can be used for various human diets (Ayaz et al. 2011; Badalyan 2015, 2016; Lalotra et al. 2016; Landi et al. 2017; Phan et al. 2018). They are characterized by high contents of water and proteins and low energy value ranging from 50 to 70 kcal/100 g (Chang and Miles 2008; Dadakova et al. 2009; Chang and Wasser 2012; Cheung 2013; Kalač 2013; Glamočlija et al. 2015). Mushrooms are also regarded as a good source of vitamins (B_1, B_2, B_3, B_5, B_6, B_{12}, D, and H), phosphorus, iron, and ergosterol (Donnini et al. 2013; Zotti et al. 2013; Kumar 2015; Kalač 2016). Macrofungi are considered a gourmet food because of their unique taste and flavor attributable to soluble sugars, eight-carbon derivatives, and various volatile organic compounds (Rapior et al. 1993, 1997; Breheret et al. 1998; Rapior et al. 2000; Nyegue et al. 2003; Badalyan and Zambonelli 2019; Kües et al. 2018).

Mushrooms contain 0.1–1 mg/g ergothioneine (2-mercaptohistidine trimethylbetaine), a water-soluble amino acid that naturally forms in some bacteria and non-yeast fungi. Mushrooms and meat are well-known sources of ergothioneine in human diet (Ey et al. 2007). The antioxidant effect of ergothioneine extracted from two *Pleurotus* species, i.e., *P. eryngii* and *P. citrinopileatus*, was recently reported (Yen et al. 2018).

The nutritive and medicinal value of ascomycetous mushroom, *M. esculenta*, is appreciated all over the world. Except for culinary properties, it has been used to treat gastric problems, to heal wounds, and reduce joint pain (Badalyan and Zambonelli 2019; Paul et al. 2018).

Based on their nutritional profile, mushrooms have been cultivated around the world as healthy dietary food (nutraceuticals) (Guillamón et al. 2010; Dutta 2013). The term "nutraceutical" is a conjunction of words "nutrition" and "pharmaceuticals"; therefore, nutraceutical is any substance which may be considered a food or part of the food that provides health-enhancing effects, including prevention and treatment of certain diseases (Barros et al. 2008; Farzana et al. 2017).

The nutritional and nutraceutical properties were evaluated in many mushroom species, e.g., *A. bisporus*, *P. ostreatus*, *P. sajor-caju*, and *S. commune* (Chang and Miles 2008; Chang and Wasser 2012, 2017; Chandrawanshi et al. 2017; Finimundy et al. 2018b; Ishara et al. 2018). The supplementation of food products with mushroom nutraceuticals (dietary fibers, proteins, and bioactive ingredients) obtained from *A. bisporus*, *Auricularia* spp., *B. edulis*, *F. velutipes*, and *L. edodes* be recommended to increase the quality and nutritional values of different types of products, such as dairy beverages, yogurts, bread, pasta, beer, etc. (Gregori 2014; Yen et al. 2015; Lu et al. 2016, 2018; Singh et al. 2016; Farzana et al. 2017; Heleno et al. 2017; Yuan et al. 2017; Francisco et al. 2018; Sknepnek et al. 2018). Mushroom-derived β-glucans possess effective health-enhancing effects and are ideal components for any baked goods, beverages, and DSs (Wasser and Didukh 2005; Zhu et al. 2015).

The potential of mushroom polysaccharides for the development of nutraceutical foods and drugs was recently summarized by Singdevsachan et al. (2016).

Among medicinal mushrooms, cultivated edible species *A. bisporus*, *A. campestris*, *A. auricula-judae*, *L. edodes*, *Pleurotus* spp., and *V. volvacea* are in high demand due to their nutritional value and pharmaceutical potential (Chang 1996; Chang and Buswell 1996; Chang and Miles 2008; Chang and Wasser 2012, 2017; Dutta 2013; Sękara et al. 2015). However, it should be noted that after consumption of *Auricularia* species, the Szechwan purpura syndrome may develop caused by the injury of thrombocytes (Hammerschmidt 1980; Giacomoni 2004; Brunelli 2009). Furthermore, the consumption of raw or undercooked *L. edodes* may cause shiitake dermatitis, a skin reaction caused by glucan lentinan (Nguyen et al. 2017).

The analysis of nutritional values (carbohydrate, crude fiber, protein, and fat contents), phenolic and flavonoid contents, and antioxidant capacity of wild mushrooms *Amanita calyptroderma*, *A. princeps*, *Astraeus odoratus*, *Heimiella retispora*, *Mycoamaranthus cambodgensis*, *Russula alboareolata*, *R. cyanoxantha*, *R. emetica*, *R. virescens*, and *T. clypeatus* and cultivated edible species *A. auricula-judae*, *Lentinus polychrous*, *L. squarrosulus*, *P. sajor-caju*, and *V. volvacea* showed that wild specimens contain higher content of fibers and bioactive compounds than cultivated specimens (Srikram and Supapvanich 2016).

The edible and medicinal mushrooms are also considered functional food or DSs ("nutriceuticals") due to their high contents of proteins, polyunsaturated fatty acids, α-tocopherol, phenols, vitamins, and other bioactive molecules (Chang 1996; Chang and Buswell 1996; Wasser and Didukh 2005; Chang and Miles 2008; Wasser 2010; Chang and Wasser 2012, 2017; Dutta 2013; Heo et al. 2014; Kumar 2015; Landi et al. 2017; Reis et al. 2017; Lu et al. 2018). Moreover, as a natural source of bioactive molecules, they may be used to develop myco-nutriceuticals and myco-pharmaceuticals.

The mushroom nutriceuticals can be extracted from fruiting bodies, mycelia, sclerotia, and spores' powder of basidiomycetous (*A. auricula-judae*, *C. cibarius*, *G. lucidum*, *L. edodes*, *Lentinus* (syn. *Pleurotus*) *tuber-regium*, *Lyophyllum decastes*, *P. eryngii*, *P. ostreatus*, *R. delica*, *S. commune*, *T. versicolor*, *T. fuciformis*, *V. volvacea*, and *Wolfiporia cocos*) and ascomycetous (*T. borchii* and *T. melanosporum*) macrofungi (Palazzolo et al. 2012; Adejumo et al. 2015; Teklit 2015; Venturella et al. 2016). The functional properties of uncooked and cooked ectomycorrhizal gasteroid fungi *Astraeus hygrometricus* have selective advantages and serve as a valuable raw material for the production of functional nutritional and pharmaceutical products with desired qualities (Pavithra et al. 2017).

A recent study addressed the chemical composition (glucans, glycerides, phospholipids, and sterols) and functional food value of cultivated culinary-medicinal mushroom *G. frondosa* possessing anti-inflammatory and antioxidant activities (Dissanayakea et al. 2018). Several compounds inhibited COX-1 and COX-2 enzymes at 100 μg/mL similar to NSAID, such as aspirin, ibuprofen, and naproxen.

The dietary supplements formulated from *Ganoderma* species are traditionally used in China, Japan, and Korea to improve life expectancy and prevent and treat many diseases, such as arthritis, asthma, bronchitis, CVD, gastritis, hypercholesterolemia, arterial hypertension, nephritis, neurasthenia, neoplasia, etc. (Paterson 2006; Cheng et al. 2010; Wang et al. 2012; Tan and Miyamoto 2015).

The screening of chemical composition, nutritional value, and bioactive properties of wild *Polyporus squamosus* mushroom revealed trehalose as the main free sugar and *p*-hydroxybenzoic acid as the main phenolic compound present in this species (Mocan et al. 2018). The highest levels of malic acid and β-tocopherol were detected in the fruiting bodies, while polyunsaturated fatty acids represented more than 57% of total fatty acids. As a source of polyunsaturated fatty acids, β-tocopherol, and bioactive compounds with antioxidant, antifungal, and antibacterial activities, *P. squamosus* can be further explored to develop new pharmaceuticals and DSs.

Several medicinal mushrooms (*A. campestris*, *A. auricula-judae*, *G. lucidum*, *O. sinensis*, and *Sanghuang porusbaumii*) have been used in traditional medicine for the prevention of diabetes (Zuomin et al. 1998). The usage of *G. lucidum*, *G. neojaponicum*, *G. frondosa*, *H. erinaceus*, *L. rhinocerotis*, *P. giganteus*, *T. camphoratus*, and many other mushroom products helps to reduce or prevent age-related neurodegenerative processes, such as Alzheimer's and Parkinson's diseases (Sabaratnam et al. 2013; Phan et al. 2015; Brandalise et al. 2017).

The genus *Pleurotus* (oyster mushrooms) comprises a group of mushrooms possessing high nutritional value and therapeutic properties. They are rich in proteins with essential amino acids, polysaccharides, dietary fibers, important minerals, and vitamins. The oyster mushrooms possess antiaging, anticancer, antimicrobial, anti-obesity, antioxidant, hypocholesterolemic, hypoglycemic, and hypotensive properties (Badalyan et al. 2008a; Baggio et al. 2010; Patel et al. 2012; Jayasuriya et al. 2015; Fu et al. 2016; Acharya et al. 2017; Debnath et al. 2017; Ebrahimi et al. 2017; Masri et al. 2017; Abidin et al. 2018; Adebayo et al. 2018; Finimundy et al. 2018a). They have been recommended as supplementary nutritional products for adjuvant therapy in patients with colorectal carcinoma. However, further clinical trials are needed to confirm the safety of described bioactive compounds of these mushrooms as alternatives to conventional drugs (Corrěa et al. 2016; Finimundy et al. 2018a, b).

Fruiting bodies and mycelia of oyster mushrooms have been used in the human diet throughout the world due to their rich content of nutrients, including proteins, fibers, carbohydrates, minerals, vitamins, and lipids, and high culinary properties (Patel et al. 2012; Corrěa et al. 2016). Several bioactive compounds have been identified in these mushrooms, including phenolics, polysaccharides, terpenes, and sterols (Corrěa et al. 2016; Wu et al. 2016). Therefore, a high nutritional value and potent medicinal usage suggest that *Pleurotus* mushrooms can be considered renewable and easily accessible sources of innovative functional foods/nutraceuticals and pharmaceuticals with antioxidant, anti-inflammatory, antimicrobial, antitumor, and immunomodulatory effects (Gunde-Cimmerman 1999; Patel et al. 2012; Fernandes et al. 2015; Khatun et al. 2015; Corrěa et al. 2016; Wu et al. 2016; Anjana and Savita 2017; Bello et al. 2017; Masri et al. 2017; Morris et al. 2017; Finimundy et al. 2018b). However, it should be noted that within 70 discovered *Pleurotus* species, only a few (*P. florida*, *P. ostreatus*, *P. pulmonarius*, and *P. sajor-caju*) are currently available in the market.

The nutraceutical and pharmaceutical potential of bioactive compounds isolated from *A. bisporus* (lectins), *A. auricula-judae*, and other *Auricularia* spp. (acidic polysaccharides), *G. frondosa* (grifolan, lectin), *Lentinus* (=*Pleurotus*) *sajor-caju* (lovastatin), and *O. sinensis* (cordycepin) (Xu et al. 2011, 2016a, b; Liu et al. 2015a; Prasad et al. 2015; Sękara et al. 2015) and general nutritional value, pharmacological properties, as well as potential for therapeutic applications of *L. edodes* (Finimundy et al. 2014; Thaper and Lakshmi 2017), *A. aegerita* (Landi et al. 2017), *Agaricus sylvaticus* (Monro 2003), and other mushrooms were also reported. Medicinal mushroom-based DSs produced from *A. brasiliensis*, *G. lucidum*, *G.*

frondosa, *L. edodes*, and *P. ostreatus* have already been approved for therapeutic use in Croatia (Jakopovich 2011).

Nutritional and nutraceutical properties of bioactive compounds (phenolics, flavonoids, antifungal proteins, immunostimulatory glucans, and lectins) derived from the wild edible mushroom *Lentinus squarrosulus*, as well as its antioxidant and wound-healing properties, were recently evaluated by Lau and Abdullah (2017). Both fruiting body and mycelium of *L. squarrosulus* were suggested as potential sources of functional ingredients. However, further clinical trials testing the mechanism of action of bioactive compounds in vivo and quality control of *L. squarrosulus* biomass obtained from different cultivation methods are recommended. The extracts obtained from fruiting bodies and sclerotia of medicinal mushrooms *P. tuber-regium* and *P. umbellatus* were suggested for the manufacturing of food supplements, pharmaceuticals, as well as cosmetic products and beverages (Badalyan et al. 2008a; Bandara et al. 2015).

Agaricomycetes coral mushrooms from genera *Ramaria* (order *Gomphales*) and *Clavaria* (order *Agaricales*) are widely distributed worldwide. However, most of these species have not been investigated for their nutritional and nutraceutical content. The information regarding the culinary status, biochemical and bioactive profiling of six *Ramaria* (*R. aurea*, *R. botrytis*, *R. flava*, *R. flavescens*, *R. rubripermanens*, and *R. stricta*) and six *Clavaria* (*C. amoena*, *C. coralloides*, *C. fragilis*, *C. purpurea*, *C. rosea*, and *C. vermicularis*) species collected in northwestern Himalayas was originally reported (Sharma and Gautam 2017). All the species were found to contain high levels of proteins, macro- and microminerals, carbohydrates, unsaturated fatty acids, essential amino acids, phenolics, tocopherols, anthocyanidins, and carotenoids. The studied mushrooms, particularly *R. botrytis* and *C. fragilis*, showed significant antioxidant and antibacterial activities against *E. coli*, *Klebsiella pneumoniae*, *P. aeruginosa*, *Streptococcus pneumoniae*, *Vibrio alginolyticus*, and *V. cholerae*. They are promising for future large-scale commercial exploitations and could be used as natural sources of antioxidant and antibacterial nutraceuticals.

Apart from basidiomes, the submerged mycelial culture of medicinal mushrooms represents a source to search for new safe and healthy myco-products with standardized quality. Currently, biotechnological cultivation of mycelia is progressing, and the production of mycelium-based biotech products from mushrooms is continuously improving (Asatiani et al. 2007; Chen et al. 2008; Elisashvili 2012; He et al. 2017a, b). In a recent study, the mycelium and cultural broth of *P. eryngii*, were reported as alternative sources of bioactive compounds with antioxidant, anti-inflammatory, and cytotoxic activities (Souilem et al. 2017). A protein-bound polysaccharide, nucleic acid, and amino acid content in mycelia and fruiting bodies of *F. velutipes* and *C. militaris* were analyzed, and the development of new products (amino acid drink or amino acid-containing food) from these mushrooms was suggested (Kim et al. 2014).

The prebiotics are food ingredients which stimulate the growth of beneficial microbiota. Oligosaccharides and fibers are major constituents of prebiotics. A recent trend in food science and technology has shown the efficacy of prebiotics to modulate gut microbiota and attenuate several diseases, such as diabetes, obesity, and cancer.

Mushrooms are regarded as a potential source for the development of prebiotics containing different nondigestible polysaccharides, such as chitin, hemicellulose, α- and β-glucans, mannans, xylans, and galactans. From all mushroom polysaccharides, β-(1 → 3)-D-glucans and their peptide/protein derivatives (polysaccharide-peptide/protein complexes and proteoglycans) are essential prebiotics and possess immunomodulatory and antitumor effects by enhancing the growth of prebiotic bacteria in the gut (Bhakta and Kumar 2013; Singdevsachan et al. 2016). The medicinal mushrooms *C. versicolor*, *G. lucidum*, *G. frondosa*, *H. erinaceus*, *I. obliquus*, and *L. edodes* were reported as prebiotics due to the presence of immunomodulatory glucans regulating gut microbiota by inhibiting exogenous pathogens conferring health benefits to the host (Jayachandran et al. 2017; Diling et al. 2017). The prebiotic effect of Mexican *G. lucidum* (Meneses et al. 2016) and polysaccharides isolated from Polish wild mushroom (Nowak et al. 2017) have also been described.

Mushrooms have the effective capacity to absorb and accumulate trace elements from substrates, such as Selenium and other minerals. Thus, they could be used to produce mineral-enriched dietary food (Nunes et al. 2012; Milovanović et al. 2013, 2015a, b; Lalotra et al. 2016). The dietary Se has been recognized as an antioxidant, and the deficiency of this element is associated with numerous chronic degenerative diseases, various types of cancer, cardiomyopathies, and endemic osteoarthropathy (Rayman 2000; Combs 2001; Pedrero and Madrid 2009; Liu et al. 2016; Klimaszewska et al. 2017). High levels of Se (10–20 mg/kg dry matter) have been detected in widely consumable edible mushrooms, including *A. campestris*, *A. cesarea*, *A. macrosporus*, *A. silvaticus*, *Boletus aereus*, *B. aestivalis*, *B. appendiculus*, *B. edulis*, *B. erythropus*, *B. pinicola*, *G. lucidum*, and *Xerocomus badius* (Kalač and Svoboda 2000; Falandysz 2008). The effect of enrichment with Se on antioxidant, antifungal, and cytotoxic properties of mycelial extracts of *G. applanatum*, *G. lucidum*, *Lenzites betulina*, and *Trametes hirsuta* has recently been studied which revealed the stimulatory effect of Se on antioxidant activity of these species (Milovanović et al. 2015a, b).

The vitamin-enriched mushroom dietary food could also play an important role in the prevention of chronic diseases (Mehrotra et al. 2014). It has been shown that vitamin D_2-enriched mushroom (VDM) *A. bisporus* may provide a dietary source of vitamin D_2 and other bioactive molecules to prevent cognitive abnormalities associated with dementia (Bennett et al. 2013). However, this study supports the need for randomized clinical trials to determine whether VDM consumption can enhance cognitive performance in the wider population.

In recent years, mycosterols (MSs) have emerged as potential functional ingredients for the development of sterol-enriched food products and DSs (Heleno et al. 2017; Corrĕa et al. 2018; Francisco et al. 2018). The consumption of MS-enriched food may significantly reduce the levels of low-density lipoprotein, a major risk factor for CVD.

The isolation of novel bioactive MSs and study of their usage in formulations of nutritional products represent challenges for current nutritional and pharmacological research. The ergosterol (ergosta-5,7,22-trien-3β-ol) has been considered the principal sterol of hyphal membranes, followed by other derivatives, i.e., ergosta-5,8,22-trien-3-ol, ergosta-7,22-dien-3-ol, ergosta-5,7-dien-3-ol, and ergosta-7-en-3-ol (fungisterol) (Corrěa et al. 2017).

Recent studies revealed new perspectives for application of mushroom-derived bioactive compounds in cosmetic industry (Taofiq et al. 2016a, b; Wu et al. 2016; Badalyan and Zambonelli 2019).

Presently, medicinal mushrooms in the Western Hemisphere are primarily used as DSs without a declaration of a medical indication. An overview of the principles of authorization, market access of herbal drugs in Europe, and the current status regarding mushrooms has been thoroughly described to support the development of legalized pharmaceutical preparations of mushroom origin in Europe (Lindequist 2013).

1.6 Current State of Epidemiological and Clinical Studies of Medicinal Mushrooms

Mushrooms have an established history of use in the traditional medicine of East Asian countries, Eastern Europe, as well as South and North America (Hobbs 1995; Stamets 2000; Wasser 2011; Guzmán 2015; Hapuarachchi et al. 2017; Gargano et al. 2017). Modern interdisciplinary research has validated the traditional knowledge which continues to demonstrate potent and unique therapeutic properties of bioactive compounds of mushroom origin.

Epidemiological and clinical studies of medicinal mushrooms conducted in Asian countries suggest that mushroom-derived natural products are used as part of different preventive and treatment strategies together with other therapies. They are mainly used as preventive agents against certain types of cancer, particularly gastrointestinal and breast cancers (Chang and Lee 2004; Sullivan et al. 2006; Bishop et al. 2015; Chaiyasut and Sivamaruthi 2017). Researchers are also focusing on the molecular mechanism behind anticancer action of mushroom-derived biomolecules (Joseph et al. 2017). Most of the available data comes from in vitro studies and in vivo experimental animal models. Therefore, systematic clinical trails should be initiated to translate the updated knowledge to clinical research.

Until now, the health-enhancing effects of various mushroom preparations in humans have been reported in more than 600 papers and reports. However, a few reviews address systematic clinical studies on the role of several medicinal mushrooms (e.g., *G. lucidum*, *G. frondosa*, *H. erinaceus*, *T. versicolor*, etc.) in the treatment of cancer, oncological immunological diseases, and diabetes (Smith et al. 2002a, b; Khatun et al. 2012, 2015; Hapuarachchi et al. 2016a, b; Wasser 2017; Rossi et al. 2018; Zmitrovich et al. 2019). Information on different clinical trials and epidemiological studies of medicinal mushrooms can also be found in several books (Hobbs 1995; Stamets 2000; Powell 2014; Rossi et al. 2018).

Epidemiological reports on the dietary consumption of mushrooms to promote health and prevent different illnesses mainly come from Brazil, China, India, Japan, and Korea. They have indicated that consumption of medicinal mushrooms in the form of fruiting bodies or extracts may prevent or delay the development of cancer. Previously reported epidemiological studies of gastric and breast cancers were mainly carried out with consumption of *G. frondosa*, *G. lucidum*, *F. velutipes*, *Hypsizygus marmoreus*, *Pholiota nameko* (Ikekawa 2001, 2005; Hara et al. 2003; Zhang et al. 2009; Shin et al. 2010; Rossi et al. 2018), *A. bisporus*, *L. edodes*, and other edible medicinal mushrooms in combination with green tea (Zhang et al. 2009). The results have shown that three times daily consumption of 3–5 g of hot water extract of *A. subrufescens* improves quality of life (QoL) by strengthening the innate healing power and immune system (Liu et al. 2008). The preclinical assessment of safety of the shiitake (*L. edodes*) on biochemical, haematological, and antioxidant parameters in rats showed that the daily intake of 100 mg/kg of mushroom could have potential health-enhancing effects (Grotto et al. 2016).

Currently, more than 130 medicinal properties (analgesic, antibacterial, antifungal, anti-obesity, antioxidant, antiparasitic, antitumor, antiviral, cardioprotective, immunomodulatory, neuroprotective, anti-allergic, etc.) were described in different mushrooms, such as *A. bisporus*, *A. blazei*, *A. cinnamomea*, *G. lucidum*, *G. frondosa*, *H. erinaceus*, *L. edodes*, *P. cornucopiae*, *P. eryngii*, *P. ostreatus*, and *T. versicolor*.

Modern clinical practices in China, Japan, Korea, Russia, and several other countries rely on medicinal mushroom-derived drugs (MMDD) and DSs to increase the tolerance to cancer therapy in cancer patients (Wasser 2010, 2017; Suzuki et al. 2013; Tangen et al. 2015; Twardowski et al. 2015; Del Buono et al. 2016; Tanaka et al. 2016; Tsai et al. 2016; Wasser 2017; Zmitrovich et al. 2019). In many clinical studies, mushrooms were used as an adjuvant treatment with conventional chemo- or radiotherapy for different types of cancer (bladder, brain, breast, liver, lung, ovary, prostate, stomach, etc.) to reduce the side effects of cancer treatment (e.g., hair loss, nausea, and loss of appetite) (Smith et al. 2002b; Sullivan et al. 2006; Suárez-Arroyo et al. 2017; Rossi et al. 2018). However, no evidence revealed the proper dosages and treatment duration for many species, sometimes due to poor study design, non-standardized mushroom preparations, statistical methods, etc. (Wasser 2017).

Currently, both edible and inedible mushrooms used in traditional medicine are undergoing clinical trials (Figs. 1.1 and 1.2). The Asian traditional medicine has proved the significance of inedible reishi (*G. lucidum*) and edible shiitake (*L. edodes*) mushrooms. Other well-known medicinal species, such as *Coriolus* (syn. *Trametes*) *versicolor*, *F. fomentarius*, *Fomitopsis officinalis*, *I. obliquus* (Chaga), *Ph. linteus*, and *P. betulinus*, are also used for the treatment of gastrointestinal disorders, various forms of cancer, bronchial asthma, etc. (Grienke et al. 2014; Chen et al. 2016).

Preclinical and clinical studies of *G. lucidum* polysaccharide (GLPS), as an approved non-hormonal drug, have been used for treating refractory myopathy and other diseases (Zeng et al. 2018). As herbal products, *A. blazei*, *Cordyceps* species,

C. versicolor, *G. lucidum*, *L. edodes*, and *T. camphoratus* are widely used in TCM together with pomegranate, green tea, garlic, turmeric, and *Artemisia* plants (Lee et al. 2012).

Although in vitro and animal studies as well as the history of usage of mushrooms in TCM confirm the ethno-mycomedicinal knowledge, there is no scientific evidence to support their efficacy for treatment of human diseases. Therefore, consumers should critically evaluate information about the medicinal properties of mushrooms (Money 2016; Mustonen et al. 2018). Thus, systematic experimental, epidemiological, and clinical studies are required to fully explore the pharmacological potential of mushrooms and develop mushroom pharmacopeia (Taofiq et al. 2016a, 2017a; Wasser 2017).

Currently, a large number of clinical trials are carried out using edible and inedible wild and cultivated mushrooms, including *A. subrufescens*, *G. lucidum*, *G. frondosa*, *H. erinaceus*, *I. obliquus*, *L. edodes*, *O. sinensis*, *Ph. linteus*, *S. commune*, and *T. versicolor* (Figs. 1.1 and 1.2). The anticancer, immunomodulatory, hypoglycemic, and hypolipidemic activities of mushrooms were mainly investigated (Xiao et al. 2004; Hobbs 2005; Oba et al. 2007, 2009; Standish et al. 2008; Eliza et al. 2012; Ren et al. 2012; Aleem 2013; Suzuki et al. 2013; Donatini 2014; Guggenheim et al. 2014; Fritz et al. 2015; Motoi et al. 2015; Del Buono et al. 2016; Frost 2016; Wasser 2017).

Different extracts obtained from fruiting bodies, mycelial biomass, spores, and pure β-glucans (e.g., lentinan from *L. edodes*, schizophyllan from culture broth of *S. commune*, PSK or PSP complex from *T. versicolor*) have been tested to treat oncological diseases in clinical trials (Ren et al. 2012; Roupas et al. 2012; Suzuki et al. 2013; Donatini 2014; Del Buono et al. 2016; Ina et al. 2016; Wasser 2017). Clinical trials, and epidemiological data evaluate the efficacy and safety of mushroom-derived β-glucans, mainly lentinan, Maitake D-fraction, and schizophyllan in the prevention and treatment of cancer (Aleem 2013; Wasser 2017). Future randomized controlled trials to reveal the clinical efficacy of mycotherapy and the mechanisms of action of mushroom-derived anticancer compounds on long-term survival, tumor response, host immune functions, inflammation, and QoL in cancer patients should also be addressed (Rossi et al. 2018).

Some clinical studies have elucidated the mechanisms responsible for the immunomodulatory effects of β-D-glucans. They have shown to activate cytotoxic macrophages, monocytes, neutrophils, NK cells, cytokines (interleukins and interferons), and colony-stimulating factors (CSF) that trigger complementary and acute phase responses (Johnson et al. 2009; Choi et al. 2014; Dai et al. 2015; Kang et al. 2015; Tanaka et al. 2016).

Contemporary studies directed to biological activities of fungal β-glucans have gained significant attention in biomedical science. However, the mechanism of biological activity of fungal β-glucans has not been completely understood. The physicochemical and biological characteristics of β-glucans isolated from different agaricoid and polyporoid edible and medicinal mushrooms have recently been discussed by Zhu et al. (2015). The polysaccharide fraction obtained from *A. bisporus*, consisting of 90% β-glucans, induced the production of the tumor necrosis factor alfa (TNF-α) in vitro. However, administered in vivo, β-glucans have lost their ability to stimulate the immune response as observed using an in vitro animal model (Volman et al. 2010).

The recent results of clinical studies involving *A. subrufescens*, *G. lucidum*, and *I. obliquus* mushrooms revealed the effects of β-glucans on cancer cells (Frost 2016).

It has been shown that the immunomodulatory effects of β-glucans are used for prevention of viral infections (Vetvicka and Vetvickova 2015). The feeding of influenza virus-infected mice for 2 weeks with a mixture of glucans obtained from fruiting bodies of *G. frondosa* and mycelia of *A. brasiliensis*, *I. obliquus*, and *L. edodes* significantly reduced the clinical symptoms of infection. These results suggest that the consumption of dietary glucan might be useful as a complementary or alternative approach for the treatment of influenza infection. Further clinical application of β-glucans and studies of structure-activity relationship to develop β-glucan-enriched functional food products should be performed.

The five major macrofungi *A. blazei*, *G. lucidum*, *G. frondosa*, *O. sinensis*, and *T. versicolor* used in the treatment of oncological diseases appear to improve the effects of chemotherapy (Guggenheim et al. 2014).

T. versicolor-derived PSK is widely used in Japan as an adjuvant immunotherapeutic agent to treat a variety of cancers, including lung cancer. The meta-analysis of *T. versicolor* effects demonstrated an increased rate of survival, particularly in breast, gastric, and colorectal cancer patients (Eliza et al. 2012). Fifteen of 17 preclinical studies supported the anticancer effects of PSK through immunomodulation and tumor-inhibiting effects in vivo which resulted in reduced tumor growth (Fritz et al. 2015). The usage of *T. versicolor* as part of immune therapy for breast cancer treatment showed that polysaccharide constituents of the mushroom as concurrent adjuvant cancer therapy might be warranted as part of comprehensive cancer treatment and secondary prevention (Standish et al. 2008). Eight randomized controlled clinical trials, including 8009 patients with gastric cancer treated with PSK in combination with chemotherapy, indicate that adjuvant immune-chemotherapy improved the survival of cancer patients after gastric resection in comparison to chemotherapy alone (Oba et al. 2007).

The purified β-glucan lentinan from *L. edodes* is a well-established drug used for the combined treatment of oncological diseases in Japan. Moreover, preparations for intravenous and per os administrations of lentinan are currently available. Five studies, involving 650 patients with non-resectable or recurrent stomach tumors, demonstrated that patients treated with chemotherapy and lentinan had a significantly higher survival rate compared to patients treated only with chemotherapy (Oba et al. 2009).

A randomized, double-blind, placebo-controlled trail in children with recurrent respiratory tract infections showed that treatment with pleuran, a β-glucan isolated from *P. ostreatus*, reduced symptoms of atopy related to these infections (Jesenak et al. 2014).

A few in vivo studies have been performed with *Ganoderma* species (Jin et al. 2012; Hapuarachchi et al. 2016a, b, 2017; Klupp et al. 2016; Wasser 2017). In many cases, it has been questioned whether *Ganoderma* is solely a DS for well-being or merely a helpful "medication" for regenerative purposes. No conclusive report is available on clinical trials using *Ganoderma* species as a direct control agent for the diseases. In addition, there is no evidence supporting the usage of *Ganoderma* as a potential supplement for oncological or other diseases since no preclinical trials have been performed to date. The purified substances isolated from *Ganoderma* have been investigated to reveal the molecular mechanisms responsible for antitumor and immunomodulatory effects. It was concluded that the standardization and quality control of *Ganoderma* species, cultivation processes, and commercial formulations are required to consider these species as natural products for the prevention and treatment of diseases (Hapuarachchi et al. 2017). Further experimental, epidemiological, and clinical studies should be carried out to identify molecular targets and investigate the association between *Ganoderma* intake and disease risk. Moreover, the efficacy, dosage, and safety, alone or in combination with chemotherapy or radiation treatment, should also be investigated (Hapuarachchi et al. 2016a, b).

A review of more than 250 clinical studies of *G. lucidum* and other *Ganoderma* species was published by Jin et al. (2012). The patients who received *G. lucidum* in addition to chemotherapy and radiation treatment revealed a stronger response to conventional treatment compared to patients with conventional treatment alone. Moreover, this combination treatment also improved their QoL. It was concluded that *G. lucidum* could be administered as an adjunct to conventional treatment considering its potential of stimulating the immune system of the host.

Reishi mushroom *G. lucidum* is also frequently used for the treatment of CVD (Klupp et al. 2015). Five medical studies comparing *G. lucidum* with placebo in 398 patients with type 2 diabetes showed that the daily usage of *G. lucidum* or a mixture of *G. lucidum* (75%) and *O. sinensis* (25%) does not support the use of *G. lucidum* for the correction of cardiovascular risk factors in patients with type 2 diabetes mellitus. Future research of *G. lucidum* should be placebo-controlled and adhere to clinical trial-reporting standards. The results of a double-blind, randomized, placebo-controlled trial found no significant effect of administration of 3 g/day *G. lucidum* or a combination of *G. lucidum* with *O. sinensis* for 16 weeks on HbA1c and fasting plasma glucose levels of a small number of patients with type 2 diabetes (Klupp et al. 2016).

Many ex vivo and in vivo studies have revealed antidiabetic, anticancer, anti-inflammatory, antimicrobial, antimutagenic, antioxidant, antiparasitic, hepatoprotective, immunomodulatory, and other pharmacological activities in the medicinal mushroom *A. blazei* Murrill sensu Heinemann (syn. *A. subrufescens*), but only 17 clinical studies and 2 case reports on *A. blazei* were found (Therkelsen et al. 2016). The nutritional and therapeutic properties of *A. blazei*, with emphasis on its chemical composition, as well as clinical trials were recently reviewed (Da Silva de Souza et al. 2017).

The findings of clinical trials involving the regular consumption of *A. brasiliensis* Wasser (syn. *A. blazei* Murrill sensu Heinemann) (Ahn et al. 2004; Talcott et al. 2007; Motoi et al. 2015) and *G. frondosa* (Kodama et al. 2002; Konno 2009; Rajamahanty et al. 2009) by healthy volunteers showed improved immunity and, in single cases, increased rate of survival and general quality of live (QoL). However, a randomized, double-blind, placebo-controlled trial showed that the extract of *A. blazei* had no stimulating effect on IL-6, IFN-γ, or TNF-α levels in elderly females (Lima et al. 2011). Further clinical trials, using reliable statistical methods and standardized preparations, are needed to establish the efficacy of *A. blazei* as a therapeutic agent (Therkelsen et al. 2016).

The results of a randomized, double-blind, placebo-controlled clinical trials suggest that the dietary supplementation with *A. sylvaticus* improves the haematological and immunological parameters and reduces blood glucose levels in colorectal cancer patients (Fortes et al. 2008, 2009).

The medicinal properties (anti-asthmatic, anticancer/antitumor/immunomodulatory, anticoagulant, anti-inflammatory, antimicrobial, anti-obesity, antioxidant, antiviral, and neuroprotective) of *L. rhinocerotis* were recently reviewed (Nallathamby et al. 2018). In vitro results suggest that fruiting bodies and sclerotia of this species can be considered as alternative therapeutic strategies in the management of non-communicable diseases. Therefore, additional studies including in vivo clinical trials are needed to scientifically validate the usage of mycochemicals of *L. rhinocerotis* and develop therapeutic products.

Numerous bioactive compounds derived from medicinal mushroom *H. erinaceus* (Yamabushitake) have been developed into food supplements and alternative medicines (Thongbai et al. 2015). A double-blind, parallel-group, placebo-controlled trial was performed on Japanese men and women diagnosed with mild cognitive impairment to highlight the efficacy of oral administration of *H. erinaceus* (250 mg tablets containing 96% of *H. erinaceus* powder three times a day for 16 weeks). The obtained results have shown that *H. erinaceus* is effective in improving mild cognitive impairment (Mori et al. 2009).

A prospective randomized trail of 25 patients comparing the efficacy of *H. erinaceus* versus essential oils against *H. pylori* infection revealed that patients who received *H. erinaceus* showed negative results for Pyloritop® test in 89.5% of cases in contrast to 33.3% for patients who started with essential oils (Donatini 2014). It was suggested that *H. erinaceus* could be considered an alternative to antibiotic therapy against *H. pylori*-associated diseases. Additional randomized studies versus referential therapy should be performed to focus on treatment without adverse effects.

In a double-blind, placebo-controlled clinical study, oral administration of *H. erinaceus* (250 mg tablets with 96% mushroom powder, three times a day, for 16 weeks) improved cognitive abilities of 50–80-year-old patients (Mori et al. 2009). Another clinical study revealed the antidepressant effect of *H. erinaceus* (Nagano et al. 2010) which may be due to its high 5-hydroxytryptamine content (Muszynska et al. 2015).

A previous clinical study of oyster mushroom has revealed a significant reduction of blood glucose, arterial pressure, and cholesterol levels of diabetic patients without any deleterious effects on the liver and kidney (Khatun et al. 2007). However, well-designed randomized controlled trials with long-term consumption of this mushroom are needed to prove the bioactivity and safety of the product for diabetic patients.

Proceeding from the results of epidemiological and clinical studies of medicinal mushrooms, it should be noted that high-quality, long-term, randomized, double-blind, placebo-controlled clinical studies should be continued in the future.

1.7 Conclusions and Future Prospects

Mushrooms can be used as nutritional products, tonics, medicines, cosmetics, and natural biocontrol agents. The multidimensional nature of the global mushroom-cultivation industry, its role in addressing critical issues faced by humankind, and its positive contributions are noteworthy. Furthermore, mushrooms can serve as agents for promoting economic growth in a society. Mushrooms produce chemically diverse compounds with a broad spectrum of biological activities. In vitro assays, animal studies, and clinical trials justify the traditional experience and suggest a great potential of mushroom-derived compounds and products for the prevention and treatment of various diseases. In view of promising results, more efforts are needed to explore the therapeutic potential of medicinal mushrooms and promote the development of drugs. The important tasks include the realization of clinical studies, development of high-quality mushroom-derived products with standardized procedures, and their sustainable production under controlled conditions.

The present review covers the potential benefits of medicinal mushrooms related to nutrition and diet, functional food supplements and bioactive ingredients for prevention of diseases, and reduction of side effects in patients receiving conventional therapies. The complementary medicine is becoming one of the options that patients use to address their distress during and after treatment. Limited information exists about the costs and benefits of adjunct non-pharmacological treatments using medicinal mushrooms. Further research is needed in this field to expand knowledge preclinical and clinical investigations based on relevant data and facilitate the comprehensive assessment of therapeutic value of mushrooms.

Acknowledgments This chapter arises from a long-standing cooperation between two authors (S.M.B. and S.R.) on fungal research directed to the identification of bioactive compounds and medicinal properties supported by the collaboration between the Institute of Pharmacy; Yerevan State University, Armenia; and Faculty of Pharmacy of the University of Montpellier/UMR 5175 CNRS, France.

We thank Philippe Callac (INRA, Villenave d'Ornon, France) for advice on the genus *Agaricus*. We are grateful to our colleagues Claudio Angelini (Pordenone, Italy), Guy Fourré (France), Jacques Guinberteau (France), Jean-Paul Maurice (Société Lorraine de Mycologie, Neufchâteau, France), Jean-Marc Moingeon (Goux-les-Usiers, France), Luigi Perrone (Roma, Italy), Jean-Philippe Rioult (EREM, Caen, France), and Peter Verstraeten (Nazareth, Belgium) for kindly providing photos of medicinal mushrooms (Figs. 1.1 and 1.2). We are also very thankful to mycologists and researchers around the world for providing literature data.

The authors have not reported any conflict of interest that would likely raise questions about their independence.

References

Abdullah N, MZD H, Lau BF et al (2013) Domestication of a wild medicinal sclerotial mushroom, *Lignosus rhinocerotis* (Cooke) Ryvarden. Ind Crop Prod 47:256–261. https://doi.org/10.1016/j.indcrop.2013.03.012

Abidin MHZ, Abdullah N, Abidin NZ (2018) Antiatherogenic potential of extracts from the gray oyster medicinal mushroom, *Pleurotus pulmonarius* (Agaricomycetes), *in vitro*. Int J Med Mushrooms 20(3):283–290. https://doi.org/10.1615/IntJMedMushrooms.2018025821

Abugri DA, McElhenney WH (2013) Extraction of total phenolic and flavonoids from edible wild and cultivated medicinal mushrooms as affected by different solvents. J Nat Prod Plant Resour 3(3):37–42

Acharya K, Bera I, Khatua S et al (2015) Pharmacognostic standardization of *Grifola frondosa*: a well-studied medicinal mushroom. Pharm Lett 7:72–78

Acharya K, Ghosh S, Kundu I (2016) Pharmacognostic standardization of a well-known edible mushroom, *Volvariella volvacea*. J App Pharm Sci 6(11):185–190. https://doi.org/10.7324/JAPS.2016.601129

Acharya K, Khatua S, Ray S (2017) Quality assessment and antioxidant study of *Pleurotus djamor* (Rumph. Ex Fr.) Boedijn. J Appl Pharm Sci 7(6):105–110. https://doi.org/10.7324/JAPS.2017.70614

Adams LS, Phung S, Wu X et al (2008) White button mushroom (*Agaricus bisporus*) exhibits antiproliferative and proapoptotic properties and inhibits prostate tumor growth in athymic mice. Nutr Cancer 60(6):744–756. https://doi.org/10.1080/01635580802192866

Adebayo EA, Martinez-Carrera D, Morales P et al (2018) Comparative study of antioxidant and antibacterial properties of the edible mushrooms *Pleurotus levis, P. ostreatus, P. pulmonarius* and *P. tuber-regium*. Int J Food Sci Technol 53(5):1316–1330. https://doi.org/10.1111/ijfs.13712

Adejumo TO, Coker ME, Akinmoladun VO (2015) Identification and evaluation of nutritional status of some edible and medicinal mushrooms in Akoko area, Ondo state, Nigeria. Int J Curr Microbiol App Sci 4(4):1011–1028

Adotey G, Quarcoo A, Holliday JC et al (2011) Effect of immunomodulating and antiviral agent of medicinal mushrooms (immune assist 24/7™) on CD4+ T-lymphocyte counts of HIV-infected patients. Int J Med Mushrooms 13(2):109–113. https://doi.org/10.1615/IntJMedMushr.v13.i2.20

Afrin S, Rakib MA, Kim BH et al (2016) Eritadenine from edible mushrooms inhibits activity of angiotensin converting enzyme *in vitro*. J Agric Food Chem 64(11):2263–2268. https://doi.org/10.1021/acs.jafc.5b05869

Ahmad MS, Noor ZM, Ariffin ZZ (2014) Isolation and identification fibrinolytic protease endophytic fungi from *Hibiscus* leaves in Shah Alam. Int J Agric Biol Eng 8(10):1104–1107. https://doi.org/scholar.org/1307-6892/9999437

Ahmad MR, Murtaza I, Bhat G et al (2016) Evaluation of some mushrooms from Kashmir valley for their potential anti-cancer activity. Int Arch BioMed Clin Res 2(4):57–60. https://doi.org/10.21276/iabcr.2016.2.4.13

Ahn WS, Kim DJ, Chae GT et al (2004) Natural killer cell activity and quality of life were improved by consumption of a mushroom extract, *Agaricus blazei* Murill Kyowa, in gynecological cancer patients undergoing chemotherapy. Int J Gynecol Cancer 14:589–594. https://doi.org/10.1111/j.1048-891X.2004.14403.x

Ahn H, Jeon E, Kim JC et al (2017) Lentinan from Shiitake selectively attenuates AIM2 and non-canonical inflammasome activation while inducing pro-inflammatory cytokine production. Sci Rep 7:1314. https://doi.org/10.1038/s41598-017-01462-4

Aleem E (2013) β-Glucans and their application in cancer therapy: focus on human studies. Anti Cancer Agents Med Chem 13:709–719. https://doi.org/10.2174/1871520611313050005

Al-Fatimi M, Schröder G, Kreisel H et al (2013) Biological activities of selected basidiomycetes from Yemen. Pharmazie 68:221–226. https://doi.org/10.1691/ph.2013.2729

Ali SM, Ling TC, Muniandy S et al (2014) Recovery and partial purification of fibrinolytic enzymes of *Auricularia polytricha* (Mont.) Sacc. by an aqueous two-phase system. Sep Purif Technol 122:359–366. https://doi.org/10.1016/j.seppur.2013.11.016

Ali SM, Tan YS, Raman J et al (2017) Do culinary mushrooms have fibrinolytic activities? Biomed Rev 28:95–103. https://doi.org/10.14748/bmr.v28.4454

Allen JL, Lendemer JC (2015) Fungal conservation in the USA. Endanger Species Res 28:33–42. https://doi.org/10.3354/esr00678

Alonso EN, Ferronato MJ, Fermento ME et al (2018) Antitumoral and antimetastatic activity of Maitake D-Fraction in triple-negative breast cancer cells. Oncotarget 9(34):23396–23412. https://doi.org/10.18632/oncotarget.25174

Alves M, Ferreira I, Martins A et al (2012) Antimicrobial activity of wild mushroom extracts against clinical isolates resistant to different antibiotics. J Appl Microbiol 113(2):466–475. https://doi.org/10.1111/j.1365-2672.2012.05347

Alves M, Ferreira I, Lourenço I et al (2014) Wild mushroom extracts as inhibitors of bacterial biofilm formation. Pathogens 3:667–679. https://doi.org/10.3390/pathogens3030667

Anjana S, Savita J (2017) Oyster mushroom: answer to human ailments. Asian J Pharm Clin Res 10(4):24–27. https://doi.org/10.22159/ajpcr.2017.v10i4.16867

Anke H, Antelo L (2011) Cyclic peptides and depsipeptides from fungi. In: Anke T, Weber D (eds) The mycota. Physiology and genetics. Selected basic and applied aspects, vol XV. Springer, Heidelberg, pp 273–296

Anwar H, Hussain G, Mustafa I (2018) Antioxidants from natural sources. In *Antioxidants in foods and its applications*, pp. 3–28. https://doi.org/10.5772/intechopen/75961.

Aras A, Khalid S, Jabeen S et al (2018) Regulation of cancer cell signaling pathways by mushrooms and their bioactive molecules: overview of the journey from benchtop to clinical trials. Food Chem Toxicol 119:206–214. https://doi.org/10.1016/j.fct.2018.04.038

Asatiani MD, Elisashvili EI, Wasser SP et al (2007) Free-radical scavenging activity of submerged mycelium extracts from higher basidiomycetes mushrooms. Biosci Biotechnol Biochem 71:3090–3092. https://doi.org/10.1271/bbb.70280

Atila F, Owaid MN, Shariati MA (2017) The nutritional and medical benefits of *Agaricus bisporus*: a review. J Microbiol Biotechnol Food Sci 7(3):281–286. https://doi.org/10.15414/jmbfs.2017/18.7.3.281-286

Awadasseid A, Hou J, Gamallat Y et al (2017) Purification, characterization, and antitumor activity of a novel glucan from the fruiting bodies of *Coriolus versicolor*. PLoS One 12(2):e0171270. https://doi.org/10.1371/journal.pone.0171270

Ayaz FA, Torun H, Colak A et al (2011) Macro- and microelement contents of fruiting bodies of wild-edible mushrooms growing in the East Black Sea region of Turkey. 2:53–59. https://doi.org/10.4236/fns.2011.22007

Baby S, Johnson AJ, Govindan B (2015) Secondary metabolites from *Ganoderma*. Phytochem 114:66–101. https://doi.org/10.1016/j.phytochem.2015.03.010

Badalian SM (2000) Antitumor and immune-modulating activities of compounds from several basidiomycete mushrooms. Probl Med Mycology 2(1):22–30. (in Russian)

Badalian SM, Serrano JJ (1999) Hypoglycemic activity of the medicinal mushroom *Hypholoma fasciculare* (Fr.) Kumm. Int J Med Mushrooms 1(3):245–250. https://doi.org/10.1615/IntJMedMushrooms.v1.i3.40

Badalian SM, Doko L, Rapior S et al (1996) Chemical and pharmacological study of higher fungi. II. Comparative investigation of the chemical composition of fruiting bodies extracts and cultural characteristics of *Nematoloma* species (*N. fasciculare*, *N. capnoides*, *N. sublateritium*, Strophariaceae). Mikol Fitopatol 30(4):79–86

Badalian SM, Mnatzakanian VA, Aroutunian LS et al (1997a) Chemical and pharmacological study of higher fungi. IV. Comparative investigation of the chemical composition from fruiting body extracts of five xylotrophic species (Agaricales). Mikol Fitopatol 31(3):61–66. (in Russian)

Badalian SM, Serrano JJ, Le Quang J et al (1997b) Chemical and pharmacological study of higher fungi. III. Acute toxicity studies from fruiting body extracts of *Nematoloma fasciculare*

(Huds.:Fr.) Karst. and *Nematoloma capnoides* (Fr.) Karst. (Strophariaceae). Mikol Fitopatol 31(1):42–45. (in Russian)

Badalian SM, Rapior S, Andary C et al (1999) Biologically active metabolites of higher fungi. In *Produits naturels d'origine végétale*. Actes of IVth Colloque. 25–29 May 1999, Ottawa, pp. 93–98.

Badalyan SM (2001) The main groups and therapeutic significance of compounds formed by macromycetes. Probl Med Mycology 3(1):16–23

Badalyan SM (2003a) Edible and medicinal Basidiomycetes mushrooms as source of natural antioxidants. Int J Med Mushrooms 5(2):153–162. https://doi.org/10.1615/InterJMedicMush.v5.i2.40

Badalyan SM (2003b) Antioxidant activity of culinary-medicinal mushroom *Flammulina velutipes* (Curt.:Fr.) P. Karst. (Agaricomycetideae). Int J Med Mushrooms 5(3):277–286. https://doi.org/10.1615/InterJMedicMush.v5.i3.50

Badalyan SM (2004a) Screening of antifungal activity of several Basidiomycetous macromycetes. Probl Med Mycology 6(1):18–26. (in Russian)

Badalyan SM (2004b) Antiprotozoal activity and mitogenic effect of mycelium of culinary-medicinal Shiitake mushroom *Lentinus edodes* (Berk.) Singer (Agaricomycetidae). Int J Med Mushrooms 6(2):131–138. https://doi.org/10.1615/IntJMedMushr.v6.i2.40

Badalyan SM (2012) Medicinal aspects of edible ectomycorrhizal mushrooms. In: Zambonelli A, Bonito G (eds) Edible ectomycorrhizal mushrooms, current knowledge and future prospects, vol 34. Springer, Berlin/Heidelberg, pp 317–334

Badalyan SM (2015) Chemical composition of mycelia of different collections of coprinoid mushrooms. In *Biodiversity and ecology of fungi and fungiform organisms of the Northern Eurasia*. Proceedings of All-Russian conference with international participation, 20–24 April 2015, Yekaterinburg, pp. 297–299.

Badalyan SM (2016) Fatty acid composition of different collections of coprinoid mushrooms (Agaricomycetes) and their nutritional and medicinal values. Int J Med Mushrooms 18(10):883–893. https://doi.org/10.1615/IntJMedMushrooms.v18.i10.40

Badalyan SM, Gharibyan NG (2017a) Characteristics of mycelial structures of different fungal collections. YSU Press (Armenian/English), Yerevan

Badalyan SM, Gharibyan NG (2017b) Antiphytopathogenic activity of mycelia polypore mushrooms (Agaricomycetes, Polyporales). In *Current mycology in Russia*. Proceedings of the IVth congress of Russian mycologists, vol 7. 12–14 April 2017, Moscow, pp 207–208 (in Russian).

Badalyan SM, Hambardzumyan LA (2001) Investigation of immune-modulating activity of medicinal mushroom *Flammulina velutipes* (Curt.:Fr.) P. Karst. *in vitro*. Cytokines induction by fruiting body extract. Int J Med Mushrooms 3(2–3):110–111. https://doi.org/10.1615/IntJMedMushr.v3.i2-3.290

Badalyan SM, Rapior S (1999) Chemical screening and biological activities of wood-decaying fungi *Flammulina velutipes* (Curt.:Fr.) Karst. (Tricholomataceae). In *Proceedings of the scientific session for the 75-year celebration of Department of Botany at Yerevan State University*, 21–23 March 1999, Yerevan, pp. 54–56 (in Russian).

Badalyan SM, Shahbazyan TA (2015) Medicinal properties of two polypore species: *Fomes fomentarius* and *Fomitopsis pinicola*. In *Current mycology in Russia*. Proceedings of IIIrd international mycological forum, vol 5. 14–15 April 2015, Moscow, pp. 277–279 (in Russian).

Badalyan SM, Zambonelli A (2019) Biotechnological exploitation of macrofungi for the production of food, pharmaceuticals and cosmeceuticals. In: Sridhar KR, Deshmukh SK (eds) Advances in macrofungi: diversity, ecology and biotechnology. CRC Press, pp 199–230. https://doi.org/10.1201/9780429504075

Badalyan SM, Rapior S, Doko L et al (1994) Investigation of primary and secondary metabolites in a chemical study of *Cortinarius armillatus* (A.-S.:Fr.) Fr. (Cortinariaceae, Telamonia). Cryptogam Mycol 15(4):223–228

Badalyan SM, Rapior S, Le Quang J et al (1995) Investigation of fungal metabolites and acute toxicity studies from fruiting bodies of *Hypholoma* species (Strophariaceae). Cryptogam Mycol 16(2):79–84

Badalyan SM, Serrano JJ, Rapior S et al (2001) Pharmacological activity of the mushrooms *Flammulina velutipes* (Curt.:Fr.) Sing., *Paxillus involutus* (Batsch.:Fr.) Fr. and *Tricholoma pardinum* Quél. (Basidiomycota). Int J Med Mushrooms 3:27–33. https://doi.org/10.1615/IntJMedMushr.v3.i1.40

Badalyan SM, Innocenti G, Garibyan NG (2002) Antagonistic activity of xylotrophic mushrooms against pathogenic fungi of cereals in dual culture. Phytopathol Mediterr 41(3):220–225. https://doi.org/10.1400/14513

Badalyan SM, Innocenti G, Gharibyan NG (2004) Interactions between xylotrophic mushrooms and mycoparasitic fungi in dual-culture experiments. Phytopathol Mediterr 43(1):44–48. http://dx.doi.org/10.14601/Phytopathol_Mediterr-1733

Badalyan SM, Isikhuemhen OS, Gharibyan NG (2008a) Antagonistic/antifungal activity of *Pleurotus tuberregium* (Fr.) Singer against selected fungal pathogens. Int J Med Mushrooms 10(2):155–162. https://doi.org/10.1615/IntJMedMushr.v10.i2.60

Badalyan SM, Melikyan LR, Navarro-González M et al (2008b) *Fibrinolytic activity of several coprinoid mushrooms*. VIth international conference of mushroom biology & mushroom production, 9 September–3 October 2008, Bonn, pp. 66–67.

Badalyan SM, Gharibyan NG, Iotti M et al (2012) *Morphological and genetic characteristics of different collections of Ganoderma P. Karst. species*. Proceedings of the 18th congress ISMS. 26–30 August 2012, Beijing, pp. 247–254.

Badalyan SM, Shnyreva AV, Iotti M et al (2015) Genetic resources and mycelial characteristics of several medicinal polypore mushrooms (Higher Basidiomycetes). Int J Med Mushrooms 17(4):371–384. https://doi.org/10.1615/IntJMedMushrooms.v17.i4.60

Badalyan SM, Gharibyan NG, Shahbazyan TA et al (2016) Milk coagulating and thrombolytic activity of red-belt conk *Fomitopsis pinicola*. In *Advances in medical mycology*. Proceedings of memorial conference of medical mycology, vol 16. 14–15 April 2016, Moscow, pp. 239–242. (in Russian).

Baggio CH, Freitas CS, Martins DF et al (2010) Antinociceptive effects of (1→3), (1→6)-linked β-glucan isolated from *Pleurotus pulmonarius* in models of acute and neuropathic pain in mice: evidence for a role for glutamatergic receptors and cytokine pathways. J Pain 11(10):965–971. https://doi.org/10.1016/j.jpain.2010.01.005

Bandara AR, Rapior S, Bhat DJ et al (2015) *Polyporus umbellatus*, an edible-medicinal cultivated mushroom with multiple developed health-care products as food, medicine and cosmetics: a review. Cryptogam Mycol 36(1):3–42. https://doi.org/10.7872/crym.v36.iss1.2015.3

Bandara AR, Karunarathna SC, Mortimer PE et al (2017) First successful domestication and determination of nutritional and antioxidant properties of the red ear mushroom *Auricularia thailandica* (Auriculariales, Basidiomycota). Mycol Prog 16(11–12):1029–1039. https://doi.org/10.1007/s11557-017-1344-7

Bao F, Yang K, Wub C et al (2018) New natural inhibitors of hexokinase 2 (HK2): steroids from *Ganoderma sinense*. Fitoterapia 125:123–129. https://doi.org/10.1016/j.fitote.2018.01.001

Barbieri A, Quagliariello V, Del Vecchio V et al (2017) Anticancer and anti-inflammatory properties of *Ganoderma lucidum* extract effects on melanoma and triple-negative breast cancer treatment. Nutrients 9:210. https://doi.org/10.3390/nu9030210

Barceloux DG (2008) Medical toxicology of natural substances. Foods, fungi, medicinal herbs, plants and venomous animals. Wiley, Hoboken. https://doi.org/10.1002/9780470330319

Barros L, Cruz T, Baptista P (2008) Wild and commercial mushrooms as source of nutrients and nutraceuticals. Food Chem Toxicol 46:2742–2747. https://doi.org/10.1016/j.fct.2008.04.030

Barsanti L, Passarelli V, Evangelista V et al (2011) Chemistry, physico-chemistry and applications linked to biological activities of β-glucans. Nat Prod Rep 28:457–466. https://doi.org/10.1039/c0np00018c

Baskaran A (2015) Suppression of lipopolysaccharide and hydrogen peroxide-induced inflammatory responses in Raw 264.7 macrophage by *Pleurotus giganteus* and *Lignosus rhinocerotis*. University of Malaya. http://studentsrepo.um.edu.my/id/eprint/7425

Baskaran A, Chua KH, Sabaratnam V et al (2017) *Pleurotus giganteus* (Berk. Karun & Hyde), the giant oyster mushroom inhibits NO production in LPS/H_2O_2 stimulated RAW 264.7 cells via

STAT 3 and COX-2 pathways. BMC Complement Altern Med 17:40. https://doi.org/10.1186/s12906-016-1546-6

Batbayar S, Lee DH, Kim HW (2012) Immunomodulation of fungal β-glucan in host defense signaling by dectin-1. Biomol Ther 20(5):433–445. https://doi.org/10.4062/biomolther.2012.20.5.433

Bedlovičová Z, Smrčová M, Strapáč I et al (2016) Mushrooms as the source of potential antimicrobial agents: a review. Curr Org Chem 20:1–13. https://doi.org/10.2174/138527282066616 0608101058

Beekman AM, Barrow R (2014) Fungal metabolites as pharmaceuticals. Aust J Chem 67:827–843. https://doi.org/10.1071/CH13639

Bello M, Oluwamukomi MO, Enujiugha VN (2017) Nutrient composition and sensory properties of biscuit from mushroom-wheat composite flours. ACRI 9(3):1–11. https://doi.org/10.1111/jfpp.13150

Bennett L, Kersaitis C, Macaulay SL et al (2013) Vitamin D_2-enriched button mushroom (*Agaricus bisporus*) improves memory in both wild type and APPswe/PS1dE9 transgenic mice. PLoS One 8(10):e76362. https://doi.org/10.1371/journal.pone.0076362

Berihuete-Azorín M, Girbal J, Piqué R et al (2018) Punk's not dead. Fungi for tinder at the Neolithic site of La Draga (NE Iberia). PLoS One 13(4):e0195846. https://doi.org/10.1371/journal.pone.0195846

Bhakta M, Kumar P (2013) Mushroom polysaccharides as potential prebiotics. Int J Health Sci Res 3:77–84. https://doi.org/10.1016/j.bcdf.2015.11.001

Bhatt RP, Singh U, Uniyal P (2018) Healing mushrooms of Uttarakhand Himalaya, India. Curr Res Environ Appl Mycol 8(1):1–23. https://doi.org/10.5943/cream/8/1/1

Bhattacharjee J, Bhattacharjee D, Paul T et al (2015) Diversity of mushrooms in indo-Bangladesh region of north-East India. J Andaman Sci Assoc 20(1):75–82

Bishop KS, Kao CHJ, Xu Y et al (2015) From 2000 years of *Ganoderma lucidum* to recent developments in nutraceuticals. Phytochemistry 114:56–65. https://doi.org/10.1016/j.phytochem.2015.02.015

Biswas G, Nandi S, Kuila D et al (2017) A comprehensive review on food and medicinal prospects of *Astraeus hygrometricus*. Pharm J 9(6):799–806. https://doi.org/10.5530/pj.2017.6.125

Blackwell M (2011) The Fungi: 1, 2, 3 ... 5.1 million species? Am J Bot 98:426–438. https://doi.org/10.3732/ajb.1000298

Blann AD, Landray MJ, Lip GY (2002) An overview of antithrombotic therapy. Br Med J 325(7367):762–765. https://doi.org/10.1136/bmj.325.7367.762

Boa ER (2004) Wild edible fungi a global overview of their use and importance to people. Food and Agriculture Organization of the United Nations, Roma. http://www.fao.org/docrep/007/y5489e/y5489e00.htm

Borchers AT, Krishnamurthy A, Keen CL et al (2008) The immunobiology of mushrooms. Exp Biol Med (Maywood) 233(3):259–276. https://doi.org/10.3181/0708-MR-227

Bouike G, Nishitani Y, Shiomi H et al (2011) Oral treatment with extract of *Agaricus blazei* Murill enhanced Th1 response through intestinal epithelial cells and suppressed OVA-sensitized allergy in mice. eCAM 2011:1–11. https://doi.org/10.1155/2011/532180

Brandalise F, Cesaroni V, Gregori A et al (2017) Dietary supplementation of *Hericium erinaceus* increases mossy fiber-CA3 hippocampal neurotransmission and recognition memory in wild-type mice. Evid Based Complement Alternat Med 2017:3864340.):13 p. https://doi.org/10.1155/2017/3864340

Breheret S, Talou T, Rapior S et al (1998) (*Z*)- and (*E*)-1,3-octadiene – new major volatile compounds in mushroom aromas (Basidiomycotina). J Essent Oil Res 10:716–718. https://doi.org/10.1080/10412905.1998.9701020

Brown GD, Gordon S (2003) Fungal β-glucans and mammalian immunity. Immunity 19:311–315. https://doi.org/10.1016/S1074-7613(03)00233-4

Brunelli E (2009) Sindrome di Szechwan e *Auricularia auricula-judae*. Pag Micol 32:53–57

Buruleanu LC, Radulescu C, Georgescu AA et al (2018) Statistical characterization of the phytochemical characteristics of edible mushroom extracts. Anal Lett 51(7):1039–1059. https://doi.org/10.1080/00032719.2017.1366499

Butkhup L, Samappito W, Jorjong S (2018) Evaluation of bioactivities and phenolic contents of wild edible mushrooms from northeastern Thailand. Food Sci Biotechnol 27(1):193–202. https://doi.org/10.1007/s10068-017-0237-5
Canli K, Akata I, Altuner EM (2016) *In vitro* antimicrobial activity screening of *Xylaria hypoxylon*. Afr J Tradit Complement Altern Med 13(4):42–46. https://doi.org/10.21010/ajtcam.v13i4.7
Cappelli A (2011) Approccio al genere *Agaricus* – IV. Riv Micol 1:2–27
Carhart-Harris RL, Roseman L, Bolstridge M et al (2017) Psilocybin for treatment-resistant depression: fMRI-measured brain mechanisms. Sci Rep 7:13187. https://doi.org/10.1038/s41598-017-13282-71
Carocho M, Ferreira ICFR, Morales P et al (2018) Antioxidants and pro-oxidants: effects on health and aging. Oxidative Med Cell Longev 2018:1472708.):2 p. https://doi.org/10.1155/2018/1472708
Castellano G, Torrens F (2015) Information entropy-based classification of triterpenoids and steroids from *Ganoderma*. Phytochem 116:305–313. https://doi.org/10.1016/j.phytochem.2015.05.008
Cerigini E, Palma F, Buffalini M et al (2007) Identification of a novel lectin from the Ascomycetes fungus *Tuber borchii*. Int J Med Mushrooms 9(3/4):287. https://doi.org/10.1615/IntJMedMushr.v9.i34.100
Chaiyasut C, Sivamaruthi BS (2017) Anti-hyperglycemic proper ty of *Hericium erinaceus* – a mini review. Asian Pac J Trop Biomed 7(11):1036–1040. https://doi.org/10.1016/j.apjtb.2017.09.024
Chan GCF, Chan WK, Sze DMY (2009) The effects of β-glucan on human immune and cancer cells. J Hematol Oncol 2:25. https://doi.org/10.1186/1756-8722-2-25
Chandrawanshi NK, Tandia DK, Jadhav SK (2017) Nutraceutical properties evaluation of *Schizophyllum commune*. Indian J Sci Res 13(2):57–62
Chang R (1996) Functional properties of edible mushrooms. Nutr Rev 54(11 pt. 2):S91–S93. https://doi.org/10.1111/j.1753-4887.1996.tb03825.x
Chang ST, Buswell JA (1996) Mushroom nutriceuticals. World J Microbiol Biotechnol 12(5):473–476. https://doi.org/10.1007/BF00419460
Chang YS, Lee SS (2004) Utilisation of macrofungi species in Malaysia. Fungal Divers 15:15–22
Chang ST, Miles PG (2008) Mushrooms: Cultivation, nutritional value, medicinal effect, and environmental impact, 2nd edn. CRC Press, Boca Raton
Chang ST, Wasser SP (2012) The role of culinary-medicinal mushrooms on human welfare with a pyramid model for human health. Int J Med Mushrooms 14(2):95–134. https://doi.org/10.1615/IntJMedMushr.v14.i2.10
Chang ST, Wasser SP (2017) The cultivation and environmental impact of mushrooms. In: Oxford research encyclopedia of environmental science – agriculture and the environment. https://doi.org/10.1093/acrefore/9780199389414.013.231
Chang JC, Hsiao G, Lin RK et al (2017) Bioactive constituents from the termite nest-derived medicinal fungus *Xylaria nigripes*. J Nat Prod 80:38–44. https://doi.org/10.1021/acs.jnatprod.6b00249
Chen J, Seviour R (2007) Medicinal importance of fungal β-(1→3), (1→6)-glucans. Mycol Res 111:635–652. https://doi.org/10.1016/j.mycres.2007.02.011
Chen CC, Shiao YJ, Lin RD et al (2006) Neuroprotective diterpenes from the fruiting body of *Antrodia camphorata*. J Nat Prod 69:689–691. https://doi.org/10.1021/np0581263
Chen W, Zhao Z, Chen SF et al (2008) Optimization for the production of exopolysaccharide from *Fomes fomentarius* in submerged culture and its antitumor effect in vitro. Bioresour Technol 99(8):3187–3194. https://doi.org/10.1016/j.biortech.2007.05.049
Chen JT, Tominaga K, Sato Y et al (2010) Maitake mushroom (*Grifola frondosa*) extract induces ovulation in patients with polycystic ovary syndrome: a possible monotherapy and a combination therapy after failure with first-line clomiphene citrate. J Altern Complement Med 16(12):1295–1299. https://doi.org/10.1089/acm.2009.0696
Chen H, Tian T, Miao H et al (2016) Traditional uses, fermentation, phytochemistry and pharmacology of *Phellinus linteus*: a review. Fitoterapia 113:6–26. https://doi.org/10.1016/j.fitote.2016.06.009

Chen B, Tiana J, Zhang J et al (2017) Triterpenes and meroterpenes from *Ganoderma lucidum* with inhibitory activity against HMGs reductase, aldose reductase and α-glucosidase. Fitoterapia 120:6–16. https://doi.org/10.1016/j.fitote.2017.05.005

Cheng CR, Yue QX, Wu ZY et al (2010) Cytotoxic triterpenoids from *Ganoderma lucidum*. Phytochem 71:1579–1585. https://doi.org/10.1016/j.phytochem.2010.06.005

Cheng PG, Phan CW, Sabaratnam V et al (2013) Polysaccharides-rich extract of *Ganoderma lucidum* (M.A. Curtis:Fr.) P. Karst. accelerates wound healing in streptozotocin-induced diabetic rats. Evid Based Complement Alternat Med 2013:671252.:9 p. https://doi.org/10.1155/2013/671252

Chepkirui C, Yuyama K, Wanga L et al (2018) Microporenic acids A-G, biofilm inhibitors, and antimicrobial agents from the basidiomycete *Microporus* species. J Nat Prod 81(4):778–784. https://doi.org/10.1021/acs.jnatprod.7b00764

Cheung PCK (2008) Mushrooms as functional foods. Wiley, Hoboken. https://doi.org/10.1002/9780470367285

Cheung PCK (2013) Mini-review on edible mushrooms as source of dietary fiber: preparation and health benefits. Food Sci Human Wellness 2:162–166. https://doi.org/10.1016/j.fshw.2013.08.001

Chiu CH, Chyau CC, Chen CC et al (2018) Erinacine A-enriched *Hericium erinaceus* mycelium produces antidepressant-like effects through modulating BDNF/PI3K/Akt/GSK-3 signaling in mice. Int J Mol Sci 19:341. https://doi.org/10.3390/ijms19020341

Choi HS, Sa YS (2000) Fibrinolytic and antithrombotic protease from *Ganoderma lucidum*. Mycologia 92:545–552. https://doi.org/10.2307/3761514

Choi HS, Shin HH (1998) Purification and partial characterization of a fibrinolytic protease in *Pleurotus ostreatus*. Mycologia 90(4):674–679. https://doi.org/10.2307/3761226

Choi BS, Sapkota K, Choi JH et al (2013) Herinase: a novel bi-functional fibrinolytic protease from the monkey head mushroom, *Hericium erinaceum*. Appl Biochem Biotechnol 170(3):609–622. https://doi.org/10.1007/s12010-013-0206-2

Choi JY, Paik DJ, Kwan DY et al (2014) Dietary supplementation with rice bran fermented with *Lentinus edodes* increases interferon-γ activity without causing adverse effects: a randomized, double-blind, placebo-controlled, parallel-group study. Nutr J 13(1):35–42. https://doi.org/10.1186/1475-2891-13-35

Choi JH, Kim DW, Kim S et al (2017) Purification and partial characterization of a fibrinolytic enzyme from the fruiting body of the medicinal and edible mushroom *Pleurotus ferulae*. Prep Biochem Biotechnol 7(6):539–546. https://doi.org/10.1080/10826068.2016.1181083

Choi YJ, Park IS, Kim MH et al (2018) The medicinal mushroom *Auricularia auricula-judae* (Bull.) extract has antioxidant activity and promotes procollagen biosynthesis in HaCaT cells. Nat Prod Res 4:1–4. https://doi.org/10.1080/14786419.2018.1468332

Ćilerdžić J, Kosanić M, Stajić M et al (2016) Species of genus *Ganoderma* (Agaricomycetes) fermentation broth: a novel antioxidant and antimicrobial agent. Int J Med Mushrooms 18(5):397–404. https://doi.org/10.1615/IntJMedMushrooms.v18.i5.30

Combs GF (2001) Selenium in global food systems. Br J Nutr 85:517–547. https://doi.org/10.1079/BJN2000280

Corrěa RCG, Brugnari T, Bracht A et al (2016) Biotechnological, nutritional and therapeutic uses of *Pleurotus* spp. (oyster mushroom) related with its chemical composition: a review on the past decade findings. Trends Food Sci Technol 50:103–117. https://doi.org/10.1016/j.tifs.2016.01.012

Corrěa RCG, Peralta RM, Bracht A et al (2017) The emerging use of mycosterols in food industry along with the current trend of extended use of bioactive phytosterols. Trends Food Sci Technol 67:19–35. https://doi.org/10.1016/j.tifs.2017.06.012

Corrěa RCG, Barros L, Fernandes A et al (2018) A natural food ingredient based on ergosterol: optimization of the extraction from *Agaricus blazei*, evaluation of bioactive properties and incorporation in yogurts. Food Funct 9(3):1465–1474. https://doi.org/10.1039/C7FO02007D

Cucuianu V, Bratanescu V, Sterian B (2004) The edible mushrooms – an organic food and its potential use for health. J Environ Prot Ecol 5(4):801–808

Da Silva de Souza AC, Corrěa VG, Goncalves GA et al (2017) *Agaricus blazei* bioactive compounds and their effects on human health: benefits and controversies. Curr Pharm Des 23(19):2807–2834. https://doi.org/10.2174/1381612823666170119093719

Da Silva de Souza AC, de Almeida Goncalves G, Soares AA et al (2018) The antioxidant action of an aqueous extract of royal sun medicinal mushroom, *Agaricus brasiliensis* (Agaricomycetes) in rats with adjuvant-induced arthritis. Int J Med Mushrooms 20(2):101–117. https://doi.org/10.1615/IntJMedMushrooms.2018025309

Da Silva Milhorini S, Smiderle FR, Biscaia SMP et al (2018) Fucogalactan from the giant mushroom *Macrocybe titans* inhibits melanoma cells migration. Carbohydr Polym 190:50–56. https://doi.org/10.1016/j.carbpol.2018.02.063

Dadakova E, Pelikanova T, Kalc P (2009) Content of biogenic amines and polyamines in some species of European wild-growing edible mushrooms. Eur Food Res Technol 230:163–117. https://doi.org/10.1007/s00217-009-1148-3

Dai YC, Yang ZL, Ui BK et al (2009) Species diversity and utilization of medicinal mushrooms and fungi in China (review). Int J Med Mushrooms 11:287–302. https://doi.org/10.1615/IntJMedMushr.v11.i3.80

Dai X, Stanilka JM, Row CA et al (2015) Consuming *Lentinula edodes* (shiitake) mushrooms daily improves human immunity: a randomized dietary intervention in healthy young adults. J Am Coll Nutr 34(6):478–487. https://doi.org/10.1080/07315724.2014.950391

De Mattos-Shipley KMJ, Ford KL, Alberti F et al (2016) The good, the bad and the tasty: the many roles of mushrooms. Stud Mycol 85:125–157. https://doi.org/10.1016/j.simyco.2016.11.002

De Silva DD, Rapior S, Fons F et al (2012a) Medicinal mushrooms in supportive cancer therapies: an approach to anti-cancer effects and putative mechanisms of action. Fungal Divers 55:1–35. https://doi.org/10.1007/s13225-012-0151-3

De Silva DD, Rapior S, Hyde KD et al (2012b) Medicinal mushrooms in prevention and control of diabetes mellitus. Fungal Divers 56:1–29. https://doi.org/10.1007/s13225-012-0187-4

De Silva DD, Rapior S, Sudarman E et al (2013) Bioactive metabolites from macrofungi: ethnopharmacology, biological activities and chemistry. Fungal Divers 62:1–40. https://doi.org/10.1007/s13225-013-0265-2

Debnath S, Upadhyay RC, Das P et al (2017) Antioxidant activities of methanolic extracts from ten *Pleurotus* species. Int Res J Pharm 8(3):44–49. https://doi.org/10.7897/2230-8407.080335

Degreef J, Demuynck L, Mukandera A et al (2016) Wild edible mushrooms, a valuable resource for food security and rural development in Burundi and Rwanda. Biotechnol Agron Soc Environ 20(4):441–452

Del Buono A, Bonucci M, Publiese S et al (2016) Polysaccharide from *Lentinus edodes* for integrative cancer treatment: immunomodulatory effects on lymphocyte population. WCRJ 3(1):1–7

Denisova NP (1982) Proteolytic activity of cultures of higher fungi. Mikol Fitopatol 16(5):458–466. (in Russian)

Denisova NP (2010) History of the study of thrombolytic and fibrinolytic enzymes of higher basidiomycetes mushrooms at the V.L. Komarov botanical institute in Saint Petersburg, Russia. Int J Med Mushrooms 12(3):317–325. https://doi.org/10.1615/IntJMedMushr.v12.i3.110

Denisova NP, Semenova IR, Sukharevich VI (1989) Biosynthesis of proteinases with fibrinolytic action by basidiomycetes in submerged culture. Mikol Fitopatol 23(4):378–381. (in Russian). https://doi.org/10.1186/1756-8722-2-25

Diling C, Chaoqun Z, Jian Y et al (2017) Immunomodulatory activities of a fungal protein extracted from *Hericium erinaceus* through regulating the gut microbiota. Front Immunol 8:666. https://doi.org/10.3389/fimmu.2017.00666

Dissanayakea AA, Zhanga CR, Mills GL et al (2018) Cultivated Maitake mushroom demonstrated functional food quality as determined by in vitro bioassays. J Funct Foods 44:79–85. https://doi.org/10.1016/j.jff.2018.02.031

Doğan HH, Karagöz S, Duman R (2018) *In vitro* evaluation of the antiviral activity of some mushrooms from Turkey. Int J Med Mushrooms 20(3):201–212. https://doi.org/10.1615/IntJMedMushrooms.2018025468

Donatini B (2011) *Hericium erinaceus*: properties mostly related to the secretion of neuronal growth factor. Phytothérapie 9:48–52. https://doi.org/10.1007/s10298-010-0601-6

Donatini B (2014) Control of oral human papillomavirus (HPV) by medicinal mushrooms, *Trametes versicolor* and *Ganoderma lucidum*: a preliminary clinical trial. Int J Med Mushrooms 16(5):497–498. https://doi.org/10.1615/IntJMedMushrooms.v16.i5.80

Donnini D, Gargano ML, Perini C et al (2013) Wild and cultivated mushrooms as a model of sustainable development. Plant Biosyst 147(1):226–236. https://doi.org/10.1080/11263504.2012.754386

Duru ME, Çayan GT (2015) Biologically active terpenoids from mushroom origin: a review. Rec Nat Prod 9(4):456–483

Dutta S (2013) Role of mushrooms as nutraceutical an overview. Int J Pharm Bio Sci 4(4):B59–B66

Dyakova MY, Kamzolkina OV, Shtaera OV et al (2011) Morphological characteristics of natural strains of certain species of Basidiomycetes and biological analysis of antimicrobial activity under submerged cultural conditions. Microbiology 80(2):274–285. https://doi.org/10.1134/S0026261711020044

Ebrahimi A, Atashi A, Soleimani M et al (2017) Anti-invasive and anti-proliferative effects of *Pleurotus ostreatus* extract on acute leukemia cell lines. J Basic Clin Physiol Pharmacol 29(1):95–102. https://doi.org/10.1515/jbcpp-2017-0088

Ehsanifard Z, Mir-Mohammadrezaei F, Safarzadeh A et al (2017) Aqueous extract of *Inocutis levis* improves insulin resistance and glucose tolerance in high sucrose-fed Wistar rats. J Herbmed Pharmacol 6(4):160–164

Eik LF, Naidu M, David P et al (2012) *Lignosus rhinocerus* (Cooke) Ryvarden: a medicinal mushroom that stimulates neurite outgrowth in PC-12 cells. Evid Based Complement Alternat Med 2012:320308.:7 p. https://doi.org/10.1155/2012/320308

El Dine RS, El Halawany AM, Ma CM et al (2009) Inhibition of the dimerization and active site of HIV-1 protease by secondary metabolites from the Vietnamese mushroom *Ganoderma colossum*. J Nat Prod 72:2019–2023. https://doi.org/10.1021/np900279u

Elisashvili V (2012) Submerged cultivation of medicinal mushrooms: bioprocesses and products. Int J Med Mushrooms 14:211–239. https://doi.org/10.1615/IntJMedMushr.v14.i3.10

Eliza WL, Fai CK, Chung LP (2012) Efficacy of Yun Zhi (*Coriolus versicolor*) on survival in cancer patients: systematic review and meta-analysis. Recent Patents Inflamm Allergy Drug Discov 6(1):78–87. https://doi.org/10.2174/187221312798889310

Ellan K, Sabaratnam V, Thayan R (2013) Antiviral activity and mode of action of mushroom extracts against dengue virus type-2. In *Proceedings of the 3rd international conference on dengue and dengue haemorrhagic fever*, 21–23 October 2012, Bangkok.

Ellertsen LK, Hetland G (2009) An extract of the medicinal mushroom *Agaricus blazei* Murill can protect against allergy. Clin Mol Allergy 7(6):1–10. https://doi.org/10.1186/1476-7961-7-6

Ey J, Schomig E, Taubert D (2007) Dietary sources and antioxidant effects of ergothioneine. J Agric Food Chem 55:6466–6474. https://doi.org/10.1021/jf071328f

Falade OE, Oyetayo VO, Awala SI (2017) Evaluation of the mycochemical composition and antimicrobial potency of wild macrofungus, *Rigidoporus microporus* (Sw.). J Phytopharmacol 6(2):115–125

Falandysz J (2008) Selenium in edible mushrooms. J Environ Sci Health C Environ Carcinog Ecotoxicol Rev 26(3):256–299. https://doi.org/10.1080/10590500802350086

Fan L, Pan H, Soccol AT et al (2006) Advances in mushroom research in the last decade. Food Technol Biotechnol 44(3):303–311

Fang N, Li Q, Yu S et al (2006) Inhibition of growth and induction of apoptosis in human cancer cell lines by an ethyl acetate fraction from shiitake mushrooms. J Altern Complement Med 12(2):125–132. https://doi.org/10.1089/acm.2006.12.125.

Farzana T, Mohajan S, Saha T et al (2017) Formulation and nutritional evaluation of a healthy vegetable soup powder supplemented with soy flour, mushroom, and moringa leaf. Food Sci Nutr 5:911–920. https://doi.org/10.1002/fsn3.476

Fernandes A, Barros L, Martins A et al (2015) Nutritional characterisation of *Pleurotus ostreatus* (Jacq. ex Fr.) P. Kumm. produced using paper scraps as substrate. Food Chem 169:396–400. https://doi.org/10.1016/j.foodchem.2014.08.027

Ferreira ICFR, Barros L, Abreu RMV (2009) Antioxidants in wild mushrooms. Curr Med Chem 16(12):1543–1560. https://doi.org/10.3390/molecules201019489

Ferreira ICFR, Vaz JA, Vasconcelos MH et al (2010) Compounds from wild mushrooms with antitumor potential. Anti Cancer Agents Med Chem 10(5):424–436. https://doi.org/10.2174/187152061109050424

Ferreira ICFR, Heleno SA, Reis FS et al (2015) Chemical features of *Ganoderma* polysaccharides with antioxidant, antitumor and antimicrobial activities. Phytochem 114:38–55. https://doi.org/10.1016/j.phytochem.2014.10.011

Ferreira-Silva V, Gusmão NB, Gibertoni TB (2017) Antibacterial activity of ethyl acetate extract of Agaricomycetes collected in Northeast Brazil. Curr Res Environ Appl Mycol J Fungal Biol 7(4):267–274. https://doi.org/10.5943/cream/7/4/3

Filipic M, Umek A, Mlinaric A (2002) Screening of Basidiomycete mushroom extracts for antigenotoxic and bio-antimutagenic activity. Pharmazie 57:416–420

Finimundy TC, Dillon AJP, Henriques JAP et al (2014) A review on general nutritional compounds and pharmacological properties of the *Lentinula edodes* mushroom. Food Nutr Sci 5:1095–1105. https://doi.org/10.4236/fns.2014.512119

Finimundy TC, Abreu RMV, Bonetto N et al (2018a) Apoptosis induction by *Pleurotus sajor-caju* (Fr.) Singer extracts on colorectal cancer cell lines. Food Chem Toxicol 112:383–392. https://doi.org/10.1016/j.fct.2018.01.015

Finimundy TC, Barros L, Calhelha RC et al (2018b) Multifunctions of *Pleurotus sajor-caju* (Fr.) Singer: a highly nutritious food and a source for bioactive compounds. Food Chem 245:150–158. https://doi.org/10.1016/j.foodchem.2017.10.088

Fortes RC, Recôva VL, Melo AL et al (2008) Effects of dietary supplementation with medicinal fungus in fasting glycemia levels of patients with colorectal cancer: a randomized, double-blind, placebo-controlled clinical study. Nutr Hosp 23(6):591–598

Fortes RC, Novaes MRCG, Recôva VL et al (2009) Immunological, hematological, and glycemia effects of dietary supplementation with *Agaricus sylvaticus* on patients' colorectal cancer. Exp Biol Med (Maywood) 234(1):53–62. https://doi.org/10.3181/0806-RM-193

Francia C, Rapior S, Courtecuisse R et al (1999) Current research findings on the effects of selected mushrooms on cardiovascular diseases. Int J Med Mushrooms 1:169–172. https://doi.org/10.1615/IntJMedMushrooms.v1.i2.60

Francisco CRL, Heleno SA, Fernandes IPM et al (2018) Functionalization of yogurts with *Agaricus bisporus* extracts encapsulated in spray-dried maltodextrin crosslinked with citric acid. Food Chem 15(245):845–853. https://doi.org/10.1016/j.foodchem.2017.11.098

Friedman M (2016) Mushroom polysaccharides: chemistry and antiobesity, antidiabetes, anticancer, and antibiotic properties in cells, rodents, and humans. Foods 5(4):80. https://doi.org/10.3390/foods5040080

Fritz H, Kennedy DA, Ishii M et al (2015) Polysaccharide K and *Coriolus versicolor* extracts for lung cancer: a systematic review. Integr Cancer Ther 14(3):201–211. https://doi.org/10.1177/1534735415572883

Frost, M. (2016). *Three popular medicinal mushroom supplements: a review of human clinical trials*. Dissertation, Brigham Young University.

Fu Z, Liu Y, Zhang Q (2016) A potent pharmacological mushroom: *Pleurotus eryngii*. Fungal Genom Biol 6:139. https://doi.org/10.4172/2165-8056.1000139

Gao Y, Padhiar AA, Wang J et al (2018) Recombinant latcripin 11 of *Lentinula edodes* C91-3 suppresses the proliferation of various cancer cells. Gene 642:212–219. https://doi.org/10.1016/j.gene.2017.10.080

Gargano ML (2018) Mycotheca of edible and medicinal mushrooms at herbarium SAF as a potential source of nutraceuticals and cultivated mushrooms. Int J Med Mushrooms 20(4):405–409. https://doi.org/10.1615/IntJMedMushrooms.2018025966

Gargano ML, van Griensven LJLD, Isikhuemhen OS et al (2017) Medicinal mushrooms: valuable biological resources of high exploitation potential. Plant Biol 151(3):548–565. https://doi.org/10.1080/11263504.2017.1301590

Gaylan CM, Estebal JC, Tantengco OA et al (2018) Anti-staphylococcal and antioxidant properties of crude ethanolic extracts of macrofungi collected from the Philippines. Pharm J 10(1):106–109. https://doi.org/10.5530/pj.2018.1.19

Giacomoni L (2004) Le Syndrome de Szechwan. Bull AEMBA 42:43–47

Giavasis I (2014) Bioactive fungal polysaccharides as potential functional ingredients in food and nutraceuticals. Curr Opin Biotechnol 26:162–173. https://doi.org/10.1016/j.copbio.2014.01.010

Gil-Ramirez A, Morales D, Soler-Rivas C (2017) Molecular actions of hypocholesterolaemic compounds from edible mushrooms. Food Funct 9(1):53–69. https://doi.org/10.1039/C7FO00835J

Glamočlija J, Soković M (2017) Fungi a source with huge potential for "mushroom pharmaceuticals". Lek Sirov 37:50–56. https://doi.org/10.5937/leksir1737050G

Glamočlija J, Stojković D, Nikolic M et al (2015) A comparative study on edible *Agaricus* mushrooms as functional foods. Food Funct 6(6):1900–1910. https://doi.org/10.1039/c4fo01135j

Gregori A (2014) Cordycepin production by *Cordyceps militaris* cultivation on spent brewery grains. ABS 57(2):45–52

Grienke U, Zöll M, Peintner U et al (2014) European medicinal polypores – a modern view on traditional uses. J Ethnopharmacol 154:564–583. https://doi.org/10.1016/j.jep.2014.04.030

Grotto D, Bueno DCR, Ramos GKA et al (2016) Assessment of the safety of the shiitake culinary-medicinal mushroom, *Lentinus edodes* (Agaricomycetes), in rats: biochemical, hematological, and antioxidative parameters. Int J Med Mushrooms 16(10):861–870. https://doi.org/10.1615/IntJMedMushrooms.v18.i10.20

Gründemann C, Arnhold M, Meier S et al (2016) Effects of *Inonotus hispidus* extracts and compounds on human immunocompetent cells. Planta Med 82:1359–1367. https://doi.org/10.1055/s-0042-111693

Guggenheim AG, Wright KM, Zwickey HL (2014) Immune modulation from five major mushrooms: application to integrative oncology. Integr Med 13(1):32–41

Guillamón E, García-Lafuente A, Lozano M et al (2010) Edible mushrooms: role in the prevention of cardiovascular diseases. Fitoterapia 81(7):715–723. https://doi.org/10.1016/j.fitote.2010.06.005

Gunde-Cimmerman N (1999) Medicinal value of the genus *Pleurotus* (Fr.) P. Karst. (Agaricales s.l., Basidiomycetes). Int J Med Mushrooms 1:69–80. https://doi.org/10.1615/IntJMedMushrooms.v1.i1.50

Gupta S, Summuna B, Gupta M et al (2018) Edible mushrooms: cultivation, bioactive molecules, and health benefits. In: Mérillon JM, Ramawat KG (eds) Bioactive molecules in food, reference series in Phytochemistry. Springer, pp 1–33. https://doi.org/10.1007/978-3-319-54528-8_86-1

Guzmán G (2015) New studies on hallucinogenic mushrooms: history, diversity, and applications in psychiatry. Int J Med Mushrooms 17:1019–1030. https://doi.org/10.1615/IntJMedMushrooms.v17.i11.10

Hadda M, Djamel C, Akila O (2015a) Production and qualitative analysis of triterpenoids and steroids of *Ganoderma* species harvested from cork oak forest of north-eastern Algeria. Res J Microbiol 10(8):366–376. https://doi.org/10.3923/jm.2015.366.376

Hadda M, Djamel C, Akila O (2015b) Screening of extracellular enzyme activities of *Ganoderma* and *Fomes* species collected from north East Algeria. RJPBCS 6(4):1455–1462

Hammerschmidt DE (1980) Szechwan purpura. N Engl J Med 302(21):1191–1193

Hapuarachchi KK, Wen TC, Jeewon R et al (2016a) Mycosphere essays 7: *Ganoderma lucidum* – are the beneficial anti-cancer properties substantiated? Mycosphere 7(3):305–332. https://doi.org/10.5943/mycosphere/7/3/6

Hapuarachchi KK, Wen TC, Jeewon R et al (2016b) Mycosphere essays 15: *Ganoderma lucidum* – are the beneficial medical properties substantiated? Mycosphere 7(6):687–715. https://doi.org/10.5943/mycosphere/7/6/1

Hapuarachchi KK, Cheng CR, Wen TC et al (2017) Mycosphere essays 20: therapeutic potential of *Ganoderma* species: insights into its use as traditional medicine. Mycosphere 8(10):1653–1694. https://doi.org/10.5943/mycosphere/8/10/5

Hara M, Hanaoka T, Kobayashi M et al (2003) Cruciferous vegetables, mushrooms, and gastrointestinal cancer risk in a multicenter, hospital-based case-control study in Japan. Nutr Cancer 46(2):138–147. https://doi.org/10.1207/S15327914NC4602_06

Hawksworth DL (1991) The fungal dimension of biodiversity magnitude, significance and conservation. Mycol Res 95:641–655. https://doi.org/10.1016/S0953-7562(09)80810-1
Hawksworth DL (2001) Mushrooms: the extent of the unexplored potential. Int J Med Mushrooms 3:333–337. https://doi.org/10.1615/IntJMedMushr.v3.i4.50
Hawskworth DL (2012) Global species numbers of fungi: are tropical studies and molecular approaches contributing to a more robust estimate? Biodivers Conserv 21:2425–2433. https://doi.org/10.1007/s10531-012-0335-x
He X, Wang X, Fang J et al (2017a) Structures, biological activities, and industrial applications of the polysaccharides from *Hericium erinaceus* (Lion's mane) mushroom: a review. Int J Biol Macromol 97:228–237. https://doi.org/10.1016/j.ijbiomac.2017.01.040
He X, Wang X, Fang J et al (2017b) Polysaccharides in *Grifola frondosa* mushroom and their health promoting properties: a review. Int J Biol Macromol 101:910–921. https://doi.org/10.1016/j.ijbiomac.2017.03.177
Heleno SA, Ferreira RC, Antonio AL et al (2015a) Nutritional value, bioactive compounds and antioxidant properties of three edible mushrooms from Poland. Food Biosci 11:48–55. https://doi.org/10.1016/j.fbio.2015.04.006
Heleno SA, Barros L, Martins A et al (2015b) Nutritional value, bioactive compounds, antimicrobial activity and bioaccessibility studies with wild edible mushrooms. LWT Food Sci Technol 63(2):799–806. https://doi.org/10.1016/j.lwt.2015.04.028
Heleno SA, Rudke AR, Calhelha RC et al (2017) Development of dairy beverages functionalized with pure ergosterol and mycosterol extracts: an alternative to phytosterol-based beverages. Food Funct 8:103. https://doi.org/10.1039/c6fo01600f
Hennicke F, Cheikh-Ali Z, Liebisch T et al (2016) Distinguishing commercially grown *Ganoderma lucidum* from *Ganoderma lingzhi* from Europe and East Asia on the basis of morphology, molecular phylogeny, and triterpenic acid profiles. Phytochem 127:29–37. https://doi.org/10.1016/j.phytochem.2016.03.012
Heo S, Jeon S, Lee S (2014) Utilization of *Lentinus edodes* mushroom β-glucan to enhance the functional properties of gluten-free rice noodles. LWT Food Sci Technol 55:627–631. https://doi.org/10.1016/j.lwt.2013.10.002
Hibbet DS, Taylor JW (2013) Fungal systematics: is a new age of enlightenment at hand? Nat Rev Microbiol 11:129–133. https://doi.org/10.1038/nrmicro2963
Hobbs CR (1995) Medicinal mushrooms: an exploration of traditions, healing, & culture. Botanica Press, Santa Cruz
Hobbs CR (2005) The chemistry, nutritional value, immune-pharmacology, and safety of the traditional food of medicinal split-gill fugus *Schizophyllum commune* Fr.:Fr. (Schizophyllaceae). A literature review. Int J Med Mushrooms 7(1–2):127–139. https://doi.org/10.1615/IntJMedMushr.v7.i12.130
Hong L, Xun M, Wutong W (2007) Anti-diabetic effect of an alpha-glucan from fruit body of Maitake (*Grifola frondosa*) on KK-Ay mice. J Pharm Pharmacol 59(4):575–582. https://doi.org/10.1211/jpp.59.4.0013
Hossain S, Hashimoto M, Choudhury EK et al (2003) Dietary mushroom (*Pleurotus ostreatus*) ameliorates atherogenic lipid in hypercholesterolaemic rats. Clin Exp Pharmacol Physiol 30(7):470–475. https://doi.org/10.1046/j.1440-1681.2003.03857.x
Hu SH, Wang JC, Lien JL et al (2006) Antihyperglycemic effect of polysaccharide from fermented broth of *Pleurotus citrinopileatus*. Appl Microbiol Biotechnol 70:107–113. https://doi.org/10.1007/s00253-005-0043-5
Ikekawa T (2001) Beneficial effects of edible and medicinal mushrooms on human health care. Int J Med Mushrooms 3(4):291–298. https://doi.org/10.1615/IntJMEDMuhsr.v3.i2-3.30
Ikekawa T (2005) Cancer risk reduction by intake of mushrooms and clinical studies on EEM. Int J Med Mushrooms 7(3):347. https://doi.org/10.1615/IntJMedMushrooms.v7.i3.110
Ina K, Furuta R, Kataoka T et al (2016) Chemo-immunotherapy using lentinan for the treatment of gastric cancer with liver metastases. Med Sci 4:8. https://doi.org/10.3390/medsci4020008

Isaka M, Sappan M, Supothina S et al (2017) Alliacane sesquiterpenoids from submerged cultures of the Basidiomycete *Inonotus* sp. BCC 22670. Phytochem 136:175–181. https://doi.org/10.1016/j.phytochem.2017.01.018

Ishara JRM, Sila DN, Kenji GM et al (2018) Nutritional and physical attributes of maize-mushroom complementary porridges as influenced by mushroom species and ratio. Am J Food Nutr 6(1):17–27. https://doi.org/10.12691/ajfn-6-1-4

Islam T, Yu X, Xu B (2016) Phenolic profiles, antioxidant capacities and metal chelating ability of edible mushrooms commonly used in China. LWT Food Sci Technol 72:423–431. https://doi.org/10.1016/j.lwt.2016.05.005

Ivanova TS, Krupodorova TA, Barshteyn VY et al (2014) Anticancer substances of mushroom origin. Exp Oncol 36(2):58–66

Jakopovich I (2011) New dietary supplements from medicinal mushrooms: Dr. Myko San – a registration report. Int J Med Mushrooms 13(3):307–313. https://doi.org/10.1615/IntJMedMushr.v13.i3.110

Jaszek M, Osińska-Jaroszuk M, Sulej J et al (2015) Stimulation of the antioxidative and antimicrobial potential of the blood red bracket mushroom *Pycnoporus sanguineus* (higher Basidiomycetes). Int J Med Mushrooms 17(8):701–712. https://doi.org/10.1615/IntJMedMushrooms.v17.i8.10

Jaworska G, Pogoń K, Bernaś E et al (2014) Vitamins, phenolics and antioxidant activity of culinary prepared *Suillus luteus* (L) Roussel mushroom. LWT Food Sci Technol 59:701–706. https://doi.org/10.1016/j.lwt.2014.07.040

Jayachandran M, Xiao J, Xu B (2017) A critical review on health promoting benefits of edible mushrooms through gut microbiota. Int J Mol Sci 18:1934. https://doi.org/10.3390/ijms18091934

Jayasuriya WJABN, Wanigatunge CA, Fernando GH et al (2015) Hypoglycaemic activity of culinary *Pleurotus ostreatus* and *P. cystidiosus* mushrooms in healthy volunteers and type 2 diabetic patients on diet control and the possible mechanisms of action. Phytother Res 29(2):303–309. https://doi.org/10.1002/ptr.5255

Jesenak M, Hrubisko M, Majtan J et al (2014) Anti-allergic effect of pleuran (β-glucan from *Pleurotus ostreatus*) in children with recurrent respiratory tract infections. Phytother Res 28:471–474. https://doi.org/10.1002/ptr.5020

Jia X, Qu L, Panpan R et al (2018) Characterization and antioxidant activity of an exopolysaccharide produced by *Rigidoporus microporus* (Agaricomycetes). Int J Med Mushrooms 20(4):311–320. https://doi.org/10.1615/IntJMedMushrooms.2018025808

Jiang J, Sliva D (2010) Novel medicinal mushroom blend suppresses growth and invasiveness of human breast cancer cells. Int J Oncol 37(6):1529–1536. https://doi.org/10.3892/ijo-00000806

Jin X, Ruiz Beguerie J, Sze DMY et al (2012) *Ganoderma lucidum* (Reishi mushroom) for cancer treatment. Cochrane Database Syst Rev 6:CD007731. https://doi.org/10.1002/14651858.CD007731.pub2

Jing X, Mao D, Geng L et al (2013) Medium optimization, molecular characterization and bioactivity of exopolysaccharides from *Pleurotus eryngii*. Arch Microbiol 195:749–757. https://doi.org/10.1007/s00203-013-0927-1

John PA, Wong KH, Naidu M et al (2013) Combination effects of curcumin and aqueous extract of *Lignosus rhinocerotis* mycelium on neurite outgrowth stimulation activity in PC-12 cells. Nat Prod Commun 8:711–714

Johnson E, Førland DT, Saetre L et al (2009) Effect of an extract based on the medicinal mushroom *Agaricus blazei* Murrill on release of cytokines, chemokines and leukocyte growth factors in human blood *ex vivo* and *in vivo*. Scand J Immunol 69(3):242–250. https://doi.org/10.1111/j.1365-3083.2008.02218.x

Jose N, Ajith TA, Janardhanan KK (2002) Antioxidant, anti-inflammatory and antitumor activities of culinary medicinal mushroom *Pleurotus pulmonarius* (Fr.) Quél. (Agaricomycetideae). Int J Med Mushrooms 4:329–335. https://doi.org/10.1615/IntJMedMushr.v4.i4.60

Joseph TP, Chanda W, Padhiar AA et al (2017) A preclinical evaluation of the antitumor activities of edible and medicinal mushrooms: a molecular insight. Integr Cancer Ther 17(2):1–10. https://doi.org/10.1177/1534735417736861

Kalač P (2013) A review of chemical composition and nutritional value of wild-growing and cultivated mushrooms. J Sci Food Agric 93(2):209–218. https://doi.org/10.1002/jsfa.5960

Kalač P (2016) Edible mushrooms. chemical composition and nutritional value. Academic, Amsterdam. https://doi.org/10.1016/B978-0-12-804455-1.00008-4

Kalač P, Svoboda L (2000) A review of trace element concentrations in edible mushrooms. Food Chem 69:273–281. https://doi.org/10.1016/S0308-8146(99)00264-2

Kang HJ, Baik HW, Kim SJ et al (2015) *Cordyceps militaris* enhances cell-immunity in heathen Korean men. J Med Food 18(10):1164–1172. https://doi.org/10.1089/jmf.2014.3350

Karun NC, Sridhar KR (2017) Edible wild mushrooms of the Western Ghats: data on the ethnic knowledge. Data Brief 14:320–328. https://doi.org/10.1016/j.dib.2017.07.067

Ke L, Chen H (2016) Homogenate extraction of crude polysaccharides from *Lentinus edodes* and evaluation of the antioxidant activity. Food Sci Technol 36(3):533–539. https://doi.org/10.1590/1678-457X.00916

Keong CY, Rashid BAA, Ing YS et al (2007) Quantification and identification of polysaccharide contents in *Hericium erinaceus*. Nutr Food Sci 37:260–271. https://doi.org/10.1108/00346650710774631

Kerrigan RW (2005) *Agaricus subrufescens*, a cultivated edible and medicinal mushroom, and its synonyms. Mycologia 97(1):12–24. https://doi.org/10.3852/mycologia.97.1.12

Kerrigan RW (2016) *Agaricus* of North America. Memoirs of the New York botanical garden, vol 114. NYBG Press, New York

Khadhri A, Aouadhi C, Aschi-Smiti S (2017) Screening of bioactive compounds of medicinal mushrooms collected on Tunisian territory. Int J Med Mushrooms 19(2):127–135. https://doi.org/10.1615/IntJMedMushrooms.v19.i2.40

Khan M, Tania M, Zhang D-z et al (2010) *Cordyceps* mushroom: a potent anticancer nutraceutical. Open Nutraceuticals J 3:179–183. https://doi.org/10.2174/1876396001003010179

Khan MS, Zhang X, You L (2014) Structure and bioactivities of fungal polysaccharides. In: Ramawat K, Mérillon JM (eds) Polysaccharides – bioactivity and biotechnology. Springer, Cham, pp 1851–1866

Khaskheli AA, Khaskheli SG, Liu Y et al (2018) Characterization and antioxidant properties of crude water soluble polysaccharides from three edible mushrooms. J Med Plant Res 12(12):133–138. https://doi.org/10.5897/JMPR2017.6441

Khatua S, Paul S, Acharya K (2013) Mushroom as the potential source of new generation of antioxidant – a review. Res J Pharm Technol 6:496–505

Khatua S, Ghosh S, Acharya K (2017) *Laetiporus sulphureus* (Bull.:Fr.) Murr. as food as medicine. Pharm J 9(6):1–15. https://doi.org/10.5530/pj.2017.6s.151

Khatun K, Mahtab H, Khanam PA et al (2007) Oyster mushroom reduced blood glucose and cholesterol in diabetic subjects. Mymensingh Med J 16:94–99. https://doi.org/10.3329/mmj.v16i1.261

Khatun S, Islam A, Cakilcioglu U et al (2012) Research on mushroom as a potential source of nutraceuticals: a review on Indian perspective. Am J Experim Agricult 2(1):47–73. https://doi.org/10.9734/AJEA/2012/492

Khatun S, Islam A, Cakilcioglu U et al (2015) Nutritional qualities and antioxidant activity of three edible oyster mushrooms (*Pleurotus* spp.). NJAS – Wageningen J Life Sci 72(73):1–5. https://doi.org/10.1016/j.njas.2012.03.003

Kim JH, Kim YS (1999) A fibrinolytic metalloprotease from the fruiting bodies of an edible mushroom, *Armillariella mellea*. Biosci Biotechnol Biochem 63(12):2130–2136. https://doi.org/10.1271/bbb.63.2130

Kim YW, Kim KH, Choi HJ et al (2005) Anti-diabetic activity of β-glucans and their enzymatically hydrolyzed oligosaccharides from *Agaricus blazei*. Biotechnol Lett 27(7):483–487. https://doi.org/10.1007/s10529-005-2225-8

Kim JS, Sapkota K, Park SE et al (2006) A fibrinolytic enzyme from the medicinal mushroom *Cordyceps militaris*. J Microbiol 44(6):622–631

Kim JS, Kim JE, Choi BS et al (2008) Purification and characterization of fibrinolytic metalloprotease from *Perenniporia fraxinea* mycelia. Mycol Res 112:990–998. https://doi.org/10.1016/j.mycres.2008.01.029

Kim YD, Kwak SH, Kim KJ et al (2014) The analysis of useful components in *Flammulina velutipes* fruit body, *Flammulina velutipes* mycelium and *Cordyceps militaris* mycelium. J Mushroom 12(3):193–200. https://doi.org/10.14480/JM.2014.12.3.193

Kirk PM, Cannon PF, Minter DW et al (2008) In: Kirk PM, Cannon PF, Minter DW, Stalpers JA (eds) Ainsworth and Bisby's dictionary of the fungi, 10th edn. CABI Europe, Wallingford

Klaus AS, Kozarski MS, Vunduk JD et al (2016) Antibacterial and antifungal potential of wild Basidiomycete mushroom *Ganoderma applanatum*. Lek Sirov 36:37–46. https://doi.org/10.5937/leksir1636037K

Klimaszewska M, Górska S, Dawidowski M et al (2017) Selective cytotoxic activity of se-methyl-seleno-L-cysteine- and se-polysaccharide–containing extracts from shiitake medicinal mushroom, *Lentinus edodes* (Agaricomycetes). Int J Med Mushrooms 19(8):709–716. https://doi.org/10.1615/IntJMedMushrooms.2017021250

Klupp NL, Chang D, Hawke F et al (2015) *Ganoderma lucidum* mushroom for the treatment of cardiovascular risk factors (review). Cochrane Database Syst Rev 2:CD007259. https://doi.org/10.1002/14651858.CD007259.pub2

Klupp NL, Kiat H, Bensoussan A et al (2016) A double-blind, randomised, placebo-controlled trial of *Ganoderma lucidum* for the treatment of cardiovascular risk factors of metabolic syndrome. Sci Rep 6:29540. https://doi.org/10.1038/srep29540

Kodama N, Komuta K, Nanba H (2002) Can Maitake MD fraction aid cancer patients? Altern Med Rev 7(3):236–239

Kolundčzić M, Stanojković T, Radović J et al (2017) Cytotoxic and antimicrobial activities of *Cantharellus cibarius* Fr. (Cantharellaceae). J Med Food 20(8):790–796. https://doi.org/10.1089/jmf.2016.0176

Konno S (2009) Synergistic potentiation of D-fraction with vitamin C as possible alternative approach for cancer therapy. Int J Gen Med 2:91–108. https://doi.org/10.2147/IJGM.S5498

Kosanić M, Ranković B, Rančić A et al (2016) Evaluation of metal concentration and antioxidant, antimicrobial, and anticancer potentials of two edible mushrooms *Lactarius deliciosus* and *Macrolepiota procera*. J Food Drug Anal 24:477–484. https://doi.org/10.1016/j.jfda.2016.01.008

Kostic M, Smiljkovic M, Petrovic J et al (2017) Chemical, nutritive composition and wide-broad bioactive properties of honey mushroom *Armillaria mellea* (Vahl.:Fr.) Kummer. Food Funct 9:3239–3249. https://doi.org/10.1039/C7FO00887B

Kovács B, Béni Z, Dékány M et al (2018) Isolation and structure determination of antiproliferative secondary metabolites from the potato earthball mushroom, *Scleroderma bovista* (Agaricomycetes). Int J Med Mushrooms 20(5):411–418. https://doi.org/10.1615/IntJMedMushrooms.2018026043

Kozarski MS, Klaus AS, Nikšić MP et al (2014) Polysaccharides of higher fungi: biological role, structure and antioxidative activity. Hem Ind 68(3):305–320. https://doi.org/10.2298/HEMIND121114056K

Kozarski M, Klaus A, Jakovljevic D et al (2015) Antioxidants of edible mushrooms. Molecules 20:19489–19525. https://doi.org/10.3390/molecules201019489

Krupodorova T, Rybalko S, Barshteyn V (2014) Antiviral activity of Basidiomycete mycelia against influenza type A (serotype H_1N_1) and herpes simplex virus type 2 in cell culture. Virol Sin 29(5):284–290. https://doi.org/10.1007/s12250-014-3486-y

Kudryavtseva OA, Dunaevsky YE, Kamzolkina OV et al (2008) Fungal proteolytic enzymes: features of the extracellular proteases of xylotrophic Basidiomycetes. Microbiology 77(6):643–653. https://doi.org/10.1134/S0026261708060015

Kües U, Badalyan SM (2017) Making use of genomic information to explore the biotechnological potential of medicinal mushrooms. In: Agrawal DC, Tsay HS, Shyur LF, Wu YC, Wang SY (eds) Medicinal plants and fungi: recent advances in research and development, Medicinal and aromatic plants of the world, vol 4. Springer, New York, pp 397–458. https://doi.org/10.1007/978-981-10-5978-0_13

Kües U, Khonsuntia W, Subba S, Dörnte B (2018) Volatiles in communication of Agaricomycetes. In: Anke T, Schüffler A (eds) Physiology and genetics. The Mycota (a comprehensive treatise on fungi as experimental systems for basic and applied research), vol 15. Springer, Cham, pp 149–212. https://doi.org/10.1007/978-3-319-71740-1_6

Kumar K (2015) Role of edible mushroom as functional foods – a review. South Asian J Food Technol Environm 1(3&4):211–218

Kumaran S, Palani P, Nishanthi R et al (2011) Studies on screening, isolation and purification of a fibrinolytic protease from an isolate (VK12) of *Ganoderma lucidum* and evaluation of its antithrombotic activity. Med Mycol J 52(2):153–162. https://doi.org/10.3314/jjmm.52.153

Lakornwong W, Kanokmedhakul K, Kanokmedhakul S et al (2014) Triterpene lactones from cultures of *Ganoderma* sp. KM01. J Nat Prod 77:1545–1153. https://doi.org/10.1021/np400846k

Lalotra P, Bala P, Kumar S, Sharma YP (2016) Biochemical characterization of some wild edible mushrooms from Jammu and Kashmir. Proc Natl Acad Sci India, Sect B Biol Sci 88(2):539–545. https://doi.org/10.1007/s40011-016-0783-2

Landi N, Pacifico S, Ragucci S et al (2017) Pioppino mushroom in southern Italy: an undervalued source of nutrients and bioactive compounds. J Sci Food Agric 97(15):5388–5397. https://doi.org/10.1002/jsfa.8428

Lau BF, Abdullah N (2017) Bioprospecting of *Lentinus squarrosulus* Mont., an underutilized wild edible mushroom, as a potential source of functional ingredients: a review. Trends Food Sci Technol 61:116–131. https://doi.org/10.1016/j.tifs.2016.11.017

Lee JS, Baik HS, Park SS (2002) Optimal production and characterization of fibrinolytic enzymes from *Fomitella fraxinea* mycelia. Kor J Microbiol Biotechnol 30:325–331

Lee JS, Baik HS, Park SS (2006) Purification and characterization of two novel fibrinolytic proteases from mushroom, *Fomitella fraxinea*. J Microbiol Biotechnol 16(2):264–271

Lee JS, Park BC, Ko YJ et al (2008) *Grifola frondosa* (Maitake mushroom) water extract inhibits vascular endothelial growth factor-induced angiogenesis through inhibition of reactive oxygen species and extracellular signal-regulated kinase phosphorylation. J Med Food 11(4):643–651. https://doi.org/10.1089/jmf.2007.06290

Lee KH, Morris-Natschke SL, Yang X et al (2012) Recent progress of research on medicinal mushrooms, foods, and other herbal products used in traditional Chinese medicine. J Tradit Complement Med 2(2):84–95. https://doi.org/10.1016/S2225-4110(16)30081-5

Lee SS, Tan NH, Fung SY et al (2014a) Anti-inflammatory effect of the sclerotium of *Lignosus rhinocerotis* (Cooke) Ryvarden, the tiger milk mushroom. BMC Complement Altern Med 14:359. https://doi.org/10.1186/1472-6882-14-359

Lee KF, Chen JW, Teng CC et al (2014b) Protective effects of *Hericium erinaceus* mycelium and its isolated erinacine a against ischemia-injury-induced neuronal cell death via the inhibition of iNOS/p38 MAPK and nitrotyrosine. Int J Mol Sci 15:15073–15089. https://doi.org/10.3390/ijms150915073

Lemieszek MK, Nunes FHFM, Sawa-Wejksza K et al (2017) A king Bolete, *Boletus edulis* (Agaricomycetes), RNA fraction stimulates proliferation and cytotoxicity of natural killer cells against myelogenous leukemia cells. Int J Med Mushrooms 19(4):347–353. https://doi.org/10.1615/IntJMedMushrooms.v19.i4.50

Lima CUJO, Souza VC, Morita MC et al (2011) *Agaricus blazei* Murrill and inflammatory mediators in elderly women: a randomized clinical trial. Scand J Immunol 75(3):336–341. https://doi.org/10.1111/j.1365-3083.2011.02656.x

Lin KW, Maitraie D, Huang AM et al (2016) Triterpenoids and an alkamide from *Ganoderma tsugae*. Fitoterapia 108:73–80. https://doi.org/10.1016/j.fitote.2015.11.003

Lindequist U (2011) The impact of ethnomycology on modern pharmacy. Curare 34(1+2):118–123

Lindequist U (2013) The merit of medicinal mushrooms from a pharmaceutical point of view. Int J Med Mushrooms 15(6):517–523. https://doi.org/10.1615/IntJMedMushr.v15.i6.10

Lindequist U, Niedermeyer THJ, Jülich WD (2005) The pharmacological potential of mushrooms. Evid Based Complement Alternat Med 2(3):285–299. https://doi.org/10.1093/ecam/neh107

Liu JK (2005) N-containing compounds of macromycetes. Chem Rev 105(7):2723–2744. https://doi.org/10.1002/chin.200542287

Liu Y, Fukuwatari Y, Okumura K et al (2008) Immunomodulating activity of *Agaricus brasiliensis* KA21 in mice and in human volunteers. Evid Based Complement Alternat Med 5(2):205–219. https://doi.org/10.1093/ecam/nem016

Liu Y, Wang J, Wang W et al (2015a) The chemical constituents and pharmacological actions of *Cordyceps sinensis*. Evid Based Complement Alternat Med 2015:575063.):12 p. https://doi.org/10.1155/2015/575063

Liu Z, Wang Q, Cui J et al (2015b) Systemic screening of strains of the Lion's mane medicinal mushroom *Hericium erinaceus* (higher Basidiomycetes) and its protective effects on Aβ-triggered neurotoxicity in PC12 cells. Int J Med Mushrooms 7(3):219–229. https://doi.org/10.1615/IntJMedMushrooms.v17.i3.20

Liu M, Jing H, Zhang J et al (2016) Optimization of mycelia selenium polysaccharide extraction from *Agrocybe cylindracea* SL-02 and assessment of their antioxidant and anti-ageing activities. PLoS One 11(8):e0160799. https://doi.org/10.1371/journal.pone.0160799

Llarena-Hernández RC, Renouf E, Vitrac X et al (2015) Antioxidant activities and metabolites in edible fungi, a focus on the almond mushroom *Agaricus subrufescens*. In: Mérillon JM, Ramawat KG (eds) Fungal metabolites. Springer, Cham, pp 1–22. https://doi.org/10.1007/978-3-319-19456-1_35-1

Lo HC, Wasser SP (2011) Medicinal mushrooms for glycemic control in diabetes mellitus: history, current status, future perspectives, and unsolved problems (review). Int J Med Mushrooms 13(5):401–426. https://doi.org/10.1615/IntJMedMushr.v13.i5.10

Lu LC, Chen SN (2012) Fibrinolytic enzymes from medicinal mushrooms. In: Biochemistry, genetics and molecular biology. Protein structure, Shanghai, pp 337–362

Lu X, Brennan MA, Serventi L et al (2016) How the inclusion of mushroom powder can affect the physicochemical characteristics of pasta. Int J Food Sci Technol 51:2433–2439. https://doi.org/10.1111/ijfs.13246

Lu X, Brennan MA, Serventi L et al (2018) Addition of mushroom powder to pasta enhances the antioxidant content and modulates the predictive glycaemic response of pasta. Food Chem 264:199–209. https://doi.org/10.1016/j.foodchem.2018.04.130

Ludwig E (2007) Pilzkompendium 2:61. Fungicon-Verlag, Berlin

Ma G, Yang W, Mariga AM et al (2014a) Purification, characterization and antitumor activity of polysaccharides from *Pleurotus eryngii* residue. Carbohydr Polym 114:297–305. https://doi.org/10.1016/j.carbpol.2014.07.069

Ma K, Ren J, Han J et al (2014b) Ganoboninketals A – C, antiplasmodial 3,4-seco-27-norlanostane triterpenes from *Ganoderma boninense*. Pat J Nat Prod 77:1847–1852. https://doi.org/10.1021/np5002863

Ma HT, Hsieh JF, Chen ST (2015) Anti-diabetic effects of *Ganoderma lucidum*. Phytochem 114:109–113. https://doi.org/10.1016/j.phytochem.2015.02.017

Martel J, Ojcius DM, Chang CJ et al (2017) Anti-obesogenic and antidiabetic effects of plants and mushrooms. Nat Rev Endocrinol 13:149–161. https://doi.org/10.1038/nrendo.2016.142

Masri HJ, Maftoun P, Malek RA et al (2017) The edible mushroom *Pleurotus* spp.: II. Medicinal values. Int J Biotech Well Indus 6:1–11

Masuda Y, Murata Y, Hayashi M et al (2008) Inhibitory effect of MD-fraction on tumor metastasis: involvement of NK cell activation and suppression of intercellular adhesion molecule (ICAM)-1 expression in lung vascular endothelial cells. Biol Pharm Bull 31(6):1104–1108. https://doi.org/10.1248/bpb.31.1104

Mehrotra A, Calvo MS, Beelman RB et al (2014) Bioavailability of vitamin D_2 from enriched mushrooms in prediabetic adults: a randomized controlled trial. Eur J Clin Nutr 68(10):1154–1160. https://doi.org/10.1038/ejcn.2014.157

Meneses ME, Martínez-Carrera D, Torres N et al (2016) Hypocholesterolemic properties and prebiotic effects of Mexican *Ganoderma lucidum* in C57BL/6 mice. PLoS One 11(7):e0159631. https://doi.org/10.1371/journal.pone.0159631

Meng X, Liang H, Luo L (2016) Antitumor polysaccharides from mushrooms: a review on the structural characteristics, antitumor mechanisms and immunomodulating activities. Carbohydr Res 424:30–41. https://doi.org/10.1016/j.carres.2016.02.008

Merdivan S, Lindequist U (2017) Medicinal mushrooms with antiallergic activities. In: Agrawal DC, Tsay HS, Shyur LF, Wu YC, Wang SY (eds) Medicinal plants and fungi: recent advances in research and development, Medicinal and aromatic plants of the world, vol 4. Springer, New York, pp 93–111. https://doi.org/10.1007/978-981-10-5978-0_4

Merdivan S, Jenett-Siems K, Siems K et al (2017) Inhibition of degranulation of RBL-2H3 cells by extracts and compounds from *Armillaria ostoyae*. Planta Med Int Open 4:e1–e7. https://doi.org/10.1055/s-0042-121608

Milovanović I, Brčeski I, Stajić M et al (2013) Potential enrichment of medicinal mushrooms with selenium to obtain new dietary supplements. Int J Med Mushrooms 15:451–457. https://doi.org/10.1615/IntJMedMushr.v15.i5.30

Milovanović I, Stajić M, Stanojković T et al (2015a) Effects of selenium presence in mycelia of *Ganoderma* species (higher Basidiomycetes) on their medicinal properties. Int J Med Mushrooms 17(1):11–20. https://doi.org/10.1615/IntJMedMushrooms.v17.i1.20

Milovanović IN, Stanojkovic TP, Stajic MM et al (2015b) Effect of selenium enrichment of *Lenzites betulinus* and *Trametes hirsuta* mycelia on antioxidant, antifungal and cytostatics potential. Curr Pharm Biotechnol 16(10):920–926. https://doi.org/10.2174/1389201016666150618152531

Mirończuk-Chodakowska I, Witkowska AM, Zujko ME et al (2017) Quantitative evaluation of 1,3-1,6-β-D-glucan contents in wild–growing species of edible Polish mushrooms. Rocz Panstw Zakl Hig 68(3):281–290

Mizuno M, Nishitani Y (2013) Immunomodulating compounds in Basidiomycetes. J Clin Biochem Nutr 52(3):202–207. https://doi.org/10.3164/jcbn.13-3

Mizuno T, Wang G, Zhang J et al (1995) Reishi, *Ganoderma lucidum* and *Ganoderma tsugae*: bioactive substances and medicinal effects. Food Rev Int 11(1):151–166. https://doi.org/10.1080/87559129509541025

Mocan A, Fernandes Â, Barros L et al (2018) Chemical composition and bioactive properties of the wild mushroom *Polyporus squamosus* (Huds.) Fr: a study with samples from Romania. Food Funct 9:160–170. https://doi.org/10.1039/C7FO01514C

Mohanarji S, Dharmalingam S, Kalusalingam A (2012) Screening of *Lignosus rhinocerus* extracts as antimicrobial agents against selected human pathogens. J Pharm Biomed Sci 18:1–4

Money NP (2016) Are mushrooms medicinal? Fungal Biol 120:449–453. https://doi.org/10.1016/j.funbio.2016.01.006

Monro JA (2003) Treatment of cancer with mushroom products. Arch Environ Health 58:533–537. https://doi.org/10.3200/AEOH.58.8.533-537

Moore D, Chiu SW (2001) Fungal products as food. In: Pointing SB, Hyde KD (eds) Bio-exploitation of filamentous fungi. Fungal Diversity Press, Hong Kong, pp 223–251

Morales D, Piris AJ, Ruiz-Rodriguez A et al (2018) Extraction of bioactive compounds against cardiovascular diseases from *Lentinula edodes* using a sequential extraction method. Biotechnol Prog 34:746. https://doi.org/10.1002/btpr.2616

Morel S, Arnould S, Vitou M et al (2018) Antiproliferative and antioxidant activities of wild Boletales mushrooms from France. Int J Med Mushrooms 20(1):13–29. https://doi.org/10.1615/IntJMedMushrooms.v17.i3.20

Mori K, Inatomi S, Ouchi K et al (2009) Improving effects of the mushroom Yamabushitake (*Hericium erinaceus*) on mild cognitive impairment: a double-blind placebo-controlled clinical trial. Phytother Res 23(3):367–372. https://doi.org/10.1002/ptr.2634

Morozova EN, Falina NN, Denisova NP et al (1982) Analysis of the component constitution and substrate specificity of a fibrinolytic preparation from the fungus *Flammulina velutipes*. Biokhimia (Moscow, Russia) 47(7):1181–1185. (in Russian)

Morris HJ, Beltrán Y, Llauradó G et al (2017) Mycelia from *Pleurotus* sp. (oyster mushroom): a new wave of antimicrobials, anticancer and antioxidant bio-ingredients. Intl J Phyto Natur Ingrd 4(1):03. https://doi.org/10.15171/ijpni.2017.03

Motoi M, Motoi A, Yamanaka D et al (2015) Open-label study of the influence of food containing the royal sun mushroom, *Agaricus brasiliensis* KA21 (higher Basidiomycetes), on the quality of life of healthy volunteers. Int J Med Mushrooms 17(9):799–817. https://doi.org/10.1615/IntJMedMushrooms.v17.i9.10

Mustonen AM, Määttänen M, Kärjä V et al (2018) Myo- and cardiotoxic effects of the wild winter mushroom (*Flammulina velutipes*) on mice. Exp Biol Med 243:639–644. https://doi.org/10.1177/1535370218762340

Muszynska B, Lojewski M, Rojowski J et al (2015) Natural products of relevance in the prevention and supportive treatment of depression. Psychiatr Pol 49:435–453. https://doi.org/10.12740/PP/29367

Muta T (2006) Molecular basis for invertebrate innate immune recognition of (1-3)-beta-D-glucan as a pathogen-associated molecular pattern. Curr Pharm Des 12(32):4155–4161. https://doi.org/10.2174/138161206778743529

Nagano M, Shimizu K, Kondo R et al (2010) Reduction of depression and anxiety by 4 weeks *Hericium erinaceus* intake. Biomed Res 31(4):231–237. https://doi.org/10.2220/biomedres.31.231

Nallathamby N, Lee GS, Raman J et al (2016) Identification and *in vitro* evaluation of lipids from sclerotia of *Lignosus rhinocerotis* for antioxidant and anti-neuroinflammatory activities. Nat Prod Commun 11:1485–1490

Nallathamby N, Phan CW, Seow SL et al (2018) A status review of the bioactive activities of tiger milk mushroom *Lignosus rhinocerotis* (Cooke) Ryvarden. Front Pharmacol 8:998. https://doi.org/10.3389/fphar.2017.00998

Nguyen AH, Gonzaga MI, Lim VM et al (2017) Clinical features of shiitake dermatitis: a systematic review. Int J Dermatol 56(6):610–616. https://doi.org/10.1111/ijd.13433

Nielsen JC, Nielsen J (2017) Development of fungal cell factories for the production of secondary metabolites: linking genomics and metabolism. Synth Syst Biotechnol 2(1):5–12. https://doi.org/10.1016/j.synbio.2017.02.002

Novak M, Vetvicka V (2008) β-Glucans, history, and the present: immunomodulatory aspects and mechanisms of action. J Immunotoxicol 5(1):47–57. https://doi.org/10.1080/15476910802019045

Novakovič A, Karaman M, Kaišarevič S et al (2017) Antioxidant and antiproliferative potential of fruiting bodies of the wild-growing king bolete mushroom, *Boletus edulis* (Agaricomycetes), from Western Serbia. Int J Med Mushrooms 19(1):27–34. https://doi.org/10.1615/IntJMedMushrooms.v19.i1.30

Nowacka N, Nowak R, Drozd M et al (2014) Analysis of phenolic constituents, antiradical and antimicrobial activity of edible mushrooms growing wild in Poland. LWT Food Sci Technol 59:689–694. https://doi.org/10.1016/j.lwt.2014.05.041

Nowak R, Nowacka-Jechalke N, Juda M et al (2017) The preliminary study of prebiotic potential of polish wild mushroom polysaccharides: the stimulation effect on *Lactobacillus* strains growth. Eur J Nutr 57(4):1511–1521. https://doi.org/10.1007/s00394-017-1436-9

Nunes RGFL, da Luz JMR, de Freitas RB et al (2012) Selenium bioaccumulation in shiitake mushrooms: a nutritional alternative source of this element. J Food Sci 77(9):983–986. https://doi.org/10.1111/j.1750-3841.2012.02837.x

Nyegue M, Amvam Zollo PH, Bessière JM et al (2003) Volatile components of fresh *Pleurotus ostreatus* and *Termitomyces shimperi* from Cameroon. J Essent Oil Bear Plants 6:153–157. https://doi.org/10.1080/0972-060X.2003.10643344

Oba K, Teramukai S, Kobayashi M et al (2007) Efficacy of adjuvant immunochemotherapy with polysaccharide K for patients with curative resections of gastric cancer. Cancer Immunol Immunother 56:905–911. https://doi.org/10.1007/s00262-006-0248-1

Oba K, Koboyashi M, Matsui T et al (2009) Individual patient based meta-analysis of lentinan for unresectable/recurrent gastric cancer. Anticancer Res 7:2739–2745

Okolo KO, Orisakwe OE, Siminialayi IM (2018) Nephroprotective and antioxidant effects of king tuber oyster medicinal mushroom, *Pleurotus tuberregium* (Agaricomycetes), on carbon tetrachloride–induced nephrotoxicity in male Sprague dawley rats. Int J Med Mushrooms 20(5):419–429. https://doi.org/10.1615/IntJMedMushrooms.2018026044

Owaid MN, AL Saeedi SS, Abed IA et al (2017) Antifungal activities of some *Pleurotus* species (higher Basidiomycetes). Walailak J Sci Tech 14(3):215–224

Özcan Ö, Ertan F (2018) Beta-glucan content, antioxidant and antimicrobial activities of some edible mushroom species. Food Sci Technol 6(2):47–55. https://doi.org/10.13189/fst.2018.060201

Palacios I, Lozano M, Moro C et al (2011) Antioxidant properties of phenolic compounds occurring in edible mushrooms. Food Chem 128:674–678. https://doi.org/10.1016/j.foodchem.2011.03.085

Palazzolo E, Gargano ML, Venturella G (2012) The nutritional composition of selected wild edible mushrooms from Sicily (southern Italy). Int J Food Sci Nutr 63(1):79–83. https://doi.org/10.3109/09637486.2011.598850

Pandya U, Dhuldhaj U, Sahay NS (2018) Bioactive mushroom polysaccharides as antitumor: an overview. Nat Prod Res 4:1–13. https://doi.org/10.1080/14786419.2018.1466129

Panthong S, Boonsathorn N, Chuchawankul S (2016) Antioxidant activity, anti-proliferative activity, and amino acid profiles of ethanolic extracts of edible mushrooms. Genet Mol Res 15(4):gmr15048886. https://doi.org/10.4238/gmr15048886

Papaspyridi LM, Aligiannis N, Christakopoulos P et al (2011) Production of bioactive metabolites with pharmaceutical and nutraceutical interest by submerged fermentation of *Pleurotus ostreatus* in a batch stirred tank bioreactor. Proc Food Sci 1:1746–1752. https://doi.org/10.1016/j.profoo.2011.09.257

Paravamsivam P, Heng CK, Malek SNA et al (2016) Giant oyster mushroom *Pleurotus giganteus* (Agaricomycetes) enhances adipocyte differentiation and glucose uptake via activation of PPAR and glucose transporters 1 and 4 in 3T3-L1 cells. Int J Med Mushrooms 18(9):821–831. https://doi.org/10.1615/IntJMedMushrooms.v18.i9.60

Park HJ (2014) CARI III inhibits tumor growth in a melanoma-bearing mouse model through induction of G0/G1 cell cycle arrest. Molecules 19:14383–14395. https://doi.org/10.3390/molecules190914383

Park KJ, Lee SY, Kim HS et al (2007) The neuroprotective and neurotrophic effects of *Tremella fuciformis* in PC12h cells. Mycobiology 35(1):11–15. https://doi.org/10.4489/MYCO.2007.35.1.011

Parra LA (2013) *Agaricus* L. *Allopsalliota* Nauta and Bas, Part II. Candusso, Alassio

Patel Y, Naraian R, Singh VK (2012) Medicinal properties of *Pleurotus* species (oyster mushroom): a review. World J Fungal Plant Biol 3:1–12. https://doi.org/10.5829/idosi.wjfpb.2012.3.1.303

Paterson RRM (2006) *Ganoderma* – a therapeutic fungal biofactory. Phytochem 67(18):1985–2001. https://doi.org/10.1016/j.phytochem.2006.07.004

Paterson RRM, Lima N (2014) Biomedical effects of mushrooms with emphasis on pure compounds. Biom J 37(6):357–368. https://doi.org/10.4103/2319-4170.143502

Patin EC, Thompson A, Orr SJ (2018) Pattern recognition receptors in fungal immunity. Semin Cell Dev Biol. (in press). https://doi.org/10.1016/j.semcdb.2018.03.003

Patra KC, Wang Q, Bhaskar PT et al (2013) Hexokinase 2 is required for tumor initiation and maintenance and its systemic deletion is therapeutic in mouse models of cancer. Cancer Cell 24(2):213–228. https://doi.org/10.1016/j.ccr.2013.06.014

Paul N, Slathia PS, Vaid A et al (2018) Traditional knowledge of Gucchi, *Morchella esculenta* (Ascomycetes), in Doda district, Jammu and Kashmir, India. Int J Med Mushrooms 20(5):445–450. https://doi.org/10.1615/IntJMedMushrooms.2018025995

Pavithra M, Sridhar KR, Ayyanna G (2017) Functional properties of edible mushroom *Astraeus hygrometricus*. KAVAKA 49:22–27

Peay KG, Kennedy PG, Talbot JM (2016) Dimension of biodiversity in the earth mycobiome. Nat Rev Microbiol 14:434–447. https://doi.org/10.1038/nrmicro.2016.59

Pedrero Z, Madrid Y (2009) Novel approaches for selenium speciation in foodstuffs and biological specimens: a review. Anal Chim Acta 634(2):135–152. https://doi.org/10.1016/j.aca.2008.12.026

Peng XR, Liu JQ, Xia JJ et al (2015) Lanostane triterpenoids from *Ganoderma hainanense* J.D. Zhao. Phytochem 114:137–145. https://doi.org/10.1016/j.phytochem.2014.10.009

Peng X, Li L, Wang X et al (2016) Antioxidant farnesylated hydroquinones from *Ganoderma capense*. Fitoterapia 111:18–23. https://doi.org/10.1016/j.fitote.2016.04.006

Perera PK, Li Y (2011) Mushrooms as a functional food mediator in preventing and ameliorating diabetes. Funct Food Health Dis 4:161–171. https://doi.org/10.31989/ffhd.v1i4.133

Pérez-Martínez AS, Acevedo-Padilla SA, Bibbins-Martínez M et al (2015) A perspective on the use of *Pleurotus* for the development of convenient fungi-made oral subunit vaccines. Vaccine 33:25–33. https://doi.org/10.1016/j.vaccine.2014.10.05

Phan CW, David P, Naidu M et al (2013) Neurite outgrowth stimulatory effects of culinary-medicinal mushrooms and their toxicity assessment using differentiating Neuro-2a and embryonic fibroblast BALB/3T3. BMC Complement Altern Med 13:261. https://doi.org/10.1186/1472-6882-13-261

Phan CW, David P, Naidu M et al (2015) Therapeutic potential of culinary-medicinal mushroom for the management of neurodegenerative diseases: diversity, metabolite, and mechanism. Crit Rev Biotechnol 35(3):355–368. https://doi.org/10.3109/07388551.2014.887649

Phan CW, David P, Sabaratnam V (2017a) Edible and medicinal mushrooms: emerging brain food for the mitigation of neurodegenerative diseases. J Med Food 20(1):1–10. https://doi.org/10.3389/fphar.2017.0021910.1089/jmf.2016.3740

Phan CW, Wang JK, Cheah SC et al (2017b) A review on the nucleic acid constituents in mushrooms: nucleobases, nucleosides and nucleotides. Crit Rev Biotechnol 10:1–16. https://doi.org/10.1080/07388551.2017.1399102

Phan CW, Tan EYY, Sabaratnam V (2018) Bioactive molecules in edible and medicinal mushrooms for human wellness. In: Mérillon JM, Ramawat K (eds) Bioactive molecules in food, Reference series in phytochemistry. Springer, Cham, pp 1–24

Piotrowski J, Jedrzejewski T, Kozak W (2015) Immunomodulatory and antitumor properties of polysaccharide peptide (PSP). Postepy Hig Med Dosw 69:91–97. https://doi.org/10.5604/17322693.1137086

Pop RM, Puia IC, Puia A et al (2018) Characterization of *Trametes versicolor*: medicinal mushroom with important health benefits. Not Bot Horti Agrobo 46(2):343–349. https://doi.org/10.15835/nbha46211132

Poucheret P, Fons F, Rapior S (2006) Biological and pharmacological activity of higher fungi: 20-year retrospective analysis. Cryptogam Mycol 27:311–333

Powell M (2014) Medicinal mushrooms: a clinical guide. Mycology Press, Somerset

Prasad S, Rathore H, Sharma S et al (2015) Medicinal mushrooms as a source of novel functional food. Int J Food Sci Nutr Diet 04(5):221–225. https://doi.org/10.19070/2326-3350-1500040

Previtali E, Bucciarelli P, Passamonti SM et al (2011) Risk factors for venous and arterial thrombosis. Blood Transfus 9(2):120–138. https://doi.org/10.2450/2010.0066-10

Pu DB, Zheng X, Gao JB et al (2017) Highly oxygenated lanostane-type triterpenoids and their bioactivity from the fruiting body of *Ganoderma gibbosum*. Fitoterapia 119:1–7. https://doi.org/10.1016/j.fitote.2017.03.007

Qiao X, Wang Q, Ji S et al (2015) Metabolites identification and multi-component pharmacokinetics of ergostane and lanostane triterpenoids in the anticancer mushroom *Antrodia cinnamomea*. J Pharm Biomed Anal 111:266–276. https://doi.org/10.1016/j.jpba.2015.04.010

Rajamahanty S, Louie B, O'Neill C et al (2009) Possible disease remission in patient with invasive bladder cancer with D-fraction regime. Int J Gen Med 2:15–17. https://doi.org/10.2147/IJGM.S4948

Ramberg JE, Nelson ED, Sinnott RA (2010) Immunomodulatory dietary polysaccharides: a systematic review of the literature. Nutr J 9:54. https://doi.org/10.1186/1475-2891-9-54

Rapior S, Pélissier Y, Vassas A et al (1993) *Termitomyces striatus* (Beeli) Heim *forma annulatus*. Identification et étude préliminaire de la composition chimique: recherche des polyols, des sucres, des alcaloïdes, des acides phénols et des toxines. Le Pharmacien d'Afrique (77):19–22

Rapior S, Fruchier A, Bessière JM (1997) Volatile aroma constituents of Agarics and Boletes. In: Pandalai SG (ed) Recent research developments in phytochemistry. Research Signpost, Trivandrum, pp 567–584

Rapior S, Mauruc MJ, Guinberteau J et al (2000) The volatile composition of *Gyrophragmium dunalii*. Mycologia 92:1043–1046. https://doi.org/10.2307/3761469

Rathee S, Rathee D, Rathee D et al (2012) Mushrooms as therapeutic agents. Bras J Pharmacognosy 22(2):459–474. https://doi.org/10.1590/S0102-695X2011005000195

Rayman MP (2000) The importance of selenium to human health. Lancet 356(9225):233–241. https://doi.org/10.1016/S0140-6736(00)02490-9

Reinoso R, Cajas-Madriaga D, Martínez M et al (2013) Biological activity of macromycetes isolated from Chilean subantarctic ecosystems. J Chil Chem Soc 58(4):2016–2019. https://doi.org/10.4067/S0717-97072013000400024

Reis FS, Stojković D, Barros L et al (2014) Can *Suillus granulatus* (L.) Roussel be classified as a functional food? Food Funct 5:2861. https://doi.org/10.1039/c4fo00619d

Reis FS, Martins A, Vasconcelos MH et al (2017) Functional foods based on extracts or compounds derived from mushrooms. Trends Food Sci Technol 66:48–62. https://doi.org/10.1016/j.tifs.2017.05.010

Ren L, Perera C, Hemar Y (2012) Antitumor activity of mushroom polysaccharides: a review. Food Funct 3(11):1118–1130. https://doi.org/10.1039/c2fo10279j

Ren L, Hemara Y, Perera CO et al (2014) Antibacterial and antioxidant activities of aqueous extracts of eight edible mushrooms. Bioact Carbohydr Dietary Fibre 3:41–51. https://doi.org/10.1016/j.bcdf.2014.01.003

Ríos JL, Andújar I, Recio MC et al (2012) Lanostanoids from fungi: a group of potential anticancer compounds. J Nat Prod 75:2016–2044. https://doi.org/10.1021/np300412h

Ros S, Schulze A (2013) Glycolysis back in the limelight: systemic targeting of HK2 blocks tumor growth. Cancer Discov 3:1105–1107. https://doi.org/10.1158/2159-8290.CD-13-0565

Rossi P, Difrancia R, Quagliariello V et al (2018) B-glucans from *Grifola frondosa* and *Ganoderma lucidum* in breast cancer: an example of complementary and integrative medicine. Oncotarget 9(37):24837–24856. https://doi.org/10.18632/oncotarget.24984

Roupas P, Keogh J, Noakes M (2012) The role of edible mushrooms in health: evaluation of the evidence. J Funct Foods 4(4):687–709. https://doi.org/10.1016/j.jff.2012.05.003

Roussel B, Rapior S, Masson CL et al (2002) *Fomes fomentarius* (L.:Fr.) Fr.: un champignon aux multiples usages. Cryptogam Mycol 23:349–366

Rubel R, Herta Santa HS, dos Santos LF et al (2018) Immunomodulatory and antitumoral properties of *Ganoderma lucidum* and *Agaricus brasiliensis* (Agaricomycetes) medicinal mushrooms. Int J Med Mushrooms 20(4):393–403. https://doi.org/10.1615/IntJMedMushrooms.2018025979

Rupcic Z, Rascher M, Kanaki S et al (2018) Two new cyathane diterpenoids from mycelial cultures of the medicinal mushroom *Hericium erinaceus* and the rare species, *Hericium flagellum*. Int J Mol Sci 19(3):740. https://doi.org/10.3390/ijms19030740

Ruthes AC, Carbonero ER, Cordova MM et al (2013) *Lactarius rufus* (1-3)-, (1-6)-β-D-glucans: structure, antinociceptive and anti-inflammatory effects. Carbohydr Polym 94(1):129–136. https://doi.org/10.1016/j.carbpol.2013.01.026

Sabaratnam V, Kah-Hui W, Naidu M et al (2013) Neuronal health – can culinary and medicinal mushrooms help? J Tradit Complement Med 3(1):62–68. https://doi.org/10.4103/2225-4110.106549

Saltarelli R, Ceccaroli P, Iotti M et al (2009) Biochemical characterisation and antioxidant activity of mycelium of *Ganoderma lucidum* from Central Italy. Food Chem 116:143–151. https://doi.org/10.1016/j.foodchem.2009.02.023

Saltarelli R, Ceccaroli P, Buffalini M et al (2015) Biochemical characterization, antioxidant and antiproliferative activities of different *Ganoderma* collections. J Mol Microbiol Biotechnol 25(1):16–25. https://doi.org/10.1159/000369212

Samberkar S, Gandhi S, Naidu M et al (2015) Lion's mane, *Hericium erinaceus* and Tiger milk, *Lignosus rhinocerotis* (higher Basidiomycetes) medicinal mushrooms stimulate neurite outgrowth in dissociated cells of brain, spinal cord, and retina: an in vitro study. Int J Med Mushrooms 17:1047–1054. https://doi.org/10.1615/IntJMedMushrooms.v17.i11.40

Sánchez C (2017a) Bioactives from mushrooms and their application. In: Puri M (ed) Food bioactives. Springer, pp 23–57. https://doi.org/10.1007/978-3-319-51639-4_2

Sánchez C (2017b) Reactive oxygen species and antioxidant properties from mushrooms. Synth Syst Biotechnol 2:13–22. https://doi.org/10.1016/j.synbio.2016.12.001

Santoyo S, Ramirez AC, Garcia LA et al (2012) Antiviral activities of *Boletus edulis*, *Pleurotus ostreatus* and *Lentinus edodes* extracts and polysaccharide fractions against *Herpes simplex* virus type 1. J Food Nutr Res 51(4):225–235

Sato M, Tai T, Nunoura Y et al (2002) Dehydrotrametenolic acid induces preadipocyte differentiation and sensitizes animal models of noninsulin-dependent diabetes mellitus to insulin. Biol Pharm Bull 25:81–86. https://doi.org/10.1248/bpb.25.81

Scarpari M, Parroni A, Zaccaria M et al (2016) *Trametes versicolor* bioactive compounds stimulate *Aspergillus flavus* antioxidant system and inhibit aflatoxin synthesis. Plant Biosyst 150(4):653–659. https://doi.org/10.1080/11263504.2014.981235

Schillaci D, Arizza V, Gargano ML et al (2013) Antibacterial activity of mediterranean oyster mushrooms, species of genus *Pleurotus* (higher Basidiomycetes). Int J Med Mushrooms 15(6):591–594. https://doi.org/10.1615/IntJMedMushr.v15.i6.70

Schillaci D, Cusimano MG, Cascioferro SM et al (2017) Antibacterial activity of desert truffles from Saudi Arabia against *Staphylococcus aureus* and *Pseudomonas aeruginosa*. Int J Med Mushrooms 19(2):121–125. https://doi.org/10.1615/IntJMedMushrooms.v19.i2.30

Schueffler A, Anke T (2014) Fungal natural products in research and development. Nat Prod Rep 31(10):1425–1448. https://doi.org/10.1039/c4np00060a

Sękara A, Kalisz A, Grabowska A et al (2015) *Auricularia* spp. – mushrooms as novel food and therapeutic agents – a review. Sydowia 67:1–10. https://doi.org/10.12905/0380.sydowia67-2015-0001

Seow SLS, Naidu M, David P et al (2013) Potentiation of neuritogenic activity of medicinal mushrooms in rat pheochromocytoma cells. BMC Complement Altern Med 13:157. http://www.biomedcentral.com/1472-6882/13/157

Seow SLS, Eik LF, Naidu M et al (2015) *Lignosus rhinocerotis* (Cooke) Ryvarden mimics the neuritogenic activity of nerve growth factor via MEK/ERK1/2 signaling pathway in PC-12 cells. Sci Rep 5:16349. https://doi.org/10.1038/srep16349

Seow LS, Naidu M, Sabaratnam V et al (2017) Tiger's milk medicinal mushroom, *Lignosus rhinocerotis* (Agaricomycetes) sclerotium inhievbits nitric oxide production in LPS-stimulated BV2 microglia. Int J Med Mushrooms 19(5):405–418. https://doi.org/10.1615/IntJMedMushrooms.v19.i5.30

Shang X, Tan Q, Liu R et al (2013) *In vitro* anti-*Helicobacter pylori* effects of medicinal mushroom extracts, with special emphasis on the Lion's mane mushroom, *Hericium erinaceus* (higher Basidiomycetes). Int J Med Mushrooms 15(2):165–174. https://doi.org/10.1615/IntJMedMushr.v15.i2.50

Sharjahan MA, Tan YS, Raman J et al (2017) Do culinary mushrooms have fibrinolytic activities? Biomed Rev 28:95–103

Sharma SK, Gautam N (2017) Chemical and bioactive profiling, and biological activities of coral fungi from Northwestern Himalayas. Sci Rep 7:46570. https://doi.org/10.1038/srep46570

Sharma D, Singh VP, Singh NK (2017) A review on phytochemistry and pharmacology of medicinal as well as poisonous mushrooms. Mini-Rev Med Chem. https://doi.org/10.2174/1389557517666170927144119

Shen HS, Shao S, Chen JC et al (2017) Antimicrobials from mushrooms for assuring food safety. Compr Rev Food Sci Food Saf 16:316–329. https://doi.org/10.1111/1541-4337.12255

Shibata T, Kudou M, Hoshi Y et al (2010) Isolation and characterization of a novel two-component hemolysin, erylysin A and B, from an edible mushroom, *Pleurotus eryngii*. Toxicon 56(8):1436–1442. https://doi.org/10.1016/j.toxicon.2010.08.010

Shibu MA, Agrawal DC, Huang CY (2017) Mushrooms: a Pandora box of cardioprotective phytochemicals. In: Agrawal DC, Tsay HS, Shyur LF, Wu YC, Wang SY (eds) Medicinal plants and fungi: recent advances in research and development, Medicinal and aromatic plants of the world, vol 4. Springer, New York, pp 337–362. https://doi.org/10.1007/978-981-10-5978-0_11
Shin A, Kim J, Lim SY et al (2010) Dietary mushroom intake and the risk of breast cancer based on hormone receptor status. Nutr Cancer 62(4):476–483. https://doi.org/10.1080/01635580903441212
Singdevsachan SK, Auroshree P, Mishra J et al (2016) Mushroom polysaccharides as potential prebiotics with their antitumor and immunomodulating properties: a review. Bioact Carbohydr Dietary Fibre 7:1–14. https://doi.org/10.1016/j.bcdf.2015.11.001
Singh J, Sindhu SC, Sindhu A et al (2016) Development and evaluation of value added biscuits from dehydrated shiitake (*Lentinula edodes*) mushroom. Int J Curr Res 8(03):27155–27159
Singh V, Bedi GK, Shri R (2017) *In vitro* and *in vivo* antidiabetic evaluation of selected culinary-medicinal mushrooms (Agaricomycetes). Int J Med Mushrooms 19(1):17–25. https://doi.org/10.1615/IntJMedMushrooms.v19.i1.20
Sivanandhan S, Khusro A, Paulraj MG et al (2017) Biocontrol properties of Basidiomycetes: an overview. J Fungi 3:2. https://doi.org/10.3390/jof3010002
Sknepnek A, Pantić M, Matijašević D et al (2018) Novel kombucha beverage from Lingzhi or Reishi medicinal mushroom, *Ganoderma lucidum*, with antibacterial and antioxidant effects. Int J Med Mushrooms 20(3):243–258. https://doi.org/10.1615/IntJMedMushrooms.2018025833
Smith JE, Rowan NJ, Sullivan R (2002a) Medicinal mushrooms: a rapidly developing area of biotechnology for cancer therapy and other bioactivities. Biotechnol Lett 24:1839–1845. https://doi.org/10.1023/A:1020994628109
Smith JF, Rowan NJ, Sullivan R (2002b) Medicinal mushrooms: their therapeutic properties and current medical usage with special emphasis on cancer treatment. University of Strathclyde Press, Glasgow
Soković M, Ciric A, Glamočlija J et al (2014) *Agaricus blazei* hot water extract shows anti quorum sensing activity in the nosocomial human pathogen *Pseudomonas aeruginosa*. Molecules 19:4189–4199. https://doi.org/10.3390/molecules19044189
Souilem F, Fernandes Â, Calhelha RC et al (2017) Wild mushrooms and their mycelia as sources of bioactive compounds: antioxidant, anti-inflammatory and cytotoxic properties. Food Chem 230:40–48. https://doi.org/10.1016/j.foodchem.2017.03.026
Souza Dias E, Abe C, Schwan RF (2004) Truths and myths about the mushroom *Agaricus blazei*. Sci Agric 61(5):545–549. https://doi.org/10.1590/S0103-90162004000500014
Srikram A, Supapvanich S (2016) Proximate compositions and bioactive compounds of edible wild and cultivated mushrooms from Northeast Thailand. Agric Nat Res 50:432–436. https://doi.org/10.1016/j.anres.2016.08.001
Stachowiak B, Reguła J (2012) Health-promoting potential of edible macromycetes under special consideration of polysaccharides: a review. Eur Food Res Technol 234(3):369–380. https://doi.org/10.1007/s00217-011-1656-9
Stadler M, Hoffmeister D (2015) Fungal natural products – the mushroom perspective. Front Microbiol 6:127. https://doi.org/10.3389/fmicb.2015.00127
Stamets P (2000) Growing gourmet and medicinal mushrooms. Ten Speed Press, Toronto
Standish LJ, Wenner CA, Sweet ES et al (2008) *Trametes versicolor* mushroom immune therapy in breast cancer. J Soc Integr Oncol 6(3):122–128
Stanikunaite R, Trappe JM, Khan SI et al (2007) Evaluation of therapeutic activity of hypogeous Ascomycetes and Basidiomycetes from North America. Int J Med Mushrooms 9:7–14. https://doi.org/10.1615/IntJMedMushr.v9.i1.20
Stojković DS, Barros L, Calhelha RC et al (2014a) A detailed comparative study between chemical and bioactive properties of *Ganoderma lucidum* from different origins. Int J Food Sci Nutr 65(1):42–47. https://doi.org/10.3109/09637486.2013.832173
Stojković DS, Reis FS, Glamočlija J et al (2014b) Cultivated strains of *Agaricus bisporus* and *A. brasiliensis*: chemical characterization and evaluation of antioxidant and antimicro-

bial properties for the final healthy product – natural preservatives in yoghurt. Food Funct 5(7):1602. https://doi.org/10.1039/c4fo00054d

Stojković DS, Kovacevic-Grujicic N, Reis FS et al (2017) Chemical composition of the mushroom *Meripilus giganteus* Karst. and bioactive properties of its methanolic extract. LWT Food Sci Technol 79:454–462. https://doi.org/10.1016/j.lwt.2017.01.045

Streith J (2011) Ergotisme, mal des ardents ou feu de Saint-Antoine du Moyen Âge aux temps modernes. Actual Chim (358):40–46

Suárez-Arroyo IJ, Loperena-Alvarez Y, Rosario-Acevedo R et al (2017) *Ganoderma* spp.: a promising adjuvant treatment for breast cancer. Medicines 4:15. https://doi.org/10.3390/medicines4010015

Suay I, Arenal F, Asensio FJ et al (2000) Screening of basidiomycetes for antimicrobial activities. Antonie Leeuwenhoek 78:129–139. https://doi.org/10.1023/A:1026552024021

Subramaniam S, Raman J, Sabaratnam V et al (2017) Functional properties of partially characterized polysaccharide from the medicinal mushroom *Ganoderma neo-japonicum* (Agaricomycetes). Int J Med Mushooms 19(10):849–859. https://doi.org/10.1615/IntJMedMushrooms.2017024355

Süfer O, Bozok F, Demir H (2016) Usage of edible mushrooms in various food products. Turkish J Agri: Food Sci Technol 4(3):144–149. https://doi.org/10.24925/turjaf.v4i3.144-149.599

Sullivan R, Smith JE, Rowan NJ (2006) Medicinal mushrooms and cancer therapy: translating a traditional practice into Western medicine. Perspect Biol Med 49(2):159–170. https://doi.org/10.1353/pbm.2006.0034

Surup F, Kuhnert E, Böhm A et al (2018) The rickiols, 20-, 22-, and 24-membered macrolides from the ascomycete *Hypoxylon rickii*. Chemistry 24(9):2200–2213. https://doi.org/10.1002/chem.201704928

Suzuki N, Takimoto Y, Suzuki R et al (2013) Efficacy of oral administration of *Lentinula edodes* mycelia extract for breast cancer patients undergoing postoperative hormone therapy. Asian Pac J Cancer Prev 14(6):3469–3472. https://doi.org/10.7314/APJCP.2013.14.6.3469

Talcott JA, Clark JA, Lee IP (2007) Measuring perceived effects of drinking an extract of basidiomycetes *Agaricus blazei* Murrill: a survey of Japanese consumers with cancer. BMC Complement Altern Med 7:246. https://doi.org/10.1186/1472-6882-7-32

Tan VP, Miyamoto S (2015) HK2/hexokinase-II integrates glycolysis and autophagy to confer cellular protection. Autophagy 11(6):963–964. https://doi.org/10.1080/15548627.2015.1042195

Tanaka A, Nishimura M, Sato Y et al (2016) Enhancement of the Th1-phenotype immune system by the intake of oyster mushroom (Tamogitake) extract in a double-blind, placebo-controlled study. J Tradit Complement Med 6:424–430. https://doi.org/10.1016/j.jtcme.2015.11.004

Tangen JM, Tierens A, Caers J et al (2015) Immunomodulatory effects of the *Agaricus blazei* Murrill-based mushroom extract AndoSan in patients with multiple myeloma undergoing high dose chemotherapy and autologous stem cell transplantation: a randomized, double blinded clinical study. Biomed Res Int 2015:718539. https://doi.org/10.1155/2015/718539

Tao QQ, Ma K, Bao L et al (2016) New sesquiterpenoids from the edible mushroom *Pleurotus cystidiosus* and their inhibitory activity against α-glucosidase and PTP1B. Fitoterapia 111:29–35. https://doi.org/10.1016/j.fitote.2016.04.007

Taofiq O, Martins A, Barreiro MF et al (2016a) Anti-inflammatory potential of mushroom extracts and isolated metabolites. Trends Food Sci Technol 50:193–210. https://doi.org/10.1016/j.tifs.2016.02.005

Taofiq O, González-Paramás AM, Martins A et al (2016b) Mushrooms extracts and compounds in cosmetics, cosmeceuticals and nutricosmetics – a review. Ind Crop Prod 90:38–48. https://doi.org/10.1016/j.indcrop.2016.06.012

Taofiq O, Heleno S, Calhelha R et al (2016c) Development of mushroom-based cosmeceutical formulations with anti-inflammatory, anti-tyrosinase, antioxidant, and antibacterial properties. Molecules 21(10):1372. https://doi.org/10.3390/molecules21101372

Taofiq O, Heleno SA, Calhelha RC et al (2017a) The potential of *Ganoderma lucidum* extracts as bioactive ingredients in topical formulations, beyond its nutritional benefits. Food Chem Toxicol 108:139–147. https://doi.org/10.1016/j.fct.2017.07.051

Taofiq O, Fernandes Â, Barros L et al (2017b) UV-irradiated mushrooms as a source of vitamin D_2: a review. Trends Food Sci Technol 70:82–94. https://doi.org/10.1016/j.tifs.2017.10.008

Tedersoo L, Bahram M, Põlme S et al (2014) Global diversity and geography of soil fungi. Science 346(6213):1256688. https://doi.org/10.1126/science.1256688

Teklit GA (2015) Chemical composition and nutritional value of the most widely used mushrooms cultivated in Mekelle Tigray Ethiopia. J Nutr Food Sci 5:408. https://doi.org/10.4172/2155-9600.1000408

Teo CP (2014) *In-vitro investigation of anticoagulant activities in edible and medicinal mushrooms*. Dissertation, University of Malaya. http://studentsrepo.um.edu.my/id/eprint/4773.

Teplyakova TV, Kosogova TA (2016) Antiviral effect of Agaricomycetes mushrooms (review). Int J Med Mushrooms 18(5):375–386. https://doi.org/10.1615/IntJMedMushrooms.v18.i5.10

Teplyakova TV, Psurtseva NV, Kosogova TA et al (2012) Antiviral activity of polyporoid mushrooms (higher Basidiomycetes) from Altai Mountains (Russia). Int J Med Mushrooms 14(1):37–45. https://doi.org/10.1615/IntJMedMushr.v14.i1.40

Thaper S, Lakshmi T (2017) Effects of mushroom on dental caries. JAPER 7(3):197–199

Therkelsen SP, Hetland G, Lyberg T et al (2016) Effect of the medicinal *Agaricus blazei* Murill-based mushroom extract, AndoSan™, on symptoms, fatigue and quality of life in patients with Crohn's disease in a randomized single-blinded placebo controlled study. PLoS One 11(7):e0159288. https://doi.org/10.1371/journal.pone.0159288

Thongbai B, Rapior S, Hyde KD et al (2015) *Hericium erinaceus*, an amazing medicinal mushroom. Mycol Prog 14:91. https://doi.org/10.1007/s11557-015-1105-4

Thongklang N, Hoang E, Rodriguez Estrada AE et al (2014) Evidence for amphithallism and broad geographical hybridization potential among *Agaricus subrufescens* isolates from Brazil, France, and Thailand. Fungal Biol 118(12):1013–1023. https://doi.org/10.1016/j.funbio.2014.10.004

Thongklang N, Chen J, Bandara AR et al (2016) Studies on *Agaricus subtilipes*, a new cultivatable species from Thailand, incidentally reveal the presence of *Agaricus subrufescens* in Africa. Mycoscience:239–250. https://doi.org/10.1016/j.myc.2016.02.003

Thongklang N, Thongbai B, Chamyuang S et al (2017) Blazeispirol A, a chemotaxonomic marker from mycelia of the medicinal mushroom *Agaricus subrufescens*. Chiang Mai J Sci 44(2):298–308

Tohtahon Z, Xue J, Han J et al (2017) Cytotoxic lanostane triterpenoids from the fruiting bodies of *Piptoporus betulinus*. Phytochem 143:98–103. https://doi.org/10.1016/j.phytochem.2017.07.013

Tsai MY, Hung YC, Chen YH et al (2016) A preliminary randomised controlled study of short-term *Antrodia cinnamomea* treatment combined with chemotherapy for patients with advanced cancer. BMC Complement Altern Med 16(1):322. https://doi.org/10.1186/s12906-016-1312-9

Tsuk S, Lev YH, Rotstein A et al (2018) Effects of a commercial supplement of *Ophiocordyceps sinensis* and *Ganoderma lucidum* on physiological responses to maximal exercise in healthy young participants. Int J Med Mushrooms 20(4):359–367. https://doi.org/10.1615/IntJMedMushrooms.2018025989

Twardowski P, Kanaya N, Frankel P et al (2015) A phase 1 trial of mushroom powder in patients with biochemically recurrent prostate cancer: roles of cytokines and myeloid-derived suppressor cells for *Agaricus bisporus*-induced prostate-specific antigen responses. Cancer 121:2949–2950. https://doi.org/10.1002/cncr.29421

Tylš F, Páleníček T, Horáček J (2014) Psilocybin – summary of knowledge and new perspectives. Eur Neuropsychopharmacol 24:342–356. https://doi.org/10.1016/j.euroneuro.2013.12.006

Uzunov BA, Stoyneva-Gärtner MP (2015) Mushrooms and lichens in Bulgarian ethnomycology. J Mycol 2015:361053.:7 p. https://doi.org/10.1155/2015/361053

Valverde ME, Hernández-Pérez T, Paredes-López O (2015) Edible mushrooms: improving human health and promoting quality life. Intern J Microbiol 2015:376387.):14 p. https://doi.org/10.1155/2015/376387

Vasdekis EP, Karkabounas A, Giannakopoulos I (2018) Screening of mushrooms bioactivity: Piceatannol was identified as a bioactive ingredient in the order Cantharellales. Eur Food Res Technol 244:861–871. https://doi.org/10.1007/s00217-017-3007-y

Venkatachalapathi A, Paulsamy S (2016) Exploration of wild medicinal mushroom species in Walayar valley, the southern Western Ghats of Coimbatore District Tamil Nadu. Mycosphere 7(2):118–130. https://doi.org/10.5943/mycosphere/7/2/3

Venturella G, Zervakis GI, Polemis E et al (2016) Taxonomic identity, geographic distribution, and commercial exploitation of the culinary-medicinal mushroom *Pleurotus nebrodensis* (Basidiomycetes). Int J Med Mushrooms 18(1):59–65. https://doi.org/10.1615/IntJMedMushrooms.v18.i1.70

Vetvicka V, Vetvickova J (2015) Glucan supplementation enhances the immune response against an influenza challenge in mice. Ann Transl Med 3(2):22. https://doi.org/10.3978/j.issn.2305-5839.2015.01.08

Vitak T, Yurkiv B, Wasser S et al (2017) Effect of medicinal mushrooms on blood cells under conditions of diabetes mellitus. World J Diabetes 8(5):187–201. https://doi.org/10.4239/wjd.v8.i5.187

Volman JJ, Mensink RP, van Griensven LJLD et al (2010) Effects of alpha-glucans from *Agaricus bisporus* on *ex vivo* cytokine production by LPS and PHA-stimulated PBMCs; a placebo-controlled study in slightly hypercholesterolemic subjects. Eur J Clin Nutr 64(7):720–726. https://doi.org/10.1038/ejcn.2010.32

Waithaka PN, Gathuru EM, Githaiga BM et al (2017) Antimicrobial activity of mushroom (*Agaricus bisporus*) and fungal (*Trametes gibbosa*) extracts from mushrooms and fungi of Egerton Main campus, Njoro Kenya. J Biomed Sci 6(3):20. https://doi.org/10.4172/2254-609X.100064

Wang XL, Yao YJ (2011) Host insect species of *Ophiocordyceps sinensis*: a review. ZooKeys 127:43–59. https://doi.org/10.3897/zookeys.127.802

Wang J, Liu YM, Cao W et al (2012) Anti-inflammation and antioxidant effect of cordymin, a peptide purified from the medicinal mushroom *Cordyceps sinensis*, in middle cerebral artery occlusion-induced focal cerebral ischemia in rats. Metab Brain Dis 27(2):159–165. https://doi.org/10.1007/s11011-012-9282-1

Wang H, Fu Z, Han C (2013) The medicinal values of culinary-medicinal royal sun mushroom (*Agaricus blazei* Murrill). Evid Based Complement Alternat Med 2013:842619.:6 p. https://doi.org/10.1155/2013/842619

Wang K, Bao L, Ma K et al (2015a) Eight new alkaloids with PTP1B and α-glucosidase inhibitory activities from the medicinal mushroom *Hericium erinaceus*. Tetrahedron 71:9557–9563. https://doi.org/10.1016/j.tet.2015.10.0.68

Wang K, Bao L, Xiong W et al (2015b) Lanostane triterpenes from the Tibetan medicinal mushroom *Ganoderma leucocontextum* and their inhibitory effects on HMG-CoA reductase and α-glucosidase. J Nat Prod 78:1977–1989. https://doi.org/10.1021/acs.jnatprod.5b00331

Wang XL, Ding ZY, Zhao Y et al (2017a) Efficient accumulation and *in vitro* antitumor activities of triterpene acids from submerged batch-cultured Lingzhi or Reishi medicinal mushroom, *Ganoderma lucidum* (Agaricomycetes). Int J Med Mushrooms 19(5):419–431. https://doi.org/10.1615/IntJMedMushrooms.v19.i5.40

Wang Q, Wang F, Xu Z, Ding Z (2017b) Bioactive mushroom polysaccharides: a review on monosaccharide composition, biosynthesis and regulation. Molecules 22:955. https://doi.org/10.3390/molecules22060955

Wang Y, Tian Y, Shao J et al (2018) Macrophage immunomodulatory activity of the polysaccharide isolated from *Collybia radicata* mushroom. Int J Biol Macromol 108:300–306. https://doi.org/10.1016/j.ijbiomac.2017.12.025

Wasser SP (2002) Medicinal mushrooms as a source of antitumor and immunomodulating polysaccharides. Appl Microbiol Biotechnol 60:258–274. https://doi.org/10.1007/s00253-002-1076-7

Wasser SP (2010) Medicinal mushroom science: history, current status, future trends, and unsolved problems. Int J Med Mushrooms 12(1):1–16. https://doi.org/10.1615/IntJMedMushr.v12.i1.10

Wasser SP (2011) Current findings, future trends, and unsolved problems in studies of medicinal mushrooms. Appl Microbiol Biotechnol 89:1323–1332. https://doi.org/10.1007/s00253-010-3067-4

Wasser SP (2014) Medicinal mushroom science: current perspectives, advances, evidences, and challenges. Biom J 37(6):345–356. https://doi.org/10.4103/2319-4170.138318

Wasser SP (2017) Medicinal mushrooms in human clinical studies. Part I. Anticancer, oncoimmunological, and immunomodulatory activities: a review. Int J Med Mushrooms 19(4):279–317. https://doi.org/10.1615/IntJMedMushrooms.v19.i4.10

Wasser SP, Didukh MY (2005) Mushroom polysaccharides in human health care. In: Deshmukh SK, Rai MK (eds) Biodiversity of fungi: their role in human life. Oxford & IBH Publishing, pp 289–328

Wasser SP, Weis AL (1999) Therapeutic effects of substances occurring in higher Basidiomycetes mushrooms: a modern perspective. Crit Rev Immunol 19:65–96. https://doi.org/10.1615/CritRevImmunol.v19.i1.30

Wasser SP, Nevo E, Sokolov D et al (2000) Dietary supplements from medicinal mushrooms: diversity of types and variety of regulations. Int J Med Mushrooms 2:1–19. https://doi.org/10.1615/IntJMedMushr.v2.i1.10

Welti S, Moreau PA, Azaroual N et al (2010) Antiproliferative activities of methanolic extracts from a neotropical *Ganoderma* species (Aphyllophoromycetideae): identification and characterization of a novel ganoderic acid. Int J Med Mushrooms 12(1):17–31. https://doi.org/10.1615/IntJMedMushr.v12.i1.20

Welti S, Moreau PA, Decock C et al (2015) Oxygenated lanostane-type triterpenes profiling in laccate *Ganoderma* chemotaxonomy. Mycol Prog 14:45. https://doi.org/10.1007/s11557-015-1066-7

Wesa KM, Cunningham-Rundles S, Klimek VM et al (2015) Maitake mushroom extract in myelodysplastic syndromes (MDS): a phase II study. Cancer Immunol Immunother 64(2):237–247. https://doi.org/10.1007/s00262-014-1628-6

Wisitrassameewong K, Karunarathna SC, Thongklang N et al (2012) *Agaricus subrufescens*: a review. Saudi J Biol Sci 19:131–146. https://doi.org/10.1016/j.sjbs.2012.01.003

Wittstein K, Rascher M, Rupcic Z et al (2016) Corallocins A–C, nerve growth and brain-derived neurotrophic factor inducing metabolites from the mushroom *Hericium coralloides*. J Nat Prod 79:2264–2269. https://doi.org/10.1021/acs.jnatprod.6b00371

Wong KH, Ng CC, Kanagasabapathy G et al (2017) An overview of culinary and medicinal mushrooms in neurodegeneration and neurotrauma research. Int J Med Mushrooms 19(3):191–202. https://doi.org/10.1615/IntJMedMushrooms.v19.i3.10

Wu Y, Choi MH, Li J et al (2016) Mushroom cosmetics: the present and future. Cosmetics 3:22. https://doi.org/10.3390/cosmetics3030022

Xiao Y, Huang XZ, Zhu JS (2004) Randomized double-blind placebo-controlled clinical trial and assessment of fermentation product of *Cordyceps sinensis* (Cs-4) in enhancing aerobic capacity and respiratory function of the healthy elderly volunteers. CJLM 10(3):187–192

Xu X, Yan H, Chen J et al (2011) Bioactive proteins from mushrooms. Biotechnol Adv 29(6):667–674. https://doi.org/10.1016/j.biotechadv.2011.05.003

Xu H, Kong YY, Chen X et al (2016a) Recombinant FIP-gat, a fungal immunomodulatory protein from *Ganoderma atrum*, induces growth inhibition and cell death in breast cancer cells. J Agric Food Chem 64:2690–2698. https://doi.org/10.1021/acs.jafc.6b00539

Xu J, Huang Y, Chen XX et al (2016b) The mechanisms of pharmacological activities of *Ophiocordyceps sinensis* fungi. Phytother Res 30:1572–1583. https://doi.org/10.1002/ptr.5673

Yamac M, Kanbak G, Zeytinoglu M et al (2010) Pancreas protective effect of button mushroom *Agaricus bisporus* (J.E. Lange) Imbach (Agaricomycetidae) extract on rats with Streptozotocin-induced diabetes. Int J Med Mushrooms 12(4):379–389. https://doi.org/10.1615/IntJMedMushr.v12.i4.50

Yamac M, Zeytinoglu M, Swenturk H et al (2016) Effects of black hoof medicinal mushroom, *Phellinus linteus* (Agaricomycetes), polysaccharide extract in streptozotocin-induced diabetic rats. Int J Med Mushrooms 18(4):301–311. https://doi.org/10.1615/IntJMedMushrooms.v18.i4.30

Yang H, Hwang I, Kim S et al (2013) *Lentinus edodes* promotes fat removal in hypercholesteremic mice. Exp Ther Med 6(6):1409–1413. https://doi.org/10.3892/etm.2013.1333

Yang YL, Tao QQ, Han JJ et al (2017) Recent advance on bioactive compounds from the edible and medicinal fungi in China. In: Agrawal DC, Tsay HS, Shyur LF, Wu YC, Wang SY (eds) Medicinal plants and fungi: recent advances in research and development, Medicinal and aromatic plants of the world, vol 4. Springer, New York, pp 253–313. https://doi.org/10.1007/978-981-10-5978-0_9

Yang H, Sun W, Zhang J et al (2018) Autophagy inhibition enhances SPCA-1 cell proliferation inhibition induced by By-1 from the stout camphor medicinal mushroom, *Taiwanofungus camphoratus* (Agaricomycetes). Int J Med Mushrooms 20(4):321–335. https://doi.org/10.1615/IntJMedMushrooms.2018025836

Yao HM, Wang G, Liu YP et al (2016) Phenolic acids isolated from the fungus *Schizophyllum commune* exerts analgesic activity by inhibiting voltage-gated sodium channels. Chin J Nat Med 14(9):661–670. https://doi.org/10.1016/S1875-5364(16)30078-4

Yap HY, Chooi YH, Firdaus-Raih M et al (2014) The genome of the tiger milk mushroom, *Lignosus rhinocerotis*, provides insights into the genetic basis of its medicinal properties. BMC Genomics 15:635. https://doi.org/10.1186/1471-2164-15-635

Yap HY, Fung SY, Ng ST et al (2015) Genome-based proteomic analysis of *Lignosus rhinocerotis* (Cooke) Ryvarden sclerotium. Int J Med Sci 12:23–31. https://doi.org/10.7150/ijms.10019

Yen MT, Tseng YH, Lee CE et al (2015) Quality of singhtake supplemented bun. Am J Adv Food Sci Technol 3(1):1–13. https://doi.org/10.7726/ajafst.2015.1001

Yen MT, Chang YH, Huang SJ et al (2018) Extraction of ergothioneine from *Pleurotus eryngii* and *P. citrinopileatus* (Agaricomycetes) and preparation of its product. Int J Med Mushrooms 20(4):381–392. https://doi.org/10.1615/IntJMedMushrooms.2018025953

Yoon TJ, Koppula S, Lee KH (2013) The effects of β-glucans on cancer metastasis. Anti Cancer Agents Med Chem 13(5):699–708. https://doi.org/10.2174/1871520611313050004

Yuan B, Zhao L, Yang W et al (2017) Enrichment of bread with nutraceutical-rich mushrooms: impact of *Auricularia auricula* (mushroom) flour upon quality attributes of wheat dough and bread. J Food Sci 82(9):2041–2050. https://doi.org/10.1111/1750-3841.13812

Yuyama KT, Fortkamp D, Abraham WR (2017a) Eremophilane-type sesquiterpenes from fungi and their medicinal potential. Biol Chem 399(1):13––28. https://doi.org/10.1515/hsz-2017-0171

Yuyama KT, Chepkirui C, Wendt L et al (2017b) Bioactive compounds produced by *Hypoxylon fragiforme* against *Staphylococcus aureus* biofilms. Microorganisms 5(80). https://doi.org/10.3390/microorganisms5040080

Zeng P, Guo Z, Zeng X et al (2018) Chemical, biochemical, preclinical and clinical studies of *Ganoderma lucidum* polysaccharide as an approved drug for treating myopathy and other diseases in China. J Cell Mol Med:1–20. https://doi.org/10.1111/jcmm.13613

Zhang M, Cui SW, Cheung PCK et al (2007) Antitumor polysaccharides from mushrooms: a review on their isolation process, structural characteristics and antitumor activity. Trends Food Sci Technol 18:4–19. https://doi.org/10.1016/j.tifs.2006.07.013

Zhang M, Huang J, Xie X et al (2009) Dietary intakes of mushrooms and green tea combine to reduce the risk of breast cancer in Chinese women. Int J Cancer 124(6):1404–1408. https://doi.org/10.1002/ijc.24047

Zhang C, Li S, Zhang J et al (2016) Antioxidant and hepatoprotective activities of intracellular polysaccharide from *Pleurotus eryngii* SI-04. Int J Biol Macromol 91:568–577. https://doi.org/10.1016/j.ijbiomac.2016.05.104

Zhang Y, Liu W, Xu C et al (2017) Characterization and antiproliferative effect of novel acid polysaccharides from the spent substrate of shiitake culinary-medicinal mushroom *Lentinus edodes* (Agaricomycetes) cultivation. Int J Med Mushrooms 19(5):395–403. https://doi.org/10.1615/IntJMedMushrooms.v19.i5.20

Zhang SB, Li ZH, Stadler M et al (2018) Lanostane triterpenoids from *Tricholoma pardinum* with NO production inhibitory and cytotoxic activities. Phytochem 152:105–112. https://doi.org/10.1016/j.phytochem.2018.05.002

Zhao XR, Huo XK, Dong PP et al (2015) Inhibitory effects of highly oxygenated lanostane derivatives from the fungus *Ganoderma lucidum* on P-glycoprotein and α-glucosidase. J Nat Prod 78:1868–1876. https://doi.org/10.1021/acs.jnatprod.5b00132

Zhao ZZ, Chen HP, Huang Y et al (2016) Lanostane triterpenoids from fruiting bodies of *Ganoderma leucocontextum*. Nat Prod Bioprospect 6:103–109. https://doi.org/10.1007/s13659-016-0089-3

Zhao J, Yang Y, Yu M et al (2018) Lanostane-type triterpenoid derivatives from the fruiting bodies of cultivated *Fomitopsis palustris*. Phytochem 152:10–21. https://doi.org/10.1016/j.phytochem.2018.04.012

Zhou LW, Cao Y, Wu SH et al (2015) Global diversity of the *Ganoderma lucidum* complex (Ganodermataceae, Polyporales) inferred from morphology and multilocus phylogeny. Phytochem 114:7–15. https://doi.org/10.1016/j.phytochem.2014.09.023

Zhu F, Dua B, Bian Z et al (2015) Beta-glucans from edible and medicinal mushrooms: characteristics, physicochemical and biological activities. J Food Compos Anal 41:165–173. https://doi.org/10.1016/j.jfca.2015.01.019

Zjawiony JK (2004) Biologically active compounds from Aphyllophorales (polypore) fungi. J Nat Prod 67(2):300–310. https://doi.org/10.1021/np030372w

Zmitrovich IV, Belova NV, Balandaykin ME, Bondartseva MA, Wasser SP (2019) Cancer without pharmacological illusions and a niche for Mycotherapy (review). Int J Med Mushrooms 21(2):105–119. https://doi.org/10.1615/IntJMedMushrooms.2019030047

Zong A, Cao H, Wang F (2012) Anticancer polysaccharides from natural resources: a review of recent research. Carbohydr Polym 90:1395–1410. https://doi.org/10.1016/j.carbpol.2012.07.026

Zotti M, Persiani AM, Ambrosio E et al (2013) Macrofungi as ecosystem resources: conservation versus exploitation. Plant Biosyst 147(1):219–225. https://doi.org/10.1080/11263504.2012.753133

Zuomin Y, Puming H, Jianhui C et al (1998) Hyperglycemic effect of water-soluble polysaccharide from *Auricularia auricularia-judae* Quél. on genetic diabetic KK-Ay mice. Biosci Biotechnol Biochem 62:1898–1903. https://doi.org/10.1271/bbb.62.1898

Edible Mushrooms as Neuro-nutraceuticals: Basis of Therapeutics

2

V. R. Remya, Goutam Chandra, and K. P. Mohanakumar

Contents

V. R. Remya · G. Chandra · K. P. Mohanakumar (✉)
Inter University Centre for Biomedical Research & Super Speciality Hospital, Mahatma Gandhi University at Thalappady Campus, Kottayam, Kerala, India
e-mail: iucbr@mgu.ac.in

D. C. Agrawal, M. Dhanasekaran (eds.), *Medicinal Mushrooms*,
https://doi.org/10.1007/978-981-13-6382-5_2

Abstract

Mushrooms, the macro-fungi with distinctive fruiting bodies and mycelia, are valued throughout the world not only for their nutritive values but also for their medicinal uses. The former desirable property of mushrooms derives out of their rich contents of vitamins, trace elements, and minerals and appetite-enhancing flavors, but the therapeutic values result from its antioxidant and immunomodulatory activities, from its ability to activate very specific neuronal receptors, and from its capacity to rejuvenate cellular metabolism. Reactive oxygen species cause extensive damage to cells and tissues by interfering with normal metabolism, which are aggravated during stressful environments, such as during infections, stress, metabolic diseases, and various degenerative disorders including neurodegenerative diseases, cardiovascular diseases, and accelerated or normal aging. Mushrooms contain many biologically dynamic compounds, some of which are neuroactive substances, free radical scavengers, anti-apoptotic factors, and nerve growth factor stimulators that exert positive effects on brain cells, all of which essentially qualify them as good neuro-nutraceuticals that help in the protection of different neural cells, in vivo and in vitro. While discussing the neuroprotective and neuromodulatory properties of mushrooms, we rationally postulate here indirect benefits of these fungi through the enhanced environment for gut microbiome that are meaningful for healthy brain functions.

Keywords

Antioxidants · Cellular respiration · Free radical scavenger · Gut-brain axis stimulants · Probiotics · Nerve growth factors · Neurotrophic factors

Abbreviations

ROS	Reactive oxygen species
LPS	Lipopolysaccharides
RNS	Reactive nitrogen species
ER	Endoplasmic reticulum
TNF	Tumor necrosis factor
CNS	Central nervous system
NGF	Nerve growth factor
BBB	Blood-brain barrier
AD	Alzheimer's disease
ICV	Intracerebroventricular
STZ	Streptozotocin
PKC	Protein kinase C
PD	Parkinson's disease
MPP^+	1-Methyl-4-phenylpyridinium
MPTP	1-Methyl-4-phenyl-1,2,3,6-tetrahydropyridine
GDNF	Glial cell-derived neurotrophic factor

BDNF	Brain-derived neurotrophic factor
OCTN1	Organic cation transporter
SLC22A4	Solute carrier family 22, member 4
IL-1β	Interleukin 1β
JNK	c-Jun N-terminal kinases
p38 MAPK	*p38* mitogen-activated protein kinases
IRE1α	Inositol-requiring protein 1alpha
TRAF2	TNF receptor-associated factor 2
C/EBP	CCAAT-enhancer-binding proteins
CHOP	C/EBP homologous protein
NF-κB	Nuclear factor-κB
Bax	Bcl-2-associated protein x
Bcl-2	B-cell lymphoma 2
PKB/Akt	Protein kinase B
mTOR	The mammalian target of rapamycin
Bid	BH3 interacting-domain death agonist

2.1 Introduction

2.1.1 The Good and Bad Mushrooms

The expanded fleshy fruiting body of a fungus that is originated over a cup or volva from underground mycelia is a mushroom and consists typically of a stipe or stem or stalk bearing a cap or pileus, underneath of which contains spore-bearing structures termed as gills or lamellae. Almost all parts of a nonpoisonous mushroom are edible and are used worldwide as food because of their unique flavor, aroma, nutritive values, and unique medicinal properties. In general, mushrooms could be classified into four categories: edible mushrooms, medicinal mushrooms, poisonous mushrooms, and mushrooms with miscellaneous quality.

2.1.1.1 Edible Mushrooms

Edible mushrooms and medicinal mushrooms cannot be distinctly classified, since most of the edible mushrooms have medicinal values, having potent bioactive molecules that provide health benefits. The nutritional value of the edible mushrooms depends upon the growth characteristics, developmental stage, and postharvest condition. As these contain all essential amino acids, they are good food supplements for vegetarians. The category of edible mushroom species includes *Agaricus*, *Lentinus*, *Pleurotus*, *Ganoderma*, and *Huitlacoche*. Their culinary values are due to high-quality proteins, vitamins, fibers, minerals, and low fat and calorie contents. These highly nutritious values together with significant biomolecules that contribute to medicinal properties help to classify mushrooms as nutraceuticals (Table 2.1).

Table 2.1 Biologically active constituents from mushrooms and their bioactive properties

Sl. No.	Mushrooms: scientific name	Mushrooms: common name	Active constituents	Bioactivity	References
1	*Agrocybe cylindracea*	Poplar mushroom	Glucans	Antioxidant	Rathee et al. (2012)
2	*Antrodia camphorata*	Stout camphor fungus	Diterpenes	Neuroprotective	Chen et al. (2006)
3	*Albatrellus ovinus*	Forest lamb mushroom	Grifolin and grifolin derivatives	Anti-inflammatory, antioxidant	Nukata et al. (2002)
4	*Boletus* spp.	Gelam mushroom	2,4,6-Trimethylacetophenone imine, glutamyl tryptophan, azatadine, lithocholic acid glycine conjugate	Antioxidant	Yuswan et al. (2015)
5	*Cordyceps sinensis*	Summer grass, winter-worn	Cordycepin	Antioxidant	Holliday et al. (2004)
6	*Cortinarius infractus*	The bitter webcap	6-Hydroxyinfractine, infractopicrine	Neuroprotective	Brondz et al. (2007) and Geissler et al. (2010)
7	*Craterellus cornucopioides*	Black chanterelle, horn of plenty, black trumpet, trumpet of the dead	Myricetin	Antioxidant	Palacios et al. (2011)
8	*Daldinia concentrica*	King Alfred's cake, cramp balls, coal fungus	1-3,4,5-Trimethoxyphenyl ethanol, caruilignan	Neuroprotective	Lee et al. (2002a)
10	*Dictyophora indusiata*	Veiled lady mushroom, bamboo mushroom	Dictyophorines A and B	Neuroprotective	Lee et al. (2002b)
11	*Hypsizygus marmoreus*	Brown beech	Ergosterol, mannitol	Antioxidant, anti-inflammatory	Jang et al. (2013)
12	*Inonotus obliquus*	Clinker polypore, cinder conk, black mass	β-D-glucans	Antioxidant	Rathee et al. (2012)
13	*Lentinula edodes*	Shiitake	Catechin, phenolic compound	Antioxidant	Chowdhury et al. (2015)
14	*Lentinula polychrous*		Catechin	Antioxidant	Attarat and Phermthai (2014)
15	*Lentinula squarrosulus*		Catechin	Antioxidant	Attarat and Phermthai (2014)

16	*Lenzites betulina*	Gilled polypore, birch maze gill, multicolor gill, polypore	Betulinan A	Antioxidant	Rathee et al. (2012)
17	*Mycoleptodonoides aitchisonii*	Bunaharitake	3-Hydroxymethyl-4-methylfuran-25H-one, 3R,4S,1'R-3-1′-hydroxyethyl-4methyldihydrofuran-23H-one, 5-hydroxy-4-1-hydroxyethyl-3-e-methylfuran-25H-one, 5-phenylpentane-1,3,4-triol	Neuroprotective	Choi et al. (2009) and Choi et al. (2014)
18	*Phellinus linteus*	Black hoof mushroom	Hispidin polyphenol	Antioxidant	Park et al. (2004)
19	*Pleurotus florida*	White oyster	β-Glucans	Antioxidant	Ganeshpurkar et al. (2015)
20	*Pleurotus ostreatus*	Oyster mushroom	Proteoglycan	Antioxidant	Tong et al. (2009)
21	*Termitomyces albuminosus*	Termite mushroom	Termitomycesphins, cerebrosides, termitomycamides, fatty acid amides	Neuroprotective	Qi et al. (2000) and Qu et al. (2012)
22	*Grifola frondosa*	Maitake	Polysaccharides	Antioxidant	Yeh et al. (2011)
23	*Pleurotus abalonus*	Tree mushrooms or oyster mushrooms	9-β-D-Ribofuranosidoadenine, 5′-deoxy-5′-methylthio adenosine, triterpenoid complex	Antioxidant	Li et al. (2007)

2.1.1.2 Medicinal Mushrooms

Medicinal mushrooms include both nonedible, poisonous mushrooms and edible mushrooms. Interventional strategies using edible mushrooms as functional foods and as preventive agents for age-related diseases, including neurodegenerative diseases, are gaining rapt recognition. Edible medicinal mushrooms have a wide variety of bioactive and nutritional components that could scavenge reactive oxygen species (ROS) and upregulate cellular antioxidant defense mechanisms. It is mainly by virtue of their strong antioxidant activity; mushrooms exert neuroprotective effects and promote neuritogenesis and neuroregeneration. It is shown that specific polysaccharides contained in mushrooms play an important medicinal role. Phenolic compounds contained in mushrooms contribute to the substantial inhibitory effect on neuronal and subcellular organelles membrane lipid peroxidation, thereby offering neuroprotection.

Hericium erinaceus or commonly known as lion's mane mushroom, *Ganoderma lucidum* long known for its antiaging effects, *Pleurotus cornucopiae* recognized for its health-promoting and health-rejuvenating effects, *P. giganteus* with great antioxidant potential, *Amanita caesarea* a mitochondrial rejuvenator with antioxidant activity, and *Polyozellus multiplex* and *Armillaria mellea* strong anti-apoptotic antioxidants rich in phenolic compounds are the major medicinal mushrooms. These mushrooms contain several biologically relevant compounds that are active at the cellular or molecular level. Some of the major molecules that contribute to the health benefits of these mushrooms are hericenones, erinacines, dilinoleoyl-phosphatidylethanolamine, several polysaccharides, lipopolysaccharides (LPS), ergothioneine, linoleic acid, oleic acid, cinnamic acid, caffeic acid, p-coumaric acid, succinic acid, benzoic acid, uridine, polyozellin phenols, flavonoids, carotenoids, and phenolic acids.

2.1.1.3 Edible Mushrooms with Toxic Effects

There are exceptions in case of edible mushrooms when mushrooms cause intoxication. For example, *P. ostreatus*, an edible mushroom, when consumed in large quantity causes acute intoxication. Ostreolysin is a cytotoxic protein isolated from *P. ostreatus*, which causes cell membrane pore formation and causes hemolysis (Schlumberger et al. 2014). This toxin induces hyperkalemia, which in turn results in cardiotoxicity. Ostreolysin can cause bradycardia, myocardial ischemia, and respiratory arrest (Frangež et al. 2017). There is a solitary report on spores of a well-known edible mushroom *Volvariella volvacea*, infecting an immunocompromised patient, causing typical invasive mycoses of the lungs and the brain culminating in patient's death (Salit et al. 2010).

2.1.1.4 Poisonous Mushrooms

On the other hand, poisonous mushrooms contain varying levels of hazardous toxins, which are harmful to the body. The intoxication depends upon the species that consume the mushroom or its products and the amount of toxin contained in it. Incorrect identification of a poisonous mushroom as edible is the major cause for severe adverse reactions in humans. These toxins can cause fatal symptoms, i.e., some are

Table 2.2 List of poisonous mushrooms, toxins, and signs of nervous system intoxication

Sl no	Mushrooms	Toxins present	Symptoms of poisoning	References
1	*Amanita muscaria*, *A. pantherina*, *A. gemmata*	Muscimol and ibotenic acid	Stupor, coma, agitation, hallucination, seizures, delirium	Curtis et al. (1979)
2	*Amanita phalloides*, *A. virosa*, *A. verna*, *A. bisporigera*, *Galerina autumnalis*, *G. marginata*, *G. venenata*, *Lepiota helveola*	Phallatoxin and amatoxin	Delayed liver toxicity, delayed gastroenteritis	Ennecker-Jans et al. (2007) and Aygul et al. (2010)
3	*Chlorophyllum molybdites* *Clitocybe nebularis* *Omphalotus illudens*	Gastrointestinal irritants	Acute gastroenteritis without liver failure	Stenklyft and Augenstein (1990) and French and Garrettson (1988)
4	*Cortinarius orellanus*, *C. speciosissimus*, *Mycena pura*, *O. orarius*	Orellanine, orellinine, cortinarin	Delayed renal failure, cellular and edematous intestinal fibrosis	Mount et al. (2002)
5	*Clitocybe dealbata*, *C. illudens*, *Boletus calopus*	Muscarine	Vomiting, diarrhea, salivation, bradycardia	Marciniak et al. (2010) and Jo et al. (2014)
6	*Gyromitra esculenta*, *G. infula*, *Sarcosphaera coronaria*	Gyromitrin	Seizures, delayed gastroenteritis, and liver toxicity	Clarke and Crews (2013)
7	*Russula* species	Sesquiterpenoids	Vomiting, diarrhea, gastrointestinal distress	Meuninck (2015)
8	*Lactarius* species		Gastroenteritis	Meuninck (2015)
9	*Pholiota* species	Amatoxin	Hepatic damage	Meuninck (2015)
10	*Psilocybe cubensis*, *P. mexicana*, *Conocybe cyanopus*, *G. aeruginosa*, *Stropharia* species	Psilocin, psilocybin	Distorted vision and tactile sensation, vomiting, increased heartbeat, hallucinations	Jo et al. (2014) and Clarke and Crews (2013)
11	*Coprinus atramentarius*	Coprine	Flushing, headache, tachycardia, chest pain, anxiety	Clarke and Crews (2013)
12	*Tricholoma equestre*		Rhabdomyolysis	Nieminen et al. (2008) and Bedry et al. (2001)
13	*Clitocybe acromelalga*	Acromelic acid	Erythromelalgia	Jo et al. (2014)
14	*Pleurocybella porrigens*		Delayed encephalopathy	Graeme (2014)
15	*Clitocybe rivulosa*		Abdominal pain, diarrhea, and intense sweating	Jo et al. (2014)

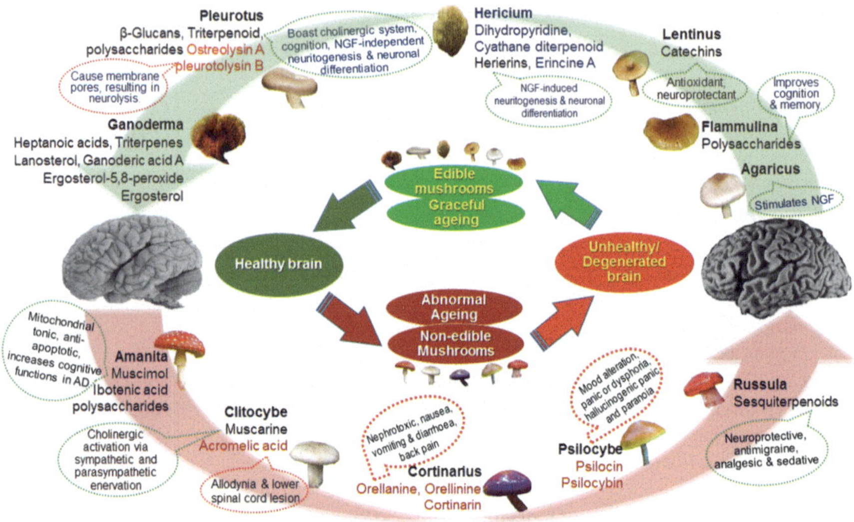

Fig. 2.1 Mushrooms and the aging brain
Note: Many of the mushrooms, as shown in the cartoons, are loaded with biochemicals having potent antioxidant and/or neuritogenic activities that may have antiaging potential. Regular consumption of those mushrooms may slow down the age-related cognitive decline in geriatric population. Some of these edible mushrooms, however, are shown to have compounds like ostreolysin A and pleurotolysin B, which can cause membrane pore resulting in neurolysis. On the other hand, nonedible mushrooms possess compounds with neuroactive and neuroprotective abilities. For examples, muscimol from *Amanita* has been found to be anti-apoptotic, cognition booster, and a mitochondrial tonic

cytotoxic; some others cause toxicity to specific cells causing nephrotoxicity or neurotoxicity. A list of inedible mushrooms and the poisonous ingredients contained therein are listed in Table 2.2. Interestingly, mushrooms that are poisonous contain several neuroactive compounds, some of which are muscarine, ibotenic acid, muscimol, psilocin, psilocybin, acromelic acid, etc. Figure 2.1 depicts the good and bad health aspects of mushrooms that make a brain healthy or degenerated, and contained therein are certain examples of edible mushrooms that are toxic and some toxic mushrooms that have components, which are beneficial to humans. In this context, the data accumulated in the literature on the pharmacological basis of medicinal and nutritive effects are classified and shown in schematic diagram (Fig. 2.2).

2.2 Mushrooms and Neural Cells

2.2.1 Neuroprotective Effects of Mushrooms

The brain utilizes a lot of energy for its metabolic processes and thereby creates a high level of a redox reaction, culminating in the production of a variety of ROS and reactive nitrogen species (RNS). Brain is having relatively low levels of antioxidant

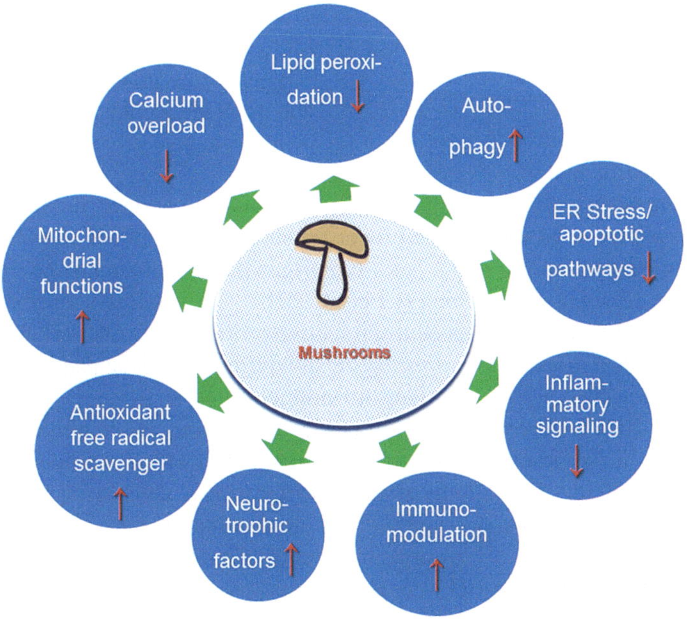

Fig. 2.2 Pharmacological basis of beneficial effects of mushroom constituents
Note: As detailed in Fig. 2.1 and Table 2.1, many of the edible and medicinal mushrooms contain bioactive molecules and are regularly used in natural medicines and as traditional home remedies. The validated neuropharmacological activities of these mushrooms depicted antioxidant, anti-apoptotic, immunomodulatory, and/or neurotrophic activities

defense mechanisms such as oxidized or reduced glutathione; antioxidant enzymes such as superoxide dismutase, catalase, glutathione reductase or glutathione oxidase, etc, as compared to other tissues of the body. Also, the nervous system will be under duress with oxidative stress even under normal levels of physiology, which are elevated substantially under disease conditions and during aging. For example, neurodegenerative diseases are shown to be resulting from uncontrolled levels of ROS and RNS in the brain. These extreme stressors are controlled by higher intake of nutritive foods containing minerals and vitamins that support enhanced synthesis of antioxidant defense molecules and enzymes in the brain.

A wide variety of antioxidant food supplements are shown to be useful in retarding or blocking the degenerative processes (Ames et al. 1993; Williams et al. 2015a, 2016a). Of very special interests are diets rich in polyphenols and potent antioxidants, traditional or natural medicines, micronutrients, and food supplements for reducing oxidative and nitrosative stresses (Williams et al. 2015b, 2016b). Mushrooms are well-known for its micronutrients, excessive vitamins, free radical scavengers, and potent antioxidants (Abdullah et al. 2012; Aruoma et al. 1999). Administration of either fresh, dried under shade, freeze-dried, or oven dried, fermented, or processed mushrooms contains a variety of phenolic compounds, free radical scavengers, ferric reducing substances, anti-carotene bleaching compounds, and a variety of polysaccharides (Wong et al. 2009; Han et al. 2013; Zhang et al. 2012b). Antiaging indications have been shown to be derived from total antioxidant molecule levels and total antioxidant enzymes activities in vitro and in vivo in experimental animals (Wang et al. 2015), and such molecular and biochemical markers could be used as determinants for neuroprotective effects of mushrooms and their constituents.

H. erinaceus is a mushroom that usually grows on old or dead, broad-leaf trees. It possesses a range of therapeutic properties such as antioxidant, hypolipidemic, hemagglutinating, antimicrobial, anti-tumorigenic, and endoplasmic reticulum (ER) stress modulatory activities (Li et al. 2014). This culinary mushroom has been extensively studied for its neuroprotective properties. Daily oral administration of aqueous extract of *H. erinaceus* fresh fruiting bodies helped to regenerate injured rat peroneal nerve following axonotmetic peroneal nerve injury in rats (Wong et al. 2011). Neuroprotective properties of *H. erinaceus* in glutamate-mediated death in differentiated PC12 cells in vitro and in vivo have been described (Zhang et al. 2016). These authors have demonstrated that the neuroprotective effect of *H. erinaceus* is associated with its eliminating power of ROS accumulation and blockade of the Ca^{2+} overload in the cells (Zhang et al. 2016). It is reported that the active ingredients of *H. erinaceus* are polysaccharides, fatty acids, amino acids, hericenones, and the diterpenoid compounds erinacines, and the latter two were isolated, respectively, from the fruiting body and mycelium of this mushroom (Mori et al. 2008). ER stress-attenuating compounds and dilinoleoyl-phosphatidylethanolamine, a phospholipid with linoleic acid, an unsaturated fatty acid, were derived from extracts of dried fruiting bodies of *H. erinaceus*. These compounds are found to reduce ER stress-induced neural death and thereby lowering the risk of neurodegenerative diseases.

G. lucidum has been used as a general preventive medicine for thousands of years in eastern Asia. Polysaccharides isolated from *G. lucidum* fruiting bodies possess antioxidant, immunomodulatory, and other health beneficial characteristics and are demonstrated to be protective against cerebral ischemic injury (Gokce et al. 2015). The extracts of *G. lucidum* are shown to exhibit immunomodulatory activity by means of inhibition of pro-inflammatory cytokines and other molecules and cell toxic factors and inhibit lipid peroxidation and protect against hypoxia/reoxygenation-mediated neuronal injury (Xuan et al. 2015; Gokce et al. 2015). *G. lucidum* polysaccharides when administered in traumatized animals with spinal cord injury; the increased levels of tumor necrosis factor-α (TNF-α), malondialdehyde, and nitric oxide; and enhanced activities of myeloperoxidase and caspase-3 were significantly attenuated to offer spinal neuronal protection (Gokce et al. 2015).

Health-promoting edible golden oyster mushroom, *P. cornucopiae* is said to be a body-energy restorer, and the major medicinal bioactive compound is ergothioneine, which is a strong antioxidant. The molecule is taken up into the brain through specific intake mechanisms, is made available in the CNS following per oral administration, and is found to be a neuroprotectant (Nakamichi et al. 2016)

P. giganteus is another mushroom from *Pleurotus* family with strong antioxidant activity and neuroprotective and neuro-regenerative potential. The chemical constituents include linoleic acid, oleic acid, cinnamic acid, caffeic acid, p-coumaric acid, succinic acid, benzoic acid, and uridine. Linoleic acid, the major fatty acid present abundantly in the ethanol extract of *P. giganteus*, helps to protect Neuro-2a and embryonic fibroblast BALB/3T3 cells in culture (Phan et al. 2013). Interestingly, it is of importance to note here that linoleic acid or its derivatives possess neuroprotective activity as have been shown in cerebral cortical neurons against glutamate-induced excitotoxicity (Hunt et al. 2010) and to prevent nitroprusside-induced cerebral cortical neuronal demise by inhibiting activation of caspase-3 and caspase-9 (Yaguchi et al. 2010).

Amanita caesarea contains a polysaccharide that exhibits antioxidant activity and protects against glutamate-induced loss in HT22 cell viability and cell death caused by activation of apoptosis (Li et al. 2017). Neuroprotection by *A. caesarea* aqueous extract is found to be working via reversal of mitochondrial dysfunction and by inhibiting excessive free radical production and calcium accumulation within the cells (Li et al. 2017).

The major bioactive constituent of *Polyozellus multiplex*, polyozellin, is reported to protect against glutamate-induced cell death in mouse hippocampal neuronal cell line, HT22 acting via its strong antioxidant property involving the inhibition of Ca^{2+} influx, intracellular ROS production, and lipid peroxidation (Yang and Song 2015).

Medicinal *Armillaria mellea* hot water or ethanolic extracts contain several potent antioxidant molecules including ascorbic acid, phenols, flavonoids, carotenoids, and phenolic acids and has been shown to be anti-inflammatory, antiaging, and neuroprotective (Lung and Chang 2011; Geng et al. 2017). These effects are probably mediated through inhibition of nitric oxide production and inflammatory cytokines TNF-α, interleukin-6, and interleukin 1β (IL-1β) (Geng et al. 2017).

2.2.2 Neuro-regenerative Effects of Mushrooms

Erinacines are shown to exert biological properties through stimulation of nerve growth factor (NGF) synthesis (Li et al. 2014). In addition to erinacines, polysaccharides from the aqueous extract of lion's mane mushroom play an important role in inducing neuronal differentiation and promote neuronal survival (Sabaratnam et al. 2013). Long-term administration of *H. erinaceus* mycelium extracts causes an increase in the number of normal neurons (Li et al. 2014). Neuronal differentiation is brought about by the increase in the release of neurotrophic factors including NGF from glial cells. Some unknown active compounds that are lipid soluble promote NGF expression other than hericenones. *H. erinaceus* extract treatment marked up NGF mRNA expression suggesting the possibility of an active ingredient getting absorbed into the blood and delivered into the central nervous system (CNS), passing through the blood-brain barrier (BBB). The active ingredient of *P. cornucopiae*, ergothioneine, is permeable to BBB following systemic administration and is shown to cause increases in the number of neuronal precursor doublecortin-positive cells implying increased neuronal precursor migration and neurogenesis in mice (Nakamichi et al. 2016).

Ethanolic extracts of *P. giganteus* with potential antioxidant activity contains linoleic acid and uridine in addition to simple sugars, sugar alcohols, and acids. While uridine results in significant increase in a number of neurites in a dose-dependent manner (Phan et al. 2015), linoleic acid displays a key role in neuritogenesis in rat PC12 cells (Phan et al. 2012).

2.2.3 Neuronal Differentiation, Stem Cell Generation, and Myelinogenesis

Neurotrophic factors including NGF, glial cell-derived neurotrophic factor (GDNF), and brain-derived neurotrophic factor (BDNF) and micronutrients such as vitamins and minerals (Fe^{2+}, Zn^{2+}, Mg^{2+}) are crucial for the health, normal physiology, survival, maintenance, and regeneration of neural population in the young, adult, and aged brain. Continuous infusion into the globus pallidus neurotrophic factor of GDNF in Parkinson's disease (PD) patients is shown to attenuate disease syndromes and initiate neuronal sprouting significantly. It is understood that treatment of Alzheimer's disease (AD) with NGF met with some degree of success, but no clinical trials are reported for spino-cerebral ataxia, stroke, or Huntington's disease. However, numerous preclinical studies are available in the literature that describe the recovery of dysfunctions in animal models following one or the other neurotrophic factor (Allen et al. 2013).

The presence of an extract of *H. erinaceus* in culture media stimulated the development of cerebellar neural cells in vitro by stimulating myelinogenesis regulatory processes (Kolotushkina et al. 2003). Traditionally used medicinal mushrooms, *H. erinaceus* and *Lignosus rhinocerotis* known as lion's mane and tiger milk, respectively, are known to have neuro-healing activity by means of regenerative capability. This is in addition to anecdotal evidences in support of several hundreds of such edible mushrooms around the world. Extracts of both these mushrooms have been demonstrated to stimulate neuronal differentiation as seen by neuritic outgrowth and neuronal elongation in dissociated cells of the brain, spinal cord, and retina from chick embryo and are comparable to positive control cultures that received BDNF (Samberkar et al. 2015). Aqueous and ethanolic extracts of the fruiting bodies of *P. giganteus*, a cultivated mushroom in Malaysia, have been shown to sprout neurite outgrowth in rat PC12 cells (Phan et al. 2012)

G. lucidum, an edible mushroom believed by Orientals to retard senescence, contains different bioactive compounds, such as triterpenoids, polysaccharides, nucleotides, sterols, and steroids, in addition to micronutrients that activate antiaging activity through immunomodulation, increased sleep, and strong anti-oxidative and radical-scavenging activities. *G. lucidum* extracts have been demonstrated to aid in neuronal differentiation (Cheung et al. 2000).

Two novel eudesmane-type sesquiterpenes, dictyophorines A and B, and a known compound, teucrenone, were isolated from the mushroom *D. indusiata*.

Dictyophorines A and B promoted NGF synthesis by astroglial cells (Kawagishi et al. 1997). Golden oyster mushroom (*P. cornucopiae*), an edible mushroom extensively used as food by the Orientals, contains a potent neurogenic antioxidant and a substrate of carnitine/organic cation transporter OCTN1/SLC22A4, ergothioneine (Gründemann et al. 2005). It is expressed in the brain and neuronal stem cells (Lamhonwah et al. 2008), and it passes through the BBB when administered externally since the brain levels of ergothioneine are increased (Nakamichi et al. 2016). It is known to be transported into the neurons by OCTN1 in a Na^+-dependent fashion and is indicted in neuronal differentiation (Nakamichi et al. 2012).

C. militaris is a rare Chinese caterpillar mushroom, and its active component, cordycepin, attenuated LPS-induced microglial activation, as evidenced by reductions in inducible nitric oxide synthase and cyclooxygenase-2 mRNA levels, inhibition of nuclear factor-κB pathway, and reduced release of TNF-α and IL-1β. These helped to recover the hippocampal neuronal loss, correct neural growth defects, neuronal growth cone extension, neurite sprouting, cell viability, and dendritic spinogenesis (Peng et al. 2015).

2.3 Mushrooms and Some Neurological Diseases

2.3.1 Fatigue and Depression

Cordyceps militaris, one of the most important traditional Chinese medicines, has been used extensively as a crude drug and a folk tonic, as well as food in East Asia (Das et al. 2010). Due to its kind of bioactive substances including 3′-deoxyadenosine (cordycepin), ergosterol, polysaccharides, certain glycoproteins, and mannitol (Ng and Wang 2005), *C. militaris* possesses anti-inflammatory, antioxidant, antihypoxic, antiaging, antitumor, antidiabetic, immunomodulatory, neuroprotective, antihypoglycemic, and antinephropathic activities (Das et al. 2010; Dong et al. 2014). It has been demonstrated that fatigue recovery following treatment with *C. militaris* fruiting body extract is mainly through attenuating mitochondrial malfunctions and by activating 5-AMP-activated protein kinase and protein kinase B/mammalian target of rapamycin pathways and by regulating serum estradiol, testosterone, and cortisol levels (Song et al. 2015). *H. erinaceus* and *Dictyophora indusiata* are well-known edible mushrooms, and hericenone A and dictyophorines A and B isolated from these stimulate synthesis and production of NGF in astroglial cells in vitro and in vivo in experimental animals to increase neuronal catecholamine synthesis, evidencing improvements in brain and autonomic neuronal functions (Kawagishi et al. 1991, 1997). In a controlled prospective clinical study, depression and anxiety were reduced, and the quality of sleep was found to be better in volunteers on a month's diet with *H. erinaceus* containing snacks (Nagano et al. 2010).

2.3.2 Beneficial Effects of Mushrooms in Alzheimer's Disease

AD is a chronic, progressive neurodegenerative disease with symptoms of loss of memory and cognitive functions. Short-term memory is affected initially, which slowly progresses to affect orientation, language, recognition, and behavioral abnormalities and to a stage of complete loss of memory and inability to take care of self. The pathology includes extracellular amyloid plaques and intraneuronal fibrillary tau tangles within the brain and in certain other areas. The cause or risk factors for AD are still speculative, but aging, oxidative stress, proteotoxicity, mitochondrial dysfunction, inflammatory response, and to a great extent genetic predisposition are suggested to be at the roots of pathology (Chakrabarti and Mohanakumar 2016). At a time when no medication is available for the disease, and no intervention has been indicated to reduce the risk of AD, it has been suggested that regular consumption of edible medicinal mushrooms could reduce the risk of AD, since these possess strong antioxidant activity and are capable of strong immunomodulation (Phan et al. 2017). In a placebo-controlled, double-blind clinical trial, tablets containing *H. erinaceus* dried powder given 36 weeks were shown to improve mild cognitive impairment in a small Japanese group of people (Mori et al. 2009). The Mediterranean diet containing oyster mushrooms of this region, under the group of *Basidiomycetes* such as *P. ostreatus* and *P. eryngii*, provides a lowered risk for AD incidence (Scarmeas et al. 2006).

Intracerebroventricular (ICV) infusion of streptozotocin (STZ) is shown to cause AD syndromes, cellular mitochondrial dysfunctions, and pathology including amyloidosis in the hippocampus of rats akin to the human disease (Paidi et al. 2015). ICV-STZ rats pre-treated with *G. lucidum* spores significantly attenuated the STZ-induced increase of malondialdehyde and decreases of antioxidant enzyme activities in the brain and glutathione and ATP levels, which accompanied a marked reversal of spatial learning and memory impairments and hippocampal neuronal loss (Zhou et al. 2012). Choi et al. (2015) have demonstrated in a sketchy, descriptive study that fermented *G. lucidum* extract could be used to enhance learning, memory, and cognitive function via influencing cholinergic dysfunction. *G. lucidum* triterpenoids have been shown to inhibit acetylcholinesterase selectively, thereby increasing acetylcholine content in the brain to improve learning and memory in animal models of AD (Lee et al. 2011; Zhang et al. 2012a).

G. lucidum extracts have potent neuroprotective activities, and it has been shown to improve immunity and memory (Zhou et al. 2012; Zhang et al. 2014). Alcoholic extract of *G. lucidum* is demonstrated to protect against hippocampal neuronal death (Lee et al. 2014; Zhang et al. 2011). *G. lucidum* water extract or polysaccharides from this mushroom promoted neural progenitor cell proliferation in transgenic AD mice and in cell culture (Huang et al. 2017). *H. erinaceus* plays a major role in improving cell viability and inverses the nuclear apoptotic alternation via mitochondrial-related pathway (Zhang et al. 2016). Neuro-2a neuronal death caused by ER stress was absolved by dilinoleoyl-phosphatidylethanolamine from the edible monkey head mushroom, *H. erinaceous*, by increasing PKC-epsilon activity (Nagai et al. 2006). *A. caesarea* has been shown to reduce cholinergic dysfunction in

$AlCl_3$-mediated neuronal apoptotic cholinergic cell death in an experimental mouse model of AD (Li et al. 2017). Aqueous extracts of *A. caesarea* enhanced acetylcholine levels in the mouse brain, by increasing the activity of choline acetyltransferase and through inhibiting acetylcholinesterase activity (Li et al. 2017).

NGF affects the activity of cholinergic neurons and the level of choline acetyltransferase and acetylcholinesterase in the CNS (Karami et al. 2015). The alcohol extract *of H. erinaceus*, but not of *P. eryngii*, *Grifola frondosa*, or *Agaricus blazei*, was found to stimulate NGF mRNA and protein levels in 1321-N1 human astrocytoma cells and promoted neurite outgrowth in PC12 pheochromocytoma cells via promoting c-Jun N-terminal kinase activity (Mori et al. 2008). The same group has also demonstrated prevention of spatial short-term and visual recognition memory loss caused by intracerebral administration of amyloid-β25–35 peptide fragments, by a diet containing *H. erinaceus* powder (Mori et al. 2011).

2.3.3 Mushrooms Are Good for Parkinson's Disease

PD is the most consistent neurodegenerative disease affecting movement, subsequent to the loss of dopamine-containing neurons of the midbrain substantia nigra and other related structures. In addition to the progressive motor neuronal disabilities resulting from bradykinesia, rigidity, tremor, and postural instability, the patients suffer from non-motor complications, including sleep, anxiety, depression, cognitive functions, disturbances in thought processes, and certain autonomic and sensory functions. In PD incidents with about 5–8% familial and more than 90% idiopathic, the undisputed risk factor for this progressive disease is aging. Pathological features of PD are the presence of intracellular Lewy bodies in the living basal ganglia neurons, severe gliosis, and loss of astrocytes in masses. Consistent reasons for the pathology are stated to be severe oxidative stress in the dopaminergic neurons, neural lysosomal and proteasomal system dysfunctions, and mitochondrial damage in terms of its bioenergetics and biogenesis (Chandra et al. 2017). Genetic predisposition and environmental factors are strongly suggested to be the foundation of PD, for which the mainstay treatment is based on management of deficient dopaminergic functions in the brain by exogenous dopaminergic stimulation in the basal ganglia circuitry to control the motor outcome. While symptomatic control of the dysfunctions is taken care by such medications, a cure for the disease is not forthcoming, yet prospects of novel therapeutic means from traditional herbals are strongly indicated (Sengupta et al. 2016). Measures for prevention of the disease is unknown, and several lines of evidence suggest regular intake of antioxidants, nonsteroidal anti-inflammatory drugs, L-type calcium channel antagonists, and immune system rejuvenators helps to reduce incidents of PD development (Chandra et al. 2017; Singh et al. 2016, 2018; Naskar et al. 2013; Madathil et al. 2013). Though literature reveals scanty, some carefully executed studies support the inclusion of mushrooms in regular diets to help to ward off certain syndromes of PD such as chronic stress, certain motor, sensory and autonomic functions; and insomnia (Phan et al. 2015).

It is understood that oxidative stress plays a fundamental role in the pathophysiology that leads to nigral dopaminergic neurodegeneration causing PD. Most of the edible mushrooms have significant levels of antioxidants and factors that stimulate enzyme protein expression or aid in the synthesis of molecules that regulate antioxidant activity in the brain. *H. erinaceus* mycelium has neuroprotective potential against the parkinsonian neurotoxin, 1-methyl-4-phenyl pyridinium (MPP+), mediated neuronal damage. The major constituent of this mushroom, erinacine, contributes to this by inactivating oxidative stress (Kuo et al. 2016). *H. erinaceus* mycelium extract or erinacine A, an active constituent of this mushroom, protects against the PD neurotoxin 1-methyl-4-phenyl-1,2,3,6 tetrahydropyridine (MPTP) induced dopaminergic apoptotic cell death by increasing the levels of reduced glutathione, by inhibiting nitrotyrosine and 4-hydroxy-2-nonenal levels, and probably by activating anti-inflammatory signaling in mice (Kuo et al. 2016). Additionally, *G. lucidum* extract reduces the expressions of pro-inflammatory and cytotoxic factors and effectively protects the dopaminergic neurons against inflammatory and oxidative damages (Chen et al. 2007; Guo et al. 2016). MPP^{+}- or rotenone-induced death of embryonic mouse mesencephalic primary dopaminergic cell cultures was prevented by *G. lucidum* polysaccharides through its strong antioxidant properties and via correcting mitochondrial membrane potential (Guo et al. 2016). *G. lucidum* extracts can help to prevent dopaminergic neuron degeneration in vitro in co-cultures of neuron-glia by boosting immunological responses via microgliosis (Zhang et al. 2011). In particular, oil from *G. lucidum* spores administered in MPTP-treated animals with neuronal loss akin to PD was effective in improving the neurotoxin-induced motor disability and dopaminergic neuronal loss in the midbrain substantia nigra region (Zhu et al. 2005).

2.4 Mechanisms by Which Mushrooms Exert Medicinal Effects

General mechanisms by which mushrooms exert medicinal effects are depicted in Fig. 2.2 and are only the cumulative pharmacological actions. Under this section, plausible mechanisms of action of mushroom extracts or active constituents are discussed, where experimental data exist. Of all the cell organelles, ER and mitochondria probably play active roles in neuronal toxicity or neuronal rejuvenation by the actions of mushroom constituents (Fig. 2.3).

Fig. 2.3 (continued) Subsequently, IRE1α oligomerizes, phosphorylates itself, and splices XBP1 mRNA, which results in the translation of the transcription factor XBP1. Upon prolonged ER stress, PERK oligomerizes and phosphorylates itself together with eIF2α, where it attenuates protein translation. It further activates the transcription factor ATF4, which carries out downstream activation of UPR genes. After the dissociation of GRP78/BiP, ATF6 translocates to the Golgi apparatus where it is cleaved into its active form. The N-terminal cytosolic fragment of ATF6, which is the active form, localizes into the nucleus to activate UPR target genes. Severe ER stress induces apoptosis through various pathways, including transcriptional induction of CHOP by the PERK and ATF6 pathways, the IRE1α-TRAF2 -ASK1 pathway, and the caspase-12 pathway. Stimulation of IRE1α-TRAF2-ASK1 complex leads to c-Jun N-terminal kinase (JNK) phosphorylation, activating Bax and Bak ER-to-mitochondria signaling to induce an apoptotic mode of cell destruction

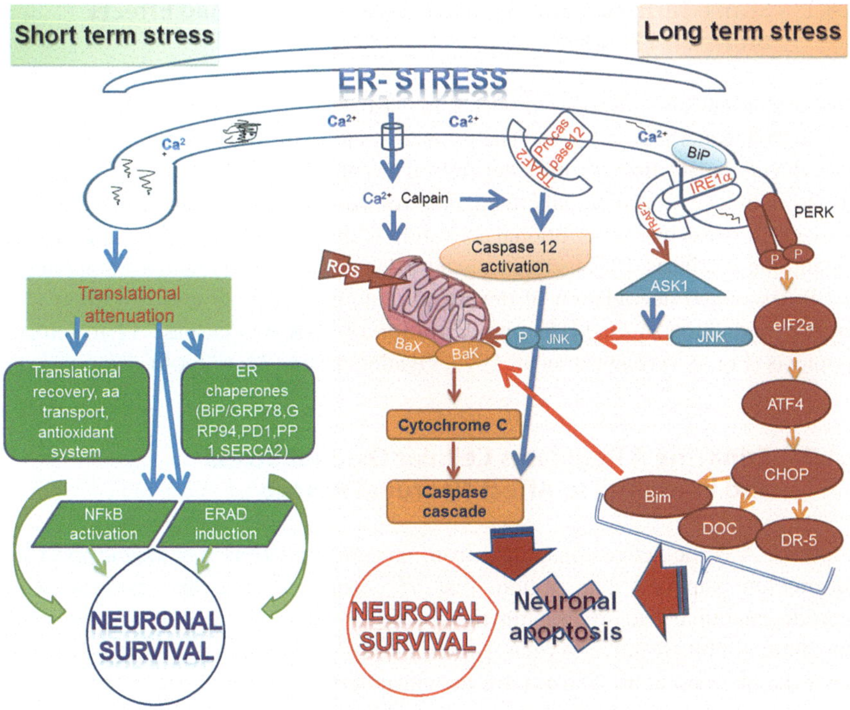

Fig. 2.3 Effects of ER stress on neuronal survival/death
Note: ER is the major site for protein synthesis and folding, and maintaining its physiological homeostasis is essential for neuronal survival. Many genetic and environmental insults that interfere with its functions lead to the accumulation of misfolded proteins in the ER lumen, causing ER stress. ER lumen is a Ca^{2+}-rich environment, and so the functions of many of the proteins located in this compartment are calcium-dependent. Thus, Ca^{2+} imbalances are one of the major causes of improper protein folding, triggering the ER stress. Short-term ER stress (*portion under light-green shade*) leads to a general downregulation of protein translation, except the translation of chaperones and other proteins involved in degradation of misfolded proteins. This translational attenuation results in lower protein-folding load to the ER and is accompanied by transcriptional induction of several groups of genes including ER chaperones for long-term adaptation to ER stress. If the initial response of induction of ER chaperones fails to refold the misfolded proteins, ER-associated degradation pathway (ERAD) components are induced to eliminate those proteins. To regain the ER homeostasis, genes coding the proteins involved in the amino acid import, glutathione biosynthesis, and ER overload response (NF-kB) are activated. ER contains three branches of stress sensors, i.e., pancreatic ER kinase (PKR)-like ER kinase (PERK), activating transcription factor 6 (ATF6) and inositol-requiring enzyme 1 (IRE1), which recognize misfolded proteins in ER and activate a complex signaling network, called the unfolded protein response (UPR). UPR is a pro-survival response to reduce the accumulation of unfolded proteins and restore normal ER functioning. However, if protein aggregation is persistent and the stress cannot be resolved, signaling switches from pro-survival to pro-apoptotic mode. During prolonged ER stress (*portion under magenta shade*), signaling through those stress sensors can trigger pro-apoptotic signals. Induction of ER stress causes the release of the molecular chaperone GRP78/BiP (glucose-regulated protein of 78 kDa/binding immunoglobulin protein), from the luminal domain of these sensors. The release of GRP78/BiP from IRE1α promotes its binding to misfolded proteins.

2.4.1 Short-Term or Long-Term ER Stress Have Varied Effects on Neural Health

Polysaccharides and unsaturated fatty acids also exert a preventative effect on certain neurodegenerative diseases. The protective effect of unsaturated fatty acids on neurodegenerative diseases may depend on attenuating the ER stress. Short-term ER stress induces translational attenuation, resulting in compensatory mechanisms to adapt for stress by transcriptional activation of amino acid transporters, increased glutathione synthesis, ER chaperons, and nuclear factor-κB to regulate components of ER-associated degradation of unwanted proteins, but long-term ER stress will result in induction of Gadd153 and activation of JNK and caspase 12 to initiate apoptosis (Fig. 2.3) (Fawcett et al. 1999; Oyadomari and Mori 2004).

2.4.2 Erinacine A Regulates Cellular Oxidant Stress and Immunity to Affect Neuronal Rescue

Erinacine A is the active constituent from mycelium of *Hericium erinaceus*. This information supports the notion that *H. erinaceus* benefits patients with neurodegenerative diseases as a functional food by reducing ER stress. *H. erinaceus* mycelium inhibits reactive oxygen species, oxidized and nitrated proteins, and lipid peroxidation. Kuo et al. (2016) have demonstrated that *H. erinaceus* extract or its bioactive compound erinacine A treatment protected against MPP^+- or MPTP-mediated dopaminergic apoptotic cell death resulted through sustained activation of ER stress responses via activation of JNK1/2, p38 MAPK, and IRE1α/TRAF2 pathways and by increased expression of C/EBP homologous protein (CHOP or Gadd153), IKB-β and NF-κB, Fas, and Bax (Fig. 2.4).

Erinacine increases neuronal growth by stimulating NGF synthesis. Brain tissue damage results from induction of free radicals. These free radicals are upregulated by pro-inflammatory cytokines. Erinacine inhibits inflammatory cytokine expression. Oxidative stress after ischemic damage causes activation of transcription factor CHOP. Erinacine inhibits neuronal cell death by inactivating p38 MAPK-dependent CHOP expression. Erinacine abolishes stroke-induced infraction volume (Li et al. 2014). Erinacine results in a reduction of Fas expression (apoptosis antigen) and inhibition of JNK1/2 and p38 phosphorylation (Kuo et al. 2016). A component from *H. erinaceus*, dilinoleoyl-phosphatidylethanolamine, an activator of protein kinase C (PKC), may also directly help in the attenuation of neuronal cell death (Nagai et al. 2006). PKC enzymes control functions of several proteins through hydroxyl group phosphorylation at serine and threonine residues of these proteins and activate signals linked to calcium, phosphatidylserine, diacylglycerol, etc. to affect smooth muscle contraction, neuronal stimulation, and glutamate activation depending on the types of cells and the endogenous neuronal activators such as acetylcholine, serotonin, adrenalin, glutamate, and prostaglandin.

It is reported that polysaccharides are the major antioxidants in *G. lucidum* similar to *H. erinaceus*, but *G. lucidum* polysaccharides act via scavenging hydroxyl

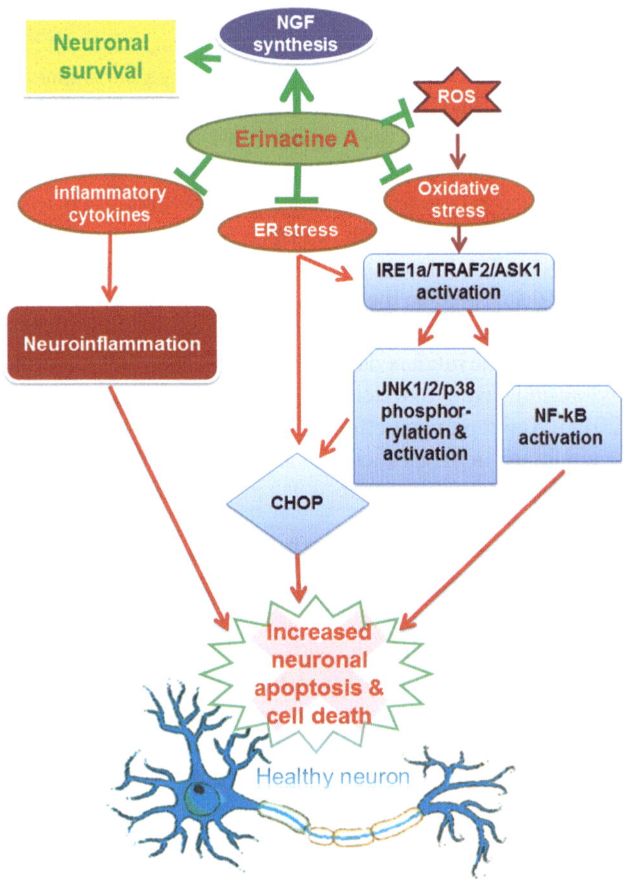

Fig. 2.4 Erinacine A scavenges ER and oxidative stresses and promotes neuronal survival
Note: Erinacine A is the active constituent from mycelium of *Hericium erinaceus*. Erinacine A scavenges ROS and attenuates oxidative and ER stress-induced neuronal apoptosis by inhibiting the IRE1α-TRAF2-ASK1 complex formation further inhibiting JNK phosphorylation and CHOP induction. Erinacine A also inhibits neuroinflammation and promotes nerve growth factor (NGF) synthesis

radicals (Chen et al. 2015). *G. lucidum* polysaccharides mediate neuronal differentiation and protect the neurons from apoptosis, which is made possible by the involvement of the Erk1/2 and the CREB signaling pathways (Gokce et al. 2015). The observed anti-inflammatory activity of *G. lucidum* polysaccharides was attributed to its capability in suppressing pro-inflammatory cytokines. Oxidative stress produces free radicals and initiates lipid peroxidation in the damaged spinal cord tissue, and *G. lucidum* polysaccharide administration actively inhibits lipid peroxidation and protects against hypoxia-/reoxygenation-induced injury in cortical neurons by reducing malondialdehyde levels, the end product of lipid peroxidation (Chen et al. 2015; Gokce et al. 2015). It has been shown that *G. lucidum* extracts act by regulating anti-inflammatory signals in mixed dopaminergic-microglia cultures treated with the parkinsonian neurotoxin, MPP^+. LPS- and MPP^+-induced increases in the mRNA and protein levels in the pro-inflammatory glial signals such as nitric oxide synthase, TNF-α, and IL-1β were significantly downregulated by the *G. lucidum* extracts and blocked cell demise in the in vitro system (Zhang et al. 2011).

Ergothioneine abundantly present in *P. cornucopiae* is membrane impermeable due to its hydrophilicity. It gets absorbed from gastrointestinal tract due to the presence of specific carrier-mediated transport system, carnitine/organic cation transporter OCTN1. This food-derived antioxidant (Aruoma et al. 1999) ergothioneine gets distributed to the brain, and the saturation of brain distribution reflects saturation of uptake into neurons after passage through the BBB (Ishimoto et al. 2014). Ergothioneine hinders damage to DNA and protein. It protects neurons against β-amyloid-induced cytotoxicity by scavenging cellular hydrogen peroxide (Yang et al. 2012). Antidepressant effect of this mushroom is predominantly due to ergothioneine contained in the extract (Nakamichi et al. 2016).

2.4.3 *Pleurotus giganteus* Blocks Nitrosative Stress to Aid in Neuronal Protection

P. giganteus, a culinary mushroom with medicinal properties, showed anti-inflammatory effects against LPS or LPS and hydrogen peroxide-induced inflammation in RAW 264.7 macrophages by inhibiting nitric oxide production and thereby limiting cyclooxygenase-2 activity and by upregulation of nuclear transcriptional signal transducer and activator of transcription 3 protein (STAT 3) pathway (Baskaran et al. 2017; Fig. 2.5).

2.4.4 *Amanita caesarea* Regulates Ca^{2+}-Mediated Apoptosis and Increases Autophagy in Cells

A. caesarea water extracts are protected against glutamate-mediated cell death in HT22 cell line and improved cholinergic functions in the brain of aluminum-treated mice by improving mitochondrial membrane potential dissipation; by reducing the increased levels of intracellular ROS and Ca^{2+}; by inhibiting the increased expression of cleaved caspase 3, calpain, and apoptosis-inducing factor, Bax; and by enhancing mitochondrial function via the modulation of Akt/mTOR signaling (Li et al. 2017; Fig. 2.6).

2.4.5 Polyozellin Tames Apoptosis During Excitotoxicity

P. multiplex one of the favorite edible mushrooms of the Orientals contains an active constituent, polyozellin, that has remarkable neuroprotective effects. Polyozellin exerts its effects at mitochondria by mediating Ca^{2+} influx, free radical production,

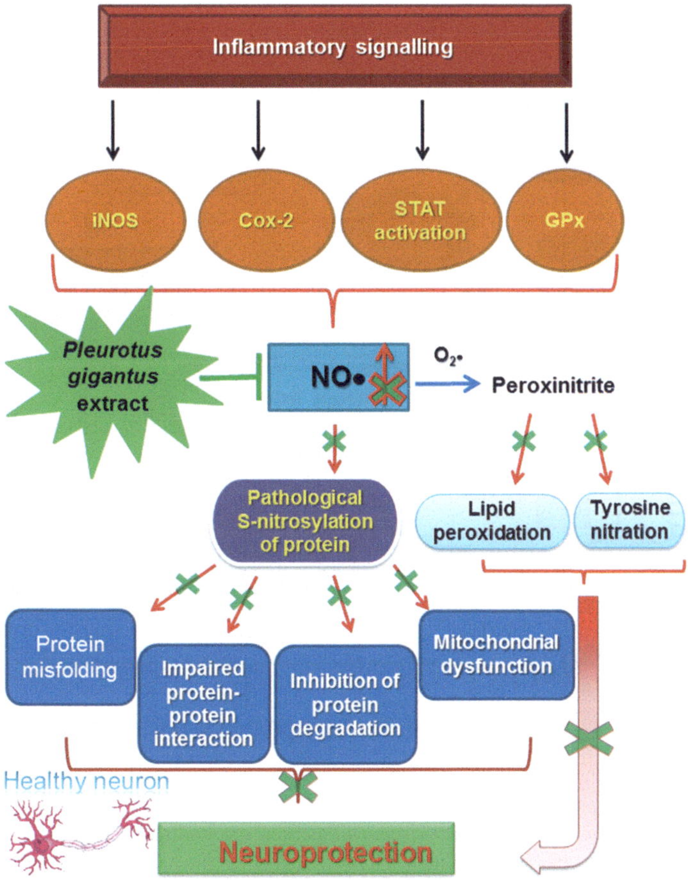

Fig. 2.5 Cartoon depicting neuroprotective potential of *Pleurotus giganteus*
Note: The ethanolic extract of this mushroom inhibits nitric oxide (NO•) production in inflammatory and oxidative stress-induced signaling. Therefore, the neuroprotective potential is mediated by inhibition of lipid peroxidation and tyrosine nitration and blockade of a host of pathological pathways mediated by S-nitrosylation of proteins by NO•

and lipid peroxidation and by regulating expressions of Bid, Bcl-2, and apoptosis-inducing factor and phosphorylation of mitogen-activated protein kinases (Yang and Song 2015; Fig. 2.7).

2.5 Mushrooms as Probiotic: The Gut-Brain Axis

It is of great interest to speculate on potential benefits of mushrooms for the peripheral nervous system or CNS disorders, chronic or otherwise. In a careful investigation, Kim et al. (2017) examined suppressive mechanisms of an extract from bioprocessed *Lentinus edodes* mycelial liquid culture supplemented with turmeric

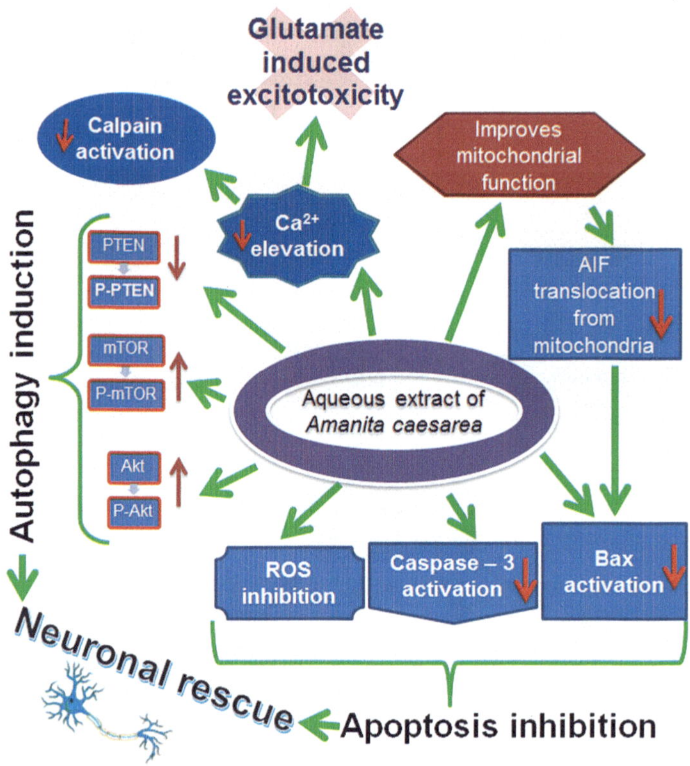

Fig. 2.6 Biological pathways influenced by *Amanita caesarea*
Note: The aqueous extract of *Amanita caesarea* has been reported to possess good antioxidant activities and mitigate glutamate-induced excitotoxicity. Experimental evidence points toward activation of signaling pathways that lead to autophagy induction and clearance of accumulated proteins by this extract. It also inhibits neuronal apoptosis by scavenging ROS, blocking caspase-3 activation, and improving mitochondrial functions

(bioprocessed *Curcuma longa* extract) against the gut microbiome, *Salmonella*. Dietary administration of the combined extract caused significant increases in the levels of Th1 cytokines and four types of interleukins and protected the infected mice from salmonellosis through increasing immunity, blockade of *Salmonella* across intestinal epithelial cells, and immunoglobulin production. Similar reports on the beneficial effects of mushrooms singularly or in combination on gut microflora have been reported by many. In this context, it is interesting to see that mushroom polysaccharides are suggested to be great prebiotics, and experimental studies have demonstrated enhanced immunity and promotion of gut microbiota following diets containing heteropolysaccharides isolated from the fruit body of *L. edodes* (Xu et al. 2015). White button mushroom (*A. bisporus*) diet changed the composition of the normal microflora and the urinary metabolome of mice resulting in better control of inflammation (Varshney et al. 2013). Also, resolution of infection effects of a polysaccharopeptide isolated from the medicinal mushroom

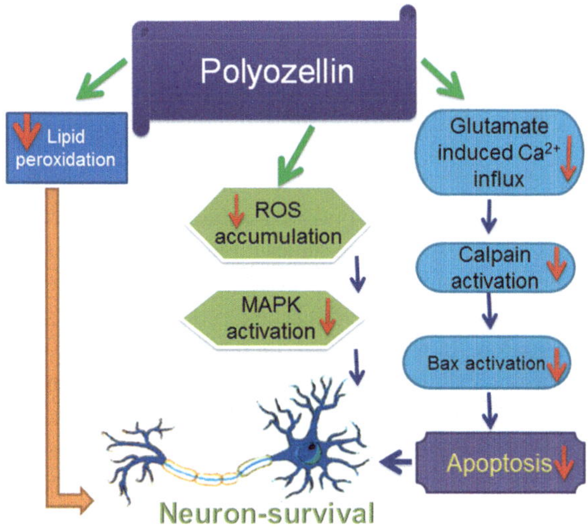

Fig. 2.7 Polyozellin neuroprotective pathways
Note: *Polyozellus multiplex*, a Korean edible mushroom, contains a key component, polyozellin, which is a prolyl endopeptidase inhibitor. Previous research showed that it might be useful for the prevention and treatment of neurodegenerative disorders since it possesses antioxidant and anti-apoptotic activities and can ameliorate glutamate excitotoxicity

Trametes versicolor showed that it enhances human intestinal microbiome regulation to better the interplay with host cells as a prebiotic (Varshney et al. 2013; Wu et al. 2016).

Under the circumstances, and in view of the recent developments in microbiome research in relation to neuro(auto)immune and neurodegenerative conditions such as AD, multiple sclerosis, PD, Huntington's disease, amyotrophic lateral sclerosis, etc. (Tremlett et al. 2017) or the stronger relationship with gut-neuronal axis in regulating immunological functions and autism spectrum disorders and other such diseases (Jaiswal et al. 2015; Israelyan and Margolis 2018), it is only prudent to suggest that mushrooms having strong effects on gut microbiome could also be befitting neuro-nutraceuticals for patients suffering from chronic neurodegenerative and development-associated neurological conditions. Especially in the cases of geriatric neurodegenerative diseases such as AD and PD, helpful probiotics are suggested to prevent many aging-associated problems, viz., decreased neurotransmitter levels, chronic inflammation, increased ROS/RNS, and several pro-apoptotic factors (Westfall et al. 2017). Since patients with neurodegenerative diseases and autism (Israelyan and Margolis 2018; O'Banion et al. 1978) suffer from a high rate of gastrointestinal comorbidities, it could be suggested that better management of helpful gut microbiota will be very beneficial to prevent or retard the progression of such chronic diseases.

2.6 Conclusions

In a recent report, Friedman (2016) has reviewed the health benefits of major homosaccharides and polysaccharides from edible mushrooms given alone or in combination with other nutritional diets. He has strongly suggested the dire need for extensive yet methodical investigation of mushrooms for health-promoting constituents. A sea of changes in food intake habits of modern population is in vogue, and that has adversely affected general mental health and the occurrence of chronic developmental disabilities such as autism, attention deficit hyperactivity disorders, and mental retardation, as well as age-related diseases such as AD and PD of the world population. In view of the increasing incidents and prevalence of developmental neurological disabilities, and geriatric neurodegenerative diseases in the modern era, it is strongly suggested that advanced research on mushrooms as neuro-nutraceuticals should be encouraged, and the results of these studies are made available to the public for reintroducing mushrooms in their daily diets. Some of us have reviewed a number of diets, diet constituents, and beverages of daily intake that have nutraceutical properties, which affect mentation, and termed them as neuro-nutraceuticals (Williams et al. 2016b). In this context, mushrooms packed with potent antioxidants, pro-inflammatory molecules, micronutrients, and constituents that are used as medicines in modern clinical practices are a welcome addition to those daily food habits. The existing scientific literature is compelling in support of the use of mushrooms as potent neuro-nutraceuticals.

Acknowledgments We acknowledge the financial support from Department of Health and Family Welfare and Department of Higher Education, Government of Kerala, India. Mahatma Gandhi University acts as our mentoring institution, support of which is acknowledged.

Conflict of Interests The authors declare no conflict of interest in publishing this manuscript.

References

Abdullah N, Ismail SM, Aminudin N, Shuib AS, Lau BF (2012) Evaluation of selected culinary-medicinal mushrooms for antioxidant and ACE inhibitory activities. Evid Based Complement Alternat Med 2012:1–12. https://doi.org/10.1155/2012/464238

Allen SJ, Watson JJ, Shoemark DK, Barua NU, Patel NK (2013) GDNF, NGF and BDNF as therapeutic options for neurodegeneration. Pharmacol Ther 138(2):155–175. https://doi.org/10.1016/j.pharmthera.2013.01.004

Ames BN, Shigenaga MK, Hagen TM (1993) Oxidants, antioxidants, and the degenerative diseases of aging. Proc Natl Acad Sci 90(17):7915–7922

Aruoma OI, Spencer JP, Mahmood N (1999) Protection against oxidative damage and cell death by the natural antioxidant ergothioneine. Food Chem Toxicol 37(11):1043–1053. https://doi.org/10.1016/S0278-6915(99)00098-8

Attarat J, Phermthai T (2014) Bioactive compounds in three edible Lentinus mushrooms. Walailak J Sci Technol 12:491–504. https://doi.org/10.2004/wjst.v12i2.842

Aygul N, Duzenli MA, Ozdemir K, Altunkeser BB (2010) A case report of an unusual complication of Amanita phalloides poisoning: development of cardiogenic shock and its successful treatment with intra-aortic balloon counterpulsation. Toxicon 55:630–632. https://doi.org/10.1016/j.toxicon.2009.10.022

Baskaran A, Chua KH, Sabaratnam V, Ravishankar Ram M, Kuppusamy UR (2017) Pleurotus giganteus (Berk. Karun & Hyde), the giant oyster mushroom inhibits NO production in LPS/ H_2O_2 stimulated RAW 264.7 cells via STAT 3 and COX-2 pathways. BMC Complement Altern Med 17(1):40. https://doi.org/10.1186/s12906-016-1546-6

Bedry R, Baudrimont I, Deffieux G, Creppy EE, Pomies JP, Ragnaud JM, Dupon M, Neau D, Gabinski C, De Witte S, Chapalain JC, Beylot J, Godeau P (2001) Wild-mushroom intoxication as a cause of Rhabdomyolysis. N Engl J Med 345:798–802. https://doi.org/10.1056/NEJMoa010581

Brondz I, Ekeberg D, Høiland K, Bell DS, Annino AR (2007) The real nature of the indole alkaloids in Cortinarius infractus: evaluation of artifact formation through solvent extraction method development. J Chromatogr A 1148:1–7. https://doi.org/10.1016/j.chroma.2007.02.074

Chakrabarti S, Mohanakumar KP (2016) Aging and neurodegeneration: a tangle of models and mechanisms. Aging Dis 7(2):111–113. https://doi.org/10.14336/AD.2016.0312

Chandra G, Shenoi RA, Anand R, Rajamma U, Mohanakumar KP (2017) Reinforcing mitochondrial functions in aging brain: an insight into Parkinson's disease therapeutics. J Chem Neuroanat S0891-0618(17):30122–30129. https://doi.org/10.1016/j.jchemneu.2017.12.004

Chen CC, Shiao YJ, Lin RD, Shao YY, Lai MN, Lin CC, Ng LT, Kuo YH (2006) Neuroprotective diterpenes from the fruiting body of *Antrodia camphorata*. J Nat Prod 69:689–691. https://doi.org/10.1021/np0581263

Chen LW, Wang YQ, Wei LC, Shi M, Chan YS (2007) Chinese herbs and herbal extracts for neuroprotection of dopaminergic neurons and potential therapeutic treatment of Parkinson's disease. CNS Neurol Disord Drug Targets 6(4):273–281

Chen P, Yong Y, Gu Y, Wang Z, Zhang S, Lu L (2015) Comparison of antioxidant and antiproliferation activities of polysaccharides from eight species of medicinal mushrooms. Int J Med Mushrooms 17(3):287–295

Cheung WWM, Hui WS, Chu PWK, Chiu SW, Ip NY (2000) Ganoderma extract activates MAP kinases and induces the neuronal differentiation of rat pheochromocytoma PC12 cells. FEBS Lett 486(3):291–296

Choi J, Horikawa M, Okumura H, Kodani S, Nagai K, Hashizume D, Koshino H, Kawagishi H (2009) Endoplasmic reticulum (ER) stress protecting compounds from the mushroom Mycoleptodonoides aitchisonii. Tetrahedron 65:221–224. https://doi.org/10.1016/J.TET.2008.10.068

Choi JH, Suzuki T, Okumura H, Noguchi K, Kondo M, Nagai K, Hirai H, Kawagishi H (2014) Endoplasmic reticulum stress suppressive compounds from the edible mushroom *Mycoleptodonoides aitchisonii*. J Nat Prod 77:1729–1733. https://doi.org/10.1021/np500075m

Choi YJ, Yang HS, Jo JH, Lee SC, Park TY, Choi BS, Seo KS, Huh CK (2015) Anti-amnesic effect of fermented Ganoderma lucidum water extracts by lactic acid bacteria on scopolamine-induced memory impairment in rats. Prev Nutr Food Sci 20(2):126–132. https://doi.org/10.3746/pnf.2015.20.2.126

Chowdhury M, Kubra K, Ahmed S (2015) Screening of antimicrobial, antioxidant properties and bioactive compounds of some edible mushrooms cultivated in Bangladesh. Ann Clin Microbiol Antimicrob 14:8. https://doi.org/10.1186/s12941-015-0067-3

Clarke D, Crews C (2013) Natural toxicants: mushrooms and toadstools. Encycl Food Saf 2:269–276. https://doi.org/10.1016/B978-0-12-378612-8.00429-7

Curtis DR, Lodge D, McLennan H (1979) The excitation and depression of spinal neurones by ibotenic acid. J Physiol 291:19–28

Das SK, Masuda M, Sakurai A, Sakakibara M (2010) Medicinal uses of the mushroom Cordyceps militaris: current state and prospects. Fitoterapia 81(8):961–968. https://doi.org/10.1016/j.fitote.2010.07.010

Dong Y, Jing T, Meng Q, Liu C, Hu S, Ma Y, Liu Y, Lu J, Cheng Y, Wang D, Teng L (2014) Studies on the antidiabetic activities of Cordyceps militaris extract in diet-streptozotocin-induced diabetic Sprague-Dawley rats. Biomed Res Int 2014:1–11. https://doi.org/10.1155/2014/160980

Ennecker-Jans SAM, van Daele PLA, Blonk MI, Varin DSE, van Laar JAM (2007) Amatoxin poisoning due to soup from personally picked deathcap mushrooms (Amanita phalloides). Ned Tijdschr Geneeskd 151:764–768

Fawcett TW, Martindal JL, Guyton KZ, Hai T, Holbrook NJ (1999) Complexes containing activating transcription factor (ATF)/cAMP-responsive element-binding protein (CREB) interact with the CCAAT/enhancer-binding protein (C/EBP)–ATF composite site to regulate Gadd153 expression during the stress response. Biochem J 339(Pt 1):135–141

Frangež R, Šuput D, Molgó J, Benoit E (2017) Ostreolysin A/pleurotolysin and equinatoxins: structure, function and pathophysiological effects of these pore-forming proteins. Toxins (Basel) 9:1–10. https://doi.org/10.3390/toxins9040128

French AL, Garrettson LK (1988) Poisoning with the North American Jack O'Lantern mushroom, Omphalotus illudens. J Toxicol Clin Toxicol 26:81–88

Friedman M (2016) Mushroom polysaccharides: chemistry and antiobesity, antidiabetes, anticancer, and antibiotic properties in cells, rodents, and humans. Foods 5(4):pii: E80. https://doi.org/10.3390/foods5040080

Ganeshpurkar A, Pardhi P, Bhadoriya SS, Jain N, Rai G, Jain AP (2015) Antioxidant potential of white oyster culinary-medicinal mushroom, Pleurotus Florida (higher basidiomycetes). Int J Med Mushrooms 17:491–498

Geissler T, Brandt W, Porzel A, Schlenzig D, Kehlen A, Wessjohann L, Arnold N (2010) Acetylcholinesterase inhibitors from the toadstool Cortinarius infractus. Bioorg Med Chem 18:2173–2177. https://doi.org/10.1016/j.bmc.2010.01.074

Geng Y, Zhu S, Cheng P, Lu ZM, Xu HY, Shi JS, Xu ZH (2017) Bioassay-guided fractionation of ethyl acetate extract from Armillaria mellea attenuates inflammatory response in lipopolysaccharides (LPS)-stimulated BV-2 microglia. Phytomedicine 26:55–61. https://doi.org/10.1016/j.phymed.2017.01.005

Gokce EC, Kahveci R, Gürer B, Aksoy N, Gokce A, Sargon MF, Cemil B, Erdogan B, Kahveci O (2015) Neuroprotective effects of Ganoderma lucidum polysaccharides against traumatic spinal cord injury in rats. Injury 46(11):2146–2155. https://doi.org/10.1016/j.injury.2015.08.017

Graeme KA (2014) Mycetism: a review of the recent literature. J Med Toxicol 10:173–189. https://doi.org/10.1007/s13181-013-0355-2

Gründemann D, Harlfinger S, Golz S, Geerts A, Lazar A, Berkels R, Jung N, Rubbert A, Schömig E (2005) Discovery of the ergothioneine transporter. Proc Natl Acad Sci USA 102(14):5256–5261

Guo SS, Cui XL, Rausch WD (2016) Ganoderma Lucidum polysaccharides protect against MPP(+) and rotenone-induced apoptosis in primary dopaminergic cell cultures through inhibiting oxidative stress. Am J Neurodegener Dis 5(2):131–144

Han ZH, Ye JM, Wang GF (2013) Evaluation of in vivo antioxidant activity of Hericium erinaceus polysaccharides. Int J Biol Macromol 52:66–71. https://doi.org/10.1016/j.ijbiomac.2012.09.009

Holliday JC, Cleaver P, Loomis-Powers M, Patel D (2004) Analysis of quality and techniques for hybridization of medicinal fungus Cordyceps sinensis (Berk.)Sacc. (Ascomycetes). Int J Med Mushrooms 6:151–164. https://doi.org/10.1615/IntJMedMushr.v6.i2.60

Huang S, Mao J, Ding K, Zhou Y, Zeng X, Yang W, Wang P, Zhao C, Yao J, Xia P, Pei G (2017) Polysaccharides from Ganoderma lucidum promote cognitive function and neural progenitor proliferation in mouse model of Alzheimer's disease. Stem Cell Rep 8:84–94. https://doi.org/10.1016/j.stemcr.2016.12.007

Hunt WT, Kamboj A, Anderson HD, Anderson CM (2010) Protection of cortical neurons from excitotoxicity by conjugated linoleic acid. J Neurochem 115(1):123–130. https://doi.org/10.1111/j.1471-4159.2010.06908.x

Ishimoto T, Nakamichi N, Hosotani H, Masuo Y, Sugiura T, Kato Y (2014) Organic cation transporter mediated ergothioneine uptake in mouse neural progenitor cells suppresses proliferation and promotes differentiation into neurons. PLoS One 9(2):e89434. https://doi.org/10.1371/journal.pone.0089434

Israelyan N, Margolis KG (2018) Serotonin as a link between the gut-brain-microbiome axis in autism spectrum disorders. Pharmacol Res 132:1–6. https://doi.org/10.1016/j.phrs.2018.03.020
Jaiswal P, Mohanakumar KP, Rajamma U (2015) Serotonin mediated immunoregulation and neural functions: complicity in the aetiology of autism spectrum disorders. Neurosci Biobehav Rev 55:413–431. https://doi.org/10.1016/j.neubiorev.2015.05.013
Jang MJ, Lee YH, Ju YC, Kim SM, Koo HM (2013) Effect of color of light emitting diode on development of fruit body in Hypsizygus marmoreus. Mycobiology 41:63–66. https://doi.org/10.5941/MYCO.2013.41.1.63
Jo WS, Hossain MA, Park SC (2014) Toxicological profiles of poisonous, edible, and medicinal mushrooms. Mycobiology 42:215–220. https://doi.org/10.5941/MYCO.2014.42.3.215
Karami A, Eyjolfsdottir H, Vijayaraghavan S, Lind G, Almqvist P, Kadir A, Linderoth B, Andreasen N, Blennow K, Wall A, Westman E, Ferreira D, Kristoffersen Wiberg M, Wahlund LO, Seiger Å, Nordberg A, Wahlberg L, Darreh-Shori T, Eriksdotter M (2015) Changes in CSF cholinergic biomarkers in response to cell therapy with NGF in patients with Alzheimer's disease. Alzh Dement 11(11):1316–1328. https://doi.org/10.1016/j.jalz.2014.11.008
Kawagishi H, Ando M, Sakamoto H, Yoshida S, Ojima F, Ukai N, Furukawa S (1991) Hericenones C, D and E, stimulators of nerve growth factor (NGF)-synthesis, from the mushroom Hericium erinaceum. Tetrahedron Lett 32(35):4561–4564. https://doi.org/10.1016/0040-4039(91)80039-9
Kawagishi H, Ishiyama D, Mori H, Sakamoto H, Ishiguro Y, Furukawa S, Li J (1997) Dictyophorines A and B, two stimulators of NGF-synthesis from the mushroom Dictyophora indusiata. Phytochemistry 45(6):1203–1205. https://doi.org/10.1016/S0031-9422(97)00144-1
Kim SP, Lee SJ, Nam SH, Friedman M (2017) Turmeric bioprocessed with mycelia from the shiitake culinary-medicinal mushroom Lentinus edodes (Agaricomycetes) protects mice against salmonellosis. Int J Med Mushrooms 19(4):363–376. https://doi.org/10.1615/IntJMedMushrooms.v19.i4.70
Kolotushkina EV, Moldavan MG, Voronin KY, Skibo GG (2003) The influence of Hericium erinaceus extract on myelination process in vitro. Fiziol Zh 49(1):38–45
Kuo H-C, Lu C-C, Shen C-H, Tung S-Y, Hsieh MC, Lee K-C, Lee L-Y, Chen C-C, Teng C-C, Huang W-S, Chen T-C, Lee K-F (2016) Hericium erinaceus mycelium and its isolated erinacine A protection from MPTP-induced neurotoxicity through the ER stress, triggering an apoptosis cascade. J Transl Med 14:78. https://doi.org/10.1186/s12967-016-0831-y
Lamhonwah AM, Hawkins CE, Tam C, Wong J, Mai L, Tein I (2008) Expression patterns of the organic cation/carnitine transporter family in adult murine brain. Brain Dev 30(1):31–42
Lee IK, Yun BS, Kim YH, Yoo ID (2002a) Two neuroprotective compounds from mushroom Daldinia concentrica. J Microbiol Biotechnol 12:692–694
Lee I-K, Yun BS, Han G, Cho DH, Kim YH, Yoo ID (2002b) Dictyoquinazols A, B, and C, new neuroprotective compounds from the mushroom *Dictyophora indusiata*. J Nat Prod 65:1769–1772. https://doi.org/10.1021/np020163w
Lee IS, Ahn BR, Choi JS, Hattori M, Min BS, Bae KH (2011) Selective cholinesterase inhibition by lanostane triterpenes from fruiting bodies of Ganoderma lucidum. Bioorg Med Chem Lett 21(21):6603–6607. https://doi.org/10.1016/j.bmcl.2011.04.042
Lee SC, Im NK, Jeong HY, Choi EH, Jeon SM, Jeong GS (2014) Neuroprotective effects of ethanol extract of Ganoderma lucidum L. on murine hippocampal cells. Kor J Pharmacogn 45(2):161–167
Li L, Ng TB, Song M, Yuan F, Liu ZK, Wang CL, Jiang Y, Fu M, Liu F (2007) A polysaccharide–peptide complex from abalone mushroom (Pleurotus abalonus) fruiting bodies increases activities and gene expression of antioxidant enzymes and reduces lipid peroxidation in senescence-accelerated mice. Appl Microbiol Biotechnol 75:863–869. https://doi.org/10.1007/s00253-007-0865-4
Li IC, Chen YL, Chen WP, Lee LY, Tsai YT, Chen CC, Chen CS (2014) A genotoxicity profile of erinacine A enriched Hericium erinaceus mycelium. Toxicol Rep 1:1195–1201. https://doi.org/10.1016/j.toxrep.2014.11.009

Li Z, Chen X, Lu W, Zhang S, Guan X, Li Z, Wang D (2017) Anti-oxidative stress activity is essential for Amanita caesarea mediated neuroprotection on glutamate-induced apoptotic HT22 cells and an Alzheimer's disease mouse model. Int J Mol Sci 18:1623. https://doi.org/10.3390/ijms18081623

Lung MY, Chang YC (2011) Antioxidant properties of the edible basidiomycete Armillaria mellea in submerged cultures. Int J Mol Sci 12:6367–6384. https://doi.org/10.3390/ijms12106367

Madathil SK, Karuppagounder SS, Mohanakumar KP (2013) Sodium salicylate protects against rotenone-induced parkinsonism in rats. Synapse 67(8):502–514. https://doi.org/10.1002/syn.21658

Marciniak B, Ferenc T, Kusowska J, Ciećwierz J, Kowalczyk E (2010) Poisoning with selected mushrooms with neurotropic and hallucinogenic effect. Med Pr 61:583–595

Meuninck J (2015) Basic illustrated edible medicinal mushrooms, 1st edn. Falcon Guides, Morris Books Publishing LLC, Globe Pequot Press, Gilford, pp 85

Mori K, Obara Y, Hirota M, Azumi Y, Kinugasa S, Inatomi S, Nakahata N (2008) Nerve growth factor-inducing activity of Hericium erinaceus in 1321N1 human astrocytoma cells. Biol Pharm Bull 31(9):1727–1732

Mori K, Inatomi S, Ouchi K, Azumi Y, Tuchida T (2009) Improving effects of the mushroom Yamabushitake (Hericium erinaceus) on mild cognitive impairment: a double-blind placebo-controlled clinical trial. Phytother Res 23(3):367–372. https://doi.org/10.1002/ptr.2634

Mori K, Obara Y, Moriya T, Inatomi S, Nakahata N (2011) Effects of Hericium erinaceus on amyloid β(25–35) peptide-induced learning and memory deficits in mice. Biomed Res 32(1):67–72

Mount P, Harris G, Sinclair R, Finlay M, Becker GJ (2002) Acute renal failure following ingestion of wild mushrooms. Intern Med J 32:187–190

Nagai K, Chiba A, Nishino T, Kubota T, Kawagishi H (2006) Dilinoleoyl- phosphatidylethanolamine from Hericium erinaceum protects against ER stress- dependent Neuro2a cell death via protein kinase C pathway. J Nutr Biochem 17(8):525–530. https://doi.org/10.1016/j.jnutbio.2005.09.007

Nagano M, Shimizu K, Kondo R, Hayashi C, Sato D, Kitagawa K, Ohnuki K (2010) Reduction of depression and anxiety by 4 weeks Hericium erinaceus intake. Biomed Res 31(4):231–237

Nakamichi N, Taguchi T, Hosotani H, Wakayama T, Shimizu T, Sugiura T, Iseki S, Kato Y (2012) Functional expression of carnitine/organic cation transporter OCTN1 in mouse brain neurons: possible involvement in neuronal differentiation. Neurochem Int 61(7):1121–1132. https://doi.org/10.1016/j.neuint.2012.08.004

Nakamichi N, Nakayama K, Ishimoto T, Masuo Y, Wakayama T, Sekiguchi H, Sutoh K, Usumi K, Iseki S, Kato Y (2016) Food-derived hydrophilic antioxidant ergothioneine is distributed to the brain and exerts antidepressant effect in mice. Brain Behav 6(6):e00477. https://doi.org/10.1002/brb3.477

Naskar A, Manivasagam T, Chakraborty J, Singh R, Thomas B, Dhanasekaran M, Mohanakumar KP (2013) Melatonin synergizes with low doses of L-DOPA to improve dendritic spine density in the mouse striatum in experimental Parkinsonism. J Pineal Res 55(3):304–312. https://doi.org/10.1111/jpi.12076

Ng TB, Wang HX (2005) Pharmacological actions of Cordyceps, a prized folk medicine. J Pharm Pharmacol 57(12):1509–1519. https://doi.org/10.1211/jpp.57.12.0001

Nieminen P, Kärjä V, Mustonen AM (2008) Indications of hepatic and cardiac toxicity caused by subchronic Tricholoma flavovirens consumption. Food Chem Toxicol 46(2):781–786

Nukata M, Hashimoto T, Yamamoto I, Iwasaki N, Tanaka M, Asakawa Y (2002) Neogrifolin derivatives possessing anti-oxidative activity from the mushroom Albatrellus ovinus. Phytochemistry 59:731–737

O'Banion D, Armstrong B, Cummings RA, Stange J (1978) Disruptive behavior: a dietary approach. J Autism Child Schizophr 8(3):325–337

Oyadomari S, Mori M (2004) Roles of CHOP/GADD153 in endoplasmic reticulum stress. Cell Death Differ 11(4):381–389. https://doi.org/10.1038/sj.cdd.4401373

Paidi RK, Nthenge-Ngumbau DN, Singh R, Kankanala T, Mehta H, Mohanakumar KP (2015) Mitochondrial deficits accompany cognitive decline following single bilateral intracerebroventricular streptozotocin. Curr Alzheimer Res 12(8):785–795

Palacios I, Lozano M, Moro C, D'Arrigo M, Rostagno MA, Martínez JA, García-Lafuente A, Guillamón E, Villares A (2011) Antioxidant properties of phenolic compounds occurring in edible mushrooms. Food Chem 128:674–678. https://doi.org/10.1016/J.FOODCHEM.2011.03.085

Park IH, Chung SK, Lee KB, Yoo YC, Kim SK, Kim GS, Song KS (2004) An antioxidant hispidin from the mycelial cultures of Phellinus linteus. Arch Pharm Res 27:615–618

Peng J, Wang P, Ge H, Qu X, Jin X (2015) Effects of cordycepin on the microglia-overactivation-induced impairments of growth and development of hippocampal cultured neurons. Plos One 10(5):e0125902. https://doi.org/10.1371/journal.pone.0125902

Phan C-W, Wong W, David P, Naidu M, Sabaratnam V (2012) *Pleurotus giganteus* (Berk.) Karunarathna & K.D. Hyde: nutritional value and in vitro neurite outgrowth activity in rat pheochromocytoma cells. BMC Complement Altern Med 12:102. https://doi.org/10.1186/1472-6882-12-102

Phan CW, David P, Naidu M, Wong KH, Sabaratnam V (2013) Neurite outgrowth stimulatory effects of culinary-medicinal mushrooms and their toxicity assessment using differentiating Neuro-2a and embryonic fibroblast BALB/3T3. BMC Complement Altern Med 13:261. https://doi.org/10.1186/1472-6882-13-261

Phan CW, David P, Naidu M, Wong KH, Sabaratnam V (2015) Therapeutic potential of culinary-medicinal mushrooms for the management of neurodegenerative diseases: diversity, metabolite, and mechanism. Crit Rev Biotechnol 35(3):355–368. https://doi.org/10.3109/07388551.2014.887649

Phan CW, David P, Sabaratnam V (2017) Edible and medicinal mushrooms: emerging brain food for the mitigation of neurodegenerative diseases. J Med Food 20(1):1–10. https://doi.org/10.1089/jmf.2016.3740

Qi J, Ojika M, Sakagami Y (2000) Termitomycesphins A–D, novel neuritogenic cerebrosides from the edible chinese mushroom Termitomyces albuminosus. Tetrahedron 56:5835–5841. https://doi.org/10.1016/S0040-4020(00)00548-2

Qu Y, Sun K, Gao L, Sakagami Y, Kawagishi H, Ojika M, Qi J (2012) Termitomycesphins G and H, additional cerebrosides from the edible Chinese mushroom *Termitomyces albuminosus*. Biosci Biotechnol Biochem 76:791–793. https://doi.org/10.1271/bbb.110918

Rathee S, Rathee D, Rathee D, Kumar V, Rathee P (2012) Mushrooms as therapeutic agents. Rev Bras Farmacogn 22:459–474. https://doi.org/10.1590/S0102-695X2011005000195

Sabaratnam V, Kah-Hui W, Naidu M, Rosie David P (2013) Neuronal health - can culinary and medicinal mushrooms help? J Tradit Complement Med 3:62–68. https://doi.org/10.4103/2225-4110.106549

Salit RB, Shea YR, Gea-Banacloche J, Fahle GA, Abu-Asab M, Sugui JA, Carpenter AE, Quezado MM, Bishop MR, Kwon-Chung KJ (2010) Death by edible mushroom: first report of Volvariella volvacea as an etiologic agent of invasive disease in a patient following double umbilical cord blood transplantation. J Clin Microbiol 48:4329–4332. https://doi.org/10.1128/JCM.01222-10

Samberkar S, Gandhi S, Naidu M, Wong KH, Raman J, Sabaratnam V (2015) Lion's Mane, Hericium erinaceus and Tiger Milk, Lignosus rhinocerotis (Higher Basidiomycetes) medicinal mushrooms stimulate neurite outgrowth in dissociated cells of brain, spinal cord, and retina: an in vitro study. Int J Med Mushrooms 17(11):1047–1054

Scarmeas N, Stern Y, Tang M-X, Mayeux R, Luchsinger JA (2006) Mediterranean diet and risk for Alzheimer's disease. Ann Neurol 59(6):912–921. https://doi.org/10.1002/ana.20854

Schlumberger S, Kristan KČ, Ota K, Frangež R, Molgó J, Sepčić K, Benoit E, Maček P (2014) Permeability characteristics of cell-membrane pores induced by ostreolysin A/pleurotolysin B, binary pore-forming proteins from the oyster mushroom. FEBS Lett 588(1):35–40. https://doi.org/10.1016/j.febslet.2013.10.038

Sengupta T, Vinayagam J, Singh R, Jaisankar P, Mohanakumar KP (2016) Plant-derived natural products for parkinson's disease therapy. Adv Neurobiol 12:415–496. https://doi.org/10.1007/978-3-319-28383-8_23

Singh A, Verma P, Balaji G, Samantaray S, Mohanakumar KP (2016) Nimodipine, an L-type calcium channel blocker attenuates mitochondrial dysfunctions to protect against 1-methyl-4-phenyl-1,2,3,6-tetrahydropyridine-induced Parkinsonism in mice. Neurochem Int 99:221–232. https://doi.org/10.1016/j.neuint.2016.07.003

Singh A, Verma P, Raju A, Mohanakumar KP (2018) Nimodipine attenuates the parkinsonian neurotoxin, MPTP-induced changes in the calcium binding proteins, calpain and calbindin. J Chem Neuroanat pii: S0891-0618(17)30148-5. https://doi.org/10.1016/j.jchemneu.2018.02.001

Song J, Wang Y, Teng M, Cai G, Xu H, Guo H, Liu Y, Wang D, Teng L (2015) Studies on the antifatigue activities of Cordyceps militaris fruit body extract in mouse model. Evid Based Complement Alternat Med 2015:174616. https://doi.org/10.1155/2015/174616

Stenklyft PH, Augenstein WL (1990) Chlorophyllum molybdites–severe mushroom poisoning in a child. J Toxicol Clin Toxicol 28:159–168

Tong H, Xia F, Feng K, Sun G, Gao X, Sun L, Jiang R, Tian D, Sun X (2009) Structural characterization and in vitro antitumor activity of a novel polysaccharide isolated from the fruiting bodies of Pleurotus ostreatus. Bioresour Technol 100:1682–1686. https://doi.org/10.1016/j.biortech.2008.09.004

Tremlett H, Bauer KC, Appel-Cresswell S, Finlay BB, Waubant E (2017) The gut microbiome in human neurological disease: a review. Ann Neurol 81(3):369–382. https://doi.org/10.1002/ana.24901

Varshney J, Ooi JH, Jayarao BM, Albert I, Fisher J, Smith RL, Patterson AD, Cantorna MT (2013) White button mushrooms increase microbial diversity and accelerate the resolution of Citrobacter rodentium infection in mice. J Nutr 143(4):526–532. https://doi.org/10.3945/jn.112.171355

Wang JH, Xu JL, Zhang JC, Liu Y, Sun HJ, Zha X (2015) Physicochemical properties and antioxidant activities of polysaccharide from floral mushroom cultivated in Huangshan Mountain. Carbohydr Polym 131:240–247. https://doi.org/10.1016/J.CARBPOL.2015.05.052

Westfall S, Lomis N, Kahouli I, Dia SY, Singh SP, Prakash S (2017) Microbiome, probiotics and neurodegenerative diseases: deciphering the gut brain axis. Cell Mol Life Sci 74(20):3769–3787. https://doi.org/10.1007/s00018-017-2550-9

Williams RJ, Mohanakumar KP, Beart PM (2015a) Neuro-nutraceuticals: the path to brain health via nourishment is not so distant. Neurochem Int 89:1–6. https://doi.org/10.1016/j.neuint.2015.08.012

Williams RJ, Mohanakumar KP, Beart PM (2015b) Nutraceuticals: molecular and functional insights into how natural products nourish the brain. Neurochem Int 89:1–280

Williams RJ, Mohanakumar KP, Beart PM (2016a) Neuro-nutraceuticals: further insights into their promise for brain health. Neurochem Int 95:1–3. https://doi.org/10.1016/j.neuint.2016.03.016

Williams RJ, Mohanakumar KP, Beart PM (2016b) Special issue: neuro-nutraceuticals: further insights into their promise for brain health. Neurochem Int 95:1–118

Wong KH, Sabaratnam V, Abdullah N, Kuppusamy UR, Naidu M (2009) Effects of cultivation techniques and processing on antimicrobial and antioxidant activities of Hericium erinaceus (Bull.:Fr.) Pers. extracts. Food Technol Biotechnol 47(1):47–55

Wong KH, Naidu M, David P, Abdulla MA, Abdullah N, Kuppusamy UR, Sabaratnam V (2011) Peripheral nerve regeneration following crush injury to rat peroneal nerve by aqueous extract of medicinal mushroom Hericium erinaceus (Bull.: Fr) Pers. (Aphyllophoromycetideae). Evid Based Complement Alternat Med 2011:580752. https://doi.org/10.1093/ecam/neq062

Wu JM, Doonan BB, Hsieh TC, Yang Q, Yang XT, Ling MT (2016) Recent advances and challenges in studies of control of cancer stem cells and the gut microbiome by the Trametes-derived polysaccharopeptide. Int J Med Mushrooms 18(8):651–660. https://doi.org/10.1615/IntJMedMushrooms.v18.i8.10

Xu X, Yang J, Ning Z, Zhang X (2015) Lentinula edodes-derived polysaccharide rejuvenates mice in terms of immune responses and gut microbiota. Food Funct 6(8):2653–2663. https://doi.org/10.1039/c5fo00689a

Xuan M, Okazaki M, Iwata N, Asano S, Kamiuchi S, Matsuzaki H, Sakamoto T, Miyano Y, Iizuka H, Hibino Y (2015) Chronic treatment with a water-soluble extract from the culture medium of Ganoderma lucidum mycelia prevents apoptosis and necroptosis in hypoxia/ischemia-induced injury of type 2 diabetic mouse brain. Evid Based Complement Alternat Med 2015:865986. https://doi.org/10.1155/2015/865986

Yaguchi T, Fujikawa H, Nishizaki T (2010) Linoleic acid derivative DCP-LA protects neurons from oxidative Stress-Induced apoptosis by inhibiting caspase-3/-9 activation. Neurochem Res 35(5):712–717. https://doi.org/10.1007/s11064-010-0124-4

Yang EJ, Song KS (2015) Polyozellin, a key constituent of the edible mushroom Polyozellus multiplex, attenuates glutamate-induced mouse hippocampal neuronal HT22 cell death. Food Funct 6(12):3678–3686

Yang NC, Lin HC, Wu JH, Ou HC, Chai YC, Tseng CY, Liao JW, Song TY (2012) Ergothioneine protects against neuronal injury induced by beta-amyloid in mice. Food Chem Toxicol 50(11):3902–3911. https://doi.org/10.1016/j.fct.2012.08.021

Yeh JY, Hsieh LH, Wu KT, Tsai CF (2011) Antioxidant properties and antioxidant compounds of various extracts from the edible Basidiomycete Grifola frondosa (Maitake). Molecules 16:3197–3211. https://doi.org/10.3390/molecules16043197

Yuswan MHMY, Al-Obaidi JR, Rahayu A, Sahidan S, Shazrul F, Fauzi D (2015) New bioactive molecules with potential antioxidant activity from various extracts of wild edible Gelam mushroom (*Boletus* spp.). Adv Biosci Biotechnol 6:320–329. https://doi.org/10.4236/abb.2015.64031

Zhang R, Xu S, Cai Y, Zhou M, Zuo X, Chan P (2011) Ganoderma lucidum protects dopaminergic neuron degeneration through inhibition of microglial activation. Evid Based Complement Alternat Med 2011:156810. https://doi.org/10.1093/ecam/nep075

Zhang Y, Luo J, Huang NH, Zhang XY (2012a) Influence of Ganoderma lucidum triterpenoids on learning memory function and ACh content in Alzheimer disease model rat. Chin J Exp Tradit Med Formulae 16(16):172–175

Zhang Z, Lv G, Pan H, Pandey A, He W, Fan L (2012b) Antioxidant and hepatoprotective potential of endo-polysaccharides from Hericium erinaceus grown on tofu whey. Int J Biol Macromol 51(5):1140–1146. https://doi.org/10.1016/j.ijbiomac.2012.09.002

Zhang W, Zhang Q, Deng W, Li Y, Xing G, Shi X, Du Y (2014) Neuroprotective effect of pretreatment with Ganoderma lucidum in cerebral ischemia/reperfusion injury in rat hippocampus. Neural Regen Res 9(15):1446–1452. https://doi.org/10.4103/1673-5374.139461

Zhang J, An S, Hu W, Teng M, Wang X, Qu Y, Liu Y, Yuan Y, Wang D (2016) The neuroprotective properties of Hericium erinaceus in glutamate-damaged differentiated PC12 cells and an Alzheimer's disease mouse model. Int J Mol Sci 17(11):pii E1810

Zhou Y, Qu ZQ, Zeng YS, Lin YK, Li Y, Chung P, Wong R, Hägg U (2012) Neuroprotective effect of pre-administration with Ganoderma lucidum spore on rat hippocampus. Exp Toxicol Pathol 64(7–8):673–680. https://doi.org/10.1016/j.etp.2010.12.011

Zhu WW, Liu Z, Xu HW, Chu WZ, Ye QY, Xie AM, Chen L, Li JR (2005) Effect of the oil from Ganoderma lucidum spores on pathological changes in the substantia nigra and behaviors of MPTP-treated mice. Di Yi Jun Yi Da Xue Xue Bao 25(6):667–671

Overview of Therapeutic Efficacy of Mushrooms

3

Sindhu Ramesh, Mohammed Majrashi, Mohammed Almaghrabi, Manoj Govindarajulu, Eddie Fahoury, Maali Fadan, Manal Buabeid, Jack Deruiter, Randall Clark, Vanisree Mulabagal, Dinesh Chandra Agrawal, Timothy Moore, and Muralikrishnan Dhanasekaran

Contents

S. Ramesh · M. Govindarajulu · E. Fahoury · M. Fadan · J. Deruiter · R. Clark · T. Moore
M. Dhanasekaran (✉)
Department of Drug Discovery and Development, Harrison School of Pharmacy, Auburn University, Auburn, AL, USA
e-mail: dhanamu@auburn.edu

M. Majrashi
Department of Drug Discovery and Development, Harrison School of Pharmacy, Auburn University, Auburn, AL, USA

Department of Pharmacology, Faculty of Medicine, University of Jeddah, Jeddah, Kingdom of Saudi Arabia

D. C. Agrawal, M. Dhanasekaran (eds.), *Medicinal Mushrooms*,
https://doi.org/10.1007/978-981-13-6382-5_3

Abstract

Mushrooms have been used globally for various nutritional and medicinal values and now are gaining worldwide recognition due to its various health benefits and potent and unique pharmaceutical properties. Researchers in different parts of the world have demonstrated different species of mushrooms possessing immunomodulatory, antitumor, anticancer, antibacterial, antiviral, anti-inflammatory, anti-atherosclerotic, neuroprotective, antioxidant, and anti-hypoglycemic properties. The chapter presents an overview of the research on the therapeutic efficacy of mushrooms.

Keywords

Ganoderma lucidum · *Hericium erinaceus* · Medicinal mushrooms · Neurodegenerative diseases · Therapeutic value

Abbreviations

AD	Alzheimer's disease
APP	Amyloid precursor protein
BACE1	Beta-secretase 1
BDNF	Brain-derived neurotrophic factor
CA3	*Cornu amonis*
CAT	Catalase
CCL4	Carbon tetrachloride
CNS	Central nervous system
COMT	Catechol-O-methyltransferase
EMM	Ectomycorrhizal mushrooms
GDNF	Glia-derived neurotrophic factor
GIT	Gastrointestinal tract

M. Almaghrabi
Department of Pharmaceutical Chemistry, College of Pharmacy, Taibah University, Almadinah Almunawwarah, Kingdom of Saudi Arabia

M. Buabeid
Department of Drug Discovery and Development, Harrison School of Pharmacy, Auburn University, Auburn, AL, USA

College of Pharmacy and Health Science, Ajman University, Ajman, UAE

V. Mulabagal
Department of Civil Engineering, Auburn University, Auburn, AL, USA

D. C. Agrawal (✉)
Department of Applied Chemistry, Chaoyang University of Technology, Taichung, Taiwan
e-mail: dcagrawal@gm.cyut.edu.tw

GSH	Glutathione
HCC	Hepatocellular carcinoma
HDL	High-density lipoprotein
HFD	High-fat diet
HIV	Human immunodeficiency virus
HMG-CoA	Hydroxymethylglutaryl-CoA
IFN-γ	Interferon-gamma
iNOS	Inducible nitric oxide synthase
JNK	c-Jun N-terminal kinases
LDL	Low-density lipoprotein
LPS	Lipopolysaccharide
MAPK	Mitogen-activated protein kinase
MPTP	1-Methyl-4-phenyl-1,2,3,6-tetrahydropyridine
MRSA	Methicillin-resistant *Staphylococcus aureus*
NDC	Non-digestible carbohydrates
NF-κB	Nuclear factor-*κB*
NGF	Nerve growth factor
NK	Natural killer cells
NO	Nitric oxide
NPC	Neural progenitor cells
PI3K	Phosphatidylinositol-4,5-bisphosphate 3-kinase
PKC	Protein kinase C
RFA	Radio frequency ablation
RNA	Ribonucleic acid
ROS	Reactive oxygen species
SOD	Superoxide dismutase
TC	Total cholesterol
TD1	Type 1 diabetes mellitus
TD2	Type 2 diabetes mellitus
Th1/Th2	T-helper cells
TNF-α	Tumor necrosis factor-alpha
UDP	Uridine diphosphate
VEGF	Vascular endothelial growth factor

3.1 Introduction

One of the great concerns of the current era is the ever-increasing population which is causing decreased healthy living and increased risk for chronic disease leading to diminished quality of life. Also, urbanization and climate forces and change have resulted in a significant loss of agricultural land. Consequently, there is a substantial reduction in the variety of crops cultivated and livestock produced. The new tendency of health-conscious consumers has led to the use of functional foods and complementary or alternative medicine. One of the alternates which humans looked for, right from the historical period, is the consumption of "edible mushrooms." Mushrooms are foods with high nutritional value, widely used in food and by the pharmaceutical industry. In addition to its nutritional and therapeutic advantages, ingestion of toxic mushrooms accidentally due to misidentification of species has been a great concern throughout the civilizations. The objective of this article is to elaborate the most common mushrooms possessing nutritional and medicinal properties.

Egyptians (4600 years ago) assumed mushrooms to be a plant of immortality, conferring to the hieroglyphics. Interestingly, the pharaohs of Egypt dictated that the mushrooms were food to be reserved for royalty and that no commoner could ever touch them. Mushroom rituals were adapted in many places including Russia, China, Greece, Mexico, and Latin America. Historically, mushrooms have been regarded to produce superhuman strength, help in finding lost objects, and lead the soul to the realm of God. In Europe, the first cultivated fungi, the mushroom, was introduced in the seventeenth century. Mushrooms were introduced into the Netherlands for the first time at the beginning of the nineteenth century, but it was not until after the 1900s that they were cultivated on a large scale in the marl mines in Limburg. The mushroom was discovered near Paris by sprinkling the waste from melon crops with leachate from ripe mushrooms. With this discovery, France became the leader in the formal cultivation of mushrooms. Later, the gardeners of England found mushrooms a very easy crop to grow requiring little labor, investment, and space. Hence, mushroom cultivation began gaining popularity in England with more experimentation and increasing publicity in journals and magazines. In the late nineteenth century, the mushroom production made its way across the Atlantic to the United States. Even with the blossoming industrial revolution, farmers still endured the hardships of poor spawn quality. This setback led the US Department of Agriculture to start its manufacturing process of producing spawn. The first producer of pure culture virgin spawn was the American Spawn Company of St. Paul, Minnesota, headed by Louis F. Lambert, a French mycologist. By 1914, marketing and other business tactics began to play a larger role in the mushroom industry. After 1930, many changes led to the industry becoming more productive. Such improvements were better spawn production, the development of synthetic manure, and improvements made to mushroom grow houses. Organizations such as the "Mushroom Growers' Cooperative Association" and the "Farm Credit Administration" were developed to help guide growers from an agricultural and economic standpoint. Pennsylvania State University helped substantially as well, becoming a major contributor to the growth of the US mushroom industry by helping farmers maximize their output of mushrooms. The founding of the "American Mushroom Institute" (AMI) by Chester County, Pennsylvania, growers coordinated the actions of independent growers and acted on behalf of the mushroom industry. The AMI had multiple goals, which included exposing the public to cultivated mushrooms through research, advertising, publicity, merchandising, consumer education, and government relations. Also, they assisted in developing better technical skills, such as mushroom growing and handling protocols.

Every form of media was used to promote mushrooms. This exposed the public to various ways to eat and prepare them. Mushroom merchandising posters were everywhere. The AMI promoted recipes that used mushrooms in casseroles, appetizers, salads, and other combinations. These were the first organized marketing efforts of the American Mushroom Institute. In 1990, the "Mushroom Promotion, Research and Consumer Information Act" was passed by the Congress to strengthen the mushroom industry. It assisted in numerous ways with regard to the marketplace, for instance, it helped its position, expansion, and development. In 1993, the "Mushroom Council" was established to perform the duties necessary for this Act. The Council started with a meager treasury but possessed a dedicated mindset to the sale and use of mushrooms. They used research as a method to define the typical mushroom user, and this turned into the bedrock of what they used for their marketing efforts. Once this bedrock was in place, a well-put-together promotion program began to take off. Mushroom recipes could be found anywhere, which exposed consumers to a way to eat them. In 1996, the Mushroom Council broke through and made the pages of multiple national magazines including *Family Circle*, *Woman's Day*, and *Good Housekeeping*. Today, the Mushroom Council still plays many important roles on the national level, like the promotion of fresh mushrooms through consumer public relations and foodservice and retail service communications. Recently, they have become creative with their methods; one example is how they work with professional chefs in developing and promoting new recipes. Behind the scenes, they work with produce department managers to maintain the highest-quality product for customers and send out thousands of brochures to people who might be interested in new mushroom ideas. Mushrooms now have their month to be honored and eaten; September is "National Mushroom Month."

The word mushroom is derived from the French word "mousseron" for fungi and molds. Mushrooms can be unicellular or multicellular organisms that do not contain cellulose and chlorophyll. Due to the absence of chlorophyll and cellulose, mushrooms cannot synthesize their nutrients. Thus, they live as a parasite on other plants or animals to absorb nutrition from them. Mushrooms are a heterogeneous group of macrofungi belonging to the *Basidiomycetes* and *Ascomycetes* divisions of the subkingdom Dikarya inside the kingdom Fungi. The cell cycle of mushroom divisions consists of sexual spore formation in a distinct structure called the basidium (for Basidiomycetes) or the ascus (for Ascomycetes). Mushrooms grow either above the soil surface (epigeous macrofungi), giving mainly umbrellalike structures which include basidiospores, or at depths of 10–20 cm below the soil surface (hypogeous macrofungi or truffles). The latter of these belong mainly to *Ascomycetes* and usually grow in a symbiotic relationship with a host plant as ectomycorrhizal mushrooms (EMM). These mushrooms have substantial nutritional value because they are rich sources of carbohydrates, proteins, amino acids, vitamins, essential minerals, and trace elements (Colak et al. 2009; Kalač 2013).

Additionally, they contain lectins, polysaccharides, phenolics, polyphenolics, terpenoids, ergosterols, and volatile organic compounds which have medicinal value. Mushrooms have been used in traditional medicine for thousands of years, and at least 270 species of mushrooms are considered to possess therapeutic properties (Ying et al. 1989; Powell 2010). Hence, the term "medicinal mushrooms" is now gaining worldwide recognition. Many edible mushrooms used in traditional folk medicine, including *Lentinus edodes*, *Grifola frondosa*, *Hericium erinaceus*, *Flammulina velutipes*, *Pleurotus ostreatus*, and *Tremella mesenterica*, are also a source of relatively pure bioactive compounds for medical usage, while other, nonedible, species, such as *Ganoderma lucidum*, *Schizophyllum commune*, and *Trametes versicolor*, are used only for their medicinal properties. Various researchers have identified many of the bioactive compounds present in the medicinal mushrooms and have provided evidence of their efficacy in a range of diseases (Wasser and Weis 1999; Ooi and Liu 2000). Recent research in medical mushrooms has demonstrated potent and unique properties of compounds extracted from a wide range of mushrooms species. Modern clinical practice in Japan, China, Korea, Russia, and several other countries relies on mushroom-derived preparations (Mizuno 1999; Wasser and Weis 1999; Chang 2007; Wasser 2010).

Edible mushrooms are also usually referred to as macrofungi. These mushrooms are a rich source of dietary fibers that have various beneficial health effects for humans. Mushroom cell walls contain a mixture of fibrillar and matrix components which include chitin (a straight chain (1→4)-β-linked polymer of N-acetylglucosamine) and the polysaccharides such as (1→3)-β-D-glucans and mannans, respectively. These mushroom cell wall components are nondigestible carbohydrates (NDCs) that are resistant to human enzymes and can be considered a source of rich dietary fiber. Carbohydrate is a major component in mushrooms with a total content ranging from 35% to 70% of dry weight with variations in different species. Most of the carbohydrates in mushrooms are NDCs including oligosaccharides such as trehalose and cell wall polysaccharides such as chitin, β-glucans, and mannans. The level of chitin found in different mushrooms usually represents only a small percentage of the total dry matter, while the content of β-glucans can be very high. While there is a large variation in the dietary fiber content of mushrooms between different species, consumption of edible mushrooms as part of a daily diet typically can provide up to 25% of the recommended dietary intake of fiber. Furthermore, various extracts of mushrooms have been used medicinally as an immunomodulator, antitumor, anticancer, antibacterial, antiviral, anti-inflammatory, anti-atherosclerotic, antioxidant, and anti-hypoglycemic therapeutic agents. Therefore, in the present chapter, we explore the compelling medicinal values of the mushrooms.

3.2 Overview of the Therapeutic Efficacy of Mushrooms

Mushrooms mainly contain fiber, carbohydrate, lipids, proteins, vitamins, and minerals (please see Tables 3.1 and 3.2) (Kurtzman 1997). These fungi have significantly low content of cholesterol, sodium, and gluten but contain valuable, beneficial compounds that can contribute to the therapeutic properties. Hence, in this section, we discuss various types of mushrooms and its beneficial role in various pathological conditions.

Table 3.1 Various species of mushrooms and its composition (g/100g)

Mushroom species	Moisture (%)	Protein	Lipid	Carbohydrate	Fiber
Agrocybe aegerita	91.4 ± 1.1	27.6 ± 2.0	5.4 ± 0.8	28.7 ± 1.8	26.7 ± 2.3
Auricularia polytricha	85.0 ± 2.7	18.3 ± 1.9	2.0 ± 0.6	18.9 ± 2.2	50.0 ± 3.4
Agaricus bisporus	89.4 ± 1.2	37.6 ± 2.1	10.0 ± 0.5	23.6 ± 2.4	21.2 ± 1.4
Coprinus comatus	90.5 ± 0.9	24.2 ± 1.4	7.3 ± 0.4	32.1 ± 2.8	27.9 ± 1.8
Caribena versicolor	58.2 ± 4.5	26.0 ± 3.1	3.7 ± 0.4	9.2 ± 1.4	50.1 ± 2.1
Ganoderma lucidum	70.2 ± 2.2	26.4 ± 2.9	2.7 ± 0.4	12.8 ± 1.1	51.5 ± 4.5
Lentinula edodes	78.5 ± 2.1	27.0 ± 1.6	7.2 ± 1.1	21.7 ± 2.1	38.3 ± 1.4
Pleurotus eryngii	85.0 ± 3.4	23.5 ± 2.2	11.9 ± 0.8	29.9 ± 1.9	25.8 ± 2.1
Volvariella volvacea	89.1 ± 0.8	20.0 ± 1.8	10.8 ± 1.3	42.0 ± 1.0	15.2 ± 4.7

Table 3.2 Minerals in mushrooms (mg/100g)

Mushroom species	Iron (Fe)	Calcium (Ca)	Zinc (Zn)	Magnesium (Mg)	Phosphorus (P)
Agrocybe aegerita	27.3	16.5	26.1	22.6	1432
Auricularia polytricha	29.3	55.0	13.2	26.0	1281
Agaricus bisporus	18.5	44.0	24.6	31.4	1289
Coprinus comatus	69.0	16.0	23.1	16.0	983
Caribena versicolor	15.4	275.3	11.1	14.0	1007
Ganoderma lucidum	53.5	27.5	14.7	16.8	685
Lentinula edodes	18.0	22.0	23.4	17.0	875
Pleurotus eryngii	37.1	17.5	23.0	17.7	1602
Volvariella volvacea	49.2	23.0	26.4	15.0	1740

3.2.1 Beneficial Role of Mushrooms or Mushroom-Derived Substances in Neurological Diseases

Alzheimer's disease (AD) is a progressive neurodegenerative disease characterized by progressive memory deficit, by the decline in activities of daily living, and, in late stages, by psychiatric symptoms which result from neurotransmission dysfunction, synaptic deficits, and loss of functional neurons (Hardy and Selkoe 2002). The pathological hallmarks of Alzheimer's disease are characterized by β-amyloid plaques and tau hyperphosphorylation (Claeysen et al. 2012). Also, neuroinflammation, cholinergic dysfunction, and free radical generation play a significant role in the neurodegeneration (Martorana et al. 2012). Currently available drugs are not able to provide therapeutic benefits for Alzheimer's disease, and hence there has been a recent upsurge of interest in complementary and alternative medicine in delaying the onset of age-associated neurodegenerative diseases. Mushrooms offer great potential as a poly-pharmaceutic drug because of the complexity of their chemical contents and different varieties of bioactivities. Aβ plaque deposition causes impaired memory and cholinergic dysfunction; hence inhibiting the production of Aβ or preventing the aggregation of Aβ into amyloid plaques is proposed to be of benefit. Also, acetylcholine, a neurotransmitter involved in the regulation of learning and memory functions, decreases dramatically in the neocortex and hippocampus in Alzheimer's disease. Oxidative stress and inflammation in the brain induced by one isoform Aβ 1–40 promote p-tau protein which is implicated in neuronal damage (Bharadwaj et al. 2010). Inhibition of beta-amyloid formation, p-tau, and acetylcholinesterase can improve the symptoms associated with cognitive impairment. *Antrodia camphorata* is a mushroom endemic to Taiwan and is rich in anticin B, antrodioxolanone, antroquinonol, antrocamphins, and other antioxidant substances. Its extracts have been shown to reverse the toxic effects of in vivo Aβ-40 infusion and in vitro Aβ-40 treatment. In addition, p-tau protein levels were significantly decreased by the treatment of *Antrodia camphorata* in rat pheochromocytoma (PC-12) cells (Wang et al. 2012a). *Hericium erinaceus* (lion's mane mushroom) contains a number of polysaccharides as well as the triterpenes hericenone and erinacine which are believed to be responsible for the neurodegenerative effects. *Hericium erinaceus* prevented impairments of spatial short-term and visual recognition memory induced by Aβ25–35 given intracerebroventricularly in mice (Mori et al. 2011). Studies on humans with *Hericium erinaceus* derivatives have shown benefits in dementia patients based on "Revised Hasegawa Dementia Scale" (HDS-R) (Mori et al. 2009). *Ganoderma lucidum* (lingzhi or reishi mushroom) contains triterpenes called the ganoderic acids (Fig. 3.1) as well as other constituents typically found in fungal cells including a beta-glucan polysaccharide, coumarin, mannitol, and alkaloids. *Ganoderma lucidum* is used in Chinese traditional medicine for its presumed immunomodulatory effects. This mushroom also has been shown to significantly attenuate Aβ-induced synaptotoxicity and apoptosis by preserving the synaptic density protein called synaptophysin (Lai et al. 2008). Also, Ganoderma extract treated senescence-accelerated mice (strain SAMP8) showed lower brain amyloid and higher antioxidant levels compared with the control mice (Wang et al. 2004). In addition, *Ganoderma lucidum* extract possesses nerve growth factor (NGF)-like properties for the processing of APP via an enhanced NGF signaling pathway resulting in promoting non-amyloidogenic protein secretion (Pinweha et al. 2008).

Cortinarius fractus is a medicinal mushroom known to possess acetylcholinesterase-inhibiting activity (Geissler et al. 2010). This mushroom

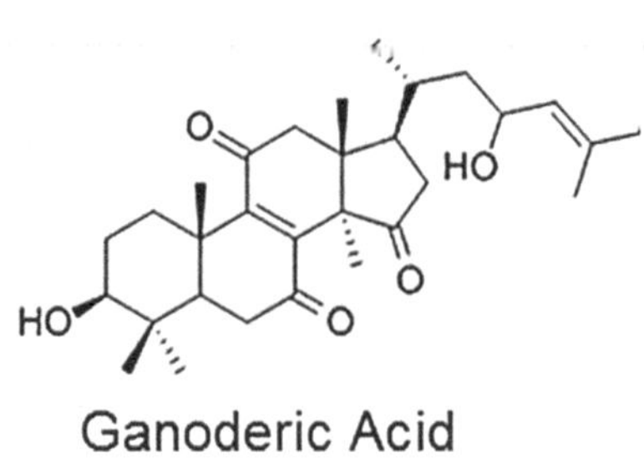

Fig. 3.1 Ganoderic acid in *Ganoderma lucidum*

contains several indole alkaloids including infractine, 10-hydroxyinfractine, and infractopicrine (Fig. 3.2), and these compounds are responsible for the inhibition of acetylcholinesterase (Brondz et al. 2007).

Beta-secretase 1 (BACE1), also known as beta-site amyloid precursor protein cleaving enzyme 1, is an aspartic acid protease important in the formation of myelin sheaths in peripheral nerve cells. Mutations in APP cause BACE1 to generate toxic amyloid. *Auricularia polytricha* (wood ear mushroom) is known to possess BACE1 inhibiting activity (Willem et al. 2006; Bennett et al. 2013).

Neurotrophic factors play a vital role in the differentiation and maintenance of neurons, and its inadequate levels lead to neurodegenerative diseases like Alzheimer's and Parkinson's disease. Few of the neurotrophic factors include nerve growth factor (NGF), brain-derived neurotrophic factor (BDNF), and glia-derived neurotrophic factor (GDNF). Various attempts to treat these conditions with neurotrophic polypeptides were hampered by their limited ability to penetrate the blood–brain barrier. Therefore, finding small molecules that show neurotrophic properties and/or enhancing the action of endogenous neurotrophic factors is important in the current situation (Shi et al. 2011). *Sarcodon* spp., also called "bitter tooth," are widely distributed in Europe, North America, and Asia. One compound called cyrneines isolated from *Sarcodon cyrneus* was shown to exhibit neuritic outgrowth in pheochromocytoma cells without any cytotoxicity (Marcotullio et al. 2006). The extracts of *Hericium erinaceus* also caused neuritic outgrowth in NG108-15 neuronal cells (Wong et al. 2007). Also, in vivo study using Sprague–Dawley rats treated with *Hericium erinaceus* showed recovery of axonotmetic peroneal nerve injury (Wong et al. 2011). In addition, the neuronal function also depends on the activity of signal transduction mechanisms of which mitogen-activated protein kinase (MAPK), phosphatidylinositol-3-kinase-Akt (PI3K-Akt), and protein kinase C (PKC) are crucial. These signal transduction pathways play an important role in regulating cell growth and differentiation (Zhang and Liu 2002). MAP kinase activity was significantly increased and induced neuronal differentiation of rat PC12 cells (Liu et al. 2006). As for the inhibitory mechanism of *Ganoderma lucidum* on Aβ25–35 neurotoxicities, the levels of stress kinases, namely, phosphorylated JNK, phosphorylated c-Jun, and phosphorylated p38, were markedly attenuated (Lai et al. 2008). Further, neurotrophic factor levels in the newborn rat's brain were enhanced after the pregnant rats were fed with *Mycoleptodonoides aitchisonii* for 7 days before delivery (Okuyama et al. 2004). A recent study showed that

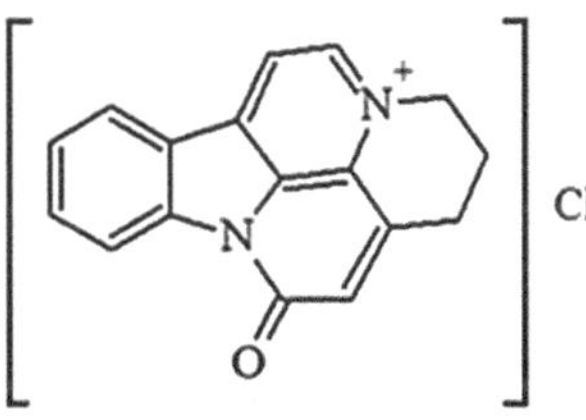

Infractopine

Fig. 3.2 Infractopicrines in *Cortinarius fractus*

Mycoleptodonoides aitchisonii treatment increased dopaminergic and serotoninergic neuronal activities following brain ischemia damage in the cerebral cortex (Okuyama et al. 2012).

Oxidative stress and reactive oxygen species play a vital role in pathogenesis and progression of neurodegenerative diseases (Chen et al. 2012b; Chu et al. 2012). The imbalance between reactive oxygen species generation and antioxidants' enzyme activities leads to lipid peroxidation, nuclear mitochondrial DNA damage resulting in neuronal death, and cognitive deficits (Biasibetti et al. 2013). Mushrooms possess the potent antioxidant property and hence find its application in offering neuroprotection, in various experimental models (Gunawardena et al. 2014). Mice treated with *Ganoderma lucidum* increased superoxide dismutase (SOD), reduced glutathione (GSH), catalase (CAT), and glutathione peroxidase in the blood and liver tissues. Further, the aqueous extracts of *Ganoderma lucidum* prevented peroxide-induced oxidative damage to cellular DNA (Shi et al. 2002). Hence, the ability of the triterpenes of *Ganoderma lucidum* to scavenge free radicals may suppress reactive oxygen damage that leads to Alzheimer's disease pathology. Considerable focus of attention has shifted to promote neurogenesis by mobilizing endogenous neural progenitor cells (NPC). In Alzheimer's disease patients and in vivo models, the proliferation and self-renewal of NPC are dysregulated and result in aberrant neurogenesis (Jin et al. 2004a, b; Rodríguez et al. 2008; Niidome et al. 2008). Growing evidence has shown that pharmaceutical approaches that promote NPC proliferation alleviate Alzheimer's-related cognitive decline. Thus this represents a feasible therapeutic strategy for Alzheimer's disease (Jin et al. 2006; Wang et al. 2010; Fiorentini et al. 2010). These studies indicate that active components of mushrooms known to possess neurotrophic and neuroprotective effects. A novel and effective neuroprotective approach to prevent dopamine neurons from progressive death in the brain (substantia nigra region) is currently being explored. *Ganoderma lucidum* decreased parkinsonism induced by dopaminergic neurotoxin 1-methyl-4-phenyl-1,2,3,6-tetrahydropyridine (MPTP). Furthermore, involuntary movement of mice was also significantly reduced (Zhu et al. 2005). Microglia are the resident innate immune cells of the central nervous system (CNS), and they play a major role in the neuroinflammatory process by stimulating production of pro-inflammatory and cytotoxic factors including tumor necrosis factor-(TNF)-α, nitric oxide (NO), superoxide radicals, interleukin-β (IL-β), and cyclooxygenase 2 (COX2) (Liu 2006). Thus, the over-activation of microglia in the CNS may contribute to neurodegenerative processes (Brown and Neher 2010). *Ganoderma lucidum* offered neuroprotective effect in co-cultures of 1-methyl-4-phenylpyridinium-(MPP+)-treated dopaminergic neuronal cell line MES23.5. It also protected against LPS-activated microglia by significantly inhibiting the production of microglia-derived pro-inflammatory and cytotoxic factors (NO, TNF-α, and IL-1β) (Zhang et al. 2011). Signal transduction cascades like the mitogen-activated protein kinase (MAPK), phosphatidylinositol-3-kinase-Akt (PI3K-Akt), and protein kinase C (PKC) pathways play important roles in neurons downstream of multiple signals including neurotrophins and neurotransmitters (Martin et al. 2012). Ganoderma extract activates MAP kinases and induces the neuronal differentiation of rat pheochromocytoma PC12 cells (Nishina et al. 2006). Another study showed that *Antrodia camphorata* prevented serum deprivation-induced PC12 cell apoptosis through a PKA-dependent pathway and by suppression of JNK and p38 activities (Lu et al. 2008). *Mycoleptodonoides aitchisonii* is a rare mushroom, and its constituent's phenylpentane is a neuroprotective agent which improved dopamine release in the brains of rats fed with the mushroom powder or aqueous extracts (Okuyama et al. 2004). *Ganoderma lucidum*, an edible, medicinal mushroom, has been used to promote health and longevity for centuries (Sanodiya et al. 2009). Recent studies have demonstrated that the water extract of *Ganoderma lucidum* induces neuronal differentiation and neurite outgrowth of PC12 cells (Cheung et al. 2000; Chu et al. 2007; Matsuzaki et al. 2013). *Ganoderma lucidum* polysaccharides (GLP), one of the major active components in this mushroom, also protect neurons from hypoxia/reoxygenation injury in vitro (Zhao et al. 2004).

Epilepsy is defined as a brain disorder characterized by an enduring predisposition to generate epileptic seizures and by the neurobiological, cognitive, psychological, and social consequences of this condition (Fisher et al. 2005). Epilepsy is a medical disorder marked by recurrent, unprovoked seizures. A seizure results when a sudden imbalance occurs between the excitatory and inhibitory forces within the network of cortical neurons in favor of a sudden-onset net excitation. They are paroxysmal manifestations of the electrical properties of the cerebral cortex. Epileptic seizures are only one manifestation of neurologic or metabolic diseases, and seizures have many causes, including a genetic predisposition for certain types of seizures, head trauma, stroke, brain tumors, alcohol or drug withdrawal, repeated episodes of metabolic insults, such as hypoglycemia, and other conditions. The goal of treatment in patients with epileptic seizures is to achieve a seizure-free status without adverse effects that could compromise patient compliance or drug efficacy. Despite adequate treatment, the risk of recurrence in 2 years after a first unprovoked seizure is 15–70%. Complementary and alternative medicine has been proposed to be beneficial in the treatment of epilepsy, and extracts of mushroom have been known to possess anti-epileptic activity. For example, *Ganoderma lucidum* extracts significantly increased the threshold for psychomotor seizures, but no changes in seizure thresholds were observed in the intravenous pentylenetetrazole and maximal electroshock seizure threshold tests after acute treatment (Socala et al. 2015). One other study showed *Ganoderma lucidum* inhibited convulsions in rats from Kainic acid-induced seizures, reduced the degeneration pattern in the CA3 region of rats, decreased astrocytic reactivity, and reduced the expression of IL-1β and TNF-α induced by Kainic acid (Tello et al. 2013). On the other hand, *Amanita phalloides*, also known as "death cap," is one of the most poisonous mushrooms, being involved in most human fatal cases of mushroom poisoning worldwide. This species contains three main groups of toxins: amatoxins, phallotoxins, and virotoxins. From these, amatoxins, especially α-amanitin, are the main responsible for the toxic effects in humans. It is recognized that α-amanitin inhibits RNA polymerase II, causing protein deficit and ultimately cell death, although other mechanisms are thought to be involved. The liver is the main target organ of toxicity, but other organs are also affected, especially the kidneys. Intoxication symptoms usually appear after a latent period and may include gastrointestinal disorders followed by jaundice, seizures, and coma, culminating in death. Neurologic manifestations, either primary due to the accumulation of ammonia or secondary due to multi-organ failure combined with hypotension, may develop in response to abnormal liver and kidney functions (Barceloux 2008). Ammonia, a by-product of protein metabolism, is neurotoxic at high concentrations (Ytrebo 2006). Rising levels of blood ammonia are known to cause encephalopathy, disorientation, confusion, lethargy, somnolence, vertigo, convulsions, and coma (Bonnet and Basson 2002).

3.2.2 Beneficial Role of Mushrooms or Mushroom-Derived Substances in Cardiovascular Diseases

Cardiovascular diseases have been one of the major causes of morbidity and mortality in our present-day society. Most important risk factors for cardiovascular diseases include high blood pressure, hyperglycemia, and hypocholesteremia. People with cardiovascular risk factors are more likely to have a heart attack or stroke than people without them. Several investigations have shown the influence of mushroom intake on some metabolic markers related to cardiovascular disease (total LDL, HDL cholesterol, fasting triacylglycerol, homocysteine, blood pressure, homeostatic function, and oxidative and inflammatory damage). For example, in the pathogenesis of atherosclerosis, reactive oxygen species and high blood cholesterols play significant roles in atherosclerotic plaque formation. Therefore, adequate control of blood cholesterol is important for reducing the risk of the development or progression of atherosclerosis. Various studies suggest that consuming mushrooms could potentially reduce the risk of cardiovascular disease (Guillamón et al. 2010). The first reported research on the cholesterol-lowering effects of mushrooms was conducted in Japan in the 1960s. *Lentinula edodes* was shown to significantly decrease cholesterol levels in rodents fed with a high-fat and high-cholesterol diet (Sun et al. 1984). Interestingly, edible mushrooms (e.g., *Hypsizygus marmoreus*) are an ideal food for the dietetic prevention of atherosclerosis due to their high fiber and low-fat content. Edible mushrooms have been a rich source of dietary fibers which have various beneficial health effects. Dietary fiber extracted from *Pleurotus cornucopiae* showed a marked anti-atherosclerotic effect in vitro and in vivo. Patients with coronary disease showed a decreased atherogenic activity (20–40%) in their sera after the consumption of *Pleurotus cornucopiae* suggesting that this mushroom contains a natural cholesterol-lowering agent (Ryong et al. 1989). In Oriental medicine, mushrooms have been prescribed because they possess natural hypocholesterolemic and antisclerotic properties (Ishikawa et al. 1984). The hypocholesterolemic actions of *Lentinus edodes*, *Auricularia polytricha*, *Flammulina velutipes*, and *Agaricus bisporus* have also been reported (Kaneda and Tokuda 1966). The adenosine derivative lentinacin or lentysine (currently known as eritadenine [2(R), 3(R)-dihydroxy-4-(9-adenyl)-butyric acid]) (Fig. 3.3) was subsequently isolated and identified to be one of the active hypocholesterolemic components in the shiitake mushroom (Kaneda and Tokuda 1966; Lin et al. 1973). In addition to

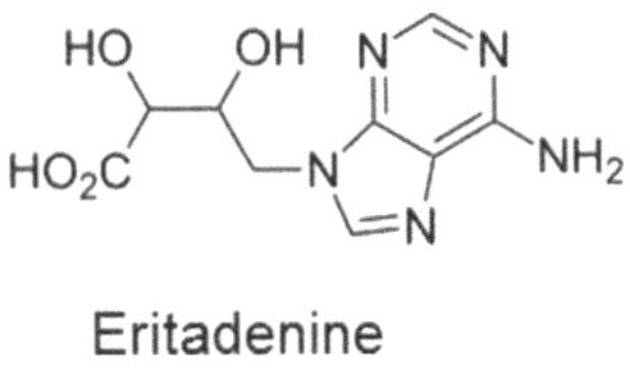

Fig. 3.3 Eritadenine present in *Lentinus edodes*

eritadenine, nucleic acid compounds extracted from *Lentinus edodes* were found to be inhibitors of platelet agglutination (Sugiyama et al. 1995).

A pronounced hypocholesterolemic effect of oyster mushroom (*Pleurotus ostreatus*), combined with inhibition of lipid peroxidation, was shown in rats and rabbits. Oyster mushroom (10% dried fruiting bodies) significantly reduced the incidence and size of atherosclerotic plaques in rabbits (Bobek and Galbavý 1999). Lovastatin, the lead compound for the statins (HMG-CoA reductase inhibitors), has been detected in this species of mushroom (Gunde-Cimerman et al. 1993). Hypocholesteremic effects of *Auricularia auricularia-judae* and *Tremella fuciformis* are also attributed to the HMG-CoA reductase inhibitory effect (Chen 1989; Hobbs 1995; Cheung 1996). Several studies on Lentinula extracts have shown to induce a significant decrease in serum cholesterol in young women and people older than 60 years of age in Japan (Tokita et al. 1972). Some triterpenes from *Ganoderma lucidum* [ganoderic acid C and derivatives] can inhibit the biosynthesis of cholesterol (Komoda et al. 1989). Other triterpenes of this fungus contribute to atherosclerosis protection by inhibition of angiotensin-converting enzyme (Morigiwa et al. 1986) or platelet aggregation (Su et al. 1999). The inhibition of low-density lipoprotein (LDL) oxidation by endothelial cells and of monocyte adhesion to endothelial cells has been demonstrated by mushrooms (Lin 2004). It has been reported that eritadenine from *Lentinus edodes* may elicit its effect by the suppression of the hyperhomocysteinemic effect of guanidinoacetic acid, which leads to the decreased production of homocysteine and increased cystathionine formation (Suzuki and Ohshima 1976). *A. auricula-judae* display anticoagulation, anti-aggregatory activity in the blood platelets of mice and rats, thus serving to lower their total cholesterol, total triacylglyceride, and lipid levels (Chen 1989; Sheng and Chen 1990). *Grifola frondosa* reduced blood pressure in rats without changing the plasma high-density lipoprotein (HDL) level or serum cholesterol level (Mizuno 1995). Various studies have shown that Lentinula mushrooms can lower both the blood pressure and the free cholesterol level in plasma and can accelerate the accumulation of lipids in the liver by removing them from circulation (Kabir and Kimura 1989). It has also been reported that dried *Agrocybe aegerita* can significantly reduce the serum total cholesterol (TC), triacylglyceride, atherogenic index, hepatic TC, and total triacylglyceride levels in rats fed with a semisynthetic high-cholesterol diet compared with the control group (Yeung and Cheung 2002). The hypocholesterolemic effect of *Agrocybe aegerita* has been suggested to be linked with its antioxidant activity (Ng 2005). The steroid ergosta-4-6-8(14),22-tetraen-3-one, isolated from many mushrooms, has been shown to possess anti-aldosteronic diuretic properties (Yuan et al. 2004) which could be of benefit in cardiovascular disease (Fig. 3.4).

Fig. 3.4 Ergosta-4-6-8(14),22-tetraen-3-one anti-aldosterone constituent of *Agrocybe aegerita*

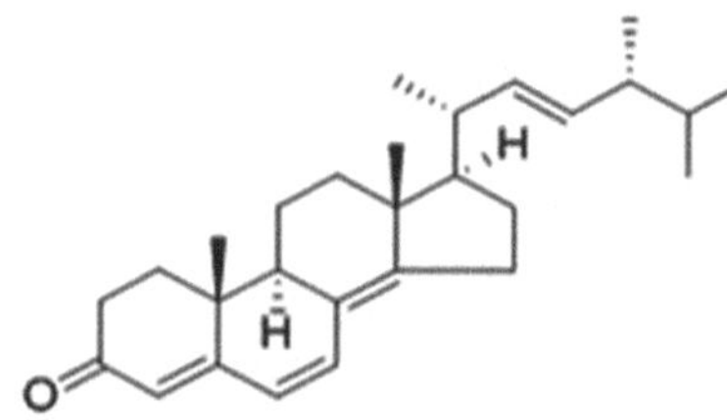

3.2.3 Beneficial Role of Mushrooms or Mushroom-Derived Substances in Cancer

Cancer is a leading cause of death worldwide. Despite vast innovative research, current oncology treatment remains very expensive with new drugs entering the market carrying ever-greater costs. In addition, most anticancer drugs cause several severe side effects which greatly complicate the clinical management of various forms of cancer. Hence, there remains an urgent need for novel effective and less-toxic therapeutic approaches. Various studies have shown therapeutic benefits when using constituents of mushroom extracts alone or with other drugs in cancer therapy. In addition, mushrooms complement chemotherapy and radiation therapy by countering the adverse effects of cancer and decreasing the incidence of developing drug resistance. Recently, several bioactive antitumor molecules have been identified from various mushrooms. Some of the bioactive antitumor constituents include alkaloids, polysaccharides, proteins, fats, glycosides, volatile oils, tocopherols, phenolics, flavonoids, carotenoids, and organic acids. The active components in mushrooms responsible for conferring anticancer potential are lentinan, krestin, hispolon, lectin, calcaelin, illudin, psilocybin, ganoderic acid, schizophyllan, laccase, etc. Polysaccharides especially β-glucan have been the most potent in having antitumor and immunomodulating properties. Numerous studies have demonstrated that mushrooms and its constituents show significant inhibitory activity against breast cancer, hepatocellular carcinoma, uterine cervix cancer, pancreatic cancer, gastric cancer, and acute leukemia. In addition, antitumor compounds have been identified in various mushrooms species (Zhang et al. 2007).

Worldwide, breast cancer is the most frequently diagnosed life-threatening cancer in women and the leading cause of cancer death among women. Surgery and radiation therapy, along with adjuvant hormone or chemotherapy when indicated, are now considered primary treatment for breast cancer. Despite improved treatment modalities, mortality due to breast cancer remains a great concern. There are various studies which have shown the beneficial effects of mushrooms on breast cancer. Moreover, a meta-analysis of various observational studies has shown dietary intake of mushrooms reduces the risk of breast cancer (Li et al. 2014). For example, *Pleurotus eryngii* has demonstrated efficacy in breast cancer. This mushroom contains three main triterpenoids (2,3,6,23-tetrahydroxy-urs-12-en-28-oic acid, 2,3,23-trihydroxy-urs-12-en-28-oic acid, and lupeol) which appear to be the primary anticancer constituents (Xue et al. 2015). This assumption is supported by research showing that 2,3,6,23-tetrahydroxy-urs-12-en-28-oic acid has significant inhibitory activity against breast cancer MCF-7 cell lines in vitro. *Ganoderma lucidum* induced cell apoptosis by exhibiting inhibitory effects on human breast adenocarcinoma MCF-7 cells (Zheng et al. 2009; Martin and Brophy 2010). Hepatocellular carcinoma (HCC) is a primary malignancy of the liver and occurs predominantly in patients with underlying chronic liver disease and cirrhosis. Tumors have been known to progress with local expansion, intrahepatic spread, and distant metastases. Treatments include resection of the tumor which may benefit certain patients, albeit mostly transiently. Many patients are not candidates given the advanced stage of their cancer at diagnosis or their degree of liver disease and, ideally, could be cured by liver transplantation. Globally, only a fraction of all patients have access to transplantation, and, even in the developed world, organ shortage remains a major limiting factor. In these patients, local ablative therapies, including radiofrequency ablation (RFA), chemoembolization, and potentially novel chemotherapeutic agents, may extend life and provide palliation. Constituents of certain mushrooms have shown promise versus liver cancer cells. Polysaccharides from *Trametes robiniophila* inhibited the proliferation of SMMC-7721 cells in vitro. In addition, SMMC-7221-bearing mice given the oral administration of extracts suppressed the growth of HCC tumor and metastatic nodules to the lung. Hence these polysaccharides might be a promising chemopreventive agent for the tumorigenesis and metastasis of HCC (Zou et al. 2015).

Coriolus versicolor (turkey tail) polysaccharides (CVPs) treated in vitro on a human hepatoma cancer (QGY) cell line showed cytotoxic activity by MTT assay. Flow cytometry analysis of cell cycle and cell apoptosis of QGY cells showed significant inhibition of proliferation of human hepatoma cancer and a decrease in the expression of the cell cycle-related genes (p53, Bcl-2, and Fas). Hence, *Coriolus versicolor* may be able to ameliorate the toxic effects in cancer therapy (Patel and Goyal 2012). Also, some experiments have demonstrated that suillin from the mushroom *Suillus placidus* might be an effective agent to treat liver cancer by inducing apoptosis in human hepatoma HepG2 cells (Liu et al. 2009).

Cervical cancer is the third most common malignancy in women worldwide, and it remains a leading cause of cancer-related death for women in developing countries. The treatment of cervical cancer varies with the stage of the disease. For early invasive cancer, surgery is the treatment of choice. In more advanced cases, radiation combined with chemotherapy is the current standard of care. In patients with disseminated disease, chemotherapy or radiation provides symptom palliation. *Cordyceps sinensis* is a traditional medicinal mushroom in China and known to have high levels of selenium. Most of the studies have demonstrated its effects on uterine and cervical cancer. One study compared the tumor incidence in mice treated with *Cordyceps sinensis*. The methylcholanthrene (MCA)-induced group showed 85.7% tumor incidence, and the animals showed 40% tumor incidence ($p < 0.05$) after administration of *Cordyceps sinensis*, suggesting that constituents of this mushroom could be beneficial in the treatment for uterine cervical cancer (Ji et al. 2014). *Ganoderma lucidum* has been used for the treatment of a variety of cancers

(Martin and Brophy 2010). It has displayed a wide range of antitumor activities, including cell cycle arrest, induction of apoptosis, and inhibition of motility, anti-angiogenesis, and anti-mutagenesis (Hu et al. 2002). Therefore, *Ganoderma lucidum* was identified as a potential source of chemopreventive agents for bladder cancer (Yuen and Gohel 2008). *Mushrooms* could also be antitumor agents of acute leukemia and gastric cancer (Sun et al. 2007; Gao et al. 2007). *Inonotus obliquus* caused the death of the cancer cell by causing G0/G1 phase arrest, inducing cell apoptosis (Song et al. 2013). *Ganoderma lucidum* polysaccharides potentiated the production of cytokines from human monocytes–macrophages and T-lymphocytes. *Ganoderma lucidum* exhibited stronger immunomodulatory activities and generation of cytokines (IFN-γ, IL-4, and IL-6) from spleen lymphocytes (Wang et al. 1997; Guggenheim et al. 2014). Most recently, proteoglycan fraction obtained from *Ganoderma lucidum* activated B-cells and improved the immune response of tumor patients as an immune-stimulatory drug (Ye et al. 2011). Apoptosis is a cascade of events regulated by a subfamily containing Bcl-2, Bcl-xL, and Bcl-w which exerts anti-apoptotic activities; a subfamily of Bax, Bad, and Bak exerting pro-apoptotic activity; and a subfamily containing Bik and Bid and exhibiting pro-apoptotic activity (Kinnally and Antonsson 2007). The polysaccharide from *Cordyceps militaris* (WECM) causes apoptosis in human lung carcinoma A549 cells, resulting in an increase in Bax in a dose-dependent manner and a decrease in Bcl-2 (Park et al. 2009). Angiogenesis is a cascade of events which triggers tumor growth, progression, and metastasis (Folkman 2004). *Ganoderma lucidum* possess anti-angiogenesis activity and inhibit the production of NO which is overexpressed in tumor cells. NO acts as an inducing agent of angiogenesis (Song et al. 2004). Moreover, another study showed that *G. lucidum* inhibits the morphogenesis of capillary by preventing the release of angiogenic factors, namely, vascular endothelial growth factor (VEGF) and transforming growth factor (TGF)-β1, constituting a key step in angiogenesis about cancer development. Similarly, polysaccharides from *Phellinus linteus* induce anti-angiogenic activity, revealing a novel inhibitor of angiogenesis, especially for tumor treatment and prevention. This finding indicates that *Phellinus linteus* acts as an immune potentiator and as a direct inhibitor of cancer cell adhesion (Lee and Hong 2010). Adjuvant therapy in cancer involves additional cancer treatment given after the primary treatment to lower the risk of recurrence. Mushroom polysaccharides are considered as adjuvant medicines with conventional chemotherapy/radiotherapy to treat various cancers. Their incorporation into treatment regimens often reduces encountered side effects by patients (Watanabe et al. 2013). Mice were treated with *Ganoderma lucidum* polysaccharides followed by administration of Con A, which would activate spleen lymphocytes to significantly express cytokines, such as IFN-γ, TNF-α, and IL-2. Also, the treatment of GL and Con A was suggested to enhance the function of T-helper cells. Also, another study reported that a GLP ameliorated nausea and vomiting induced by cisplatin (CDDP), as well as improved food intake, in a dose-dependent manner in a rat pica model based on measured kaolin intake (Wang et al. 2005). *Phellinus linteus* antitumor effects are attributed to immunomodulating activity and prevention of metastasis owing to its b(1–3)-linked glycan (Baker et al. 2008). The phenolic ketone

Fig. 3.5 Hispolon

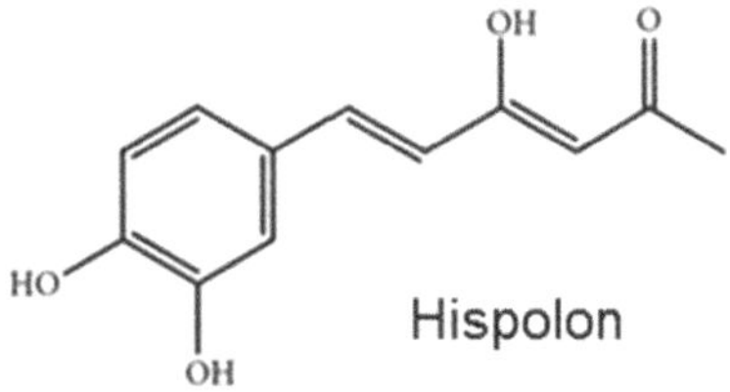

compound hispolon induced apoptosis of breast and bladder cancer cell (Fig. 3.5) (Lu et al. 2009). A protein-bound polysaccharide from this mushroom induces G2/M phase arrest and apoptosis in SW480 human colon cancer cells (Li et al. 2004). It also has anti-inflammatory and anti-angiogenic activities (Kim et al. 2004). These findings indicate the potential for the use of the mushroom extract in stimulated angiogenesis, such as inflammation and tumor development (Lee et al. 2010b).

3.2.4 Beneficial Role of Mushrooms or Mushroom-Derived Substances in Diabetes Mellitus

Mushrooms are considered to have fewer calories and low levels of cholesterol and sodium, thereby promoting good health, especially in disease states such as obesity, diabetes mellitus, dyslipidemia, and hypertension (Kundaković and Kolundžić 2013; Wasser 2014). Diabetes mellitus is a disorder characterized by high blood glucose levels due to deregulation of insulin hormone. Type 1 diabetes (IDDM) is an autoimmune disorder characterized by insufficient or complete lack of insulin production by the pancreatic β-cells. In contrast, type 2 diabetes (NIDDM) is due to insulin resistance leading to failure of glucose utilization by the cells and, over time, leads to a buildup of glucose in the blood. Despite significant research in developing an adequate pharmacological therapy, people with diabetes have an increased risk of developing several serious health problems. Consistently high blood glucose levels can lead to serious diseases affecting the heart and blood vessels, eyes, kidneys, nerves, and brain. Also, people with diabetes have a higher risk of developing infections. Mushroom contains β-glucan polysaccharides which have numerous biological functions that may be of benefit in diabetes (Rahar et al. 2011). For example, α-glycosidase enzymes present in the gastrointestinal tract are known to induce postprandial hyperglycemia by breaking down ingested complex carbohydrates and disaccharides, thereby facilitating the absorption of glucose. A polysaccharide from oyster mushrooms has been shown to inhibit the enzyme α-glucosidase and might benefit in controlling post-meal hyperglycemia in diabetes mellitus (Zhu et al. 2014a, b). GLUT4 transporter plays a key role in regulating whole-body glucose homeostasis by insulin-mediated uptake of glucose into skeletal and adipose tissue. Polysaccharide-rich β-glucans from *Pleurotus sajor-caju* mushrooms prevented the onset of hyperglycemia as well as hyperinsulinemia/insulin resistance in mice that were fed a diet high in fat. The mechanisms include upregulation of insulin-responsive glucose-transporter (GLUT-4) and hormone adiponectin genes and downregulation of nuclear factor κB (NF-κB) and interleukin-6 (IL-6), the key regulators of inflammatory responses. The noted decrease of insulin resistance was similar to that observed with the antidiabetic drug metformin (Vincent et al. 2013; Kanagasabapathy et al. 2014). Diabetes mellitus is associated with a decrease in the anti-inflammatory and metabolic protein adiponectin and an increase in pro-inflammatory cytokines. This imbalance leads to the formation of abnormally high levels of cytokines in adipose tissues. Consequently, the expression of inflammatory cytokines tumor necrosis factor-α (TNF-α) and IL-6 amplifies insulin resistance in regions dense with adipose tissues. This was noticed in high-fat diet (HFD) animals where the expression of NF-κB in the HFD groups treated with polysaccharide and metformin was significantly downregulated; the levels were lower than those of the untreated HFD group (Kanagasabapathy et al. 2012). Diabetic mellitus mice treated with α-glucan from Grifola frondosa mushrooms had decreased body weight and led to low levels of plasma glucose, glycosylated serum protein, serum insulin, triglycerides, cholesterol, free fatty acids, and malondialdehyde levels in the liver. The treatment also increased the content of hepatic glycogen, glutathione, superoxide dismutase, and glutathione peroxidase enzymes. This suggests that beneficial biochemical changes can be due to the effect of the α-glucan on insulin receptors that results in increased insulin sensitivity and a decrease in insulin resistance of peripheral target tissues (Hong et al. 2007). Also, α-glucan of *Grifola frondosa* mushrooms treated on high-fat diet mice showed improved immune function and also decreased the levels of mediators that destroyed β-pancreatic cells (Lei et al. 2012). Ob/Ob (obese) mice are genetically modified mice carrying mutations in gene leptin and hence are profoundly obese and serve as animal models of Type 2 diabetes mellitus. In one study, the mice treated with exopolysaccharides isolated from mycelial cultures of *Phellinus baumii* and *Tremella fuciformis* mushrooms showed a significant reduction in fasting blood glucose levels and an improvement in glucose tolerance. It should also be noted that this was observed without significant weight gain. This can be attributed to the increase of PPAR-γ transcription messenger expression in response to treatment. This seems to be correlated to the regulation of hyperglycemia and dyslipidemia; it is possible this is regulated via lipid metabolism (Cho et al. 2007). Diabetic rats treated for 7 days with a crude exopolysaccharide of the edible *Lentinus strigosus* mushroom were shown to have up to a 21.1% reduction in the serum glucose level, hypoglycemia, and regeneration in the pancreatic islets of Langerhans and amelioration of the destruction of the microvasculature of the islets (Yamac et al. 2008). Altered lipid metabolism and increased levels of oxidative stress are other associated pathologies of diabetes. One study evaluated the potential of a saponin-containing broth extract and polysaccharides from *Tremella aurantialba mycelia* and its ability to alleviate these specific conditions in diabetic rodents. This extract was more effective than the polysaccharides in decreasing blood glucose, cholesterol, phospholipids, and triglyceride serum levels. In addition, the results showed an increase in superoxide dismutase, catalase, glutathione peroxidase, and reductase levels that are closely associated with oxidative stress (Zhang et al. 2009). Furthermore, the destruction of pancreatic β-cells in mice was prevented by treatment with a polysaccharide from *Agrocybe chaxingu* mushrooms. These benefits were supplemented by reduced plasma glucose and nitric oxide (NO) production and inducible nitric oxide synthase (iNOS) expression in the cells, which suggests the important value of the polysaccharide to treat diabetes mellitus (Lee et al. 2010a). Homogeneous water-and alkali-extracted polysaccharides from *Pleurotus tuber-regium* mushrooms were shown to have strong antioxidant properties in a variety of antioxidant chemical and cell assays (oxygen and hydroxyl radical and DPPH scavenging activities; inhibition on liver peroxidation, liver mitochondria swelling, and red blood cell hemolysis) (Wu et al. 2014). An investigation into the study of the mechanism of action of polysaccharides in diabetes found that immune cells could acknowledge β-glucans using a cell surface pathogen recognition receptor called dectin-1, suggesting that the perceived innate immune response that is induced by low-dose β-glucan is regulatory. This can be used to modulate T-cell responses to pancreatic β-cells which induce protection from type 1 diabetes (T1DM) (Karumuthil-Melethil et al. 2014). A retrospective study of 37 human subjects indicates that dietary consumption of *Agaricus bisporus* mushrooms can reduce diabetes risk factors. This suggests that the mushrooms contain compounds that have potential anti-inflammatory and antioxidant health benefits that can occur over time in adults that are predisposed to type 2 diabetes mellitus (Calvo et al. 2016). Surveillance related to the study shows that the *Agaricus bisporus* mushrooms have dual hypoglycemic and hypolipidemic activity in rats and that *Agaricus bisporus* lectins (glycoproteins) regenerated pancreatic β-cells in mice following 70% partial pancreatectomy, which suggests that induction of islet β-cell proliferation has potential therapeutic benefits for diabetes mellitus (Jeong et al. 2010; Wang et al. 2012b). There have not been many studies conducted in humans on the advantageous properties of mushroom polysaccharides. *Pleurotus ostreatus* and *Pleurotus cystidiosus* mushrooms (consumed as freeze-dried powders at a dose of 50 mg/kg/body weight) in healthy human volunteers and patients with type 2 diabetes showed a reduction in the ability to fast and postprandial serum glucose levels of healthy volunteers (Jayasuriya et al. 2015). Also, a reduction in postprandial serum glucose levels and increased serum insulin levels of type 2 diabetic mellitus patients were observed (Jayasuriya et al. 2015). Thus the mechanism of action increased glucokinase activity and promotion of insulin secretion by the pancreas, which increased the utilization of glucose by peripheral tissues, inhibited glycogen synthase kinase activity, and promoted glycogen synthesis.

3.2.5 Beneficial Role of Mushrooms or Mushroom-Derived Substances Against Infectious Diseases

Mushrooms possess intrinsic antibacterial and antifungal constituents to survive in their natural environment. Hence, they are a potential source of natural antimicrobials. Most of the secondary metabolites secreted extracellularly by the mycelium have demonstrated activity against some bacterial cell lines (Benedict and Brady 1972; Kupka et al. 1979; Lindequist et al. 1990). The terpenoid hydroquinones present in *Ganoderma pfeifferi* inhibited the growth of methicillin-resistant *Staphylococcus aureus* (MRSA) as well as several other bacterial pathogens. Also, whole extracts from this mushroom inhibited the growth of microorganisms responsible for skin infections including *Pityrosporum ovale*, *Staphylococcus epidermis*, and *Propionibacterium acnes* (Mothana et al. 2000). β-Glucan from *Pleurotus ostreatus* is used to protect athletes against respiratory tract infections (Bergendiova et al. 2011). Silver nanoparticles prepared using glucan isolated from *Pleurotus florida* blue variant mushrooms constrained the multiple antibiotic-resistant (MAR) bacterium *Klebsiella pneumoniae* in a dose-dependent manner. Also, it acted synergistically with four medicinal antibiotics to inhibit nearly all other bacterial growth, which eludes to the premise that certain combinations might control the MAR bacteria, which causes pneumonia (Sen et al. 2013). A related study completed by Manna et al. (Manna et al. 2015) reported that nanoparticles synthesized using a heteropolysaccharide isolated from *Lentinus squarrosulus* mushrooms showed promising antibacterial activity against MAR *Escherichia coli* bacteria. These nanoparticles also exhibited synergistic effects with four antibiotics that inhibited all bacterial growth. The *Lentinus edodes* polysaccharide was used to protect mice against a *Salmonella* lipopolysaccharide-induced endotoxemia (septic shock), which often results in death in humans (Kim et al. 2013). The same polysaccharide, as well as *Hericium erinaceus* mushroom extract, shielded mice from infections by the deadly foodborne pathogen *Salmonella typhimurium*, via stimulation of the immune system (Kim et al. 2012, 2014). Furthermore, a rough polysaccharide from *Auricularia auricula-judae* exhibited in vitro activity against the foodborne pathogens *Escherichia coli* and *Staphylococcus aureus*, (Cai et al. 2015) and sulfated polysaccharides from *Pleurotus eryngii* (oyster) mushrooms stunted the growth of the same bacterial pathogens. A purified *Lentinus edodes* extract exhibited antimicrobial properties against many bacterial pathogens found in the oral cavity. This points to their obvious value in oral hygiene improvement (anticaries and antigingivitis) (Signoretto et al. 2014). Biofilm is a slick outer layer of bacteria that can assist in the bacteria's resistance to phagocytosis and antibiotics (de Carvalho et al. 2011). *Coprinus comatus* can be used against biofilm infections due to its ability to block both quorum sensing and MurA. Scientists isolated the compound *Coprinus lactone* [(3R,4S)-2-methylene-3,4-dihydroxy pentanoic acid 1,4-lactone] (Fig. 3.6) from the edible mushroom *Coprinus comatus*. It obstructed quorum sensing and dissipated the biofilms of *Pseudomonas aeruginosa* and reduced the formation of pathogenicity factors pyocyanin and rhamnolipid B. Also, *Coprinus lactone* damaged *Staphylococcus aureus* cells in biofilms at subtoxic levels and inhibited UDP-acetyl glucosamine enolpyruvyl transferase, which is essential to its ability to synthesize a cell wall. Studies related to this one showed that extracts of *Agaricus* mushroom species inhibited the biofilm forming ability of *Pseudomonas aeruginosa* (Soković

Fig. 3.6 *Coprinus lactone*

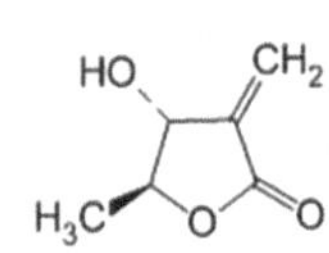

Coprinuslactone

et al. 2014). Extracts of *Ganoderma lucidum* and *Phellinus igniarius* inhibited quorum sensing in *Chromobacterium violaceum* (Zhu et al. 2011). These acute observations suggest that mushrooms produce compounds that can serve as a source of antimicrobial and anti-quorum sensing agents (Kim et al. 2016).

Viral diseases have different morphology, and hence conventional antibiotics do not play a role in containing the infections. Drugs with specific antiviral mechanisms are required such as direct inhibition of viral enzymes, interference with nucleic acid synthesis or inhibiting entry of the virus into mammalian cells. Such direct antiviral effects have been noted for smaller molecular constituents of mushrooms, while indirect effects are due to the activity of polysaccharides or other complex molecules (Brandt and Piraino 2000). Small molecular compounds with antiviral activities include several triterpenes ganodermanontriol and ganoderic acid from *Ganoderma lucidum*. These are active against human immunodeficiency virus type 1 (El-Mekkawy et al. 1998). Ganodermadiol, lucidadiol, and applanoxidic acid G isolated from *Gibberula pfeifferi* possess in vitro antiviral activity against influenza virus type A. Furthermore, ganodermadiol is active against herpes simplex virus type 1 (Mothana et al. 2003).

A pioneering study discovered that non-sulfated and sulfated polysaccharides completely inhibit the cell-to-cell infection capabilities of the human immunodeficiency virus HIV-1 and HIV-2, as well as human T-cell lymphotropic virus type 1 (Tochikura et al. 1989). Polysaccharides from *Pleurotus abalonus* mushrooms can also be seen to inhibit the HIV (Wang et al. 2011). Also, *Fomes fomentarius* mushroom constituents are shown to have a higher anti-HIV activity potential in comparison to the drug zidovudine in vitro, and in rats, and also exhibited antimicrobial properties that were detrimental to *Helicobacter pylori* (associated with gastritis and peptic ulcers) and with greater efficiency than medicinal antibiotics (Seniuk et al. 2011). The polysaccharide from *Agaricus brasiliensis* and its sulfated derivative exhibited strong anti-herpes simplex virus activities (Minari et al. 2011; Yamamoto et al. 2013; Cardozo et al. 2014). Moreover, a polysaccharide extract of *Lentinula edodes* (shiitake) mushrooms exhibited viricidal activity. This activity was focused against the bovine herpes simplex type 1 and poliovirus type 1 viruses (Rincão et al. 2012). Aqueous extracts from a multitude of mushroom varieties protected mice against lethality induced by the herpes simplex type 2 virus (Razumov et al. 2013). Mushroom lentinan and its sulfated product protected tobacco seedlings against a viral infection from the tobacco mosaic virus (Wang et al. 2015). The studies mentioned show that mushroom polysaccharides and other elements protect against viral diseases that affect plants, animals, and humans.

3.2.6 Cytoprotective (Antioxidant, Immunostimulating, and Anti-inflammatory) Properties of Mushrooms

Antioxidants are substances which protect the body from damage caused by reactive oxygen species (ROS). The antioxidants show its benefit by counteracting ROS which is known to damage DNA and essential proteins. Polysaccharides from mushrooms possess antioxidant properties which are known to increase the activity of liver enzymes (catalase, glutathione peroxidase, superoxide dismutase) and increase glutathione and malondialdehyde levels (Xiao et al. 2012; Kozarski et al. 2015). Polysaccharide from *Cordyceps militaris* mushrooms suppressed in vivo growth of melanoma in mouse models, and this was attributed to antioxidant and immune-stimulating properties (Lee and Hong 2011; Zhu et al. 2014a, b). Therapeutic and nutritional properties of polysaccharides from *Inonotus oblique* mushrooms can be looked upon as providing its antioxidant properties (Mu et al. 2012). *Hericium erinaceus* was shown to bring forth strong in vitro antioxidant activity and protected mice against potential liver damage caused by carbon tetrachloride (Zhang et al. 2012). A fucogalactan isolated from *Macrolepiota dolichaula* demonstrated antioxidant and immunostimulating properties (Samanta et al. 2015). Near identical results were observed with a β-glucan isolated from *Russula albonigra* (Nandi et al. 2014). A purified polysaccharide from *Pleurotus nebrodensis* improved immunity and coordinated innate immunity and inflammatory responses by activating macrophages (Cui et al. 2015). An exopolysaccharide from the medicinal mushroom *Clitocybe maxima* helped in maximizing the immune systems response and inhibited tumor cell growth in mice (Hu et al. 2015). A polysaccharide from *Hericium erinaceus* exhibited antioxidant in addition to neuroprotective effects on Aβ-induced neurotoxicity in neurons (Cheng et al. 2016). A polysaccharide from *Agaricus brasiliensis* induced immune-stimulation in mice (increased spleen and thymus indexes) and increased RAW 264.7 cell proliferation in vitro (Fang et al. 2016). A similar study found that the polysaccharide from this mushroom was observed to have strong in vitro free radical scavenging activity (Yin et al. 2015). Ganoderma β-D-glucans have higher molecular weights and better in vitro antioxidant activity in comparison with conventional extraction methods (Alzorqi et al. 2017). A water-soluble β-glucan isolated from the edible mushroom *Entoloma lividoalbum* encouraged the production of macrophages, splenocytes, and thymocytes and demonstrated good hydroxyl and superoxide radical scavenging activities in addition to reducing properties. A fucogalactomannan from *Tylopilus ballouii* mushroom inhibited superoxide and hydroxyl radicals with IC50 values of 1.25 and 1.6 mg/mL, respectively, and reduced edema by up to 56% in an anti-inflammatory assay (Lima et al. 2016).

In addition, a β-glucan-rich mushroom preparation AndoSan™ that consists of a mixture of an extract of *Agaricus blazei Murill* (82.4%), *Hericium erinaceus* (14.7%), and *Grifola frondosa* (2.9%) significantly inhibited the activity of the tumor-associated protease, legumain, in RAW 264.7 macrophage cells (Berven et al. 2015). Anti-allergic/anti-asthma, anti-infective, and antitumor properties in mice and anti-inflammatory effects in inflammatory bowel disease were attributed to immunomodulating polysaccharides (β-glucans) from *Agaricus blazei Murill* which is due to stimulation of the innate immune cells, natural killer (NK) cells, and dendritic cells, which leads to changes in the T-helper cell (Th1/Th2) ratio (balance) and inflammation (Hetland et al. 2011). Surprisingly, administration of AndoSan™ to 40 patients with multiple myeloma planned to go through high-dose chemotherapy for 7 weeks did not result in statistically significant differences in treatment response and overall survival (Tangen et al. 2015). It should also be noted that an *Agaricus blazei* extract enhanced the immune response elicited by the DNA vaccine for foot-and-mouth disease (Chen and Shao 2006). *Hericium erinaceus* extracts shielded mice against *Salmonella typhimurium* mortality (Kim et al. 2012). A polysaccharide isolated from a liquid culture of *Lentinus edodes* mushroom mycelia, which contained black rice bran, shielded mice against both *Salmonella*-induced endotoxemia and salmonellosis by means of upregulating the immune response (Kim et al. 2014). Mushrooms are a good potential source of anti-inflammatory compounds (Chan et al. 2015). Attenuation of inflammatory mediators (TNF-α and nitric oxide) and upregulation of IL-10 have been demonstrated by basidiocarps of *Amauroderma rugosum* (Mishra et al. 2012; Elsayed et al. 2014; Chan et al. 2015). The biological activities of polysaccharides seem to be related to their specific structure or conformation properties (Huang and Nie 2015). Nevertheless, the studies mentioned above suggest the strong potential value of polysaccharides derived from mushrooms as antioxidant, immunostimulatory, and anti-inflammatory agents.

3.2.7 Hepatoprotective Properties of Mushrooms

The liver is a large, complex organ involved in excretion and metabolism of a host of endogenous substances (carbohydrate, protein, and fat) as well as xenobiotics. Also, it detoxifies toxic compounds produced during metabolism and plays an important role in the elimination of drugs. Liver damage can be caused by many occurrences, but the central pathogenesis involves some level of oxidative stress and is characterized by a progressive evolution from steatosis to chronic hepatitis, fibrosis, cirrhosis, and finally hepatocellular carcinoma. Mushrooms and their constituents have been reported to possess hepatoprotective properties and are reported to be beneficial in hepatic diseases (Ooi 1996). For example, the extracts from the *Lentinus edodes and Grifola frondosa* exerted a positive and noteworthy hepatoprotective effect by reducing the paracetamol-induced acute elevation of AST and ALT levels. The mechanisms proposed include preservation of structural integrity as it relates to hepatocytic membranes; prevention of the fall of the GSH levels by acting on enzymes involved in the GSH redox cycle; and scavenging of free radicals originating from the paracetamol metabolism. All of these actions have the potential to result from antioxidant compounds of the mushroom extracts (Ooi 1996). Also, *Antrodia cinnamomea* significantly inhibited the ethanol-induced elevations in AST, ALT, ROS, NO, and induced GSH depletion in human hepatoma cell lines (HepG2 cells). Ethanolic extract of *Antrodia cinnamomea prevented* elevation of serum ALT and AST levels, hepatocellular lipid peroxidation, and GSH depletion in a dose-dependent manner in an in vivo acute ethanol-intoxicated mouse model (Kumar et al. 2011). *Ganoderma lucidum* is considered to be the most studied of all mushrooms in respect to its hepatoprotective effects. Predominantly, the polysaccharide and triterpenoid components in *G. lucidum* are responsible for the protective activities against toxin-induced liver injury (Zhou et al. 2002; Gao et al. 2003; Wasser 2005).

Various in vitro and in vivo studies show that *Ganoderma lucidum* extracts, largely polysaccharides or triterpenoids, reveal some protective effects against potential liver injury induced by toxic chemicals, (e.g., CCl4) Bacillus Calmette–Guerin (BCG) and lipopolysaccharide (LPS). Ganoderenic acid, one of the triterpenoids found in *G. lucidum*, exhibited acute inhibition of β-glucuronidase activity, which is considered to be an indicator of hepatic damage (Kim et al. 1999). Correspondingly, polysaccharide administration alone protected hepatocytes from the damage induced by carbon tetrachloride (Nada et al. 2010). The polysaccharide prevented the increased activities of serum ALT and AST, reduced the formation of malondialdehyde, and enriched the activities of superoxide dismutase and glutathione peroxidase (Gan et al. 2012). Carbon tetrachloride-induced liver injury was evaluated in an in vivo mouse model by treating with *Pleurotus eryngii* rich in the polysaccharide, and it was found to be beneficial as a food additive for hypolipidemic and hepatoprotective treatments (Chen et al. 2012a). *Hericium erinaceus* polysaccharides act as antioxidant and can be considered as supplements in the prevention of hepatic diseases (Zhang et al. 2012). Hence various mushrooms and its constituents are considered to be potential hepatoprotective agents.

3.3 Conclusions

There are numerous therapeutic applications of mushrooms (Fig. 3.7, Table 3.3). Edible mushrooms have significant beneficial effects in the treatment of neurodegenerative (Alzheimer's and Parkinson's), neurological (epilepsy), cardiovascular (hypertension, diabetes mellitus), hepatological, and infectious diseases. Nevertheless, with the evolving and emerging new evidence of health benefits, further detailed mechanisms of various benefits of mushrooms in humans still require intensive investigation. There seems to be a need to express further current knowledge about the health benefiting properties, which have engrossed prevalent interest. The assessment of afresh cultivated mushrooms and segregation of their vigorous ingredients with a mechanism based potential therapeutic value remains a challenge, and hence mushrooms will continue to be the notable limelight of research in the upcoming prospect as well.

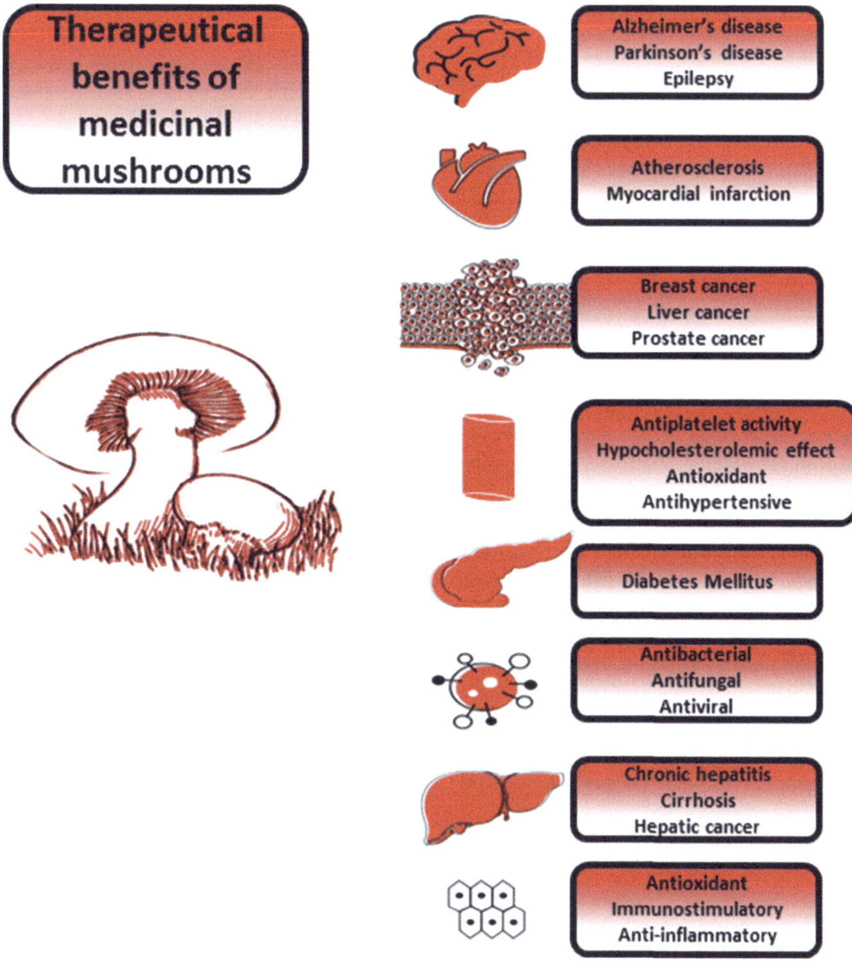

Fig. 3.7 Therapeutical benefits of medicinal mushrooms

Table 3.3 Beneficial effects of mushrooms

Mushroom species	Medical benefits	References
Agaricus bisporus	Infectious diseases	Soković et al. (2014)
	Hypocholesterolemic action	Kaneda and Tokuda (1966)
	Diabetic mellitus	Calvo et al. (2016)
Agaricus blazei Murill	Cytoprotective	Hetland et al. (2011)
Agaricus brasiliensis	Infectious diseases	Minari et al. (2011), Yamamoto et al. (2013), and Cardozo et al. (2014)
	Cytoprotective	Fang et al. (2016)
Agaricus blazei	Cytoprotective	Chen and Shao (2006)
Agrocybe aegerita	Decreases total cholesterol, triglycerides, and lipid levels	Ng (2005)
Agrocybe chaxingu	Diabetic mellitus	Lee et al. (2010a)
Amantia phylloides	Toxic, increases ammonia	Ytrebo (2006)
Amauroderma rugosum	Cytoprotective	Mishra et al. (2012), Elsayed et al. (2014), and Chan et al. (2015)
Antrodia camphorate	Alzheimer's disease	Wang et al. (2012b)
Antrodia cinnamomea	Hepatological protective properties	Kumar et al. (2011)
Auricularia auricula-judae	Decreases total cholesterol, triglycerides, and lipid levels	Chen (1989) and Sheng and Chen (1990)
	Infectious diseases	Cai et al. (2015)
Auricularia polytricha	Alzheimer's disease	Willem et al. (2006) and Bennett et al. (2013)
	Hypocholesterolemic action	Kaneda and Tokuda (1966)
Clitocybe maxima	Cytoprotective	Hu et al. (2015)
Cordyceps militaris	Cytoprotective	Lee and Hong (2011) and Zhu et al. (2014a, b)
	Lung cancer	Park et al. (2009)
Cordyceps sinensis	Cervical cancer	Ji et al. (2014)
Coriolus versicolor	Hepatocellular carcinoma	Patel and Goyal (2012)
Cortinarius infractus	Alzheimer's disease	Brondz et al. (2007) and Geissler et al. (2010)
Entoloma lividoalbum	Cytoprotective	Lima et al. (2016)
Flammulina velutipes	Hypocholesterolemic action	Kaneda and Tokuda (1966)

(continued)

Table 3.3 (continued)

Mushroom species	Medical benefits	References
Fomes fomentarius	Infectious diseases	Seniuk et al. (2011)
Ganoderma lucidum	Alzheimer's disease	Wang et al. (2004), Jin et al. (2004a), and Lai et al. (2008)
	Parkinson's disease	Pinweha et al. (2008)
	Longevity	Sanodiya et al. (2009)
	Epilepsy	Socala et al. (2015)
	Anticancer	Wang et al. (1997) and Guggenheim et al. (2014)
	Inhibits biosynthesis of cholesterol	Komoda et al. (1989)
	Inhibits platelet aggregation	Su et al. (1999)
	Bladder cancer	Yuen and Gohel (2008)
	Infectious diseases	Zhu et al. (2011)
	Hepatoprotective properties	Zhu et al. (2011), Zhou et al. (2002), Gao et al. (2003), and Wasser (2005)
Ganoderma pfeifferi	Infectious diseases	Mothana et al. (2000, 2003)
Grifola frondosa	Decreases blood pressure	Mizuno (1995)
	Cytoprotective	Berven et al. (2015)
	Diabetic mellitus	Hong et al. (2007)
	Hepatoprotective	Ooi (1996)
Hericium erinaceus	Alzheimer's disease	Wong et al. (2007) and Mori et al. (2009)
	Parkinson's disease	Wong et al. (2011)
	Infectious diseases	Kim et al. (2012)
	Cytoprotective	Kim et al. (2012), Zhang et al. (2012), Berven et al. (2015), and Cheng et al. (2016)
	Hepatoprotective properties	Zhang et al. (2012)
Hypsizygus marmoreus	Prevents atherosclerosis	Ryong et al. (1989)
Inonotus obliquus	Cytoprotective	Mu et al. (2012)
	Tumor apoptosis	Song et al. (2013)
Lentinula edodes	Decreases cholesterol and homocysteine	Suzuki and Ohshima (1976), Sun et al. (1984), and Sugiyama et al. (1995)
	Infectious diseases	Li and Shah (2014) and Signoretto et al. (2014)
	Hypocholesterolemic action	Tokita et al. (1972)
	Cytoprotective	Kim et al. (2014)
	Hepatoprotective properties	Ooi (1996)
Lentinus squarrosulus	Infectious diseases	Manna et al. (2015)
Lentinus strigosus	Diabetic mellitus	Yamac et al. (2008)

(continued)

Table 3.3 (continued)

Mushroom species	Medical benefits	References
Macrolepiota dolichaula	Cytoprotective	Samanta et al. (2015)
Mycoleptodonoides aitchisonii	Improves brain ischemic damage	Okuyama et al. (2012)
	Parkinson's disease	Okuyama et al. (2004)
Phellinus igniarius	Infectious diseases	El-Mekkawy et al. (1998)
Phellinus linteus	Anti-angiogenic	Baker et al. (2008)
Pleurotus abalonus	Infectious diseases	Wang et al. (2011)
Pleurotus cornucopiae	Anti-atherosclerotic	Bobek and Galbavý (1999)
Pleurotus eryngii	Hepatoprotective properties	Chen et al. (2012a)
	Breast cancer	Xue et al. (2015)
Pleurotus florida	Infectious diseases	Sen et al. (2013)
Pleurotus nebrodensis	Cytoprotective	Cui et al. (2015)
Pleurotus ostreatus	Infectious diseases	Zhu et al. (2014a, b) and Bergendiova et al. (2011)
Pleurotus tuber-regium	Diabetic mellitus	Wu et al. (2014)
Pleurotus sajor-caju	Diabetic mellitus	Vincent et al. (2013) and Kanagasabapathy et al. (2014)
Russula albonigra	Cytoprotective	Nandi et al. (2014)
Sarcodon cyrneus	Neuritic growth	Marcotullio et al. (2006)
Suillus placidus	Hepatocellular carcinoma	Liu et al. (2009)
Trametesro biniophila	Inhibit tumorigenesis	Zou et al. (2015)
Tremella aurantialba mycelia	Diabetic mellitus	Zhang et al. (2009)
Tremella fuciformis	HMG COA reductase inhibitor	Chen (1989)
Tylopilus ballouii	Cytoprotective	Lima et al. (2016)

Acknowledgment We would like to thank Mrs. Fatimah Almaghrabi and Mrs. Bhavanee Samyuktha for their technical assistance with the manuscript.

References

Alzorqi I, Sudheer S, Lu TJ, Manickam S (2017) Ultrasonically extracted β-d-glucan from artificially cultivated mushroom, characteristic properties and antioxidant activity. Ultrason Sonochem 35:531–540. https://doi.org/10.1016/j.ultsonch.2016.04.017

Baker JR, Kim JS, Park SY (2008) Composition and proposed structure of a water-soluble glycan from the Keumsa Sangwhang Mushroom (*Phellinus linteus*). Fitoterapia 79:345–350. https://doi.org/10.1016/j.fitote.2008.03.002

Barceloux DG (2008) Medical toxicology of natural substances. Wiley, Hoboken

Benedict RG, Brady LR (1972) Antimicrobial activity of mushroom metabolites. J Pharm Sci 61:1820–1822. https://doi.org/10.1002/jps.2600611130

Bennett L, Sheean P, Zabaras D, Head R (2013) Heat-stable components of wood ear mushroom, *Auricularia polytricha* (higher Basidiomycetes), inhibit in vitro activity of beta secretase (BACE1). Int J Med Mushrooms 15:233–249

Bergendiova K, Tibenska E, Majtan J (2011) Pleuran (β-glucan from *Pleurotus ostreatus*) supplementation, cellular immune response and respiratory tract infections in athletes. Eur J Appl Physiol 111:2033–2040. https://doi.org/10.1007/s00421-011-1837-z

Berven L, Karppinen P, Hetland G, Samuelsen ABC (2015) The polar high molecular weight fraction of the *Agaricus blazei* Murill extract, AndoSan™, reduces the activity of the tumor-associated protease, legumain, in RAW 264.7 cells. J Med Food 18:429–438. https://doi.org/10.1089/jmf.2014.0018

Bharadwaj P, Martins R, Macreadie I (2010) Yeast as a model for studying Alzheimer's disease. FEMS Yeast Res 10:961–969. https://doi.org/10.1111/j.1567-1364.2010.00658.x

Biasibetti R, Tramontina AC, Costa AP, Dutra MF, Quincozes-Santos A, Nardin P, Bernardi CL, Wartchow KM, Lunardi PS, Gonçalves CA (2013) Green tea (−)epigallocatechin-3-gallate reverses oxidative stress and reduces acetylcholinesterase activity in a streptozotocin-induced model of dementia. Behav Brain Res 236:186–193. https://doi.org/10.1016/j.bbr.2012.08.039

Bobek P, Galbavý Š (1999) Hypocholesterolemic and antiatherogenic effect of oyster mushroom (*Pleurotus ostreatus*) in rabbits. Nahrung/Food 43:339–342. https://doi.org/10.1002/(SICI)1521-3803(19991001)43:5<339::AID-FOOD339>3.0.CO;2-5

Bonnet MS, Basson PW (2002) The toxicology of Amanita phalloides. Homeopathy 91:249–254

Brandt C, Piraino F (2000) Mushroom antivirals. Recent Res Dev Antimicrob Agents Chemother 4:11–26

Brondz I, Ekeberg D, Høiland K, Bell DS, Annino AR (2007) The real nature of the indole alkaloids in *Cortinarius infractus*: evaluation of artifact formation through solvent extraction method development. J Chromatogr A 1148:1–7. https://doi.org/10.1016/j.chroma.2007.02.074

Brown GC, Neher JJ (2010) Inflammatory neurodegeneration and mechanisms of microglial killing of neurons. Mol Neurobiol 41:242–247. https://doi.org/10.1007/s12035-010-8105-9

Cai M, Lin Y, Luo Y, Liang H, Sun P (2015) Extraction, antimicrobial, and antioxidant activities of crude polysaccharides from the wood ear medicinal mushroom *Auricularia auricula-judae* (higher basidiomycetes). Int J Med Mushrooms 17:591–600

Calvo MS, Mehrotra A, Beelman RB, Nadkarni G, Wang L, Cai W, Goh BC, Kalaras MD, Uribarri J (2016) A retrospective study in adults with metabolic syndrome: diabetic risk factor response to daily consumption of *Agaricus bisporus* (white button mushrooms). Plant Foods Hum Nutr 71:245–251. https://doi.org/10.1007/s11130-016-0552-7

Cardozo FTGS, Camelini CM, Leal PC, Kratz JM, Nunes RJ, de Mendonça MM, Simões CMO (2014) Antiherpetic mechanism of a sulfated derivative of *Agaricus brasiliensis* fruiting bodies polysaccharide. Intervirology 57:375–383. https://doi.org/10.1159/000365194

Chan PM, Tan YS, Chua KH, Sabaratnam V, Kuppusamy UR (2015) Attenuation of inflammatory mediators (TNF-α and nitric oxide) and up-regulation of IL-10 by wild and domesticated basidiocarps of *Amauroderma rugosum* (Blume T. Nees) torrend in LPS-stimulated RAW264.7 cells. PLoS One 10:e0139593. https://doi.org/10.1371/journal.pone.0139593

Chang ST (2007) Mushrooms and mushroom cultivation. In: Van Nostrand's scientific encyclopedia. Wiley, Hoboken

Chen Q (1989) Antilipemic effect of polysaccharides from *Auricularia auricular*, *Tremella fuciformis*, and *Tremella fuciformis* spores. Zhongguo Yaoke Daxue Xuebao 20:344–347

Chen L, Shao H (2006) Extract from *Agaricus blazei* Murill can enhance immune responses elicited by DNA vaccine against foot-and-mouth disease. Vet Immunol Immunopathol 109:177–182. https://doi.org/10.1016/j.vetimm.2005.08.028

Chen J, Mao D, Yong Y, Li J, Wei H, Lu L (2012a) Hepatoprotective and hypolipidemic effects of water-soluble polysaccharidic extract of *Pleurotus eryngii*. Food Chem 130:687–694. https://doi.org/10.1016/j.foodchem.2011.07.110

Chen X, Guo C, Kong J (2012b) Oxidative stress in neurodegenerative diseases. Neural Regen Res 7:376–385. https://doi.org/10.3969/j.issn.1673-5374.2012.05.009

Cheng JH, Tsai CL, Lien YY, Lee MS, Sheu SC (2016) High molecular weight of polysaccharides from *Hericium erinaceus* against amyloid beta-induced neurotoxicity. BMC Complement Altern Med 16:170. https://doi.org/10.1186/s12906-016-1154-5

Cheung PCK (1996) Dietary fiber content and composition of some cultivated edible mushroom fruiting bodies and mycelia. J Agric Food Chem 44:468–471. https://doi.org/10.1021/jf9504551

Cheung WM, Hui WS, Chu PW, Chiu SW, Ip NY (2000) *Ganoderma* extract activates MAP kinases and induces the neuronal differentiation of rat pheochromocytoma PC12 cells. FEBS Lett 486:291–296

Cho EJ, Hwang HJ, Kim SW, Oh JY, Baek YM, Choi JW, Bae SH, Yun JW (2007) Hypoglycemic effects of exopolysaccharides produced by mycelial cultures of two different mushrooms *Tremella fuciformis* and *Phellinus baumii* in ob/ob mice. Appl Microbiol Biotechnol 75:1257–1265. https://doi.org/10.1007/s00253-007-0972-2

Chu QP, Wang LE, Cui XY, Fu HZ, Lin ZB, Lin SQ, Zhang YH (2007) Extract of *Ganoderma lucidum* potentiates pentobarbital-induced sleep via a GABAergic mechanism. Pharmacol Biochem Behav 86:693–698. https://doi.org/10.1016/j.pbb.2007.02.015

Chu YF, Chang WH, Black RM, Liu JR, Sompol P, Chen Y, Wei H, Zhao Q, Cheng IH (2012) Crude caffeine reduces memory impairment and amyloid β1–42 levels in an Alzheimer's mouse model. Food Chem 135:2095–2102. https://doi.org/10.1016/j.foodchem.2012.04.148

Claeysen S, Cochet M, Donneger R, Dumuis A, Bockaert J, Giannoni P (2012) Alzheimer culprits: cellular crossroads and interplay. Cell Signal 24:1831–1840. https://doi.org/10.1016/j.cellsig.2012.05.008

Colak A, Kolcuoglu Y, ES AJ of, 2007 U (2009). Biochemical composition of some Turkish fungi. https://search.proquest.com/

Cui HY, Wang CL, Wang YR, Li ZJ, Chen MH, Li FJ, Sun YP (2015) *Pleurotus nebrodensis* polysaccharide (PN-S) enhances the immunity of immunosuppressed mice. Chin J Nat Med 13:760–766. https://doi.org/10.1016/S1875-5364(15)30076-5

De Carvalho RSM, Duarte FS, De Lima TCM (2011) Involvement of GABAergic non-benzodiazepine sites in the anxiolytic-like and sedative effects of the flavonoid baicalein in mice. Behav Brain Res 221:75–82. https://doi.org/10.1016/j.bbr.2011.02.038

El-Mekkawy S, Meselhy MR, Nakamura N, Tezuka Y, Hattori M, Kakiuchi N, Shimotohno K, Kawahata T, Otake T (1998) Anti-HIV-1 and anti-HIV-1-protease substances from *Ganoderma lucidum*. Phytochemistry 49:1651–1657

Elsayed EA, El Enshasy H, Wadaan MAM, Aziz R (2014) Mushrooms: a potential natural source of anti-inflammatory compounds for medical applications. Mediators Inflamm 2014:1–15. https://doi.org/10.1155/2014/805841

Fang L, Zhang Y, Xie J, Wang L, Zhang H, Wei W, Li Y (2016) Royal sun medicinal mushroom, *Agaricus brasiliensis* (Agaricomycetidae), derived polysaccharides exert immunomodulatory activities in vitro and in vivo. Int J Med Mushrooms 18:123–132. https://doi.org/10.1615/IntJMedMushrooms.v18.i2.30

Fiorentini A, Rosi MC, Grossi C, Luccarini I, Casamenti F (2010) Lithium improves hippocampal neurogenesis, neuropathology and cognitive functions in APP mutant mice. PLoS One 5:e14382. https://doi.org/10.1371/journal.pone.0014382

Fisher RS, Boas W, van E, Blume W, Elger C, Genton P, Lee P, Engel J (2005) Epileptic seizures and epilepsy: definitions proposed by the International League Against Epilepsy (ILAE) and the International Bureau for Epilepsy (IBE). Epilepsia 46:470–472. https://doi.org/10.1111/j.0013-9580.2005.66104.x

Folkman J (2004) Endogenous angiogenesis inhibitors. APMIS 112:496–507. https://doi.org/10.1111/j.1600-0463.2004.apm11207-0809.x

Gan D, Ma L, Jiang C, Wang M, Zeng X (2012) Medium optimization and potential hepatoprotective effect of mycelial polysaccharides from *Pholiota dinghuensis Bi* against carbon tetrachloride-induced acute liver injury in mice. Food Chem Toxicol 50:2681–2688. https://doi.org/10.1016/j.fct.2012.05.003

Gao Y, Huang M, Lin ZB, Zhou S (2003) Hepatoprotective activity and the mechanisms of action of *Ganoderma lucidum* (Curt.:Fr.) P. Karst. (Ling Zhi, Reishi Mushroom) (Aphyllophoromycetideae) (Review). Int J Med Mushrooms 5:22. https://doi.org/10.1615/InterJMedicMush.v5.i2.20

Gao L, Sun Y, Chen C, Xi Y, Wang J, Wang Z (2007) Primary mechanism of apoptosis induction in a leukemia cell line by fraction FA-2-b-ss prepared from the mushroom *Agaricus blazei* Murill. Brazilian J Med Biol Res = Rev Bras Pesqui medicas e Biol 40:1545–1555

Geissler T, Brandt W, Porzel A, Schlenzig D, Kehlen A, Wessjohann L, Arnold N (2010) Acetylcholinesterase inhibitors from the toadstool *Cortinarius infractus*. Bioorg Med Chem 18:2173–2177. https://doi.org/10.1016/j.bmc.2010.01.074

Guggenheim AG, Wright KM, Zwickey HL (2014) Immune modulation from five major mushrooms: Application to integrative oncology. Integr Med (Encinitas) 13:32–44

Guillamón E, García Lafuente A, Lozano M, D'Arrigo M, Rostagno MA, Villares A, Martínez JA (2010) Edible mushrooms: role in the prevention of cardiovascular diseases. Fitoterapia 81:715–723. https://doi.org/10.1016/j.fitote.2010.06.005

Gunawardena D, Shanmugam K, Low M, Bennett L, Govindaraghavan S, Head R, Ooi L, Münch G (2014) Determination of anti-inflammatory activities of standardised preparations of plant- and mushroom-based foods. Eur J Nutr 53:335–343. https://doi.org/10.1007/s00394-013-0531-9

Gunde-Cimerman N, Friedrich J, Cimerman A, BeniÄki N (1993) Screening fungi for the production of an inhibitor of HMG CoA reductase: production of mevinolin by the fungi of the genus Pleurotus. FEMS Microbiol Lett 111:203–206. https://doi.org/10.1111/j.1574-6968.1993.tb06386.x

Hardy J, Selkoe DJ (2002) The amyloid hypothesis of Alzheimer's disease: progress and problems on the road to therapeutics. Science (80-) 297:353–356. https://doi.org/10.1126/science.1072994

Hetland G, Johnson E, Lyberg T, Kvalheim G (2011) The mushroom *Agaricus blazei Murill* elicits its medicinal effects on tumor, infection, allergy, and inflammation through its modulation of innate immunity and amelioration of Th1/Th2 imbalance and inflammation. Adv Pharmacol Sci 2011:1–10. https://doi.org/10.1155/2011/157015

Hobbs C (1995) Medicinal mushrooms: an exploration of tradition, healing, & culture. Book Publishing Company, Christopher Hobbs, p. 251. 9781570671432

Hong L, Xun M, Wutong W (2007) Anti-diabetic effect of an α-glucan from fruit body of maitake (*Grifola frondosa*) on KK-Ay mice. J Pharm Pharmacol 59:575–582. https://doi.org/10.1211/jpp.59.4.0013

Hu H, Ahn NS, Yang X, Lee YS, Kang KS (2002) *Ganoderma lucidum* extract induces cell cycle arrest and apoptosis in MCF-7 human breast cancer cell. Int J Cancer 102:250–253. https://doi.org/10.1002/ijc.10707

Hu SH, Cheung PCK, Hung RP, Chen YK, Wang JC, Chang SJ (2015) Antitumor and immunomodulating activities of exopolysaccharide produced by big cup culinary- medicinal mushroom *Clitocybe maxima* (Higher Basidiomycetes) in liquid submerged culture. Int J Med Mushrooms 17:891–901

Huang X, Nie S (2015) The structure of mushroom polysaccharides and their beneficial role in health. Food Funct 6:3205–3217. https://doi.org/10.1039/c5fo00678c

Ishikawa Y, Morimoto K, Hamasaki T (1984) Flavoglaucin, a metabolite of *Eurotium chevalieri*, its antioxidation and synergism with tocopherol. J Am Oil Chem Soc 61:1864–1868. https://doi.org/10.1007/BF02540819

Jayasuriya W, Wanigatunge CA, Fernando GH, Abeytunga DTU, Suresh TS (2015) Hypoglycaemic activity of culinary *Pleurotus ostreatus* and *P. cystidiosus* mushrooms in healthy volunteers and type 2 diabetic patients on diet control and the possible mechanisms of action. Phyther Res 29:303–309. https://doi.org/10.1002/ptr.5255

Jeong SC, Jeong YT, Yang BK, Islam R, Koyyalamudi SR, Pang G, Cho KY, Song CH (2010) White button mushroom (*Agaricus bisporus*) lowers blood glucose and cholesterol levels in diabetic and hypercholesterolemic rats. Nutr Res 30:49–56. https://doi.org/10.1016/j.nutres.2009.12.003

Ji J, Liu J, Liu H, Wang Y (2014) Effects of fermented mushroom of *Cordyceps sinensis*, rich in selenium, on uterine cervix cancer. Evidence-Based Complement Altern Med 2014:1–7. https://doi.org/10.1155/2014/173180

Jin K, Galvan V, Xie L, Mao XO, Gorostiza OF, Bredesen DE, Greenberg DA (2004a) Enhanced neurogenesis in Alzheimer's disease transgenic (PDGF-APPSw,Ind) mice. Proc Natl Acad Sci 101:13363–13367. https://doi.org/10.1073/pnas.0403678101

Jin K, Peel AL, Mao XO, Xie L, Cottrell BA, Henshall DC, Greenberg DA (2004b) Increased hippocampal neurogenesis in Alzheimer's disease. Proc Natl Acad Sci 101:343–347. https://doi.org/10.1073/pnas.2634794100

Jin K, Xie L, Mao XO, Greenberg DA (2006) Alzheimer's disease drugs promote neurogenesis. Brain Res 1085:183–188. https://doi.org/10.1016/j.brainres.2006.02.081

Kabir Y, Kimura S (1989) Dietary mushrooms reduce blood pressure in spontaneously hypertensive rats (SHR). J Nutr Sci Vitaminol (Tokyo) 35:91–94

Kalač P (2013) A review of chemical composition and nutritional value of wild-growing and cultivated mushrooms. J Sci Food Agric 93:209–218. https://doi.org/10.1002/jsfa.5960

Kanagasabapathy G, Kuppusamy UR, Abd Malek SN, Abdulla MA, Chua KH, Sabaratnam V (2012) Glucan-rich polysaccharides from *Pleurotus sajor-caju* (Fr.) Singer prevents glucose intolerance, insulin resistance and inflammation in C57BL/6J mice fed a high-fat diet. BMC Complement Altern Med 12:1234. https://doi.org/10.1186/1472-6882-12-261

Kanagasabapathy G, Chua KH, Malek SNA, Vikineswary S, Kuppusamy UR (2014) AMP-activated protein kinase mediates insulin-like and lipo-mobilising effects of β-glucan-rich polysaccharides isolated from *Pleurotus sajor-caju* (Fr.), Singer mushroom, in 3T3-L1 cells. Food Chem 145:198–204. https://doi.org/10.1016/j.foodchem.2013.08.051

Kaneda T, Tokuda S (1966) Effect of various mushroom preparations on cholesterol levels in rats. J Nutr 90:371–376

Karumuthil Melethil S, Gudi R, Johnson BM, Perez N, Vasu C (2014) Fungal β-glucan, a dectin-1 ligand, promotes protection from type 1 diabetes by inducing regulatory innate immune response. J Immunol 193:3308–3321. https://doi.org/10.4049/jimmunol.1400186

Kim DH, Shim SB, Kim NJ, Jang IS (1999) Beta-glucuronidase-inhibitory activity and hepatoprotective effect of *Ganoderma lucidum*. Biol Pharm Bull 22:162–164

Kim SH, Song YS, Kim SK, Kim BC, Lim CJ, Park EH (2004) Anti-inflammatory and related pharmacological activities of the n-BuOH subfraction of mushroom *Phellinus linteus*. J Ethnopharmacol 93:141–146. https://doi.org/10.1016/j.jep.2004.03.048

Kim SP, Moon E, Nam SH, Friedman M (2012) *Hericium erinaceus* mushroom extracts protect infected mice against Salmonella typhimurium-induced liver damage and mortality by stimulation of innate immune cells. J Agric Food Chem 60:5590–5596. https://doi.org/10.1021/jf300897w

Kim SP, Park SO, Lee SJ, Nam SH, Friedman M (2013) A polysaccharide isolated from the liquid culture of *Lentinus edodes* (Shiitake) mushroom mycelia containing black rice bran protects mice against a salmonella lipopolysaccharide-induced endotoxemia. J Agric Food Chem 61:10987–10994. https://doi.org/10.1021/jf403173k

Kim SP, Park SO, Lee SJ, Nam SH, Friedman M (2014) A polysaccharide isolated from the liquid culture of *Lentinus edodes* (Shiitake) mushroom mycelia containing black rice bran protects mice against Salmonellosis through upregulation of the Th1 immune reaction. J Agric Food Chem 62:2384–2391. https://doi.org/10.1021/jf405223q

Kim SP, Lee SJ, Nam SH, Friedman M (2016) Turmeric (*Curcuma longa*) bioprocessed with mycelia of shiitake (*Lentinus edodes*) mushrooms: Composition and mechanism of protection against salmonellosis in mice. Int J Med Mushrooms 19(4):363–376

Kinnally KW, Antonsson B (2007) A tale of two mitochondrial channels, MAC and PTP, in apoptosis. Apoptosis 12:857–868. https://doi.org/10.1007/s10495-007-0722-z

Komoda Y, Shimizu M, Sonoda Y, Sato Y (1989) Ganoderic acid and its derivatives as cholesterol synthesis inhibitors. Chem Pharm Bull (Tokyo) 37:531–533

Kozarski M, Klaus A, Jakovljevic D, Todorovic N, Vunduk J, Petrović P, Niksic M, Vrvic M, van Griensven L (2015) Antioxidants of edible mushrooms. Molecules 20:19489–19525. https://doi.org/10.3390/molecules201019489

Kumar KJS, Chu FH, Hsieh HW, Liao JW, Li WH, Lin JCC, Shaw JF, Wang SY (2011) Antroquinonol from ethanolic extract of mycelium of *Antrodia cinnamomea* protects hepatic

cells from ethanol-induced oxidative stress through Nrf-2 activation. J Ethnopharmacol 136:168–177. https://doi.org/10.1016/j.jep.2011.04.030
Kundaković T, Kolundžić M (2013) Therapeutic properties of mushrooms in managing adverse effects in the metabolic syndrome. Curr Top Med Chem 13:2734–2744
Kupka J, Anke T, Oberwinkler F, Schramm G, Steglich W (1979) Antibiotics from basidiomycetes. VII. Crinipellin, a new antibiotic from the basidiomycetous fungus *Crinipellis stipitaria* (Fr.) Pat. J Antibiot (Tokyo) 32:130–135
Kurtzman RH (1997) Nutrition from mushrooms, understanding and reconciling available data. Mycoscience 38:247–253. https://doi.org/10.1007/BF02460860
Lai CSW, Yu MS, Yuen WH, So KF, Zee SY, Chang RCC (2008) Antagonizing β-amyloid peptide neurotoxicity of the anti-aging fungus *Ganoderma lucidum*. Brain Res 1190:215–224. https://doi.org/10.1016/j.brainres.2007.10.103
Lee JS, Hong EK (2010) *Hericium erinaceus* enhances doxorubicin-induced apoptosis in human hepatocellular carcinoma cells. Cancer Lett 297:144–154. https://doi.org/10.1016/j.canlet.2010.05.006
Lee JS, Hong EK (2011) Immunostimulating activity of the polysaccharides isolated from *Cordyceps militaris*. Int Immunopharmacol 11:1226–1233. https://doi.org/10.1016/j.intimp.2011.04.001
Lee BR, Lee YP, Kim DW, Song HY, Yoo KY, Won MH, Kang TC, Lee KJ, Kim KH, Joo JH, Ham HJ, Hur JH, Cho S-W, Han KH, Lee KS, Park J, Eum WS, Choi SY (2010a) Amelioration of streptozotocin-induced diabetes by *Agrocybe chaxingu* polysaccharide. Mol Cells 29:349–354. https://doi.org/10.1007/s10059-010-0044-9
Lee YS, Kim YH, Shin EK, Kim DH, Lim SS, Lee JY, Kim JK (2010b) Anti-angiogenic activity of methanol extract of *Phellinus linteus* and its fractions. J Ethnopharmacol 131:56–62. https://doi.org/10.1016/j.jep.2010.05.064
Lei H, Guo S, Han J, Wang Q, Zhang X, Wu W (2012) Hypoglycemic and hypolipidemic activities of MT-α-glucan and its effect on immune function of diabetic mice. Carbohydr Polym 89:245–250. https://doi.org/10.1016/j.carbpol.2012.03.003
Li S, Shah NP (2014) Antioxidant and antibacterial activities of sulphated polysaccharides from *Pleurotus eryngii* and *Streptococcus thermophilus* ASCC 1275. Food Chem 165:262–270. https://doi.org/10.1016/j.foodchem.2014.05.110
Li G, Kim DH, Kim TD, Park BJ, Park HD, Park JI, Na MK, Kim HC, Hong ND, Lim K, Hwang BD, Yoon WH (2004) Protein-bound polysaccharide from *Phellinus linteus* induces G2/M phase arrest and apoptosis in SW480 human colon cancer cells. Cancer Lett 216:175–181. https://doi.org/10.1016/j.canlet.2004.07.014
Li J, Zou L, Chen W, Zhu B, Shen N, Ke J, Lou J, Song R, Zhong R, Miao X (2014) Dietary mushroom intake may reduce the risk of breast cancer: evidence from a meta-analysis of observational studies. PLoS One 9:e93437. https://doi.org/10.1371/journal.pone.0093437
Lima ATM, Santos MN, de Souza LAR, Pinheiro TS, Paiva AAO, Dore CMPG, Costa MSSP, Santos ND, Baseia YG, Araújo RM, Leite EL (2016) Chemical characteristics of a heteropolysaccharide from *Tylopilus ballouii* mushroom and its antioxidant and anti-inflammatory activities. Carbohydr Polym 144:400–409. https://doi.org/10.1016/j.carbpol.2016.02.050
Lin Z (2004) Focus on anti-oxidative and free radical scavenging activity of *Ganoderma lucidum*. J Appl Pharmacol 12:133–137
Lin JY, Jeng TW, Chen CC, Shi GY, Tung TC (1973) Isolation of a new cardiotoxic protein from the edible mushroom, *Volvariella volvacea*. Nature 246:524–525. https://doi.org/10.1038/246524a0
Lindequist U, Teuscher E, Narbe G (1990) Neue Wirkstoffe aus Basidiomyceten. Phytotherapie 11:139–149
Liu B (2006) Modulation of microglial pro-inflammatory and neurotoxic activity for the treatment of Parkinson's disease. AAPS J 8:E606–E621. https://doi.org/10.1208/aapsj080369
Liu J, Zheng X, Yin F, Hu Y, Guo L, Deng X, Chen G, Jiajia J, Zhang H (2006) Neurotrophic property of geniposide for inducing the neuronal differentiation of PC12 cells. Int J Dev Neurosci 24:419–424. https://doi.org/10.1016/J.IJDEVNEU.2006.08.009

Liu F, Luo K, Yu Z, Co N, Wu S, Wu P, Fung K, Kwok T (2009) Suillin from the mushroom *Suillus placidus* as potent apoptosis inducer in human hepatoma HepG2 cells. Chem Biol Interact 181:168–174. https://doi.org/10.1016/j.cbi.2009.07.008

Lu MK, Cheng JJ, Lai WL, Lin YJ, Huang NK (2008) Fermented *Antrodia cinnamomea* extract protects rat PC12 cells from serum deprivation-induced apoptosis: the role of the MAPK family†. J Agric Food Chem 56:865–874. https://doi.org/10.1021/jf072828b

Lu TL, Huang GJ, Lu TJ, Wu JB, Wu CH, Yang TC, Iizuka A, Chen YF (2009) Hispolon from *Phellinus linteus* has antiproliferative effects via MDM2-recruited ERK1/2 activity in breast and bladder cancer cells. Food Chem Toxicol 47:2013–2021. https://doi.org/10.1016/j.fct.2009.05.023

Manna DK, Nandi AK, Pattanayak M, Maity P, Tripathy S, Mandal AK, Roy S, Tripathy SS, Gupta N, Islam SS (2015) A water soluble β-glucan of an edible mushroom *Termitomyces heimii*: structural and biological investigation. Carbohydr Polym 134:375–384. https://doi.org/10.1016/j.carbpol.2015.07.099

Marcotullio M, Pagiott R, Maltese F, Obara Y, Hoshino T, Nakahata N, Curini M (2006) Neurite outgrowth activity of cyathane diterpenes from *Sarcodon cyrneus*, cyrneines A and B. Planta Med 72:819–823. https://doi.org/10.1055/s-2006-946681

Martin KR, Brophy SK (2010) Commonly consumed and specialty dietary mushrooms reduce cellular proliferation in MCF-7 human breast cancer cells. Exp Biol Med 235:1306–1314. https://doi.org/10.1258/ebm.2010.010113

Martin KJ, Simon J, Arthur C (2012) Selective kinase inhibitors as tools for neuroscience research. Neuropharmacology 63:1227–1237. https://doi.org/10.1016/j.neuropharm.2012.07.024

Martorana A, Bulati M, Buffa S, Pellicanò M, Caruso C, Candore G, Colonna-Romano G (2012) Immunosenescence, inflammation and Alzheimer's disease. Longev Health 1:8. https://doi.org/10.1186/2046-2395-1-8

Matsuzaki H, Shimizu Y, Iwata N, Kamiuchi S, Suzuki F, Iizuka H, Hibino Y, Okazaki M (2013) Antidepressant-like effects of a water-soluble extract from the culture medium of *Ganoderma lucidum mycelia* in rats. BMC Complement Altern Med 13:370. https://doi.org/10.1186/1472-6882-13-370

Minari MC, Rincão VP, Soares SA, Ricardo NM, Nozawa C, Linhares RE (2011) Antiviral properties of polysaccharides from *Agaricus brasiliensis* in the replication of bovine herpesvirus 1. Acta Virol 55:255–259

Mishra SK, Kang JH, Kim DK, Oh SH, Kim MK (2012) Orally administered aqueous extract of *Inonotus obliquus* ameliorates acute inflammation in dextran sulfate sodium (DSS)-induced colitis in mice. J Ethnopharmacol 143:524–532. https://doi.org/10.1016/j.jep.2012.07.008

Mizuno T (1995) Bioactive biomolecules of mushrooms: food function and medicinal effect of mushroom fungi. Food Rev Int 11:5–21. https://doi.org/10.1080/87559129509541017

Mizuno T (1999) The extraction and development of antitumor-active polysaccharides from medicinal mushrooms in Japan (Review). Int J Med Mushrooms 1:9–29. https://doi.org/10.1615/IntJMedMushrooms.v1.i1.20

Mori K, Inatomi S, Ouchi K, Azumi Y, Tuchida T (2009) Improving effects of the mushroom Yamabushitake (*Hericium erinaceus*) on mild cognitive impairment: a double-blind placebo-controlled clinical trial. Phyther Res 23:367–372. https://doi.org/10.1002/ptr.2634

Mori K, Obara Y, Moriya T, Inatomi S, Nakahata N (2011) Effects of Hericium erinaceus on amyloid β(25-35) peptide-induced learning and memory deficits in mice. Biomed Res 32:67–72

Morigiwa A, Kitabatake K, Fujimoto Y, Ikekawa N (1986) Angiotensin converting enzyme-inhibitory triterpenes from *Ganoderma lucidum*. Chem Pharm Bull (Tokyo) 34:3025–3028

Mothana RA, Jansen R, Jülich WD, Lindequist U (2000) Ganomycins A and B, new antimicrobial farnesyl hydroquinones from the basidiomycete *Ganoderma pfeifferi*. J Nat Prod 63:416–418

Mothana RAA, Awadh Ali NA, Jansen R, Wegner U, Mentel R, Lindequist U (2003) Antiviral lanostanoid triterpenes from the fungus *Ganoderma pfeifferi*. Fitoterapia 74:177–180

Mu H, Zhang A, Zhang W, Cui G, Wang S, Duan J (2012) Antioxidative properties of crude polysaccharides from *Inonotus obliquus*. Int J Mol Sci 13:9194–9206. https://doi.org/10.3390/ijms13079194

Nada SA, Omara EA, Abdel Salam OME, Zahran HG (2010) Mushroom insoluble polysaccharides prevent carbon tetrachloride-induced hepatotoxicity in rat. Food Chem Toxicol 48:3184–3188. https://doi.org/10.1016/j.fct.2010.08.019

Nandi AK, Samanta S, Maity S, Sen IK, Khatua S, Devi KSP, Acharya K, Maiti TK, Islam SS (2014) Antioxidant and immunostimulant β-glucan from edible mushroom *Russula albonigra* (Krombh.) Fr. Carbohydr Polym 99:774–782. https://doi.org/10.1016/j.carbpol.2013.09.016

Ng Y (2005) In vitro and in vivo antioxidant activity and hypocholesterolemic effect in extracts of *Agrocybe aegerita*. The Chinese University of Hong Kong, Hong Kong

Niidome T, Taniuchi N, Akaike A, Kihara T, Sugimoto H (2008) Differential regulation of neurogenesis in two neurogenic regions of APPswe/PS1dE9 transgenic mice. Neuroreport 19:1361–1364. https://doi.org/10.1097/WNR.0b013e32830e6dd6

Nishina A, Kimura H, Sekiguchi A, Fukumoto R, Nakajima S, Furukawa S (2006) Lysophosphatidylethanolamine in *Grifola frondosa* as a neurotrophic activator via activation of MAPK. J Lipid Res 47:1434–1443. https://doi.org/10.1194/jlr.M600045-JLR200

Okuyama S, Lam NV, Hatakeyama T, Terashima T, Yamagata K, Yokogoshi H (2004) *Mycoleptodonoides aitchisonii* affects brain nerve growth factor concentration in newborn rats. Nutr Neurosci 7:341–349. https://doi.org/10.1080/10284150400020490

Okuyama S, Tashiro K, Lam NV, Hatakeyama T, Terashima T, Yokogoshi H (2012) Effect of the edible mushroom *Mycoleptodonoides aitchisonii* on transient global ischemia-induced monoamine metabolism changes in rat cerebral cortex. J Med Food 15:96–99. https://doi.org/10.1089/jmf.2011.1636

Ooi VEC (1996) Hepatoprotective effect of some edible mushrooms. Phyther Res 10:536–538. https://doi.org/10.1002/(SICI)1099-1573(199609)10:6<536::AID-PTR893>3.0.CO;2-Y

Ooi VE, Liu F (2000) Immunomodulation and anti-cancer activity of polysaccharide-protein complexes. Curr Med Chem 7:715–729

Park SE, Kim J, Lee YW, Yoo HS, Cho CK (2009) Antitumor activity of water extracts from *Cordyceps Militaris* in NCI-H460 cell xenografted nude mice. J Acupunct Meridian Stud 2:294–300. https://doi.org/10.1016/S2005-2901(09)60071-6

Patel S, Goyal A (2012) Recent developments in mushrooms as anti-cancer therapeutics: a review. 3 Biotech 2:1–15. https://doi.org/10.1007/s13205-011-0036-2

Pinweha S, Wanikiat P, Sanvarinda Y, Supavilai P (2008) The signaling cascades of *Ganoderma lucidum* extracts in stimulating non-amyloidogenic protein secretion in human neuroblastoma SH-SY5Y cell lines. Neurosci Lett 448:62–66. https://doi.org/10.1016/j.neulet.2008.10.028

Powell M (2010) Medicinal mushrooms: a clinical guide, 1st edn. Mycology Press, East Sussex

Rahar S, Swami G, Nagpal N, Nagpal M, Singh G (2011) Preparation, characterization, and biological properties of β-glucans. J Adv Pharm Technol Res 2:94. https://doi.org/10.4103/2231-4040.82953

Razumov IA, Kazachinskaia EI, Puchkova LI, Kosogorova TA, Gorbunova IA, Loktev VB, Tepliakova TV (2013) Protective activity of aqueous extracts from higher mushrooms against Herpes simplex virus type-2 on albino mice model. Antibiot i khimioterapiia = Antibiot chemoterapy [sic] 58:8–12

Rincão VP, Yamamoto KA, Ricardo NMPS, Soares SA, Meirelles LDP, Nozawa C, Linhares REC (2012) Polysaccharide and extracts from *Lentinula edodes*: structural features and antiviral activity. Virol J 9:37. https://doi.org/10.1186/1743-422X-9-37

Rodríguez JJ, Jones VC, Tabuchi M, Allan SM, Knight EM, LaFerla FM, Oddo S, Verkhratsky A (2008) Impaired adult neurogenesis in the dentate gyrus of a triple transgenic mouse model of Alzheimer's disease. PLoS One 3:e2935. https://doi.org/10.1371/journal.pone.0002935

Ryong LH, Tertov VV, Vasil'ev AV, Tutel'yan VA, Orekhov AN (1989) Antiatherogenic and anti-atherosclerotic effects of mushroom extracts revealed in human aortic intima cell culture. Drug Dev Res 17:109–117. https://doi.org/10.1002/ddr.430170203

Samanta S, Nandi AK, Sen IK, Maity P, Pattanayak M, Devi KSP, Khatua S, Maiti TK, Acharya K, Islam SS (2015) Studies on antioxidative and immunostimulating fucogalactan of the edible mushroom *Macrolepiota dolichaula*. Carbohydr Res 413:22–29. https://doi.org/10.1016/j.carres.2015.05.006

Sanodiya BS, Thakur GS, Baghel RK, Prasad GBKS, Bisen PS (2009) *Ganoderma lucidum*: a potent pharmacological macrofungus. Curr Pharm Biotechnol 10:717–742

Sen IK, Mandal AK, Chakraborti S, Dey B, Chakraborty R, Islam SS (2013) Green synthesis of silver nanoparticles using glucan from mushroom and study of antibacterial activity. Int J Biol Macromol 62:439–449. https://doi.org/10.1016/j.ijbiomac.2013.09.019

Seniuk OF, Gorovoj LF, Beketova GV, Savichuk HO, Rytik PG, Kucherov II, Prilutskay AB, Prilutsky AI (2011) Anti-infective properties of the melanin-glucan complex obtained from medicinal tinder bracket mushroom, Fomes fomentarius (L.: Fr.) Fr. (Aphyllophoromycetideae). Int J Med Mushrooms 13:7–18

Sheng J, Chen Q (1990) Effects of polysaccharides from *Auricularia auricula*, T*remella fuciformis*, and *Tremella fuciformis* spores on experimental thrombin formation. Zhongguo Yaoke Daxue Xuebae 21:39–42

Shi Y, James AE, Benzie IFF, Buswell JA (2002) Mushroom-derived preparations in the prevention of H_2O_2-induced oxidative damage to cellular DNA. Teratog Carcinog Mutagen 22:103–111

Shi JY, Liu GS, Liu LF, Kuo SM, Ton CH, Wen ZH, Tee R, Chen CH, Huang HT, Chen CL, Chao D, Tai MH (2011) Glial cell line–derived neurotrophic factor gene transfer exerts protective effect on axons in sciatic nerve following constriction-induced peripheral nerve injury. Hum Gene Ther 22:721–731. https://doi.org/10.1089/hum.2010.036

Signoretto C, Marchi A, Bertoncelli A, Burlacchini G, Papetti A, Pruzzo C, Zaura E, Lingström P, Ofek I, Pratten J, Spratt DA, Wilson M, Canepari P (2014) The anti-adhesive mode of action of a purified mushroom (*Lentinus edodes*) extract with anticaries and antigingivitis properties in two oral bacterial pathogens. BMC Complement Altern Med 14:75. https://doi.org/10.1186/1472-6882-14-75

Socala K, Nieoczym D, Grzywnowicz K, Stefaniuk D, Wlaz P (2015) Evaluation of anticonvulsant, antidepressant-, and anxiolytic-like effects of an aqueous extract from cultured mycelia of the lingzhi or reishi medicinal mushroom *Ganoderma lucidum* (higher basidiomycetes) in mice. Int J Med Mushrooms 17:209–218. https://doi.org/10.1615/IntJMedMushrooms.v17.i3.10

Soković M, Ćirić A, Glamočlija J, Nikolić M, Van Griensven L (2014) *Agaricus Blazei* hot water extract shows anti quorum sensing activity in the nosocomial human pathogen Pseudomonas aeruginosa. Molecules 19:4189–4199. https://doi.org/10.3390/molecules19044189

Song YS, Kim SH, Sa JH, Jin C, Lim CJ, Park EH (2004) Anti-angiogenic and inhibitory activity on inducible nitric oxide production of the mushroom *Ganoderma lucidum*. J Ethnopharmacol 90:17–20

Song FQ, Liu Y, Kong XS, Chang W, Song G (2013) Progress on understanding the anticancer mechanisms of medicinal mushroom: *Inonotus obliquus*. Asian Pac J Cancer Prev 14:1571–1578

Su CY, Shiao MS, Wang CT (1999) Predominant inhibition of ganodermic acid S on the thromboxane A2-dependent pathway in human platelets response to collagen. Biochim Biophys Acta Mol Cell Biol Lipids 1437:223–234. https://doi.org/10.1016/S1388-1981(98)00012-2

Sugiyama K, Akachi T, Yamakawa A (1995) Hypocholesterolemic action of eritadenine is mediated by a modification of hepatic phospholipid metabolism in rats. J Nutr 125:2134–2144

Sun M, Xiao J, Zhang S, Liu Y, Li S (1984) Therapeutic effect of some foods on hyperlipidemia in man. Acta Nutr Sin 6:127–133

Sun YQ, Guo TK, Xi YM, Chen C, Wang J, Wang ZR (2007) Effects of AZT and RNA-protein complex (FA-2-b-beta) extracted from *Liang Jin* mushroom on apoptosis of gastric cancer cells. World J Gastroenterol 13:4185–4191

Suzuki S, Ohshima S (1976) Influence of shiitake (*Lentinus edodes*) on human serum cholesterol. Mushroom Sci 9:463–470

Tangen JM, Tierens A, Caers J, Binsfeld M, Olstad OK, Trøseid AMS, Wang J, Tjønnfjord GE, Hetland G (2015) Immunomodulatory effects of the *Agaricus blazei* murill-based mushroom extract andoSan in patients with multiple myeloma undergoing high dose chemotherapy and autologous stem cell transplantation: a randomized, double blinded clinical study. Biomed Res Int 2015:1–11. https://doi.org/10.1155/2015/718539

Tello I, Campos Pena V, Montiel E, Rodriguez V, Aguirre Moreno A, Leon Rivera I, Del Rio Portilla F, Herrera Ruiz M, Villeda Hernandez J (2013) Anticonvulsant and neuroprotective effects of oligosaccharides from Lingzhi or Reishi medicinal mushroom, *Ganoderma lucidum* (Higher Basidiomycetes). Int J Med Mushrooms 15:555–568

Tochikura TS, Nakashima H, Yamamoto N (1989) Antiviral agents with activity against human retroviruses. J Acquir Immune Defic Syndr 2:441–447

Tokita F, Shibukawa N, Yasumoto T, Kaneda T (1972) Isolation and chemical structure of the plasma-cholesterol reducing substance from shiitake mushroom. Mushroom Sci 8:783–788

Vincent M, Philippe E, Everard A, Kassis N, Rouch C, Denom J, Takeda Y, Uchiyama S, Delzenne NM, Cani PD, Migrenne S, Magnan C (2013) Dietary supplementation with *Agaricus blazei* murill extract prevents diet-induced obesity and insulin resistance in rats. Obesity 21:553–561. https://doi.org/10.1002/oby.20276

Wang SY, Hsu ML, Hsu HC, Tzeng CH, Lee SS, Shiao MS, Ho CK (1997) The anti-tumor effect of *Ganoderma lucidum* is mediated by cytokines released from activated macrophages and T lymphocytes. Int J Cancer 70:699–705

Wang MF, Chan YC, Wu CL, Wong YC, Hosoda K, Yamamoto S (2004) Effects of Ganoderma on aging and learning and memory ability in senescence accelerated mice. Int Congr Ser 1260:399–404. https://doi.org/10.1016/S0531-5131(03)01682-0

Wang CZ, Basila D, Aung HH, Mehendale SR, Chang WT, McEntee E, Guan X, Yuan CS (2005) Effects of *Ganoderma lucidum* extract on chemotherapy-induced nausea and vomiting in a rat model. Am J Chin Med 33:807–815. https://doi.org/10.1142/S0192415X05003429

Wang JM, Singh C, Liu L, Irwin RW, Chen S, Chung EJ, Thompson RF, Brinton RD (2010) Allopregnanolone reverses neurogenic and cognitive deficits in mouse model of Alzheimer's disease. Proc Natl Acad Sci 107:6498–6503. https://doi.org/10.1073/pnas.1001422107

Wang CR, Ng TB, Li L, Fang JC, Jiang Y, Wen TY, Qiao WT, Li N, Liu F (2011) Isolation of a polysaccharide with antiproliferative, hypoglycemic, antioxidant and HIV-1 reverse transcriptase inhibitory activities from the fruiting bodies of the abalone mushroom *Pleurotus abalonus*. J Pharm Pharmacol 63:825–832. https://doi.org/10.1111/j.2042-7158.2011.01274.x

Wang HY, Bakshi K, Frankfurt M, Stucky A, Goberdhan M, Shah SM, Burns LH (2012a) Neurobiology of disease reducing amyloid-related Alzheimer's disease pathogenesis by a small molecule targeting filamin A. J Neurosci 32:9773–9784. https://doi.org/10.1523/JNEUROSCI.0354-12.2012

Wang LC, Wang SE, Wang JJ, Tsai TY, Lin CH, Pan TM, Lee CL (2012b) In vitro and in vivo comparisons of the effects of the fruiting body and mycelium of *Antrodia camphorata* against amyloid β-protein-induced neurotoxicity and memory impairment. Appl Microbiol Biotechnol 94:1505–1519. https://doi.org/10.1007/s00253-012-3941-3

Wang J, Yu G, Li Y, Shen L, Qian Y, Yang J, Wang F (2015) Inhibitory effects of sulfated lentinan with different degree of sulfation against tobacco mosaic virus (TMV) in tobacco seedlings. Pestic Biochem Physiol 122:38–43. https://doi.org/10.1016/j.pestbp.2014.12.027

Wasser SP (2005) Reishi or Ling Zhi (*Ganoderma lucidum*). https://doi.org/10.1081/E-EDS-120022119

Wasser SP (2010) Medicinal mushroom science: History, current status, future trends, and unsolved problems. Int J Med Mushrooms 12:1–16. https://doi.org/10.1615/IntJMedMushr.v12.i1.10

Wasser S (2014) Medicinal mushroom science: current perspectives, advances, evidences, and challenges. Biomed J 37:345. https://doi.org/10.4103/2319-4170.138318

Wasser SP, Weis AL (1999) Therapeutic effects of substances occurring in higher Basidiomycetes mushrooms: a modern perspective. Crit Rev Immunol 19:65–96

Watanabe H, Kashimoto N, Ushijima M, Tamura K (2013) Effects of a water-soluble extract of *Ganoderma lucidum* mycelia on aberrant crypt foci induced by azoxymethane and small-intestinal injury by 5-FU in F344 rats. Med Mol Morphol 46:97–103. https://doi.org/10.1007/s00795-013-0012-5

Willem M, Garratt AN, Novak B, Citron M, Kaufmann S, Rittger A, DeStrooper B, Saftig P, Birchmeier C, Haass C (2006) Control of peripheral nerve myelination by the -secretase BACE1. Science (80-) 314:664–666. https://doi.org/10.1126/science.1132341

Wong KH, Vikineswary S, Abdullah N, Naidu M, Keynes R (2007) Activity of aqueous extracts of lion's mane mushroom *Hericium erinaceus* (Bull.: Fr.) Pers. (Aphyllophoromycetideae) on the neural cell line NG108-15. Int J Med Mushrooms 9:57–65. https://doi.org/10.1615/IntJMedMushr.v9.i1.70

Wong K-H, Naidu M, David P, Abdulla MA, Abdullah N, Kuppusamy UR, Sabaratnam V (2011) Peripheral nerve regeneration following crush injury to rat peroneal nerve by aqueous extract of medicinal mushroom *Hericium erinaceus* (Bull.: Fr) Pers. (Aphyllophoromycetideae). Evid Based Complement Alternat Med 2011:580752. https://doi.org/10.1093/ecam/neq062

Wu GH, Hu T, Li ZY, Huang ZL, Jiang JG (2014) In vitro antioxidant activities of the polysaccharides from Pleurotus tuber-regium (Fr.) Sing. Food Chem 148:351–356. https://doi.org/10.1016/j.foodchem.2013.10.029

Xiao JH, Xiao DM, Chen DX, Xiao Y, Liang ZQ, Zhong JJ (2012) Polysaccharides from the medicinal mushroom *Cordyceps taii* show antioxidant and immunoenhancing activities in a *D*-galactose-induced aging mouse model. Evidence-Based Complement Altern Med 2012:1–15. https://doi.org/10.1155/2012/273435

Xue Z, Li J, Cheng A, Yu W, Zhang Z, Kou X, Zhou F (2015) Structure identification of triterpene from the mushroom *Pleurotus eryngii* with inhibitory effects against breast cancer. Plant Foods Hum Nutr 70:291–296. https://doi.org/10.1007/s11130-015-0492-7

Yamac M, Kanbak G, Zeytinoglu M, Bayramoglu G, Senturk H, Uyanoglu M (2008) Hypoglycemic effect of *Lentinus strigosus* (Schwein.) Fr. crude exopolysaccharide in streptozotocin-induced diabetic rats. J Med Food 11:513–517. https://doi.org/10.1089/jmf.2007.0551

Yamamoto KA, Galhardi LCF, Rincão VP, Soares S de A, Vieira ÍGP, Ricardo NMPS, Nozawa C, Linhares REC (2013) Antiherpetic activity of an *Agaricus brasiliensis* polysaccharide, its sulfated derivative and fractions. Int J Biol Macromol 52:9–13. https://doi.org/10.1016/j.ijbiomac.2012.09.029

Ye L, Zheng X, Zhang J, Tang Q, Yang Y, Wang X, Li J, Liu Y, Pan Y (2011) Biochemical characterization of a proteoglycan complex from an edible mushroom *Ganoderma lucidum* fruiting bodies and its immunoregulatory activity. Food Res Int 44:367–372. https://doi.org/10.1016/j.foodres.2010.10.004

Yeung M, Cheung P (2002) *The hypolipidemic effect of some lesser-known edible and medicinal mushrooms*. In Abstracts of the Institute of Food Technologists' Annual Meeting. Institute of Food Technologists, Anaheim

Yin X, You Q, Zhou X (2015) Complex enzyme-assisted extraction, purification, and antioxidant activity of polysaccharides from the button mushroom, *Agaricus bisporus* (higher basidiomycetes). Int J Med Mushrooms 17:987–996

Ying C, Wang Y, Tang H (1989) Icons of medicinal fungi from China. Science Press, Beijing

Ytrebo LM (2006) Interorgan ammonia, glutamate, and glutamine trafficking in pigs with acute liver failure. AJP Gastrointest Liver Physiol 291:G373–G381. https://doi.org/10.1152/ajpgi.00440.2005

Yuan D, Mori J, Komatsu K, Makino T, Kano Y (2004) An anti-aldosteronic diuretic component (drain dampness) in Polyporus sclerotium. Biol Pharm Bull 27:867–870

Yuen JWM, Gohel MDI (2008) The dual roles of Ganoderma antioxidants on urothelial cell DNA under carcinogenic attack. J Ethnopharmacol 118:324–330. https://doi.org/10.1016/J.JEP.2008.05.003

Zhang W, Liu HT (2002) MAPK signal pathways in the regulation of cell proliferation in mammalian cells. Cell Res 12:9–18. https://doi.org/10.1038/sj.cr.7290105

Zhang M, Cui SW, Cheung PCK, Wang Q (2007) Antitumor polysaccharides from mushrooms: a review on their isolation process, structural characteristics and antitumor activity. Trends Food Sci Technol 18:4–19. https://doi.org/10.1016/j.tifs.2006.07.013

Zhang ZC, Lian B, Huang DM, Cui FJ (2009) Compare activities on regulating lipid-metabolism and reducing oxidative stress of diabetic rats of *Tremella Aurantialba* Broth's Extract (TBE) with its mycelia polysaccharides (TMP). J Food Sci 74:H15–H21. https://doi.org/10.1111/j.1750-3841.2008.00989.x

Zhang R, Xu S, Cai Y, Zhou M, Zuo X, Chan P (2011) *Ganoderma lucidum* protects dopaminergic neuron degeneration through inhibition of microglial activation. Evidence-Based Complement Altern Med 2011:1–9. https://doi.org/10.1093/ecam/nep075

Zhang Z, Lv G, Pan H, Pandey A, He W, Fan L (2012) Antioxidant and hepatoprotective potential of endo-polysaccharides from *Hericium erinaceus* grown on tofu whey. Int J Biol Macromol 51:1140–1146. https://doi.org/10.1016/j.ijbiomac.2012.09.002

Zhao HB, Lin SQ, Liu JH, Lin ZB (2004) Polysaccharide extract isolated from *ganoderma lucidum* protects rat cerebral cortical neurons from hypoxia/reoxygenation injury. J Pharmacol Sci 95:294–298

Zheng L, Jia D, Fei X, Luo X, Yang Z (2009) An assessment of the genetic diversity within Ganoderma strains with AFLP and ITS PCR-RFLP. Microbiol Res 164:312–321. https://doi.org/10.1016/j.micres.2007.02.002

Zhou C, Jia W, Yang Y, Bai Y (2002) Experimental studies on prevention of several kinds of fungi polysaccharides against alcohol-induced hepatic injury. Edible Fungi 24:36–37

Zhu W, Liu Z, Xu H, Chu W, Ye Q, Xie A, Chen L, Li J (2005) Effect of the oil from *ganoderma lucidum* spores on pathological changes in the substantia nigra and behaviors of MPTP-treated mice. Di Yi Jun Yi Da Xue Xue Bao 25:667–671

Zhu H, Liu W, Tian B, Liu H, Ning S (2011) Inhibition of quorum sensing in the opportunistic pathogenic bacterium Chromobacterium violaceum by an extract from fruiting bodies of Lingzhi or Reishi medicinal mushroom, *Ganoderma lucidum* (W.Curt.:Fr.) P. Karst. (higher Basidiomycetes). Int J Med Mushrooms 13:559–564

Zhu L, Tang Q, Zhou S, Liu Y, Zhang Z, Gao X, Wang S, Wang Z (2014a) Isolation and purification of a polysaccharide from the caterpillar medicinal mushroom *Cordyceps militaris* (Ascomycetes) fruit bodies and its immunomodulation of RAW 264.7 macrophages. Int J Med Mushrooms 16:247–257

Zhu ZY, Zhang JY, Chen LJ, Liu XC, Liu Y, Wang WX, Zhang YM (2014b) Comparative evaluation of polysaccharides isolated from Astragalus, oyster mushroom, and yacon as inhibitors of α-glucosidase. Chin J Nat Med 12:290–293. https://doi.org/10.1016/S1875-5364(14)60056-X

Zou Y, Liao D, Huang H, Li T, Chi H (2015) A systematic review and meta-analysis of beta-glucan consumption on glycemic control in hypercholesterolemic individuals. Int J Food Sci Nutr 66:355–362. https://doi.org/10.3109/09637486.2015.1034250

Mushrooms as Potential Natural Cytostatics

4

Mirjana Stajić, Jelena Vukojević, and Jasmina Ćilerdžić

Contents

M. Stajić (✉) · J. Vukojević · J. Ćilerdžić
Faculty of Biology, University of Belgrade, Belgrade, Serbia
e-mail: stajicm@bio.bg.ac.rs

D. C. Agrawal, M. Dhanasekaran (eds.), *Medicinal Mushrooms*,
https://doi.org/10.1007/978-981-13-6382-5_4

Abstract

Cancer is the second cause of morbidity and mortality worldwide, i.e., half of the men and more than a third of women of the world population get sick with some type of cancer during a lifetime, and one-quarter of all adults die of this disease. Common treatments, chemo- and radiotherapy, are not highly effective, give satisfactory results only in the treatment of early cancer development stages or have no any effect on some cancer types, and commonly cause numerous side effects. Therefore, alternative medicine based on various natural sources attracts great attention nowadays. Although mushrooms, their extracts, and isolated metabolites cannot be considered drugs, they are a type of important dietary supplement, i.e., functional food or nutraceuticals, and could be used as auxiliary natural cytostatics. They are highly selective, i.e., not toxic or almost nontoxic to normal cells, do not cause any side effects, and even reduce harmful effects caused by conventional treatments, and finally, resistance to them cannot be developed. Mushroom extracts or biologically active compounds isolated from them affect cytotoxic activity on a few mechanisms: stimulation of immune system; antioxidative, antimutagenic, and anti-inflammatory activity; regulation of expression of regulators of some cell processes; cell cycle arrest and apoptosis; disturbance of DNA synthesis and structure; changes in morphology and mobility of malignant cells; and antiangiogenic activity.

Keywords

Anticancer activity · Mechanisms of action · Mushrooms · Polysaccharides · Proteins · Terpenoids

Abbreviations

ABL	Lectin from *Agaricus bisporus*
AP-1	Transcription factor
APP	Protein from *Agaricus polytricha*
Bad, Bax, Bcl-2, and Bid	Apoptosis regulators
Cdk-1	Enzyme that catalyzes transition G_2 to M phase of cell cycle
Cdk-2	Enzyme that catalyzes transition G_1 to S phase of cell cycle
CNL	Immunomodulatory protein from *Clitocybe nebularis*
FIP-fve	Immunomodulatory protein from *Flammulina velutipes*
FIP-gts	Immunomodulatory protein from *Ganoderma tsugae*
GISP1b	Polysaccharide-peptide complex from *Grifola frondosa*
GPx	Glutathione peroxidase
IFN	Interferon
IL	Interleukin

LZ-D	Sulfate derivate of glycopeptide from *Ganoderma lucidum*
LZ-D-4	Glycopeptide from *Ganoderma lucidum*
NF-kB	Nuclear transcription factor
reFIP-gts	Recombinant of immunomodulatory protein from *Ganoderma tsugae*
SOD	Superoxide dismutase
TML-1 and TML-2	Lectins from *Tricholoma mongolicum*
TNF	Tumor necrosis factor

4.1 Introduction

High growth rate of the world population (about 80 million people per year), the lack of food in underdeveloped countries or irregular diet in development ones, environmental pollution, and climate changes are associated with a life quality decline, a large number of diseases and disorders, and a shortening of lifespan and present the main problems which humanity of the twenty-first century faces with. Today, the second causal agent of death in the world, after cardiovascular diseases, is cancer or neoplasm. According to Parker (2014), half of the men and more than a third of women of the world population get sick with some type of cancer during a lifetime, and one-quarter of all adults die of this disease. The World Health Organization reported that only in the United States, 1,596,670 persons were new sufferers from cancer and even 571,950 ones died in 2011. The International Agency for Research on Cancer estimates that even over 21 million persons worldwide will be new cancer–diagnosed each year until 2030. If diversity of cancers, their curability rate, and treatment cost are added to these data, it is clear that strategies for prevention as well as new more efficient, healthy, and economically friendly curing have to be developed. However, to successfully combat against this "fourth apocalyptic rider" requires a good knowledge of causal agents of its appearance as well as the mechanism of its development.

4.2 Cancer Causal Agents and Development Mechanism

Numerous chemical, physical, and biological agents can be potential carcinogens, i.e., can cause oxidative stress and/or inflammation and consequently direct and irreversible changes in the genome transforming the normal cell into the neoplastic one (Reuter et al. 2010). In the case of mechanism, any cancer type develops through three phases: *(i)* initiation when mutagenic agent binds to DNA and causes damage, *(ii)* activation when tumor promoter is activated and small benign tumors are formed, and *(iii)* progression when control of the cell cycle is lost and cells begin to divide uncontrollably and unlimitedly, i.e., to proliferate (Reuter et al. 2010).

Oxidative stress, i.e., reactive oxygen and nitrogen species (O_2^-, H_2O_2, •OH, NO), can initiate changes in structure, function, and activity of numerous cell molecules and processes, such as (1) breaking DNA chain; (2) increased frequency of point mutations, mutations in proto-oncogenes and tumor suppressor genes; (3) disrupted gene expression; (4) activation of the nuclear transcription factor κB (NF-κB) and in such a way induction of immunosuppressive, inflammatory, and anti-apoptosis gene transcription; (5) modification of cell morphology and adhesion; (6) transition of normal cell to a more mobile and invasive phenotype; (7) loss of normal cell polarism and differentiation; (8) blocking the communication among cells; (9) modification of intracellular signaling molecules and pathways; (10) activation of metalloproteinases; (11) increased rate of cell migration; and *(xii)* intensive production of angiogenic factors (Folkman 1985; Shacter et al. 1988; Liotta et al. 1991; Kundu et al. 1995; Kim and Kim 1999; Westermarck and Kahari 1999; Jackson and Loeb 2001; Storz 2005; Kollmar et al. 2007; Yuecheng and Xiaoyan 2007; Petrova et al. 2008; Diers et al. 2010; Reuter et al. 2010). The main consequences of these changes are uncontrolled cell division, i.e., proliferation, the absence of apoptosis, and finally cancer development, aggressive neovascularization, and metastases (Fig. 4.1).

More than 20% of all human cancers (cancer of the prostate, cervix, esophagus, stomach, liver, colon, pancreas, and bladder) are caused by chronic inflammations or infections, which could be caused by radicals, allergens, various chemicals, radiation, bacterial and viral infections, autoimmune and chronic diseases, high-calorie

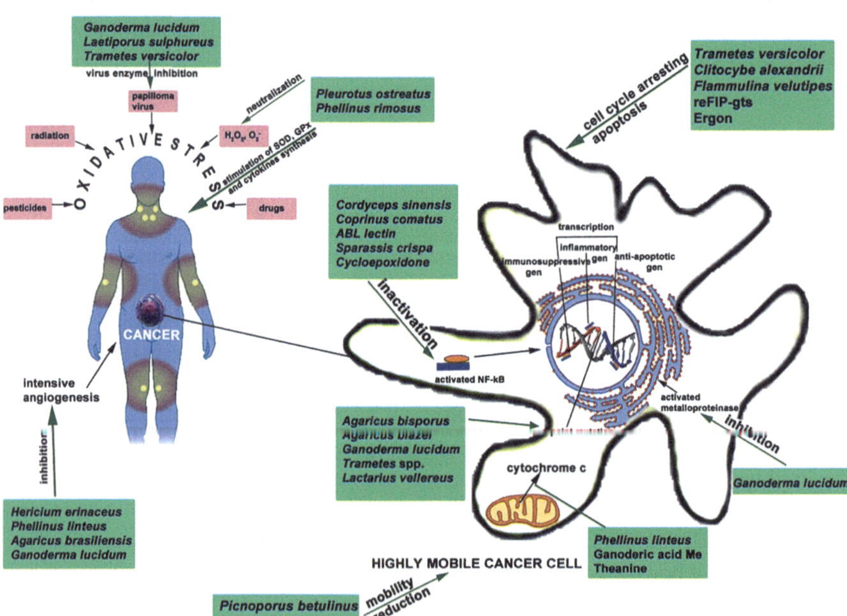

Fig. 4.1 Mechanisms of mushroom anticancer effect (schematic presentation)

diet and obesity, alcohol consumption, and smoking (Kundu and Surh 2008; Grivennikov and Karin 2010). The traditional Indian Ayurvedic medical system emphasized, even 5000 years ago, that continuous irritation during a long period can cause cancer, while Rudolf Virchow, in the nineteenth century, noted the presence of inflammatory cells into the tumor as well as the development of tumor at places of chronic inflammation (Garodia et al. 2007; Schetter et al. 2010). Modern medicine confirmed that fact, i.e., showed that Crohn's disease and ulcerative colitis, common inflammatory intestine diseases, increase the risk of colon adenocarcinoma; chronic pancreatitis and esophagitis are associated with increased rate of the pancreas and esophagus cancer, while bronchitis can lead to lung cancer (Reuter et al. 2010). Coussens and Werb (2002) and Federico et al. (2007) reported that leukocytes and immune cells containing histamine- and heparin-rich granules group during inflammation at the site of the lesion, and owing to their intensive respiration reactive oxygen species release, damage neighboring healthy cells and can cause carcinogenesis. Additionally, Estrela et al. (2006) demonstrated that the inflammatory response can also induce changes in the expression and activity of proteins involved in the transportation of drugs and in such a way cause occurrence of resistance to conventional anticancer therapies.

Generally, carcinogenesis can be initiated directly by the radicals, i.e., by oxidation, nitration, and/or halogenation of nuclear DNA, RNA, and lipids, or indirectly by activation of certain metabolic pathways.

4.3 Cancer Treatments

No matter what causes cancer, nowadays patients suffering from cancer can be treated in several ways depending on cancer type. The main treatment of majority cancer type is surgery commonly combined with chemo- and/or radiotherapy. However, chemo- and radiotherapy give satisfactory results only in the treatment of early cancer development stages or are not effective for some cancer types and also can cause numerous side effects (Chen et al. 2006). Likewise, it should be emphasized that cytostatics that are currently available on the world market are not tumor-specific and cause numerous harmful effects in patients, which indicate that new effective and less toxic preparations are necessary. Therefore, the American Cancer Society and National Cancer Institute developed and introduced a few new treatments, such as targeted, hormone, and photodynamic therapies, blood transfusion and stem cell transplant, hyperthermia, precision medicine, laser treatment, and particularly immunotherapy. Currently, scientists pay more and more attention to integrative medicine, i.e., a combination of conventional medicine with complementary and alternative ones. Today, besides massage, acupuncture, tai chi, and various diets, preparations based on plants and mushrooms or their active compounds attract the great attention of the researchers.

There are numerous data that demonstrated very important place of mushrooms in folk medicines of Far Eastern and Eastern European countries. In China, use of mushrooms in treatments of various diseases has a history long-lasting about

2000 years. Thus, *Piptoporus betulinus* and *Inonotus obliquus* have been used traditionally for the treatment of stomach diseases and rectal cancer (Lindequist et al. 2005). In 1858, Doctor Froben from the Moscow University Clinic described a case of a healing patient who suffered from the cancer of the glands around the ear with the juice from *P. betulinus*, and in 1862, Doctor Furht from the Velikogo Luka near St. Petersburg reported the success of this juice in the treatment of lip cancer (Stajić 2015).

Nowadays, mushrooms are mostly used as food and in the form of mycelium and fruiting body extracts, while a relatively small number of their active compounds are isolated and characterized. Therefore, mushrooms cannot be considered drugs, but a type of dietary supplement, i.e., functional food or nutraceuticals (Chang and Wasser 2012).

4.3.1 Anticancer Activity of Mushrooms and Various Extracts

During a few last decades, numerous in vitro studies of cytostatic activity of various macromycetes, i.e., their fresh and dry fruiting bodies as well as various extracts, have been done, and considerable cytotoxic capacity of some of them has been demonstrated (Table 4.1). Thus, *Agaricus sylvaticus* fruiting bodies are highly efficient in the treatment of patients with colon cancer, i.e., they significantly improved immunological parameters in those that take them in a daily dose of 30 mg/kg for 6 months (Fortes et al. 2009). The significant anticancer effect was also reported for *A. blazei* extracts by Patel and Goyal (2012). Namely, numerous chemotherapy side effects in patients who suffered from uterus, ovary, or prostate cancer, such as loss of appetite, alopecia, emotional instability, and weakness, were reduced by combined *A. blazei* extract and chemotherapy treatment, and in such a way, life quality was significantly raised. These extracts also showed selective antitumor activity against two leukemia cell lines, in vitro (Kim et al. 2009). The various extracts of *A. bisporus* fruiting bodies have shown a cytotoxic effect on many cancer cell lines, too. Thus, for example, its water extract is highly active against breast cancer cell line (Grube et al. 2001). Likewise, strong antiproliferative and pro-apoptotic effects on various cancer cell lines can be caused by species of the genus *Pleurotus*. *P. ostreatus* water extract induces these effects on the colon cancer cells, *P. ferulae* extracts are active against lung and cervix cancer cells, *P. eryngii* against colon and cervix cancer cells, and *P. pulmonarius* ones against hepatoma cells (Choi et al. 2004; Lavi et al. 2006; Wasonga et al. 2008; Milovanović et al. 2014). High efficiency against leukemia and skin cancer cells was reported for extracts of *Lentinus edodes* fruiting bodies (Gu and Belury 2005), while *Flammulina velutipes* extracts are effective in suppression of breast, colon, and cervix cancers (Gu and Leonard 2006; Milovanović et al. 2015a).

The centuries-old experience in the usage of *Ganoderma* spp. fruiting bodies for the treatment of various diseases has led scientists to study their anticancer potential. The obtained results confirmed the assumption that fruiting bodies and mycelia of these species and their extracts and compounds have a strong cytostatic effect on

Table 4.1 Biological activity of selected mushroom extracts

Extract sources	Target cancer cells	Mechanism of action	References
Agaricus bisporus	Breast	Inhibition of aromatase activity, stimulation of caspase-3	Grube et al. (2001)
Agaricus sylvaticus	Colon	Stimulation of immune system	Fortes et al. (2009)
Agaricus blazei	Uterus, ovary, prostate, leukemia	Inhibition of telomerase activity, induction of cytochrome c release, caspase activity, regulation of Bcl-2 synthesis	Kim et al. (2009) and Patel and Goyal (2012)
Coprinus comatus	Breast, hepatoma	Inhibition of NF-κB factor	Asatiani et al. (2011)
Cordyceps militaris	Breast, hepatoma, leukemia	Induction of IL-18, IFN-γ	Rao et al. (2010) and Patel and Goyal (2012)
Cordyceps sinensis	Colon, hepatoma	Suppression of NF-κB factor	Wang et al. (2016)
Flammulina velutipes	Breast, cervix, colon	Arresting cell cycle	Gu and Leonard (2006) and Milovanović et al. (2015a)
Ganoderma lucidum	Cervix, colon, lung adenocarcinoma, prostate, stomach	Stimulation of TNF- γ, caspase activation, inhibition of transcription factor AP-1, inhibition of matrix metalloproteinase expression	Hong et al. (2004), Sadava et al. (2009), Ćilerdžić et al. (2014), and Milovanović et al. (2015b)
Hericium erinaceus	Colon	Activation of natural killer cells and macrophages, angiogenesis inhibition	Kim et al. (2011)
Lenzites betulinus	Cervix, colon	Antioxidative	Milovanović et al. (2015c)
Phellinus linteus	Bladder, breast, hepatoma, liver, lung, stomach	Stimulation of IL-12, IFN-γ, and TNF-α synthesis; macrophage and T-, CD^{4+}-, and dendritic cell proliferation; natural killer cell activation; increasing of the spleen and thymus mass and plasma immunoglobulin receptors; reduction of epidermal growth factor receptors; stimulation of NAD(P)H:quinone oxidoreductase, glutathione S-transferase	Guo et al. (2007), Sliva (2010), and Patel and Goyal (2012)
Pleurotus eryngii	Cervix, colon	Antioxidative activity	Milovanović et al. (2014)

(continued)

Table 4.1 (continued)

Extract sources	Target cancer cells	Mechanism of action	References
Pleurotus ferulae	Lung, uterus		Choi et al. (2004)
Pleurotus ostreatus	Cervix, colon	Antioxidative activity, increase level of cytokine	Lavi et al. (2006) and Milovanović et al. (2014)
Pleurotus pulmonarius	Cervix, hepatoma	Antioxidative activity	Milovanović et al. (2014)
Trametes versicolor	Cervix, colon, leukemia, lung, lymphoma, melanoma, prostate, stomach	Antimutagenic activity, antioxidative activity	Hsieh and Wu (2001), Chu et al. (2002), and Milovanović et al. (2015c)

numerous cancer cell lines, and therefore it is suggested their usage in combination with chemo- and radiotherapy. It was shown that basidiocarp extracts of various *G. lucidum* strains are significantly efficient in the inhibition of human papillomavirus 16 and in such a way in the prevention of cervix cancer appearance, than in the inhibition of colon, cervix, and lung adenocarcinoma cell line growth, in the reduction of viability of the stomach carcinoma cells, as well as in the combat against prostate cancer (Hong et al. 2004; Sadava et al. 2009; Ćilerdžić et al. 2014; Milovanović et al. 2015b). Strong antiproliferative effect on colon adenocarcinoma cells was also caused by *G. applanatum* and *G. tsugae* fruiting body and mycelium extracts (Hsu et al. 2008; Milovanović et al. 2015b), as well as by *Hericium erinaceus*, *Cordyceps militaris*, *Clitocybe alexandri*, and *Inonotus obliquus* fruiting bodies as well as *Lenzites betulinus* and *Funalia trogii* mycelia extracts (Hu et al. 2009; Rao et al. 2010; Vaz et al. 2010; Kim et al. 2011; Rashid et al. 2011; Milovanović et al. 2015c). *I. obliquus* extracts are also effective against melanoma; *C. sinensis* against hepatoma; *C. alexandri* against lung, breast, and stomach cancers; and *F. trogii* against prostate and breast cancers (Youn et al. 2009; Vaz et al. 2010; Rashid et al. 2011; Wang et al. 2016).

The extracts of *Trametes versicolor* fruiting bodies and mycelium have the extraordinary high anticancer capacity. Namely, numerous studies showed that the extracts could suppress prostate, stomach, cervix, colon, and lung cancers, as well as leukemia, lymphoma, and melanoma cell lines (Hsieh and Wu 2001; Chu et al. 2002; Milovanović et al. 2015c). *Phellinus linteus*, *Hypsizygus marmoreus*, and *Fomes fomentarius* extracts also possess strong cytotoxic activity against stomach and lung cancer cells, while *Cordyceps militaris* is highly active against leukemia (Saitoh et al. 1997; Ikekawa 2005; Guo et al. 2007; Sliva 2010; Patel and Goyal 2012). Thus, for example, Saitoh et al. (1997) reported that daily treatment of mice with lung carcinoma with *H. marmoreus* water extract prevented the development of metastases and significantly prolonged their life. Besides these effects, *Ph. linteus* extracts are also very efficient in the inhibition of proliferation of liver, breast, bladder, stomach, and lung cancer cells, in the reduction of tumor size, and in the prevention of metastases (Sliva 2010). Inhibition of hepatoma and breast cancer can be also caused by *Cordyceps militaris*, *Coprinus comatus*, and *Antrodia camphorata* (Chang et al. 2008; Rao et al. 2010; Asatiani et al. 2011; Yang et al. 2011).

At the end of that brief review of results on cytotoxic activity of various mushroom extracts, it also should be presented data on the possibility of the activity enhancement. Namely, our group has done numerous in vitro studies on the effect of mushrooms enriched with selenium, an essential trace element for normal functioning of the organism, on their anticancer activity. Thus, the obtained results showed that enrichment of *Pleurotus ostreatus*, *P. eryngii*, *P. pulmonarius*, *Ganoderma lucidum*, *G. applanatum*, *Flammulina velutipes*, *Lenzites betulinus*, and *Trametes versicolor* mycelia with selenium caused significant increase of cytotoxic activity of their ethanol extracts against cervix and colon adenocarcinoma cell lines (Milovanović et al. 2014, 2015a, b, c).

During a last few decades, numerous anticancer compounds have been isolated from mushrooms basidiocarps and mycelia and characterized (Table 4.2). Polysaccharides are the most active antitumor compounds (Zhang et al. 2007; Ren et al. 2012). However, proteins, lectins, terpenoids, and numerous low molecular weight compounds are also highly effective though their activities in vivo are still unclear (Lindequist et al. 2005; Xu et al. 2011; Varrot et al. 2013).

4.3.2 Anticancer Activity of Mushroom Polysaccharides

The ability of polysaccharides to prevent cancer was noticed in growers of *Flammulina velutipes* in Japan and *Agaricus blazei* in Brazil, where the cancer mortality rate was about 40% lower than in the rest of the population (Zhang et al. 2007). Polysaccharides are used in the treatment of various cancer types and can induce patient remission. They act non-specifically on the organism and regulatory functions and do not cause damages and stress but help an organism to adapt to different conditions (Ren et al. 2012). Zhang et al. (2007) assumed that polysaccharides could prevent oncogenesis, enhance immunity, and/or directly inhibit tumor growth by cell cycle arrest and induction of cell necrosis/apoptosis. However, it should be emphasized that the anticancer activity of mushroom polysaccharides is mainly based on the improvement of the host immune system. Namely, on the surface of the immune cells, there are receptors known as "pattern recognition receptors" that detect mushroom polysaccharides as foreign molecules and initiate immune responses (Ren et al. 2012). Binding of mushroom polysaccharides to the receptors is a basis for organism defense against microorganisms and spontaneously generated malignant cells. Lentinan, a polysaccharide from *Lentinus edodes*, schizophyllan from *Schizophyllum commune*, grifolan from *Grifola frondosa*, a polysaccharide from *Ganoderma lucidum*, and many others act in such a way. However, mushroom polysaccharides can also directly inhibit different types of cancers. Namely, the incubation of malignant cells with polysaccharides can induce their secondary necrosis as well as cell cycle arrest and apoptosis (Ren et al. 2012). These authors also emphasized that mushroom polysaccharides are able to inhibit

Table 4.2 Mushrooms' biologically active compounds and their activities

Active compound	Fungal source	Chemical structure	Cancer type	Mechanism of action	References
Lentinan	*Lentinus edodes*	β-(1 → 3)-Glucan with B-(1 → 6)-branches	Colon	Stimulation of TNF-α, ILs, IFN production, and T-cell and macrophage proliferation and differentiation	Taki et al. (1995)
Lentin		Protein	Leukemia	Antiproliferative activity	Ngai and Ng (2003)
Grifolan	*Grifola frondosa*	Polysaccharide	Bladder, prostate, stomach	Stimulation of macrophage proliferation and increasing levels of IL-6, IL-1, and TNF-α	Louie et al. (2010), Patel and Goyal (2012), and Roupas et al. (2012)
GFPPS1b		Polysaccharide-peptide complex	Stomach	Induction of apoptotic bodies; chromatin condensation; reduction of cell volume	
Krestin	*Trametes versicolor*	β-(1 → 3)-Glucan with B-(1 → 6)-branches +25%–38% proteins	Breast, colon, liver, rectum	Activation of T cells and natural killer cells; induction of IFN-γ, IL-2, TNF-α, IL-1, IL-6, IL-8	Taki et al. (1995), Wasser (2002), and Sakamoto et al. (2006)
Cordlan	*Cordyceps militaris*	Polysaccharide	Bladder, leukemia	Induction of maturation of dendritic cells; caspase activation and mitochondria function disruption	Patel and Goyal (2012)
FIP-gts	*Ganoderma tsugae*	Protein	Lung adenocarcinoma	Induction of cytokines and IFN-γ secretion	Liao et al. (2008)
FIP-fve	*Flammulina velutipes*	Proteins	Liver	Stimulation of mitogenesis in human peripheral blood lymphocytes	Chang et al. (2010) and Patel and Goyal (2012)
Flamulin			Leukemia	Antiproliferative activity	
CNL	*Clitocybe nebularis*	Protein	Leukemia	Antiproliferative activity	Pohleven et al. (2009)

Active compound	Fungal source	Chemical structure	Cancer type	Mechanism of action	References
Laccase	*Clitocybe maxima*	Protein	Breast, liver	Antimitotic activity	Patel and Goyal (2012)
Calcelin	*Calvatia utriformis*	Protein	Breast	Antiproliferative activity	Patel and Goyal (2012)
Hispolon	*Phellinus linteus*	Phenolics	Bladder, breast, lung, stomach	Pro-apoptotic activity	Patel and Goyal (2012) and Lu et al. (2009)
Agaritine	*Agaricus blazei*	Hydrazine derivate	Leukemia	Antiproliferative and pro-apoptotic activity	Patel and Goyal (2012)
Blazein		Steroid	Lung, stomach	Initiation of morphological changes in cancer cells and apoptosis	
ABL	*Agaricus bisporus*	Lectin	Colon	Antiproliferative and pro-apoptotic activity	Wang et al. (1998)
Ganoderic acids T and Me	*Ganoderma lucidum*	Terpenoids	Colon	Stimulation of expression of antitumor proteins and apoptosis regulators and cytochrome c release	Patel and Goyal (2012)
Ganoderic acids Mf and S			Cervix	Antiproliferative and pro-apoptotic activity	
Lucidenic acid			Liver		
Ergon	*Russula cyanoxantha*	Steroid	Liver	Antiproliferative and pro-apoptotic activity	Patel and Goyal (2012)
Anthraquinone	*Antrodia camphorata*	Ubiquinone derivate	Pancreas	Arresting cell cycle in G_1 phase	Yu et al. (2012) and Patel and Goyal (2012)
Methylantacinate A		Terpenoid	Oral	Chromatin condensation; DNA fragmentation	
Marmorin	*Hypsizygus marmoreus*	Protein	Breast, liver	Antiproliferative activity	Wong et al. (2008)
Cordycepin	*Cordyceps sinensis*	Adenosine derivate	Bladder, colon, leukemia, liver	Inhibition of NF-κB, antiproliferative and pro-apoptotic activity	Wang et al. (2016)

angiogenesis, i.e., inhibit proliferation and migration of endothelial cells necessary for the formation of tumor blood vessels. However, it has been proven that polysaccharide antitumor activity is never based on only one mechanism, but on a combination of several ones (Ren et al. 2012).

Polysaccharides of several mushrooms have been studied in detail, i.e., their chemical structure and features as well as the mechanism of action, and production of some of them is even commercialized (Table 4.2). One of them is lentinan, which has the ability to inhibit the growth of various cancer cell lines efficiently. Thus, Taki et al. (1995) reported that lentinan inhibits liver metastases in mice with colon cancer by activation of Kupffer's cells, reduces tumor size from 50% to 90%, prolongs life, and in some cases leads to complete regression. Strong activity against colon and rectal cancers was also reported for krestin (Taki et al. 1995; Wasser 2002; Sakamoto et al. 2006). Namely, clinical trials have shown that this polysaccharide isolated from *Trametes versicolor* prolonged the lifespan of patients suffering from these cancer types, reduced the mortality rate by 29%, and increased the rate of healing by 28% by restoration of immunity that was impaired by chemotherapy. Krestin is also very efficient in the treatment of breast and liver cancers, as well as polysaccharides from *Pleurotus tuber-regium* and *Lycium barbarum* (Zhang et al. 2001, 2005; Roupas et al. 2012). Wong et al. (2007) and Tong et al. (2009) observed that the polysaccharide isolated from the *P. tuber-regium* fruiting bodies and mycelium also inhibits proliferation and induces apoptosis of leukemia cells in vitro and that POPS-1 isolated from *P. ostreatus* fruiting bodies possess considerable activity against cervix and slightly lower against kidney cancer cells. A polysaccharide isolated from the *Grifola frondosa* mycelium, as well as its sulfate derivative and some fractions, can cause strong antiproliferative and pro-apoptotic effects on the stomach, prostate, and bladder cancer cells (Louie et al. 2010; Patel and Goyal 2012; Roupas et al. 2012). It can be demonstrated by datum that D fraction inhibited the growth of even 75% bladder cancer cells. Cordlan from *Cordyceps militaris* is highly active against leukemia as well as bladder cancer cells and polysaccharides from *Phellinus linteus* against melanoma cells, while *Lactarius flavidulus* polysaccharide efficiently inhibits the growth of Sarcoma 180, even up to 100% (Han et al. 2006; Patel and Goyal 2012). Contrary, sulfated extracellular polysaccharides obtained from *Ganoderma lucidum* fermentation broth showed moderate cytotoxic activity against six tumor cell lines. Zhang et al. (2012) reported that they inhibited the growth of tumor cells in a range from 34.46% to 74.79% when presented in a concentration of 500.0 μg/mL.

Generally, the mushroom polysaccharides are primarily excellent immunostimulators as well as inducers of cancer cell apoptosis and therefore could be potential candidates for the treatment of patients suffering from various cancer types.

4.3.3 Anticancer Activity of Mushroom Proteins

Peptides, immunomodulatory and ribosome-inactivating proteins, as well as some enzymes produced by numerous mushroom species have also shown significant cytotoxic activity against various cancer cell lines (Table 4.2). Immunomodulatory proteins possess particularly high anticancer activity. Thus, a recombinant of the first characterized immunomodulatory protein FIP-gts isolated from *Ganoderma tsugae* (reFIP-gts) acts on lung adenocarcinoma cells, i.e., slows down their growth and causes their rapid aging (Liao et al. 2008). Immunomodulatory protein FIP-fve synthesized by *Flammulina velutipes* inhibits the growth of hepatoma in mice, and CNL from *Clitocybe nebularis* has an antiproliferative effect on the leukemia cells and other hematopoietic malignancies (Pohleven et al. 2009; Chang et al. 2010). Suppression of leukemia cell proliferation was also caused by formalin, the ribosome-inactivating protein produced by *F. velutipes*, while calcaelin, the same protein from *Calvatia utriformis*, was highly effective against breast cancer cells (Patel and Goyal 2012). These authors reported that peptide from *C. utriformis*, which is chemically similar to ubiquitin, is also active against breast cancer cells as well as laccase from *Clitocybe maxima*, which also has antimitotic activity against hepatoma cells. The main carrier of antiproliferative activity of *Phellinus linteus* against breast, colon, liver, and lung cancer cells is proteoglycan, and risk of stomach cancer significantly can be decreased by calvatic acid synthesized by *C. utriformis* that inhibits the growth of *Helicobacter pylori*, the causal agent of chronic inflammation.

4.3.4 Anticancer Activity of Mushroom Polysaccharide-Protein Complexes

The main carriers of cytotoxic activity in some mushroom species are polysaccharide-protein complexes (Table 4.2). Polysaccharide-peptide complex GFPPS1b produced by *Grifola frondosa* causes strong antiproliferative and pro-apoptotic effect on the stomach cancer cells, while polysaccharide-peptide complex and glycopeptide LZ-D-4 and its sulfate derivative (LZ-D) from *Ganoderma lucidum* are highly active against lung cancer and leukemia cells (Patel and Goyal 2012). Hemagglutinin isolated from *Flammulina velutipes* also has high cytotoxic activity against leukemia cells, while proflamin suppresses the development of melanoma (Ng and Ngai 2006; Roupas et al. 2012).

4.3.5 Anticancer Activity of Mushroom Lectins

Fungi are rich sources of lectins; even 82% of all lectins are originated from mushrooms, 15% from filamentous micromycetes, and only 3% from yeasts (Varrot et al. 2013). Many mushroom lectins have anticancer activity (Table 4.2). Thus, lectins isolated from *Volvariella volvacea* and *Tricholoma mongolicum* (TML-1 and TML-2), as well as *Schizophyllum commune* and *Marasmius oreades*, showed strong antitumor activity both in vitro and in vivo. *Agrocybe aegerita*, *Inocybe umbrinella*, *Russula lepida*, and *Lactarius flavidulus* lectins are highly active against hepatoma cells, and *Clitocybe nebularis* one is highly active against leukemia cells, while both these activities are characteristic of lectins synthesized by *Ganoderma capense*, *Russula lepida*, *Hericium erinaceus*, and *Pholiota adiposa* (Wang et al. 1998). These authors reported that *I. umbrinella* and *R. lepida* lectins are also excellent inhibitors of proliferation of breast cancer cells, *Grifola frondosa* lectin of cervix cancer cells, and ABL from *Agaricus bisporus* of colon cancer cells. Interesting and useful data was reported for the activity of *Pleurotus citrinopileatus* lectin against Sarcoma 180 by Patel and Goyal (2012). Namely, these authors noted the reduction of tumor size by about 80% after 20 days of the intraperitoneal mice treated with a daily dose of 5.0 mg/kg.

4.3.6 Anticancer Activity of Mushroom Terpenoids

Species of the genus *Ganoderma* are especially rich in triterpenoids, which are carriers of many bioactivities, among which cytotoxic one takes important place (Table 4.2). Thus, ganoderic acids T and Me from *G. lucidum* inhibit the invasion of colon tumor and appearance of metastases; ganoderic acids Mf and S possess selective antiproliferative and pro-apoptotic effect on cervix cancer cells and lucidenic acids on hepatocellular carcinoma cells (Patel and Goyal 2012). *G. concinnum* triterpenoids induce apoptosis of leukemia cells, while applanoxidic acids A–H from *G. applanatum* are effective in the treatment of skin tumors (Lindequist et al. 2005). The significant antitumor potential is also the property of sesquiterpene iludin S isolated from *Suillus placidus* basidiocarps, as well as agrocibine and drimane, sesquiterpenoids produced by *Schizophyllum commune* and *Marasmius oreades* (Petrova et al. 2008).

4.3.7 Anticancer Activity of Other Mushroom Compounds

Many other intra- and extracellular compounds produced by mushrooms are characterized with significant cytotoxic capacity (Table 4.2). If these compounds are separated based on cancer cell lines which growth and proliferation inhibit, it could be defined a few groups. Highly active compounds against stomach and lung cancer cells are hispolon, *Phellinus linteus* phenolic compound, as well as agaritine and blazein, a hydrazine derivate and a steroid synthesized by *Agaricus blazei* (Lu et al. 2009; Patel and Goyal 2012). These authors reported that hispolon also causes an antiproliferative effect on breast and bladder cancer cells and agaritine and blazein cause an antiproliferative effect on leukemia cells. Development of pancreas cancer is successfully suppressed by anthraquinone, *Antrodia camphorata* ubiquinone derivative, as well as by irofulven, *Omphalotus olearius* metabolite, while ergon, steroid from *Russula cyanoxantha*, good inhibits hepatoma (Yu et al. 2012; Patel and Goyal 2012; Roupas et al. 2012). Triterpenoid methylantacinate A, synthesized by *A. camphorate*, is characterized by activity against oral cancer cells and grifolin, *Albatrellus confluens* secondary metabolite, against pharynx cancer cells (Luo et al. 2011).

However, some mushroom compounds can have auxiliary effects in cancer treatment. Thus, theanine (γ-L-glutamethylethylamide), extracellular *Boletus badius* metabolite, has a synergistic effect with commercial cytostatics, while alkaloid psilocybin from *Inocybe umbrinella* fruiting bodies significantly reduces anxiety and improves mood in patients with cancer at late phase (Patel and Goyal 2012).

4.4 Mechanisms of Anticancer Activity

Mushroom extracts or biologically active compounds isolated from them affect cytotoxic activity on a few mechanisms: *(i)* stimulation of immune system; *(ii)* neutralization of free radicals, i.e., antioxidative activity; *(iii)* antimutagenic activity; *(iv)* anti-inflammatory activity; *(v)* inhibition or stimulation of expression of proteins that are regulators of some cell processes; *(vi)* cell cycle arrest and apoptosis; *(vii)* disturbance of DNA synthesis and structure; *(viii)* changes in morphology and mobility of malignant cells; and *(ix)* antiangiogenic activity (Fig. 4.1).

4.4.1 Stimulation of the Immune System

Fresh and dry fruiting bodies, various extracts, as well as many metabolites of numerous mushroom species possess significant immunostimulatory effect (Tables 4.1, 4.2). For example, *Agaricus sylvaticus* basidiocarps, *A. brasiliensis* polysaccharides, *A. bisporus* proteoglycans, and *A. polytricha* protein APP have this effect (Patel and Goyal 2012; Roupas et al. 2012). *Ganoderma lucidum* basidiocarp extract bases activity against stomach carcinoma on stimulation of tumor necrosis factor-γ (TNF-γ) synthesis, while *Cordyceps militaris* extracts inhibit the proliferation of hepatoma and leukemia cells by induction of interleukin-18 (IL-18) transcription and interferon-γ (IFN-γ) production (Patel and Goyal 2012). These authors also reported that reduction of hepatoma in mice can be caused by *Phellinus linteus* extract that stimulates IL-12, IFN-γ, and TNF-α synthesis; macrophage and T-, $CD4^+$-, and dendritic cell proliferation; as well as natural killer cell activity, while its inhibition effect on colon carcinoma is based on increasing of spleen and thymus mass and level of certain plasma immunoglobulin receptors. Activation of natural killer cells and macrophages is a mechanism of colon cancer inhibition by *Hericium erinaceus* fruiting body water extract.

Numerous studies have shown that β-D-glucans non-specifically activate the cellular and humoral immune system, but increase the activity of macrophages, mononuclear cells, and neutrophils as well as the number of antibodies and stimulate the production of cytokines (Ren et al. 2012). Thus, Chihara (1992) reported that lentinan, after recognition of tumor cells, can stimulate certain groups of lymphocytes to produce higher amounts of cytokines (TNF-α, IL-1, IL-3, and IFN) and consequently cause mature differentiation and proliferation of T cells and macrophages. This β-glucan from *Lentinus edodes* also has the ability to restore the suppressed activity of T-helper cells to a normal level in organisms with developed tumor resulting in a complete renewal of immune response. Grifolan from *Grifola frondosa* as well as *Ganoderma lucidum* polysaccharide fractions can also stimulate macrophage proliferation and increase levels of some cytokines (IL-6, IL-1, and TNF-α). Krestin from *Trametes versicolor* bases activity against breast and liver cancer cells on initiation of T1 helper cells, IL-2 and INF-γ production, and activation of natural killer cells, while cordlan from *Cordyceps militaris* induces the maturation of dendritic cells (Patel and Goyal 2012; Ren et al. 2012; El Enshasy and Hatti-Kaul 2013).

Polysaccharide-protein complexes of macromycetes base antitumor effect on increasing the number and activity of natural killer cells, as well as on stimulation of synthesis of cytokines (INF-γ and ILs) and expression of nitric oxide synthase gene which is followed by the production of nitric oxide in macrophages (Roupas et al. 2012). These authors observed that increased synthesis of IFN-γ and interleukins and consequently increased number and activity of natural killer cells are results of the treatment with *A. bisporus* proteoglycans.

Lectin from *Clitocybe nebularis* activates dendritic cells and stimulates the synthesis of several pro-inflammatory cytokines (IL-6, IL-8, and TNF-α) and in such a way inhibits the proliferation of leukemia cells (El Enshasy and Hatti-Kaul 2013).

4.4.2 Antioxidative Activity

Reactive oxygen and nitrogen species can damage nucleic acids, proteins, carbohydrates, and lipids and in such a way change their function that can lead to appearance and development of various cancer types (Limón-Pacheco and Gonsebatt 2009). Numerous mushroom intra- and extracellular metabolites, singly or synergistically in the basidiocarp or mycelium extracts or fermentation broth, have the ability to reduce, i.e., neutralize, free radicals, which is presented in a few papers (Stajić et al. 2013; Ćilerdžić et al. 2014, 2015a, b, 2016a). This activity is one of the mechanisms of mushroom anticancer effect. Thus, *Pleurotus ostreatus* dry fruiting bodies and their extracts as well as an ethyl acetate extract of *Phellinus rimosus* basidiocarps base high cytotoxic capacity on strong antioxidative activity (Stajić 2015).

4.4.3 Antimutagenic Activity

DNA is the most susceptible macromolecule to oxidative damage that can be induced by various agents among which H_2O_2 has significant genotoxic potential. In vitro studies have shown that water extracts of *Agaricus bisporus*, *Ganoderma lucidum*, and *Agrocybe cylindracea*, as well as ethanol extracts of *Trametes versicolor*, *T. hirsuta*, and *T. gibbosa*, have protective effects against H_2O_2-induced DNA damage (Roupas et al. 2012; Ćilerdžić et al. 2016b; Knežević et al. 2018). Mlinrič et al. (2004) and Roupas et al. (2012) showed that *A. bisporus* and *G. lucidum* water extracts as well as *Lactarius vellereus* methanol extract base antitumor activity on antimutagenic effect. Thermolabile protein isolated from *A. bisporus*, β-glucans from *A. brasiliensis* and *A. blazei*, and extracts of Se-enriched *G. lucidum* basidiocarps have a similar effect (Angeli et al. 2006, 2009; Zhao et al. 2008). These authors reported that mechanism of action of *Agaricus* spp. glucan is based on the binding of benzopyrene which induces damage or on the neutralization of free radicals and that of *G. lucidum* on the inhibition of lipid peroxidation. Roupas et al. (2012) showed that *A. blazei* crude extracts significantly reduce DNA damage in rat liver induced by either diethylnitrosamine or radicals formed from 4-hydroxymethyl benzenediazonium salts.

4.4.4 Anti-inflammatory Activity

Anti-inflammatory effect of mushroom extracts and compounds is mainly associated with their significant antioxidative and immunomodulatory capacity. Thus, Joseph et al. (2012) and El Enshasy and Hatti-Kaul (2013) observed that polysaccharide-protein complex isolated from *Phellinus rimosus* significantly increases the level of superoxide dismutase (SOD) and glutathione peroxidase (GPx) and decreases the level of reduced glutathione, while β-(1,3/1,6)-D-glucan from *Pleurotus ostreatus* changes the level of cytokines in plasma.

4.4.5 Regulation of Some Cell Processes

The antiproliferative and pro-apoptotic activity of *Agaricus blazei* extracts and agaritine against some leukemia cell lines is based on strong inhibition of telomerase activity, induction of cytochrome c release and caspase activity, as well as regulation of Bcl-2 synthesis (Roupas et al. 2012). Inhibition of breast cancer growth with water *A. bisporus* extract and colon cancer cells with its lectin ABL is based on inhibition of aromatase activity and stimulation of caspase-3 activity, respectively, while suppression of the NF-κB factor activity is based on the mechanism of antitumor activity of *A. brasiliensis* extracts (Grube et al. 2001; Hong et al. 2004; Petrova et al. 2008). Inhibition of NF-κB is also the mechanism of antiproliferative and pro-apoptotic effects of *Cordyceps sinensis*, *Coprinus comatus*, *Sparassis crispa*, and *Phallus impudicus* extracts on various cancer cell lines (Petrova et al. 2008). Cytotoxic activity of *Ganoderma lucidum* extract against stomach cancer cells is based on caspase activation, while its activity against prostate cancer cells is based on inhibition of transcription factor AP-1 (Patel and Goyal 2012; Roupas et al. 2012). However, Chen et al. (2010) reported that anticancer activity of *G. lucidum* extracts could also be based on inhibition of matrix metalloproteinase expression. Reduction of expression of Cdk2 and Cdk1, the enzymes that catalyze transition G_1 to S phase and G_2 to M phase, respectively, is based on the mechanism of antiproliferative effect of *Pleurotus tuber-regium* basidiocarp and mycelium extracts, while stimulation of NAD(P)H:quinone oxidoreductase and glutathione S-transferase is based on the antitumor activity of *Phellinus linteus* extracts (Wong et al. 2007; Patel and Goyal 2012).

Polysaccharide cordlan inhibits the growth of leukemia and bladder cancer cells by caspase activation and mitochondrial function disruption, while D fraction of *Grifola frondosa* β-glucan bases cytotoxic effect on the bladder cancer cells on activation of DNA-dependent protein kinase and consequently on cell cycle arrest (Louie et al. 2010). Regulation of pro-apoptotic protein synthesis, i.e., Bax stimulation and Bcl-2 inhibition, and the activation of caspase-3 are mechanisms of pro-apoptotic activity of *Grifola frondosa* polysaccharide-peptide complex against stomach cancer cells, while this complex from *Phellinus linteus* inhibits colon cancer cells by reduction of expression of apoptosis regulators (cyclin B1 and Bcl-2) and stimulation of cytochrome c release (Li et al. 2004; Patel and Goyal 2012).

Triterpenoid ganoderic acid Me from *G. lucidum* bases cytotoxic activity against colon cancer cells on stimulation of expression of antitumor proteins p53, Bax, and caspase-3 and cytochrome c release, while methylantacinate A from *Antrodia camphorata* inhibits oral cancer by activation of caspase-3 (Patel and Goyal 2012). These authors also reported that cytochrome c release and caspase activation are mechanisms of *Boletus badius* theanine activity against breast cancer cells and that ergon from *Russula cyanoxantha* and suilin from *Suillus placidus* base activity against liver cancer cells not only on caspase activation but also on regulation of level of apoptosis regulators, i.e., Bax, Bcl-2, Bad, Bid, and Fas receptor. Antitumor activities of cycloepoxidone from *Panus conchatus* are based on suppresses of the NF-κB activity (Petrova et al. 2008).

4.4.6 Cell Cycle Arrest and Apoptosis

Cytotoxic activity of numerous mushroom extracts on various neoplastic cells is based on the arrest of their cell cycle in certain phase and apoptosis. Thus, *Trametes versicolor* and *Inonotus obliquus* extracts reduce proliferation of melanoma cells by arrest of their cycle in G_0/G_1 phase, while *Ganoderma tsugae* and *Clitocybe alexandri* extracts arrest colon adenocarcinoma cell cycle in G_2/M and S phase, respectively, and in such a way significantly decrease tumor size, from 44.2% to even 74.6% (Hsu et al. 2008; Youn et al. 2009; Vaz et al. 2010; Roupas et al. 2012). Activities of *Clitocybe alexandri* extracts on lung, breast, and stomach cancers are also based on cell cycle arrest in S phase and apoptosis induction. Arrest of cell cycle in the certain phase and induction of rapid cell apoptosis are also cytotoxic mechanisms of extracts of *Lentinus edodes* on leukemia and skin cancers, *Coprinus comatus* on prostate cancer, and *Flammulina velutipes* on breast cancer, as well as *Phellinus linteus* on liver, bladder, stomach, and lung neoplasms (Gu and Belury 2005; Guo et al. 2007; Zaidman et al. 2008; Patel and Goyal 2012).

Carboxymethylated polysaccharide from *Pleurotus tuber-regium* arrests the cell cycle of breast cancer cells in G_0/G_1 phase, while the polysaccharide from *Lycium barbarum* terminates the hepatoma cell cycle in the S phase causing apoptosis (Zhang et al. 2005, 2007). Inhibition of colon, stomach, and lung cancers' growth and protection of metastases development can be achieved by treatment with polysaccharide-peptide complex from *Phellinus linteus* which also arrests the cell cycle in the G_1/M phase (Li et al. 2004; Guo et al. 2007; Sliva 2010). Patel and Goyal (2012) observed that polysaccharide-peptide complexes from *Grifola frondosa*, polysaccharide cordlan from *Cordyceps militaris*, as well as ergon from *Russula cyanoxantha* base anticancer activity against stomach and liver cancer, leukemia, and bladder neoplasms, respectively, on arresting the cell cycle in the G_2/M phase. Recombinant of *Ganoderma tsugae* immunomodulatory protein (reFIP-gts), *Albatrellus confluens* grifolin, *Boletus badius* theanine, as well as *Antrodia camphorata* anthraquinone suppress lung, pharynx, breast, and pancreas cancer, respectively, by terminating cell cycles in the G_1 phase (Liao et al. 2008; Luo et al. 2011; Yu et al. 2012; Patel and Goyal 2012).

4.4.7 Disturbance of DNA Synthesis and Structure

Some mushroom metabolites have the potential to disturb DNA synthesis and affect the level of chromatin condensation and on these activities base cytotoxic effects. Thus, polysaccharide-peptide complex from *Grifola frondosa* causes the appearance of apoptotic bodies on the surface of stomach cancer cells, the reduction in their volume, and chromatin condensation (Cui et al. 2007). Inhibition of cervix cancer by bioactive components from *Ganoderma lucidum* spores is based on prevention of the onset of DNA synthesis, while chromatin condensation and DNA fragmentation are mechanisms of cytotoxic activity of *Russula cyanoxantha* ergon, *Boletus badius* fermentation broth, and methylantacinate A from *Antrodia camphorata* against liver, breast, and oral cancer cells, respectively (Patel and Goyal 2012).

4.4.8 Changes in Morphology and Mobility of Malignant Cells

Agaricus blazei steroid blazein affects activity against lung and stomach cancers on cell morphology and mobility changes (Roupas et al. 2012). Patel and Goyal (2012) reported that reduction of mobility of colon and lung carcinoma and glioma cells is one of the mechanisms of anticancer activity of fractions obtained from dry *Piptoporus betulinus* fruiting bodies. However, strong activity against pancreas cancer cells of anthraquinone from *Antrodia camphorata* is based on induction of their rapid aging and degradation of unnecessary or dysfunctional cellular components (Yu et al. 2012).

4.4.9 Antiangiogenic Activity

Kim et al. (2011) reported that *Hericium erinaceus* water extract has the ability to inhibit angiogenesis within colon tumor and in such a way reduce the tumor mass by about 40%. Anticancer activities of *Phellinus linteus* methanol extract and its fractions, as well as its proteoglycan, are based on inhibition of endothelial cells proliferation and reduction of epidermal growth factor receptors, respectively (Lee et al. 2010). The antiangiogenic effect also can be caused by polysaccharides isolated from *Agaricus brasiliensis* and polysaccharide-peptide complex from *Ganoderma lucidum* that inhibits vascular cell proliferation and endothelial growth factor secretion in patients suffering from lung cancer (Ren et al. 2012).

4.5 Conclusion

Today, cancer is the second major cause of morbidity and mortality worldwide. According to data from the World Health Organization, 8.8 million persons died in 2015, mostly of lung, liver, colon, rectal, stomach, and breast cancer. Additionally, this organization estimates that a number of new cases will rise by about 70% in the next two decades, i.e., nearly one in six deaths will be due to some cancer type. Conventional cancer treatments are very expensive and not highly effective. High-development, i.e., high-income, countries adopt extremely high budget not only for treatment of patients suffering from cancer but also for the development of new cytostatics and methods. The World Health Organization reported that 1.16 trillion US$ was spent for cancer treatments in 2010 and that Pfizer alone, the world leader among pharmaceutical companies, spend about 1.4 billion US$ for development of new anticancer drugs. However, common chemo- and radiotherapies are characterized by limited efficiency and numerous side effects. Therefore, alternative medicine, based on various natural sources, attracts the great attention of the researchers, peculiarly due to several features. Firstly, mushrooms and their extracts and metabolites primarily enhance the immune system and thus can suppress various cancers in all phases of their development without any side effects. Secondly, they are characterized with high selectivity, i.e., they are not toxic or almost nontoxic to normal cells. Thirdly, they can reduce side effects caused by conventional treatments, and finally, resistance to those preparations does not develop.

Acknowledgments This study was carried out with the financial support of the Ministry of Education, Science, and Technological Development of the Republic of Serbia, Project No. 173032.

References

Angeli JP, Ribeiro LR, Gonzaga ML, Soares Sde A, Ricardo MP, Tsuboy MS, Stidl R, Knasmueller S, Linhares RE, Mantovani MS (2006) Protective effects of beta-glucan extracted from *Agaricus brasiliensis* against chemically induced DNA damage in human lymphocytes. Cell Biol Toxicol 22:285–291. https://doi.org/10.1007/s10565-006-0087-z

Angeli JP, Ribeiro LR, Bellini MF, Mantovani MS (2009) Beta-glucan extracted from the medicinal mushroom *Agaricus blazei* prevents the genotoxic effects of benzo[a]pyrene in the human hepatoma cell line HepG2. Arch Toxicol 83:81–86. https://doi.org/10.1007/s00204-008-0319-5

Asatiani MD, Wasser SP, Nevo E, Ruimi N, Mahajna JA, Reznick A (2011) The Shaggy Inc Cap medicinal mushroom, *Coprinus comatus* (O.F.Mull.: Fr.) Pers. (Agaricomycetideae) substances

interfere with H_2O_2 induction of the NF-jB pathway through inhibition of IjBa phosphorylation in MCF7 breast cancer cells. Int J Medi Mushrooms. https://doi.org/10.1615/IntJMedMushr.v13.i1.30
Chang ST, Wasser SP (2012) The role of culinary-medicinal mushrooms on human welfare with a pyramid model for human health. Int J Medi Mushrooms 14(2):95–134. https://doi.org/10.1615/IntJMedMushr.v14.i2.10
Chang CY, Huang ZN, Yu HH, Chang LH, Li SL, Chen YP, Lee KY, Chuu JJ (2008) The adjuvant effects of *Antrodia camphorata* extracts combined with anti-tumor agents on multidrug resistant human hepatoma cells. J Ethnopharmacol 118:387–395. https://doi.org/10.1016/j.jep.2008.05.001
Chang HH, Hsieh KY, Yeh CH, Tu YP, Sheu F (2010) Oral administration of an Enoki mushroom protein FVE activates innate and adaptive immunity and induces anti-tumor activity against murine hepatocellular carcinoma. Int Immunopharmacol 20:239–246. https://doi.org/10.1016/j.intimp.2009.10.017
Chen X, Hu ZP, Yang XX, Huang M, Gao Y, Tang W, Chan SY, Dai X, Ye J, Ho PC, Duan W, Yang HY, Zhu YZ, Zhou SF (2006) Monitoring of immune responses to a herbal immuno-modulator in patients with advanced colorectal cancer. Int Immunopharmacol 6(3):499–508. https://doi.org/10.1016/j.intimp.2005.08.026
Chen X, Wang W, Li S, Xue J, Fan L, Sheng Z, Chen Y (2010) Optimization of ultrasound-assisted extraction of Lingzhi polysaccharides using response surface methodology and its inhibitory effect on cervical cancer cells. Carbohydr Polym 80:944–948. https://doi.org/10.1016/j.carbpol.2010.01.010
Chihara G (1992) Immunopharmacology of lentinan, a polysaccharide isolated from *Lentinus edodes*: its application as a host defense potentiator. Int J Oriental Med 17:57–77
Choi DB, Cha WS, Kang SH, Lee BR (2004) Effect of *Pleurotus ferulae* extracts on viability of human lung cancer and cervical cancer cell lines. Biotechnol Bioprocess Eng 9:356–361. https://doi.org/10.1007/BF02933057
Chu KK, Ho SS, Chow AH (2002) *Coriolus versicolor*: a medicinal mushroom with promising immunotherapeutic values. J Clin Pharmacol 42:976–984
Ćilerdžić J, Vukojević J, Stajić M, Stanojković T, Glamočlija J (2014) Biological activity of *Ganoderma lucidum* basidiocarps cultivated on alternative and commercial substrate. J Ethnopharmacol 155:312–319. https://doi.org/10.1016/j.jep.2014.05.036
Ćilerdžić J, Stajić M, Vukojević J, Milovanović I, Muzgonja N (2015a) Antioxidant and antifungal potential of *Pleurotus ostreatus* and *Agrocybe cylindracea* basidiocarps and mycelia. Curr Pharm Biotechnol 16(2):179–186. https://doi.org/10.2174/1389201015666141202152023
Ćilerdžić J, Stajić M, Vukojević J (2015b) Potential of submergedly cultivated mycelia of *Ganoderma* spp. as antioxidant and antimicrobial agents. Curr Pharm Biotechnol 16(17):275–282. https://doi.org/10.2174/1389201016666150930115944
Ćilerdžić J, Kosanić M, Stajić M, Vukojević J, Ranković B (2016a) Species of genus *Ganoderma* (Agaricomycetes) fermentation broth – a novel antioxidant and antimicrobial agent. Int J Medi Mushrooms 18(5):397–404. https://doi.org/10.1615/intJMedMushrooms.v18.i5.30
Ćilerdžić J, Stajić M, Živković L, Vukojević J, Bajić V, Spremo-Potparević B (2016b) Genoprotective capacity of alternatively cultivated Lingzhi or Reishi Medicinal Mushroom, *Ganoderma lucidum* (Agaricomycetes), basidiocarps. Int J Medi Mushrooms 18(12):1061–1069. https://doi.org/10.1615/IntJMedMushrooms.v18.i12.10
Coussens LM, Werb Z (2002) Inflammation, and cancer. Nature 420:860–867. https://doi.org/10.1038/nature01322
Cui FJ, Li Y, Xu YY, Liu ZQ, Huang DM, Zhang ZC, Tao WY (2007) Induction of apoptosis in SGC-7901 cells by polysaccharide-peptide GFPS1b from the cultured mycelia of *Grifola frondosa* GF9801. Toxicol in Vitro 21:417–427. https://doi.org/10.1016/j.tiv.2006.10.004
Diers AR, Dranka BP, Ricart KC, Oh JY, Johnson MS, Zhou F, Pallero MA, Bodenstine TM (2010) Modulation of mammary cancer cell migration by 15-deoxy-delta(12,14)-prostaglandin J(2): implications for anti-metastatic therapy. Biochem J 430:69–78. https://doi.org/10.1042/BJ20091193

El Enshasy HA, Hatti-Kaul R (2013) Mushroom immunomodulators: Unique molecules with unlimited applications. Trends Biotechnol 31(12):668–677. https://doi.org/10.1016/j.tibtech.2013.09.003
Estrela JM, Ortega A, Obrador E (2006) Glutathione in cancer biology and therapy. Crit Rev Clin Lab Sci 43:143–181. https://doi.org/10.1080/10408360500523878
Federico A, Morgillo F, Tuccillo C, Ciardiello F, Loguercio C (2007) Chronic inflammation and oxidative stress in human carcinogenesis. Int J Cancer 121:2381–2386. https://doi.org/10.1002/ijc.23192
Folkman J (1985) Tumor angiogenesis. Adv Cancer Res 43:175–203
Fortes RC, Novaes M, Recova VL, Melo AL (2009) Immunological, hematological, and glycemia effects of dietary supplementation with *Agaricus sylvaticus* on patients' colorectal cancer. Exp Biol Med 234:53–62. https://doi.org/10.3181/0806-RM-193
Garodia P, Ichikawa H, Malani N, Sethi G, Aggarwal BB (2007) From ancient medicine to modern medicine: ayurvedic concepts of health and their role in inflammation and cancer. J Soc Integr Oncol 5:25–37
Grivennikov SI, Karin M (2010) Inflammation and oncogenesis: a vicious connection. Curr Opin Genet Dev 20:65–71. https://doi.org/10.1016/j.gde.2009.11.004
Grube BJ, Eng ET, Kao YC, Kwon A, Chen S (2001) White button mushroom phytochemicals inhibit aromatase activity and breast cancer cell proliferation. J Nutr 131:3288–3293. https://doi.org/10.1093/jn/131.12.3288
Gu YH, Belury MA (2005) Selective induction of apoptosis in murine skin carcinoma cells (CH72) by an ethanol extract of *Lentinula edodes*. Cancer Lett 220:21–28. https://doi.org/10.1016/j.canlet.2004.06.037
Gu YH, Leonard J (2006) In vitro effects on proliferation, apoptosis and colony inhibition in ER-dependent and ER-independent human breast cancer cells by selected mushroom species. Oncol Rep 15:417–423
Guo JJ, Zhu TB, Collins L, Xiao ZXJ, Kim SH, Chen CY (2007) Modulation of lung cancer growth arrest and apoptosis by *Phellinus linteus*. Mol Carcinog 46:144–154. https://doi.org/10.1002/mc.20275
Han SB, Lee CW, Kang JS, Yoon YD, Lee KH, Lee K, Park SK, Kim HM (2006) Acidic polysaccharide from *Phellinus linteus* inhibits melanoma cell metastasis by blocking cell adhesion and invasion. Int Immunopharmacol 6:697–702. https://doi.org/10.1016/j.intimp.2005.10.003
Hong KJ, Dunn DM, Shen CL, Pence BC (2004) Effects of *Ganoderma lucidum* on apoptotic and anti-inflammatory function in HT-29 human colonic carcinoma cells. Phytother Res 18:768–770. https://doi.org/10.1002/ptr.1495
Hsieh TC, Wu JM (2001) Cell growth and gene modulatory activities of Yunzhi (Windsor Wunxi) from mushroom *Trametes versicolor* in androgen-dependent and androgen-insensitive human prostate cancer cells. Int J Oncol 18:81–88
Hsu SC, Ou CC, Li JW, Chuang TC, Kuo HP, Liu JY, Chen CS, Lin SC, Su CH, Kao MC (2008) *Ganoderma tsugae* extracts inhibit colorectal cancer cell growth via G2/M cell cycle arrest. J Ethnopharmacol 120:394–401. https://doi.org/10.1016/j.jep.2008.09.025
Hu H, Zhang Z, Lei Z, Yang Y, Sugiura N (2009) Comparative study of antioxidant activity and antiproliferative effect of hot water and ethanol extracts from the mushroom *Inonotus obliquus*. J Biosci Bioeng 107:42–48. https://doi.org/10.1016/j.jbiosc.2008.09.004
Ikekawa T (2005) Cancer risk reduction by intake of mushrooms and clinical studies on EEM. Int J Medi Mushrooms 7:347. https://doi.org/10.1615/IntJMedMushrooms.v7.i3.110
Jackson AL, Loeb LA (2001) The contribution of endogenous sources of DNA damage to the multiple mutations in cancer. Mutat Res 477:7–21
Joseph J, Panicker SN, Janardhanan KK (2012) Protective effect of polysaccharide-protein complex from a polypore mushroom, *Phellinus rimosus* against radiation-induced oxidative stress. Redox Rep 17:22–27. https://doi.org/10.1179/1351000211Y.0000000018
Kim KC, Kim IG (1999) *Ganoderma lucidum* extract protects DNA from strand breakage caused by hydroxyl radical and UV irradiation. Int J Mol Med 4:273–277

Kim CF, Jiang JJ, Leung KN, Fung KP, BS Lau C (2009) Inhibitory effects of *Agaricus blazei* extracts on human myeloid leukemia cells. J Ethnopharmacol 122:320–326. https://doi.org/10.1016/j.jep.2008.12.025

Kim SP, Kang MY, Kim JH, Nam SH, Friedman M (2011) Composition and mechanism of antitumor effects of *Hericium erinaceus* mushroom extracts in tumor-bearing mice. J Agric Food Chem 59:9861–9869. https://doi.org/10.1021/jf201944n

Knežević A, Stajić M, Sofrenić I, Stanojković T, Milovanović I, Tešević V, Vukojević J (2018) Extracts of Trametes species as new antioxidative, antifungal, cytostatic and antineurodegenerative agents. PLoS One. (in press)

Kollmar O, Rupertus K, Scheuer C, Junker B, Tilton B, Schilling MK, Menger MD (2007) Stromal cell-derived factor-1 promotes cell migration and tumor growth of colorectal metastasis. Neoplasia 9:862–870. https://doi.org/10.1371/journal.pone.0131946

Kundu JK, Surh YJ (2008) Inflammation: gearing the journey to cancer. Mutat Res 659:15–30. https://doi.org/10.1016/j.mrrev.2008.03.002

Kundu N, Zhang S, Fulton AM (1995) Sublethal oxidative stress inhibits tumor cell adhesion and enhances experimental metastasis of murine mammary carcinoma. Clin Exp Metastasis 13:16–22

Lavi I, Friesem D, Geresh S, Hadar Y, Schwartz B (2006) An aqueous polysaccharide extract from the edible mushroom *Pleurotus ostreatus* induces anti-proliferative and pro-apoptotic effects on HT-29 colon cancer cells. Cancer Lett 244:61–70. https://doi.org/10.1016/j.canlet.2005.12.007

Lee YS, Kim YH, Shin EK, Kim DH, Lim SS, Lee JY, Kim JK (2010) Anti-angiogenic activity of methanol extract of *Phellinus linteus* and its fractions. J Ethnopharmacol 131:56–62. https://doi.org/10.1016/j.jep.2010.05.064

Li G, Kim DH, Kim TD, Park BJ, Park HD, Park JI, Na MK, Kim HC, Hong ND, Lim K, Hwang BD, Yoon WH (2004) Protein-bound polysaccharide from *Phellinus linteus* induces G2/M phase arrest and apoptosis in SW480 human colon cancer cells. Cancer Lett 216:175–181. https://doi.org/10.1016/j.canlet.2004.07.014

Liao CH, Hsiao YM, Lin CH, Yeh CS, Wang JCH, Ni CH, Hsu CP, Ko JL (2008) Induction of premature senescence in human lung cancer by fungal immunomodulatory protein from *Ganoderma tsugae*. Food Chem Toxicol 46:1851–1859. https://doi.org/10.1016/j.fct.2008.01.044

Limón-Pacheco J, Gonsebatt ME (2009) The role of antioxidants and antioxidant-related enzymes in protective responses to environmentally induced oxidative stress. Mutat Res 674:137–147. https://doi.org/10.1016/j.mrgentox.2008.09.015

Lindequist U, Niedermeyer THJ, Jülich WD (2005) The pharmacological potential of mushrooms. eCAM 2(3):285–299. https://doi.org/10.1093/ecam/neh107

Liotta LA, Steeg PS, Stetler-Stevenson WG (1991) Cancer metastasis and angiogenesis: an imbalance of positive and negative regulation. Cell 64:327–336

Louie B, Rajamahanty S, Won J, Choudhury M, Konno S (2010) Synergistic potentiation of interferon activity with Maitake mushroom D-fraction on bladder cancer cells. BJU Int 105:1011–1015. https://doi.org/10.1111/j.1464-410X.2009.08870.x

Lu TL, Huang GJ, Lu TJ, Wu JB, Wu CH, Yang TC, Iizuka A, Chen YF (2009) Hispolon from *Phellinus linteus* has antiproliferative effects via MDM2-recruited ERK1/2 activity in breast and bladder cancer cells. Food Chem Toxicol 47:2013–2021. https://doi.org/10.1016/j.fct.2009.05.023

Luo XJ, Li LL, Deng QP, Yu XF, Yang LF, Luo FJ, Xiao LB, Chen XY, Ye M, Liu JK, Cao Y (2011) Grifolin, a potent antitumour natural product upregulates death-associated protein kinase 1 DAPK1 via p53 in nasopharyngeal carcinoma cells. Eur J Cancer 47:316–325. https://doi.org/10.1016/j.ejca.2010.09.021

Milovanović I, Stajić M, Ćilerdžić J, Stanojković T, Knežević A, Vukojević J (2014) Antioxidant, antifungal and anticancer activities of Se-enriched *Pleurotus* spp. mycelium extracts. Arch Biol Sci 66(4):1379–1388. https://doi.org/10.2298/ABS1404379M

Milovanović I, Stanojković T, Stajić M, Vukojević J, Knežević A (2015a) Se effect on biological activity of *Flammulina velutipes*. Ital J Food Sci 27(1):1–7. https://doi.org/10.14674/1120-1770/IJFS.V27I1.74

Milovanović I, Stajić M, Stanojković T, Knežević A, Vukojević J (2015b) Effects of selenium presence in mycelia of *Ganoderma* species (Higher Basidiomycetes) on their medicinal properties. Int J Medi Mushrooms 17(1):11–20. https://doi.org/10.1615/IntJMedMushrooms.v17.i1.20

Milovanović I, Stanojković T, Stajić M, Brčeski I, Knežević A, Ćilerdžić J, Vukojević J (2015c) Effect of selenium enrichment of *Lenzites betulinus* and *Trametes hirsuta* mycelia on antioxidant, antifungal and cytostatics potential. Curr Pharm Biotechnol 16(10):920–926. https://doi.org/10.2174/1389201016666150618152531

Mlinrič A, Kac J, Fatur T, Filini M (2004) Anti-genotoxic activity of the mushroom *Lactarius vellereus* extract in bacteria and in mammalian cells in vitro. Pharmazie 59:217–221

Ng TB, Ngai PHK (2006) An agglutinin with mitogenic and antiproliferative activities from the mushroom *Flammulina velutipes*. Mycologia 98:167–171

Ngai PH, Ng TB (2003) Lentin, a novel and potent antifungal protein from shitake mushroom with inhibitory effects on activity of human immunodeficiency virus-1 reverse transcriptase and proliferation of leukemia cells. Life Sci 73:3363–3374. https://doi.org/10.1016/j.lfs.2003.06.023

Parker L (2014) The impact of the environment on cancer genomics. In: Dellaire G, Berman J, Arceci R (eds) Cancer genomics from bench to personalized medicine. Elsevier Inc, London, pp 449–465

Patel S, Goyal A (2012) Recent developments in mushroom as anti-cancer therapeutics: a review. 3 Biotech 2:1–15. https://doi.org/10.1007/s13205-011-0036-2

Petrova RD, Reznick AZ, Wasser SP, Denchev CM, Nevo E, Mahajna J (2008) Fungal metabolites modulating NF-κB activity: an approach to cancer therapy and chemoprevention (Review). Oncol Rep 19:299–308

Pohleven J, Obermajer N, Sabotic J, Anzlovar S, Sepcic K, Kos J, Kralj B, Sˇtrukelj B, Brzin J (2009) Purification, characterization and cloning of a ricin B-like lectin from mushroom *Clitocybe nebularis* with antiproliferative activity against human leukemic T cells. Biochim Biophys Acta 1790:173–181. https://doi.org/10.1016/j.bbagen.2008.11.006

Rao YK, Fang SH, Wu WS, Tzeng YM (2010) Constituents isolated from *Cordyceps militaris* suppress enhanced inflammatory mediator's production and human cancer cell proliferation. J Ethnopharmacol 131:363–367. https://doi.org/10.1016/j.jep.2010.07.020

Rashid S, Unyayar A, Mazmanci MA, McKeown SR, Banat IM, Worthington J (2011) A study of anti-cancer efforts of *Funalia trogii in vitro* and *in vivo*. Food Chem Toxicol 49:1477–1483. https://doi.org/10.1016/j.fct.2011.02.008

Ren L, Perera C, Hemar Y (2012) Antitumor activity of mushroom polysaccharides: a review. Food Funct 3:1118–1130. https://doi.org/10.1039/c2fo10279j

Reuter S, Gupta SC, Chaturvedi MM, Aggarwal BB (2010) Oxidative stress, inflammation, and cancer: How are they linked? Free Radic Biol Med 49:1603–1616. https://doi.org/10.1016/j.freeradbiomed.2010.09.006

Roupas P, Keogh J, Noakes M, Margetts C, Taylor P (2012) The role of edible mushrooms in health: evaluation of the evidence. J Funct Foods 4:687–709. https://doi.org/10.1016/j.jff.2012.05.003

Sadava D, Still DW, Mudry RR, Kane SE (2009) Effect of *Ganoderma* on drug-sensitive and multidrug-resistant small-cell lung carcinoma cells. Cancer Lett 277:182–189. https://doi.org/10.1016/j.canlet.2008.12.009

Saitoh H, Feng W, Matsuzawa T, Ikekawa T (1997) Antitumor activity of *Hypsizigus marmoreus*. II. Preventive effect against lung metastasis of Lewis lung carcinoma. Yakugaku Zasshi J Pharmaceut Soc Jpn 117:1006–1010

Sakamoto J, Morita S, Oba K, Matsui T, Kobayashi M, Nakazato H, Ohashi Y (2006) Efficacy of adjuvant immunochemotherapy with polysaccharide K for patients with curatively resected colorectal cancer: a meta-analysis of centrally randomized controlled clinical trials. Cancer Immunol Immunother 55:404–411. https://doi.org/10.1007/s00262-005-0054-1

Schetter AJ, Heegaard NH, Harris CC (2010) Inflammation and cancer: interweaving microRNA, free radical, cytokine and p53 pathways. Carcinogenesis 31:37–49. https://doi.org/10.1093/carcin/bgp272

Shacter E, Beecham EJ, Covey JM, Kohn KW, Potter M (1988) Activated neutrophils induce prolonged DNA damage in neighboring cells. Carcinogenesis 9:2297–2304

Sliva D (2010) Medicinal mushroom *Phellinus linteus* as an alternative cancer therapy (Review). Exp Therapeut Med 1:407–411. https://doi.org/10.3892/etm_00000063

Stajić M (2015) *Nutritivna svojstva i medicinski potencijal makromiceta*. University of Belgrade Faculty of Biology. ISBN 978-86-7078-122-1.

Stajić M, Vukojević J, Knežević A, Duletić Laušević S, Milovanović I (2013) Antioxidant protective effects of mushroom metabolites. Curr Top Med Chem 13(21):2660–2676

Storz P (2005) Reactive oxygen species in tumor progression. Front Biosci 10:1881–1896

Taki H, Ohishi K, Okano A, Suga T, Akiyama Y, Yamashita A (1995) Antimetastatic effect of lentinan on liver metastasis of colon carcinoma (Colon 26): possible role of activated Kupffer cells. Int J Immunother 11:29–38

Tong H, Xia F, Feng K, Sun G, Gao X, Sun L, Jiang R, Tian D, Sun X (2009) Structural characterization and in vitro antitumor activity of a novel polysaccharide isolated from the fruiting bodies of *Pleurotus ostreatus*. Bioresour Technol 100:1682–1686. https://doi.org/10.1016/j.biortech.2008.09.004

Varrot A, Basheer SM, Imberty A (2013) Fungal lectins: structure, function and potential applications. Curr Opin Struct Biol 23:678–685. https://doi.org/10.1016/j.sbi.2013.07.007

Vaz JA, Heleno SA, Martins A, Almeida GM, Vasconcelos MH, Ferreira ICFR (2010) Wild mushrooms *Clitocybe alexandri* and *Lepista inversa*: in vitro antioxidant activity and growth inhibition of human tumour cell lines. Food Chem Toxicol 48:2881–2884. https://doi.org/10.1016/j.fct.2010.07.021

Wang H, Ng TB, Ooi VC (1998) Lectins from mushrooms. Mycol Res 102:897–906. https://doi.org/10.1017/S0953756298006200

Wang PW, Hung YC, WenTai Li WT, Chau-Ting Yeh CT, Tai-Long Pan TL (2016) Systematic revelation of the protective effect and mechanism of *Cordyceps sinensis* on diethylnitrosamine-induced rat hepatocellular carcinoma with proteomics. Oncotarget 7:60270–60289. https://doi.org/10.18632/oncotarget.11201

Wasonga CG, Okoth SA, Mukuria JC, Omwandho CO (2008) Mushroom polysaccharide extracts delay progression of carcinogenesis in mice. J Exp Ther Oncol 7:147–152

Wasser SP (2002) Medicinal mushrooms as a source of antitumor and immunomodulating polysaccharides. Appl Microbiol Biotechnol 60:258–274. https://doi.org/10.1007/s00253-002-1076-7

Westermarck J, Kahari VM (1999) Regulation of matrix metalloproteinase expression in tumor invasion. FASEB J 13:781–792

Wong SM, Wong KK, Chiu LCM, Cheung PCK (2007) Non-starch polysaccharides from different developmental stages of *Pleurotus tuber-regium* inhibited the growth of human acute promyelocytic leukaemia HL-60 cells by cell-cycle arrest and/or apoptotic induction. Carbohydr Polym 68:206–217. https://doi.org/10.1016/j.carbpol.2006.12.018

Wong JH, Wang HX, Ng TB (2008) Marmorin, a new ribosome inactivating protein with antiproliferative and HIV-1 reverse transcriptase inhibitory activities from the mushroom *Hypsizigus marmoreus*. Appl Microbiol Biotechnol 81(4):669–674. https://doi.org/10.1007/s00253-008-1639-3

Xu X, Yan H, Chen J, Zhang X (2011) Bioactive proteins from mushrooms. Biotechnol Adv 29:667–674. https://doi.org/10.1016/j.biotechadv.2011.05.003

Yang HL, Kuo YH, Tsai CT, Huang YT, Chen SC, Chang HW, Lin E, Lin WH, Hseu YC (2011) Anti-metastatic activities of *Antrodia camphorata* against human breast cancer cells mediated through suppression of the MAPK signaling pathway. Food Chem Toxicol 49:290–298. https://doi.org/10.1016/j.fct.2010.10.031

Youn MJ, Kim JK, Park SY, Kim Y, Park C, Kim ES, Park KI, So HS, Park R (2009) Potential anticancer properties of the water extract of *Inonotus obliquus* by induction of apoptosis in melanoma B16–F10 cells. J Ethnopharmacol 121:221–228. https://doi.org/10.1016/j.jep.2008.10.016

Yu CC, Chiang PC, Lu PH, Kuo MT, Wen WC, Chen P, Guh JH (2012) Antroquinonol, a natural ubiquinone derivative, induces a cross talk between apoptosis, autophagy and senescence in human pancreatic carcinoma cells. J Nutr Biochem 23(8):900–907. https://doi.org/10.1016/j.jnutbio.2011.04.015

Yuecheng Y, Xiaoyan X (2007) Stromal-cell derived factor-1 regulates epithelial ovarian cancer cell invasión by activating matrix metalloproteinase-9 and matrix metalloproteinase-2. Eur J Cancer Prev 16:430–435. https://doi.org/10.1097/01.cej.0000236259.88146.a4

Zaidman BZ, Wasser SP, Nevo E, Mahajan J (2008) *Coprinus comatus* and *Ganoderma lucidum* interfere with androgen receptor function in LNCaP prostate cancer cells. Mol Biol Rep 35:107–117. https://doi.org/10.1007/s11033-007-9059-5

Zhang M, Cheung PCK, Zhang L (2001) Evaluation of mushroom dietary fiber (nonstarch polysaccharides) from sclerotia of *Pleurotus tuber-regium* (Fries) Singer as a potential antitumor agent. J Agric Food Chem 49:5059–5062

Zhang M, Chen H, Huang J, Li Z, Zhu C, Zhang S (2005) Effect of *Lycium barbarum* polysaccharide on human hepatoma QGY7703 cells: Inhibition of proliferation and induction of apoptosis. Life Sci 76:2115–2124. https://doi.org/10.1016/j.lfs.2004.11.009

Zhang M, Cui SW, Cheung PCK, Wang Q (2007) Antitumor polysaccharides from mushrooms: a review on their isolation process, structural characteristics and antitumor activity. Trends Food Sci Technol 18:4–19. https://doi.org/10.1016/j.tifs.2006.07.013

Zhang J, Liu YJ, Park HS, Xia YM, Kim GS (2012) Antitumor activity of sulfated extracellular polysaccharides of *Ganoderma lucidum* from the submerged fermentation broth. Carbohydr Polym 87:1539–1544. https://doi.org/10.1016/j.carbpol.2011.09.051

Zhao L, Zhao GH, Du M, Zhao ZD, Xiao LX, Hu XS (2008) Effect of selenium on increasing free radical scavenging activities of polysaccharide extracts from a Se-enriched mushroom species of the genus *Ganoderma*. Eur Food Res Technol 226:499–505. https://doi.org/10.1007/s00217-007-0562-7

Immunomodulatory Aspects of Medicinal Mushrooms

5

Seema Patel

Contents

Abstract

Mushrooms, the macrofungi, are enigmatic in their composition, which renders some of them umami-flavored edible, some medicinal, some hallucinogenic/psychoactive, while some lethal. In current times, appreciation and demand of the medicinal mushrooms is rising, with the validation of their efficacy as antioxidant, anti-inflammatory, immunomodulatory, anticancer, endocrine restorative, etc. Containing a rich repertoire of bioactive components as β-glucan, phenolics, peptides, and sterols, they have attracted the attention of researchers. *Ganoderma lucidum*, *Hericium erinaceus*, *Grifola frondosa*, *Lentinus edodes*, *Inonotus*

S. Patel (✉)
Bioinformatics and Medical Informatics Research Center, San Diego State University, San Diego, CA, USA

D. C. Agrawal, M. Dhanasekaran (eds.), *Medicinal Mushrooms*,
https://doi.org/10.1007/978-981-13-6382-5_5

obliquus, and *Cordyceps sinensis* are some of the most popular medicinal mushrooms. The immunomodulation, anti-inflammation, and anticancer attributes of the mycochemicals mostly rely on the inhibition of mTOR translational pathway, cell apoptosis, G1 phase cell cycle arrest, downregulation of cyclins A and B1, upregulation of p21 and p27, and induction of cytokines (TNF-α, IL-2, and IFN-γ), among others. Further attention can facilitate the identification of components of biological interest from mushrooms. In this regard, this chapter discusses the biological effects of some major mushroom mycochemicals.

Keywords
Mushrooms · Polysaccharides · Beta-glucan · Lectins · Peptides · Immunomodulation

Abbreviations

ChtBDs	Chitin-binding domains
HUVECs	Human umbilical vein endothelial cells
MMP	Matrix metalloproteinase
PSA	Prostate-specific antigen
RIP	Ribosome-inactivating proteins
TIMP	Tissue inhibitor of metalloproteinase
TLR	Toll-like receptors

5.1 Introduction

Edible mushrooms are deemed as gourmet food. These macrofungi are regarded for their umami flavor, sotolon (3-hydroxy-4, 5-dimethyl-2(5H)-furanone) (Colin Slaughter 1999; Ma et al. 2008) (Lizárraga-Guerra et al. 1997). *Agaricus bisporus*, *Boletus edulis*, *Volvariella volvacea*, *Lentinus edodes*, *Pleurotus* spp., *Grifola frondosa*, and *Flammulina velutipes* are some of the valued edible mushrooms (Guo et al. 2012; Suárez Arango and Nieto 2013; Valverde et al. 2015). Some edible as well as nonedible mushrooms possess medicinal relevance. Pharmacological aspects of mushrooms such as anticancer, cholesterol lowering, antidiabetic, antioxidant, nephroprotective, antiviral, antibacterial, and immunomodulation have been intense areas of research since decades, continuing till now (Lindequist et al. 2005; Wu et al. 2014). Some notable medicinal mushrooms belong to *Ganoderma*, *Cordyceps*, *Clitocybe*, *Antrodia*, *Trametes*, *Xerocomus*, *Phellinus*, *Pleurotus*, *Agaricus*, *Calvatia*, *Schizophyllum*, *Flammulina*, *Suillus*, *Inonotus*, *Inocybe*, *Funlia*, *Lactarius*, *Albatrellus*, *Russula*, and *Fomes genus*. The bioactive constituents in these mushrooms are polysaccharides, oligosaccharides, dietary fibers, triterpenoids, peptides, proteins (lectin), fats (stearic acid, palmitic acid), phenols (hispolon), steroids (ergosta-7,22-dien-3-one, ergosta-7,22-dien-3-ol), triterpenes (lanostane, ergostane-type), flavonoids, carotenoids, glycosides, alkaloids, volatile oils, vitamins (tocopherols, folates, ascorbic acid), enzymes (laccase), organic acids (ganoderic acid, calvatic acid), and minerals (zinc, iodine, selenium, copper, and iron). Calcaelin, illudin S, suillin, cordycepin, psilocybin, grifolin, etc. are some well-studied bioactive mycochemicals.

5.2 Immunomodulatory Properties of Diverse Mushrooms

The mycocomponents attempt to restore immune balance by acting as a mitotic kinase inhibitor, angiogenesis inhibitor, topoisomerase inhibitor, etc. An insightful review on the medicinal value of mushroom has been published previously (Patel and Goyal 2012). This chapter focusses on the immunomodulatory potential of diverse mushrooms and their bioactive components.

5.2.1 Polysaccharide

Beta-glucan, a polysaccharide, consists of a backbone of glucose residues linked by β-(1 → 3)-glycosidic bonds, often with attached side-chain glucose residues joined by β-(1 → 6) linkages (Rop et al. 2009). Mushroom polysaccharides as lentinan, krestin, schizophyllan, cordlan, *Hericium* polysaccharide, etc. have been validated as immunomodulatory and anticancer agents (Patel and Goyal 2012). Shiitake mushroom *Lentinula edodes* lentinan, a β-glucan, can suppress gastric cancer (Ina et al. 2013). Beta-glucan from *Grifola frondosa* enhanced the efficacy of anticancer agent cisplatin, by lowering the nephrotoxicity and myelosuppression caused by the drug (Masuda et al. 2009). Beta-glucan binds to dectin-1, a non-TLR (Toll-like receptors) lectin receptor, which induces an intracellular signaling cascade, leading to the stimulation of macrophages and dendritic cells (Batbayar et al. 2012) and elaboration of pro-inflammatory cytokine production (Mansour et al. 2013). This mechanism has been illustrated in Fig. 5.1.

The polysaccharide of *Fomes fomentarius* could intervene in human gastric cancer cell lines SGC-7901 and MKN-45. Further, the exopolysaccharide sensitized doxorubicin and induced growth inhibition of SGC-7901 cells at 0.25 mg/ml following 24 h incubation (Chen et al. 2008a). *Schizophyllum commune* homopolysaccharide schizophyllan has proven its immunomodulatory and antineoplastic properties. In mice model, it inhibited mammary tumors (Mansour et al. 2012). The polysaccharide *Trametes versicolor* reduces the growth of hormone-responsive breast cancer as well as prostate cancer LNCaP cell growth (Córdoba and Ríos 2012). A protein-bound polysaccharide of *P. linteus* induces G2/M phase arrest and apoptosis in SW480 human colon cancer cells (Li et al. 2004). *Cordyceps militaris* water-soluble polysaccharide has immunostimulating polysaccharide (Zhu et al. 2014). The polysaccharide cordlan from this mushroom induces dendritic cell maturation through TLR4 signaling pathways (Kim et al. 2010). The maturation of

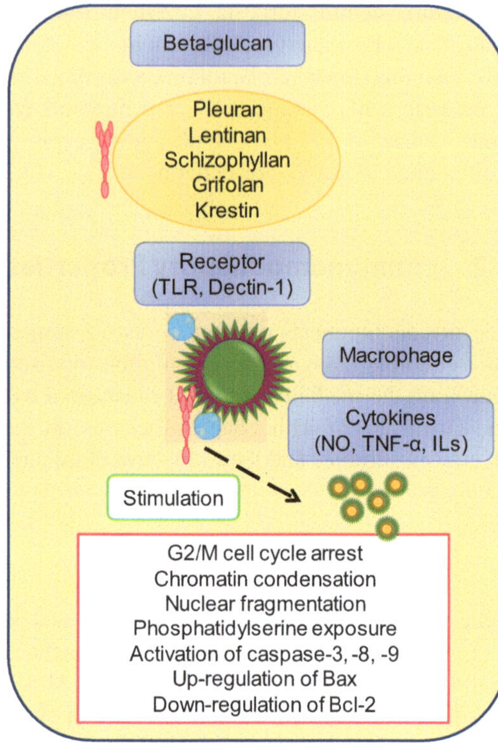

Fig. 5.1 The binding of β-glucan to dectin-1 leads to macrophage activation and pro-inflammatory cytokine elaboration

dendritic cell is expected to make cancer immunotherapy more effective. HEG-5, a novel polysaccharide-protein purified from *Hericium erinaceus* CZ-2, can evoke apoptosis by caspase-8/caspase-3-dependent, p53-dependent mitochondrial-mediated, and PI3k/Akt signaling pathways, which might be of relevance to gastric cancer treatment (Zan et al. 2015). The water-soluble polysaccharide extract from *Pleurotus tuber-regium* was antiproliferative toward human acute promyelocytic leukemia cells (HL-60) (Zhang et al. 2004). The aqueous polysaccharide extract of *Pleurotus ostreatus* induced antiproliferative and pro-apoptotic effects on HT-29 colon cancer cells (Tong et al. 2009).

Cordlan isolated from *Cordyceps militaris* induced the phenotypic maturation of dendritic cells as demonstrated by the elevated expressions of CD40, CD80, CD86, MHC-I, and MHC-II molecules (Kim et al. 2010). A chemically-sulfated polysaccharide (S-GAP-P) derived from water insoluble polysaccharide of *Grifola frondosa* had anticancer effects on human gastric carcinoma SGC-7901 cells, which occur by apoptotic induction and by augmenting the efficacy of the drug 5-FU (Shi et al. 2007). *Inonotus obliquus* polysaccharide induced apoptosis in human hepatoma HepG2 cells (Youn et al. 2008), and U251 human neurogliocytoma cells (Ning et al. 2014), by arresting cell cycle at G_0/G_1 phase.

Mushrooms contain chitins, like other fungi (Vetter 2007). The immunogenicity of chitin has been amply reported in recent times (Patel and Goyal 2017).

5.2.2 Proteoglycans and Other Glycol-Protein Conjugates

P. linteus proteoglycan has an antiproliferative effect on human hepatocellular liver carcinoma (HepG2), human colon adenocarcinoma (HT-29), human lung cancer (NCI-H 460), and human breast adenocarcinoma (MCF-7) cells (Li et al. 2011). The treatment of HT-29-bearing mice with 100 mg/kg proteoglycan led to an increment in spleen and thymus mass and IgA levels. The proteoglycan protected T cells from plasmatic prostaglandin E2 and disrupted the Reg IV/EGFR/Akt signaling pathway (Li et al. 2011). The administration of *P. linteus* extract daily for 8 weeks to human hepatoma (Hep3B) cell-transplanted mice led to a reduction in the size of the tumor and enhanced T cell numbers; IL-12, IFN-γ, and TNF-α secretion; NK cell activity; and phagocytic ability (Huang et al. 2011). *P. linteus* methanol extract and its fractions have the potential for anti-angiogenic effects through the inhibition of human umbilical vein endothelial cell (HUVEC) proliferation, migration, and assembly into capillary-like structures (Lee et al. 2010).

Cui et al. (2007) investigated the biological function of a novel polysaccharide-peptide GFPPS1b from *Grifola frondosa* GF9801 towards human gastric adenocarcinoma (SGC-7901 cells) (Cui et al. 2007). A native glycopeptide, LZ-D-4, purified from the fruiting bodies of *Ganoderma lucidum*, and its sulfated derivative, LZ-D, showed antitumor effect in an in vitro test against mouse lymphocytic leukemia L1210 cell line (Ye et al. 2009).

5.2.3 Protein, Peptides, and Amino Acids

The laccase enzyme from *Clitocybe maxima* was antiproliferative toward Hep G2 and MCF-7 tumor cells (Zhang et al. 2010c). Serine protease has been isolated from insect-pathogenic fungi *Metarhizium anisopliae*, *Cordyceps militaris* (Kim et al. 2006), *Termitomyces albuminosus* (Zheng et al. 2011), *Pholiota nameko* (Guan et al. 2011), *Conocybe apala* (Walton et al. 2010), *Lignosus rhinocerotis* (Yap et al. 2015), and *Helvella lacunosa* (Zhang et al. 2010b). In fact, this enzyme being pivotal to survival across all living organisms (Patel 2017) is expected to be presented in all mushrooms. Chitin-binding domains (ChtBDs) are present in fungal serine proteases (Merzendorfer and Zimoch 2003; Javed et al. 2013). It has come forth that poisonous mushrooms have a high level of proteases. Mushrooms not only contain serine proteases but serine protease inhibitors as well. Some notable inhibitors include cnispin (*Clitocybe nebularis*), LeSPI (*Lentinus edodes*), cospin (*Coprinopsis cinerea*), etc. (Sabotič et al. 2012). Both cnispin and cospin are β-trefoil proteins, highly specific for trypsin (Avanzo Caglič et al. 2014).

Clitocybe nebularis has ricin B-like lectin, which binds to the carbohydrate receptors on human leukemic T cells, which might be exploited for therapeutic purpose (Pohleven et al. 2009). A lectin from *Russula lepida*, when administered intraperitoneally, was antiproliferative towards Hep G2 and MCF-7 cells. Also, the injections of the lectin at 5.0 mg/kg dosage for 20 days, caused 67.6% reductions in the weight of S-180 tumor (Zhang et al. 2010a). A hemagglutinin from *Flammulina velutipes* inhibited the proliferation of leukemia L1210 cells (Ng et al. 2006). Wu et al. (2011) isolated a lectin from dried *Lactarius flavidulus* fruit bodies which suppressed the proliferation of HepG2 and L1210 cells with an IC_{50} of 8.90 μM and 6.81 μM, respectively (Wu et al. 2011). A lectin from *Pleurotus citrinopileatus* exerted potent antitumor activity in mice bearing sarcoma 180, inhibiting 80% of the tumor growth, when administered intraperitoneally at 5 mg/kg daily for 20 days (Li et al. 2008). Lectin from *Agaricus bisporus* inhibited cell proliferation of some ocular and cancer cell lines (Cheung et al. 2012). A novel lectin from the *Inocybe umbrinella* inhibited the proliferation of tumor HepG2 and MCF7 cells (Zhao et al. 2009).

FIP-fve from *Flammulina velutipes* is an immunomodulatory protein, which can stimulate lymphocytes and suppress systemic anaphylaxis reactions (Ko et al. 1995). Purified recombinant fungal immunomodulatory protein, *re*FIP-gts, from *Ganoderma tsugae* possessed anti-telomerase effect which retarded human lung adenocarcinoma (A549) cell growth (Liao et al. 2008). *Flammulina velutipes*-derived flammulin and velutin, the ribosome-inactivating proteins (RIP), have antitumor effects (Zhou et al. 2003). Other mushrooms with RIP include pleuturegin (*Pleurotus tuber-regium*) (Wang and Ng 2001), *Hypsizygus marmoreus* (hypsin), and *Lyophyllum shimeji* (lyophyllin). Some of the RIPs are devoid of ribonuclease activity. A RIP calcaelin from *Calvatia caelata* reduced the viability of breast cancer cells (Ng et al. 2003).

Calvatia utriformis ubiquitin-like peptide possessed antiproliferative and antioncogenic property towards breast cancer cells (Lam et al. 2001). *Gymnopus fusipes* cyclopeptides (gymnopeptides A and B) exerted an antiproliferative effect on several human cancer cell lines at nanomolar IC50 values (Ványolós et al. 2016). The peptide eryngin from the fruiting bodies of *Pleurotus eryngii* possessed antifungal property (Wang and Ng 2004). Pleurostrin, a 7 kDa peptide from oyster mushroom (Chu et al. 2005), and agrocybin, a 9 kDa peptide from *Agrocybe cylindracea* (Ngai et al. 2005), possessed antifungal activity.

L-theanine (γ-glutamylethylamide) is a non-protein amino acid, which provides umami flavor. *Boletus badius* or *Xerocomus badius* L-theanine potentiates the anticancer effect of drugs such as doxorubicin, anthracyclines, cisplatin, and irinotecan. Ergothioneine from mushroom could inhibit MPO activity and inflammatory responses (Asahi et al. 2016).

5.2.4 Phenolic Acids, Phenolic Compounds, and Terpenoids

Cinnamic acid-rich ethanolic extract of *Clitocybe alexandri* could inhibit lung, breast, colon, and gastric cancer cell lines by inducing S-phase cell cycle arrest (Vaz et al. 2012). Hispolon, a phenolic compound in several *Phellinus* species, is known to possess cytotoxicity and antitumor property (Chen et al. 2008b), and it can regulate estrogenic ingredient. It exhibited antiproliferative effect against estrogen-sensitive ER (+) MCF-7 (Wang et al. 2014). Hispolon induced ROS-mediated apoptosis in gastric cancer cells by causing oxidative damage to the mitochondrial membranes, leading to apoptosis. Also, hispolon potentiated the cytotoxicity of chemotherapeutic agents used in the study (Chen et al. 2008b).

Grifolin, a secondary metabolite isolated from *Albatrellus confluens*, could upregulate death-associated protein kinase 1 DAPK1 in nasopharyngeal carcinoma cells (Luo et al. 2011). Grifolin treatment inhibited cyclin D1, cyclin E, and CDK4 expression and upregulated CKI ($p19^{INK4D}$), while reducing pRB phosphorylation (Ye et al. 2007). Grifolin could inhibit the proliferation of gastric cancer cells by inducing apoptosis and suppressing the ERK1/2 pathway (Wu and Li 2017).

Antibiotic activity of calvatic acid (p-carboxyphenylazoxycyanide) and its analogues against *Helicobacter pylori* has been reported. This acid can inhibit the enzyme human placenta glutathione transferase (GST) P1–1 (Antonini et al. 1997). Ganoderic acid, a lanostane triterpenoid, from *Ganoderma lucidum* could inhibit the replication of hepatitis B virus (HBV) in HepG2215 cells and protected infected mice against *Mycobacterium bovis* BCG-caused hepatic injuries (Li and Wang 2006). Ganoderic acid T prevented tumor invasion and metastasis by promoting cell aggregation, inhibiting cell adhesion, and migration in human colon cancer HCT-116 cells (Chen et al. 2010). Ganoderic acid Me exerted cytotoxicity on HCT-116 cells in a dose-dependent manner (Chen and Zhong 2009). The expression of antitumor protein p53 in ganoderic acid Me-treated tumor cells increased, and pro-apoptotic proteins Bax was upregulated (Chen and Zhong 2009).

Suillin isolated from the ethyl acetate extract of the mushroom *Suillus placidus* induced apoptosis of HepG2 cells, by DNA fragmentation, phosphatidylserine (PS) externalization, activation of caspases (3, 8, and 9), depolarization of mitochondrial membrane potential, and the release of cytochrome c into the cytosol (Liu et al. 2009). Iso-suillin from *Suillus luteus* inhibited the proliferation and caused apoptosis in human hepatoma cell line SMMC-7721, by G1 phase arrest (Jia et al. 2014), whereas the iso-suillin from *Suillus flavus* inhibited human small cell lung cancer H446 cell line by mitochondrial as well as the death receptor pathway (Zhao et al. 2016).

Illudin S is a sesquiterpenoid toxin isolated from *Omphalotus*. Its derivative acylfulvenes has cytotoxicity and genotoxicity which occurs by metabolic sulfation. A semisynthetic derivative of hydroxymethylacylfulvene (Irofulven) is under clinical trial for antitumor efficacy (Jaspers et al. 2002).

Polyozellus multiplex-derived polyozellin, a p-terphenyl, prevented cell proliferation in stomach cancer and leukemia cells. The mechanism was by increasing quinone reductase, and glutathione level, and by upregulating the expression of p53 proteins (Kim et al. 2004). It suppressed iNOS expression by inhibiting the activation of NF-κB and SAPK/JNK (stress-activated protein/ c-Jun N-terminal kinase) (Jin et al. 2006). In another study, polyozellin suppressed TGFBIp (transforming growth factor β-induced protein)-mediated septic risks (Jung et al. 2016).

Fomitopsis betulina (*Piptoporus betulinus*)-derived lanostane triterpenoids (piptolinic acids A–E) were cytotoxic towards human promyelocytic leukemia cell line HL-60 and human acute monocytic leukemia cell line THP-1 (Tohtahon et al. 2017). Methylantcinate A, an ergostane-type triterpenoid, from *Antrodia camphorata*, inhibited the growth of oral cancer (OEC-M1 and OC-2) cell lines in a dose-dependent manner, without cytotoxicity to normal oral gingival fibroblast cells. The mechanism of cancer inhibition was apoptosis induction, resultant of caspase-3 activation, and DNA fragmentation (Tsai et al. 2010).

5.2.5 Steroids

Ergosterol gives rise to ergone (ergosta-4,6,8(14),22-tetraen-3-one), a steroid. Ergone from *Russula cyanoxantha* showed cytotoxicity toward HepG2 cells, causing their apoptosis (Zhao et al. 2011). Ergone from *Polyporus umbellatus* has an anti-aldosterone and cytotoxic effect (Lee et al. 2007). Ergosterol peroxide (5α, 8α-epidioxy-22E-ergosta-6,22-dien-3β-ol), the steroidal derivative, has antimicrobial effects. The sterol cerevisterol was isolated from the fruiting body of *Pseudoinonotus dryadeus* (Cateni et al. 2015).

5.2.6 Ceramide

A new phytosphingosine-type ceramide, suillumide, from *Suillus luteus* exerted cytotoxicity toward human melanoma cell line SK-MEL-1 (León et al. 2008). A new cerebroside, glycosphingolipid (monoglycosylceramides cerebroside E), was isolated from the fruiting bodies of *Hericium erinaceus*. The cerebroside alleviated cisplatin-induced nephrotoxicity in LLC-PK1 cells and inhibited angiogenesis in HUVECs (Lee et al. 2015).

5.2.7 Nucleoside

Cordycepin, 3′-deoxyadenosine, a derivative of adenosine, from *Cordyceps militaris* inhibited human leukemia cell growth, by inducing apoptosis through ROS-mediated caspase pathway (Jeong et al. 2011). Also, this metabolite inhibited human HepG2 cell proliferation and induced their apoptosis (Lu et al. 2014).

5.2.8 Ubiquinone Derivative

Antrodia camphorata can promote cell cycle arrest and apoptosis of human estrogen-nonresponsive breast cancer (MDA-MB-231) and colon cancer cell lines (HT-29, HCT-116, and SW-480), by suppressing the MAPK signaling pathway (Yeh et al. 2009). Antroquinonol, a ubiquinone derivative, isolated from this mushroom inhibited the cell proliferation in pancreatic cancer (PANC-1 and AsPC-1), as well as lung cancer (NSCLC A549) cells (Kumar et al. 2011). Antroquinonol can also inhibit hepatocellular carcinoma (HCC) through AMPK activation and inhibition of mTOR translational pathway, which leads to G1 phase arrest of cell cycle and subsequent cell apoptosis (Chiang et al. 2010).

5.2.9 Alkaloids

The alkaloid psilocybin from *Inocybe umbrinella* is psychoactive and hallucinogenic (Tylš et al. 2014). At a dose of 0.2 mg/kg, psilocybin lowered anxiety and improved mood in cancer patients (Grob et al. 2011). Its usage in the treatment of depression is being pursued (Patra 2016). Other mushroom alkaloids include agrocybenine (*Agrocybe cylindracea*) and laccarin (*Laccaria vinaceoavellanea*).

5.2.10 Phenylhydrazine Derivative

Agaricus blazei Murrill agaritine (N-(gamma-L(+)-glutamyl)-4-hydroxymethylphenylhydrazine), a hydrazine-containing compound, exhibits antitumor activity toward leukemic cells by inducing apoptosis (Akiyama et al. 2011).

5.3 Mushroom Extracts for Immunomodulation

In some preliminary studies, where the bioactive mycochemical has not been isolated, the roles of the mushroom extracts have been reported. The ethanol extracts of *Ganoderma lucidum* reduced the viability of human gastric carcinoma (AGS) cell line (Jang et al. 2010). In mice model, it showed 5α-reductase inhibitory activity, which holds hope for prostate cancer therapy (Fujita et al. 2005). Its function is mediated by both triterpenes and polysaccharides. Hsu et al. (2007) reported the anti-invasive effect of ethyl acetate extract from *Antrodia cinnamomea* fruiting bodies in the human liver cancer (PLC/PRF/5) cell line characterized by the fall in either the level or activity of VEGF, matrix metalloproteinases (MMP-2, MMP-9, and MT1-MMP), and an increase in the expression of tissue inhibitor of metalloproteinase (TIMP-1 and TIMP-2) (Hsu et al. 2007). The extract inhibited constitutively activated and inducible NF-κB DNA-binding and transcriptional activity, also inhibiting the TNF-α-activated NF-κB-dependent reporter gene expression of MMP-9 and VEGF (Hsu et al. 2007). The dichloromethane extract *Ganoderma lucidum* had human papillomavirus 16 (HPV 16) E6 oncoprotein inhibitory activity has been studied using epidermoid cervical carcinoma (CaSki) cells (Lai et al. 2010). *Hericium erinaceus* sensitizes doxorubicin-mediated apoptotic signaling by reducing c-FLIP expression via JNK activation and enhancing the intracellular level of the drug by the inhibition of NF-κB activity (Lee and Hong 2010). The ethanol and ethyl acetate extract of shaggy ink cap mushroom *Coprinus comatus* selectively inhibited malignant estrogen-independent breast cancer by affecting IκBα (nuclear factor of kappa light polypeptide gene enhancer in B-cell inhibitor, alpha) phosphorylation (Gu and Leonard 2006). Also, it inhibited dihydrotestosterone-induced prostate cancer LNCaP cells, by arresting at G1 phase, which occurred by the reduction in the levels of androgen receptors and prostate-specific antigen (PSA) (Dotan et al. 2011).

5.4 Discussion

Medicinal mushrooms have become an important part of pharmaceutical and alternative drugs, as well as dietary supplement sector. Mushroom as biotherapeutics attract immense interest for their easy accessibility. Some proprietary dietary supplement prepared from mushrooms include Yunzhi (*Trametes versicolor*) (Wong et al. 2005; Szeto et al. 2013), ReishiMax, (*Ganoderma lucidum*) (Thyagarajan-Sahu et al. 2011), etc.

Patients suffering from terminal diseases look up to medical mushrooms as palliative and anxiolytic therapy (Patel and Goyal 2012; Kozarski et al. 2015). Mushrooms have secured a place in the league of superfoods like spirulina, green tea, cereal bran, acai berry, seaweeds, and others (Patel and Goyal 2012). *Inonotus obliquus* (Chaga) (Patel 2015), *Ganoderma lucidum* (Reishi mushroom, Ling Zhi) (Wachtel-Galor et al. 2011), *Cordyceps* (Dworecka-Kaszak 2014), etc. are some of the most popular edible-medicinal mushrooms. However, in vitro, or even in vivo results do not translate well when applied to the human system. Each cancer or for that matter each individual's immune system is unique. Also, in the studies, a particular biological function is attributed to one bioactive component. But, it is likely that multiple components coordinate to elicit an immunomodulation property.

Amanita, *Clitocybe*, *Psilocybe*, *Cortinarius*, and *Gyromitra* are some well-studied poisonous mushrooms (Lima et al. 2012). Among the mushroom toxins, amatoxin, amanitin, phalloidin, phallacidine, gyromitrin, psilocybin, muscarine, ibotenic acid, orellanine, muscimol, and coprine have been well-characterized (Bas and Yurttagul 2004; Jo et al. 2014). They can cause gastrointestinal, hepatic, nephrological, and neurological afflictions (Bas and Yurttagul 2004) or death (Erden et al. 2013). Agaritine induces DNA damage by imposing oxidative stress as verified from the marker 8-OHdG in hydrazine-treated mouse urine (Kondo et al. 2008). A carboxylic acid cycloprop-2-ene from *Russula subnigricans* can cause fatal rhabdomyolysis, by elevating serum creatine phosphokinase (Matsuura et al. 2009). Also, this mushroom, a trove of poisonous components, can cause hypocalcemia, respiratory failure, ventricular tachycardia, cardiac shock, and death (Cho and Han 2016). Recent studies have found that Chaga mushrooms have very high content of oxalate which on regular ingestion, can cause oxalate nephropathy (Kikuchi et al. 2014). However, now it has been acknowledged that the difference between drugs and toxins are blurry. So, the mushroom toxins might be manipulated to derive therapeutics. In fact, psilocybin is already being studied for therapeutic applications (Passie et al. 2002).

Finally, there are about 5,000 species of mushrooms globally. Not all of them have been characterized well, and each of them is expected to have a unique repertoire of mycochemicals. As our medicinal arsenal is depleting, they ought to be tapped for healthcare. However, it must be understood by humankind that in the face of abusive lifestyle, a torrent of oxidative assaults are imposed on the body, which the best of panaceas may not cancel. So, the immunomodulating benefits of mushrooms can be best taken advantage of, if an inflammation-free life is followed.

5.5 Conclusion

The reported literature on mushrooms is exhaustive, yet it is a fraction of the yet-to-be mined information. This chapter has attempted to mention only selective data of biomedical relevance of medicinal mushrooms. The mycochemicals spanning diverse chemical groups ought to be exploited for healthcare. As chemotherapeutics are being linked to side effects, and antibiotics are being associated with bacterial resistance, mushroom-derived compounds might be a solution to the clinical conundrum.

Conflict of Interest There is no conflict of interest in the submission of this manuscript.

References

Akiyama H, Endo M, Matsui T et al (2011) Agaritine from *Agaricus blazei* Murrill induces apoptosis in the leukemic cell line U937. Biochim Biophys Acta Gen Subj 1810:519–525. https://doi.org/10.1016/j.bbagen.2011.02.010

Antonini G, Pitari G, Caccuri AM et al (1997) Inhibition of human placenta glutathione transferase P1-1 by the antibiotic calvatic acid and its diazocyanide analogue. Eur J Biochem 245:663–667. https://doi.org/10.1111/j.1432-1033.1997.00663.x

Asahi T, Wu X, Shimoda H et al (2016) A mushroom-derived amino acid, ergothioneine, is a potential inhibitor of inflammation-related DNA halogenation. Biosci Biotechnol Biochem 80:313–317. https://doi.org/10.1080/09168451.2015.1083396

Avanzo Caglič P, Renko M, Turk D et al (2014) Fungal β-trefoil trypsin inhibitors cnispin and cospin demonstrate the plasticity of the β-trefoil fold. Biochim Biophys Acta Proteins Proteomics 1844:1749–1756. https://doi.org/10.1016/j.bbapap.2014.07.004

Bas M, Yurttagul M (2004) Mushroom toxins and mushroom poisoning. SENDROM 16:60–65

Batbayar S, Lee DH, Kim HW (2012) Immunomodulation of fungal β-glucan in host defense signaling by dectin-1. Biomol Ther (Seoul) 20:433–445. https://doi.org/10.4062/biomolther.2012.20.5.433 (Title in lower case)

Cateni F, Zacchigna M, Altieri T et al (2015) Antioxidant properties of oak bracket mushroom, *Pseudoinonotus dryadeus* (higher basidiomycetes): a mycochemical study. Int J Med Mushrooms 17:627–637. https://doi.org/10.1615/IntJMedMushrooms.v17.i7.30

Chen NH, Zhong JJ (2009) Ganoderic acid Me induces G1 arrest in wild-type p53 human tumor cells while G1/S transition arrest in p53-null cells. Process Biochem 44:928–933. https://doi.org/10.1016/j.procbio.2009.03.018

Chen W, Zhao Z, Chen SF, Li YQ (2008a) Optimization for the production of exopolysaccharide from *Fomes fomentarius* in submerged culture and its antitumor effect in vitro. Bioresour Technol 99:3187–3194. https://doi.org/10.1016/j.biortech.2007.05.049

Chen W, Zhao Z, Li L et al (2008b) Hispolon induces apoptosis in human gastric cancer cells through a ROS-mediated mitochondrial pathway. Free Radic Biol Med 45:60–72. https://doi.org/10.1016/j.freeradbiomed.2008.03.013

Chen NH, Liu JW, Zhong JJ (2010) Ganoderic acid T inhibits tumor invasion in vitro and in vivo through inhibition of MMP expression. Pharmacol Rep 62:150–163. https://doi.org/10.1016/S1734-1140(10)70252-8

Cheung Y-H, Sheridan CM, Lo ACY, Lai WW (2012) Lectin from *Agaricus bisporus* inhibited S phase cell population and Akt phosphorylation in human RPE cells. Invest Ophthalmol Vis Sci 53:7469–7475. https://doi.org/10.1167/iovs.12-10589

Chiang PC, Lin SC, Pan SL et al (2010) Antroquinonol displays anticancer potential against human hepatocellular carcinoma cells: a crucial role of AMPK and mTOR pathways. Biochem Pharmacol 79:162–171. https://doi.org/10.1016/j.bcp.2009.08.022

Cho JT, Han JH (2016) A case of mushroom poisoning with *Russula subnigricans*: development of rhabdomyolysis, acute kidney injury, cardiogenic shock, and death. J Korean Med Sci 31:1164–1167. https://doi.org/10.3346/jkms.2016.31.7.1164 (Title in lower case)

Chu KT, Xia L, Ng TB (2005) Pleurostrin, an antifungal peptide from the oyster mushroom. Peptides 26:2098–2103. https://doi.org/10.1016/j.peptides.2005.04.010

Colin Slaughter J (1999) The naturally occurring furanones: formation and function from pheromone to food. Biol Rev Camb Philos Soc 74:259–276

Córdoba M, Ríos AH (2012) Biotechnological applications and potential uses of the mushroom *Trametes versicolor*. Vitae Rev la Fac Quim Farm 19:70–76

Cui FJ, Li Y, Xu YY et al (2007) Induction of apoptosis in SGC-7901 cells by polysaccharide-peptide GFPS1b from the cultured mycelia of *Grifola frondosa* GF9801. Toxicol Vitr 21:417–427. https://doi.org/10.1016/j.tiv.2006.10.004

Dotan N, Wasser SP, Mahajna J (2011) Inhibition of the androgen receptor activity by *Coprinus comatus* substances. Nutr Cancer 63:1316–1327. https://doi.org/10.1080/01635581.2011.607542

Dworecka-Kaszak B (2014) *Cordyceps* fungi as natural killers, new hopes for medicine and biological control factors. Ann Parasitol 60:151–158

Erden A, Esmeray K, Karagöz H et al (2013) Acute liver failure caused by mushroom poisoning: a case report and review of the literature. Int Med Case Rep J 6:85–90. https://doi.org/10.2147/IMCRJ.S53773

Fujita R, Liu J, Shimizu K et al (2005) Anti-androgenic activities of *Ganoderma lucidum*. J Ethnopharmacol 102:107–112. https://doi.org/10.1016/j.jep.2005.05.041

Grob CS, Danforth AL, Chopra GS et al (2011) Pilot study of psilocybin treatment for anxiety in patients with advanced-stage cancer. Arch Gen Psychiatry 68:71–78. https://doi.org/10.1001/archgenpsychiatry.2010.116

Gu YH, Leonard J (2006) In vitro effects on proliferation, apoptosis and colony inhibition in ER-dependent and ER-independent human breast cancer cells by selected mushroom species. Oncol Rep 15:417–423. https://doi.org/10.3892/or.15.2.417

Guan G-P, Zhang G-Q, Wu Y-Y et al (2011) Purification and characterization of a novel serine protease from the mushroom *Pholiota nameko*. J Biosci Bioeng 111:641–645. https://doi.org/10.1016/j.jbiosc.2011.02.009

Guo Y-J, Deng G-F, Xu X-R et al (2012) Antioxidant capacities, phenolic compounds and polysaccharide contents of 49 edible macro-fungi. Food Funct 3:1195–1205. https://doi.org/10.1039/c2fo30110e

Hsu YL, Kuo PL, Cho CY et al (2007) *Antrodia cinnamomea* fruiting bodies extract suppresses the invasive potential of human liver cancer cell line PLC/PRF/5 through inhibition of nuclear factor kappaB pathway. Food Chem Toxicol 45:1249–1257. https://doi.org/10.1016/j.fct.2007.01.005

Huang HY, Chieh SY, Tso TK et al (2011) Orally administered mycelial culture of *Phellinus linteus* exhibits antitumor effects in hepatoma cell-bearing mice. J Ethnopharmacol 133:460–466. https://doi.org/10.1016/j.jep.2010.10.015

Ina K, Kataoka T, Ando T (2013) The use of lentinan for treating gastric cancer. Anticancer Agents Med Chem 13:681–688. https://doi.org/10.2174/1871520611313050002

Jang K-J, Han M-H, Lee B-H et al (2010) Induction of apoptosis by ethanol extracts of *Ganoderma lucidum* in human gastric carcinoma cells. J Acupunct Meridian Stud 3:24–31. https://doi.org/10.1016/S2005-2901(10)60004-0

Jaspers NGJ, Raams A, Kelner MJ et al (2002) Anti-tumour compounds illudin S and Irofulven induce DNA lesions ignored by global repair and exclusively processed by transcription- and replication-coupled repair pathways. DNA Repair (Amst) 1:1027–1038. https://doi.org/10.1016/S1568-7864(02)00166-0

Javed S, Ahmad M, Ahmad M et al (2013) Chitinases: an update. J Pharm Bioallied Sci 5:21. https://doi.org/10.4103/0975-7406.106559

Jeong JW, Jin CY, Park C et al (2011) Induction of apoptosis by cordycepin via reactive oxygen species generation in human leukemia cells. Toxicol Vitr 25:817–824. https://doi.org/10.1016/j.tiv.2011.02.001

Jia Z-Q, Chen Y, Yan Y-X, Zhao J-X (2014) Iso-suillin isolated from *Suillus luteus*, induces G1 phase arrest and apoptosis in human hepatoma SMMC-7721 cells. Asian Pac J Cancer Prev 15:1423–1428. https://doi.org/10.7314/APJCP.2014.15.3.1423

Jin XY, Lee SH, Kim JY et al (2006) Polyozellin inhibits nitric oxide production by down-regulating LPS-induced activity of NF-kappaB and SAPK/JNK in RAW 264.7 cells. Planta Med 72:857–859. https://doi.org/10.1055/s-2006-946640

Jo W-S, Hossain MA, Park S-C (2014) Toxicological profiles of poisonous, edible, and medicinal mushrooms. Mycobiology 42:215–220. https://doi.org/10.5941/MYCO.2014.42.3.215

Jung B, Yang E-J, Bae J-S (2016) Suppressive effects of polyozellin on TGFBIp-mediated septic responses in human endothelial cells and mice. Nutr Res 36:380–389. https://doi.org/10.1016/j.nutres.2015.12.009

Kikuchi Y, Seta K, Ogawa Y et al (2014) Chaga mushroom-induced oxalate nephropathy. Clin Nephrol 81:440–444. https://doi.org/10.5414/CN107655

Kim JH, Lee JS, Song K-S et al (2004) Polyozellin isolated from *Polyozellus multiplex* induces phase 2 enzymes in mouse hepatoma cells and differentiation in human myeloid leukaemic cell lines. J Agric Food Chem 52:451–455. https://doi.org/10.1021/jf034748n

Kim J-S, Sapkota K, Park S-E et al (2006) A fibrinolytic enzyme from the medicinal mushroom *Cordyceps militari*s. J Microbiol 44:622–631

Kim HS, Kim JY, Kang JS et al (2010) Cordlan polysaccharide isolated from mushroom *Cordyceps militaris* induces dendritic cell maturation through toll-like receptor 4 signalings. Food Chem Toxicol 48:1926–1933. https://doi.org/10.1016/j.fct.2010.04.036

Ko J-L, Hsu C-I, Lin R-H et al (1995) A new fungal immunomodulatory protein, FIP-fve isolated from the edible mushroom, *Flammulina velutipes* and its complete amino acid sequence. Eur J Biochem 228:244–249. https://doi.org/10.1111/j.1432-1033.1995.0244n.x

Kondo K, Watanabe A, Akiyama H, Maitani T (2008) The metabolisms of agaritine, a mushroom hydrazine in mice. Food Chem Toxicol 46:854–862. https://doi.org/10.1016/j.fct.2007.10.022

Kozarski M, Klaus A, Jakovljevic D et al (2015) Antioxidants of edible mushrooms. Molecules 20:19489–19525

Kumar VB, Yuan TC, Liou JW et al (2011) Antroquinonol inhibits NSCLC proliferation by altering PI3K/mTOR proteins and miRNA expression profiles. Mutat Res Fundam Mol Mech Mutagen 707:42–52. https://doi.org/10.1016/j.mrfmmm.2010.12.009

Lai LK, Abidin NZ, Abdullah N, Sabaratnam V (2010) Anti-Human Papillomavirus (HPV) 16 E6 activity of Ling Zhi or Reishi medicinal mushroom, *Ganoderma lucidum* (W. Curt.: Fr.) P. Karst. (Apyllophoromycetideae) extracts. Int J Med Mushroom 12:279–286. https://doi.org/10.1615/IntJMedMushr.v12.i3.70

Lam YW, Ng TB, Wang HX (2001) Antiproliferative and antimitogenic activities in a peptide from puffball mushroom *Calvatia caelata*. Biochem Biophys Res Commun 289:744–749. https://doi.org/10.1006/bbrc.2001.6036

Lee JS, Hong EK (2010) Hericium erinaceus enhances doxorubicin-induced apoptosis in human hepatocellular carcinoma cells. Cancer Lett 297:144–154. https://doi.org/10.1016/j.canlet.2010.05.006

Lee WY, Park Y, Ahn JK (2007) Improvement of ergone production from mycelial culture of *Polyporus umbellatus*. Mycobiology 35:82. https://doi.org/10.4489/MYCO.2007.35.2.082

Lee YS, Kim YH, Shin EK et al (2010) Anti-angiogenic activity of methanol extract of *Phellinus linteus* and its fractions. J Ethnopharmacol 131:56–62. https://doi.org/10.1016/j.jep.2010.05.064

Lee SR, Jung K, Noh HJ et al (2015) A new cerebroside from the fruiting bodies of *Hericium erinaceu*s and its applicability to cancer treatment. Bioorganic Med Chem Lett 25:5712–5715. https://doi.org/10.1016/j.bmcl.2015.10.092

León F, Brouard I, Torres F et al (2008) A new ceramide from *Suillus luteus* and its cytotoxic activity against human melanoma cells. Chem Biodivers 5:120–125. https://doi.org/10.1002/cbdv.200890002

Li YQ, Wang SF (2006) Anti-hepatitis B activities of ganoderic acid from *Ganoderma lucidum*. Biotechnol Lett 28:837–841. https://doi.org/10.1007/s10529-006-9007-9

Li G, Kim DH, Kim TD et al (2004) Protein-bound polysaccharide from *Phellinus linteus* induces G 2/M phase arrest and apoptosis in SW480 human colon cancer cells. Cancer Lett 216:175–181. https://doi.org/10.1016/j.canlet.2004.07.014

Li YR, Liu QH, Wang HX, Ng TB (2008) A novel lectin with potent antitumor, mitogenic and HIV-1 reverse transcriptase inhibitory activities from the edible mushroom *Pleurotus citrinopileatus*. Biochim Biophys Acta Gen Subj 1780:51–57. https://doi.org/10.1016/j.bbagen.2007.09.004

Li YG, Ji DF, Zhong S et al (2011) Anti-tumor effects of proteoglycan from *Phellinus linteus* by immunomodulating and inhibiting Reg IV/EGFR/Akt signaling pathway in colorectal carcinoma. Int J Biol Macromol 48:511–517. https://doi.org/10.1016/j.ijbiomac.2011.01.014

Liao CH, Hsiao YM, Lin CH et al (2008) Induction of premature senescence in human lung cancer by fungal immunomodulatory protein from *Ganoderma tsugae*. Food Chem Toxicol 46:1851–1859. https://doi.org/10.1016/j.fct.2008.01.044

Lima ADL, Costa Fortes R, Carvalho Garbi Novaes MR, Percário S (2012) Poisonous mushrooms: a review of the most common intoxications. Nutr Hosp 27:402–408

Lindequist U, Niedermeyer THJ, Jülich W-D (2005) The pharmacological potential of mushrooms. Evid Based Complement Alternat Med 2:285–299. https://doi.org/10.1093/ecam/neh107

Liu F yan, Luo K wang, Yu Z ming, et al (2009) Suillin from the mushroom *Suillus placidus* as potent apoptosis inducer in human hepatoma HepG2 cells. Chem Biol Interact 181:168–174. https://doi.org/10.1016/j.cbi.2009.07.008

Lizárraga-Guerra R, Guth H, López MG (1997) Identification of the most potent odorants in Huitlacoche (*Ustilago maydis*) and Austern Pilzen (*Pleurotus* sp.) by aroma extract dilution analysis and static head-space samples. J Agric Food Chem 45:1329–1332. https://doi.org/10.1021/jf960650f (Title in lower case)

Lu H, Li X, Zhang J et al (2014) Effects of cordycepin on HepG2 and EA.hy926 cells: potential antiproliferative, antimetastatic and anti-angiogenic effects on hepatocellular carcinoma. Oncol Lett 7:1556–1562. https://doi.org/10.3892/ol.2014.1965

Luo XJ, Li LL, Deng QP et al (2011) Grifolin, a potent antitumour natural product upregulates death-associated protein kinase 1 DAPK1 via p53 in nasopharyngeal carcinoma cells. Eur J Cancer 47:316–325. https://doi.org/10.1016/j.ejca.2010.09.021

Ma B-J, Ruan Y, Liu J-K (2008) Chemical constituents study on the fruiting bodies of *Lactarius rufus*. Zhong Yao Cai 31:233–234

Mansour A, Daba A, Baddour N et al (2012) Schizophyllan inhibits the development of mammary and hepatic carcinomas induced by 7,12 dimethylbenz(α)anthracene and decreases cell proliferation: comparison with tamoxifen. J Cancer Res Clin Oncol 138:1579–1596. https://doi.org/10.1007/s00432-012-1224-0

Mansour MK, Tam JM, Khan NS et al (2013) Dectin-1 activation controls maturation of β-1,3-glucan-containing phagosomes. J Biol Chem 288:16043–16054. https://doi.org/10.1074/jbc.M113.473223

Masuda Y, Inoue M, Miyata A et al (2009) Maitake β-glucan enhances therapeutic effect and reduces myelosuppression and nephrotoxicity of cisplatin in mice. Int Immunopharmacol 9:620–626. https://doi.org/10.1016/j.intimp.2009.02.005

Matsuura M, Saikawa Y, Inui K et al (2009) Identification of the toxic trigger in mushroom poisoning. Nat Chem Biol 5:465–467. https://doi.org/10.1038/nchembio.179

Merzendorfer H, Zimoch L (2003) Chitin metabolism in insects: structure, function and regulation of chitin synthases and chitinases. J Exp Biol 206:4393–4412

Ng TB, Lam YW, Wang H (2003) Calcaelin, a new protein with translation-inhibiting, antiproliferative and antimitogenic activities from the mosaic puffball mushroom *Calvatia caelata*. Planta Med 69:212–217. https://doi.org/10.1055/s-2003-38492

Ng TB, Ngai PHK, Xia L (2006) An agglutinin with mitogenic and antiproliferative activities from the mushroom *Flammulina velutipes*. Mycologia 98:167–171. https://doi.org/10.3852/mycologia.98.2.167

Ngai PHK, Zhao Z, Ng TB (2005) Agrocybin, an antifungal peptide from the edible mushroom *Agrocybe cylindracea*. Peptides 26:191–196. https://doi.org/10.1016/j.peptides.2004.09.011

Ning X, Luo Q, Li C et al (2014) Inhibitory effects of a polysaccharide extract from the Chaga medicinal mushroom, *Inonotus obliquus* (higher Basidiomycetes), on the proliferation of human neurogliocytoma cells. Int J Med Mushrooms 16:29–36

Passie T, Seifert J, Schneider U, Emrich HM (2002) The pharmacology of psilocybin. Addict Biol 7:357–364

Patel S (2015) Emerging bioresources with nutraceutical and pharmaceutical prospects. Springer, Cham

Patel S (2017) A critical review on serine protease: key immune manipulator and pathology mediator. Allergol Immunopathol (Madr). https://doi.org/10.1016/j.aller.2016.10.011

Patel S, Goyal A (2012) Recent developments in mushrooms as anti-cancer therapeutics: a review. 3 Biotech 2:1–15. https://doi.org/10.1007/s13205-011-0036-2

Patel S, Goyal A (2017) Chitin and chitinase: role in pathogenicity, allergenicity and health. Int J Biol Macromol 97:331–338. https://doi.org/10.1016/j.ijbiomac.2017.01.042

Patra S (2016) Return of the psychedelics: psilocybin for treatment resistant depression. Asian J Psychiatr 24:51–52. https://doi.org/10.1016/j.ajp.2016.08.010

Pohleven J, Obermajer N, Sabotič J et al (2009) Purification, characterization and cloning of a ricin B-like lectin from mushroom *Clitocybe nebularis* with antiproliferative activity against human leukemic T cells. Biochim Biophys Acta Gen Subj 1790:173–181. https://doi.org/10.1016/j.bbagen.2008.11.006

Rop O, Mlcek J, Jurikova T (2009) Beta-glucans in higher fungi and their health effects. Nutr Rev 67:624–631. https://doi.org/10.1111/j.1753-4887.2009.00230.x

Sabotič J, Bleuler-Martinez S, Renko M et al (2012) Structural basis of trypsin inhibition and entomotoxicity of cospin, serine protease inhibitor involved in defense of *Coprinopsis cinerea* fruiting bodies. J Biol Chem 287:3898–3907. https://doi.org/10.1074/jbc.M111.285304

Shi BJ, Nie XH, Chen LZ et al (2007) Anticancer activities of a chemically sulfated polysaccharide obtained from *Grifola frondosa* and its combination with 5-Fluorouracil against human gastric carcinoma cells. Carbohydr Polym 68:687–692. https://doi.org/10.1016/j.carbpol.2006.08.003

Suárez Arango C, Nieto IJ (2013) Biotechnological cultivation of edible macrofungi: an alternative for obtaining nutraceutics. Rev Iberoam Micol 30:1–8. https://doi.org/10.1016/j.riam.2012.03.011

Szeto YT, Lau PC, Kalle W, Pak SC (2013) Direct human DNA protection by *Coriolus versicolor* (Yunzhi) extract. Pharm Biol 51:851–855. https://doi.org/10.3109/13880209.2013.767839

Thyagarajan-Sahu A, Lane B, Sliva D (2011) ReishiMax, mushroom based dietary supplement, inhibits adipocyte differentiation, stimulates glucose uptake and activates AMPK. BMC Complement Altern Med. https://doi.org/10.1186/1472-6882-11-74

Tohtahon Z, Xue J, Han J et al (2017) Cytotoxic lanostane triterpenoids from the fruiting bodies of *Piptoporus betulinus*. Phytochemistry 143:98–103. https://doi.org/10.1016/j.phytochem.2017.07.013

Tong H, Xia F, Feng K et al (2009) Structural characterization and in vitro antitumor activity of a novel polysaccharide isolated from the fruiting bodies of *Pleurotus ostreatus*. Bioresour Technol 100:1682–1686. https://doi.org/10.1016/j.biortech.2008.09.004

Tsai WC, Rao YK, Lin SS et al (2010) Methylantcinate A induces tumor specific growth inhibition in oral cancer cells via Bax-mediated mitochondrial apoptotic pathway. Bioorganic Med Chem Lett 20:6145–6148. https://doi.org/10.1016/j.bmcl.2010.08.006

Tylš F, Páleníček T, Horáček J (2014) Psilocybin – summary of knowledge and new perspectives. Eur Neuropsychopharmacol 24:342–356

Valverde ME, Hernández-Pérez T, Paredes-López O (2015) Edible mushrooms: improving human health and promoting quality life. Int J Microbiol 2015:376387. https://doi.org/10.1155/2015/376387

Ványolós A, Dékány M, Kovács B et al (2016) Gymnopeptides A and B, cyclic octadecapeptides from the mushroom *Gymnopus fusipes*. Org Lett 18:2688–2691. https://doi.org/10.1021/acs.orglett.6b01158

Vaz JA, Almeida GM, Ferreira ICFR et al (2012) *Clitocybe alexandri* extract induces cell cycle arrest and apoptosis in a lung cancer cell line: identification of phenolic acids with cytotoxic potential. Food Chem 132:482–486. https://doi.org/10.1016/j.foodchem.2011.11.031

Vetter J (2007) Chitin content of cultivated mushrooms *Agaricus bisporus*, *Pleurotus ostreatus* and *Lentinula edodes*. Food Chem 102:6–9. https://doi.org/10.1016/j.foodchem.2006.01.037

Wachtel-Galor S, Yuen J, Buswell JA, Benzie IFF (2011) Ganoderma lucidum (Lingzhi or Reishi). In: Benzie IFF, Wachtel-Galor S, editors. Herbal Medicine: Biomolecular and Clinical Aspects. 2nd edition. Boca Raton (FL): CRC Press/Taylor & Francis; 2011. Chapter 9.

Walton JD, Hallen-Adams HE, Luo H (2010) Ribosomal biosynthesis of the cyclic peptide toxins of *Amanita* mushrooms. Biopolymers 94:659–664. https://doi.org/10.1002/bip.21416

Wang HX, Ng TB (2001) Isolation of pleuturegin, a novel ribosome-inactivating protein from fresh sclerotia of the edible mushroom *Pleurotus tuber-regium*. Biochem Biophys Res Commun 288:718–721. https://doi.org/10.1006/bbrc.2001.5816

Wang H, Ng TB (2004) Eryngin, a novel antifungal peptide from fruiting bodies of the edible mushroom *Pleurotus eryngii*. Peptides 25:1–5. https://doi.org/10.1016/j.peptides.2003.11.014

Wang J, Hu F, Luo Y et al (2014) Estrogenic and anti-estrogenic activities of hispolon from *Phellinus lonicerinus* (Bond.) Bond. et sing. Fitoterapia 95:93–101. https://doi.org/10.1016/j.fitote.2014.03.007

Wong C-K, Bao Y-X, Wong EL-Y et al (2005) Immunomodulatory activities of Yunzhi and Danshen in post-treatment breast cancer patients. Am J Chin Med 33:381–395. https://doi.org/10.1142/S0192415X05002990

Wu Z, Li Y (2017) Grifolin exhibits anti-cancer activity by inhibiting the development and invasion of gastric tumor cells. Oncotarget 8:21454–21460. https://doi.org/10.18632/oncotarget.15250

Wu Y, Wang H, Ng TB (2011) Purification and characterization of a lectin with antiproliferative activity toward cancer cells from the dried fruit bodies of *Lactarius flavidulus*. Carbohydr Res 346:2576–2581. https://doi.org/10.1016/j.carres.2011.09.005

Wu H-T, Lu F-H, Su Y-C et al (2014) In vivo and in vitro anti-tumor effects of fungal extracts. Molecules 19:2546–2556. https://doi.org/10.3390/molecules19022546

Yap H-YY, Fung S-Y, Ng S-T et al (2015) Shotgun proteomic analysis of tiger milk mushroom (*Lignosus rhinocerotis*) and the isolation of a cytotoxic fungal serine protease from its sclerotium. J Ethnopharmacol 174:437–451. https://doi.org/10.1016/j.jep.2015.08.042

Ye M, Luo X, Li L et al (2007) Grifolin, a potential antitumor natural product from the mushroom *Albatrellus confluens*, induces cell-cycle arrest in G1 phase via the ERK1/2 pathway. Cancer Lett 258:199–207. https://doi.org/10.1016/j.canlet.2007.09.001

Ye L, Zhang J, Zhou S et al (2009) Preparation of a novel sulfated glycopeptide complex and inhibiting L1210 cell lines property in vitro. Carbohydr Polym 77:276–279. https://doi.org/10.1016/j.carbpol.2008.12.031

Yeh CT, Rao YK, Yao CJ et al (2009) Cytotoxic triterpenes from *Antrodia camphorata* and their mode of action in HT-29 human colon cancer cells. Cancer Lett 285:73–79. https://doi.org/10.1016/j.canlet.2009.05.002

Youn M-J, Kim J-K, Park S-Y et al (2008) Chaga mushroom (*Inonotus obliquus*) induces G0/G1 arrest and apoptosis in human hepatoma HepG2 cells. World J Gastroenterol 14:511–517 https://doi.org/10.3748/wjg.14.511

Zan X, Cui F, Li Y et al (2015) *Hericium erinaceus* polysaccharide-protein HEG-5 inhibits SGC-7901 cell growth via cell cycle arrest and apoptosis. Int J Biol Macromol 76:242–253. https://doi.org/10.1016/j.ijbiomac.2015.01.060

Zhang M, Zhang L, Cheung PCK, Ooi VEC (2004) Molecular weight and anti-tumor activity of the water-soluble polysaccharides isolated by hot water and ultrasonic treatment from the sclerotia and mycelia of *Pleurotus tuber-regium*. Carbohydr Polym 56:123–128. https://doi.org/10.1016/j.carbpol.2004.01.005

Zhang G, Sun J, Wang H, Ng TB (2010a) First isolation and characterization of a novel lectin with potent antitumor activity from a *Russula mushroom*. Phytomedicine 17:775–781. https://doi.org/10.1016/j.phymed.2010.02.001
Zhang G, Wang H, Zhang X, Ng T (2010b) Helvellisin, a novel alkaline protease from the wild ascomycete mushroom *Helvella lacunosa*. J Biosci Bioeng 109:20–24. https://doi.org/10.1016/j.jbiosc.2009.06.022
Zhang GQ, Wang YF, Zhang XQ et al (2010c) Purification and characterization of a novel laccase from the edible mushroom *Clitocybe maxima*. Process Biochem 45:627–633. https://doi.org/10.1016/j.procbio.2009.12.010
Zhao JK, Wang HX, Ng TB (2009) Purification and characterization of a novel lectin from the toxic wild mushroom *Inocybe umbrinella*. Toxicon 53:360–366. https://doi.org/10.1016/j.toxicon.2008.12.009
Zhao Y, Shen X, Chao X et al (2011) Ergosta-4,6,8(14),22-tetraen-3-one induces G2/M cell cycle arrest and apoptosis in human hepatocellular carcinoma HepG2 cells. Biochim Biophys Acta 1810:384–390. https://doi.org/10.1016/j.bbagen.2010.12.005
Zhao J-X, Zhang Q-S, Chen Y et al (2016) Iso-suillin from *Suillus flavus* induces apoptosis in human small cell lung cancer H446 cell line. Chinese Med J (English Ed) 129:1215–1223. https://doi.org/10.4103/0366-6999.181961
Zheng S, Wang H, Zhang G (2011) A novel alkaline protease from wild edible mushroom *Termitomyces albuminosus*. Acta Biochim Pol 58:269–273
Zhou K, Peng J, Chang N et al (2003) Purification and crystallization of flammulin, a basic protein with anti-tumor activities from *Flammulina velutipes*. CHINESE Chem Lett 14:713–716
Zhu L, Tang Q, Zhou S et al (2014) Isolation and purification of a polysaccharide from the caterpillar medicinal mushroom *Cordyceps militaris* (Ascomycetes) fruit bodies and its immunomodulation of RAW 264.7 macrophages. Int J Med Mushrooms 16:247–257

Mushrooms: A Wealth of Resource for Prospective Stem Cell-Based Therapies

6

Marthandam Asokan Shibu, Tamilselvi Shanmugam, Dinesh Chandra Agrawal, and Chih-Yang Huang

Contents

M. A. Shibu · T. Shanmugam
Medical Research Center for Exosome and Mitochondria Related Diseases, China Medical University and Hospital, Taichung, Taiwan

Graduate Institute of Basic Medical Science, School of Chinese Medicine, China Medical University and Hospital, Taichung, Taiwan

D. C. Agrawal (✉)
Department of Applied Chemistry, Chaoyang University of Technology, Taichung, Taiwan
e-mail: dcagrawal@gm.cyut.edu.tw

C.-Y. Huang (✉)
Medical Research Center for Exosome and Mitochondria Related Diseases, China Medical University and Hospital, Taichung, Taiwan

Department of Biological Science and Technology, Asia University, Taichung, Taiwan

Graduate Institute of Basic Medical Science, School of Chinese Medicine, China Medical University and Hospital, Taichung, Taiwan
e-mail: cyhuang@mail.cmu.edu.tw

D. C. Agrawal, M. Dhanasekaran (eds.), *Medicinal Mushrooms*,
https://doi.org/10.1007/978-981-13-6382-5_6

Abstract

Mushrooms are rich in nutraceutical compounds, which provide outstanding health benefits. Mounting evidences have proved that consumption of mushrooms decreases health complications associated with obesity, diabetes, and heart disease. However, mushrooms are mostly consumed as dietary supplements rather than as medicines. Nowadays, research efforts are focused on bioactive compounds extracted from mushrooms and their usage in regenerative medicine. Regenerative medicine is an interdisciplinary approach which deals with the replacement, engineering, or the regeneration of damaged human cells, tissues, or organs to regain their usual functions. It comprises the production and usage of beneficial stem cells, tissue engineering, and also generating artificial organs. Stem cell therapy is a fascinating field of research which targets myriad of disorders. Stem cells are undifferentiated cells which ultimately divide (mitosis) and develop into specialized organs. Researches are now keen on the development of stem cell-based disease management especially for neurodegenerative diseases, diabetes, and cardiovascular diseases. Numerous chemical compounds have been reported to precondition stem cells. Therapeutic agents derived from natural compounds are much preferred to prime or to regulate the activation of various types of stem cells. Although the medicinal properties of compounds extracted from mushrooms are well-documented, the comprehensive information on the compounds present in mushrooms which target different types of stem cells is still lacking. This chapter deals with various compounds present in mushrooms which mediate the activation of various stem cells, especially under pathological conditions.

Keywords

Cancer · Inflammation mushrooms · Neurodegeneration stem cells · Tissue engineering

Abbreviations

ANQ	Antroquinonol
BMSCs	Bone marrow-derived mesenchymal stem cells
CSCs	Cancer stem cells
ECM	Extracellular matrix
ESCs	Embryonic stem cells
HSCs	Hematopoietic stem cells
iPSCs	Induced pluripotent stem cells
MHC	Major histocompatibility complex
MSCs	Mesenchymal stem cells
ND	Neurodegenerative diseases
NSCs	Neural stem cells
NSCs	Neural stem cells

PSP	Polysaccharopeptides
Pten	Phosphatase and tensin homolog
SOD	Superoxide dismutase
TMP	Tetramethylpyrazine
TSPSCs	Tissue-specific progenitor stem cells

6.1 Introduction

Mushrooms belong to the kingdom fungi which are different from plants and animals. Plant cells possess totipotency, which is a rare phenomenon in animals. In fungi, almost all cells could be called "stem cells." In fact, all hypha of mature mushroom are totipotent in nature (Money 2002). Mushrooms are fungi with a stipe, a pileus, and a lamella, which is present inside the pileus. Mushrooms have long been consumed directly by human beings as food and medicine. In fact, the first reported reproductions of fungi by man could be before 3000 years. The "mushroom stones" of Guatemala are known for mushroom cultivation, which are for medicinal rituals (Molitoris 1994). *Ganoderma lucidum* or the bracket fungus (lingzhi in Chinese and reishi in Japanese) is a well-known medicinal mushroom, which has been cited in a 2000-year-old poem documented by the Han dynasty referring the mushroom as "Mushroom of Immortality" (Wachtel-Galor et al. 2004). Interest on mushrooms has been growing in various fields including food and biopharmaceuticals because of their nutritional and medicinal properties. Approximately, 140, 000 species of mushrooms have been estimated to be present on earth, among which only about 10% of mushrooms have been identified and documented by the scientific community. Around 2000 mushrooms are reported as harmless, and nearly 700 of them are considered to have significant medicinal properties (Lull et al. 2005; Manzi et al. 1999; Mattila et al. 2000). Mushrooms have been divided into three groups based on their nutritional aspects; (1) mycorrhizal or symbiotic species, (2) saprophytic species, and (3) parasitic species (Kalac 2013).

6.1.1 Mushroom Compounds and Their Uses

Mushrooms contain almost all essential amino acids in higher quantities than many common vegetables. Mushrooms contain moisture, total carbohydrates, dietary fibers, crude fat, ash, N-acetyl glucosamine, nitrogen, phosphorus, calcium, and proteins (Mattila et al. 2002). Mushrooms are low in fat and calorie. However, they are a good source of polyunsaturated fatty acid, which makes them a healthy diet (Kalac 2009). Also, mushrooms are rich in vitamins like riboflavin (Vit B2), niacin (Vit B3), folates (Vit B9), and traces of vitamin C, B1, B12, D, and E.

Mushrooms are rich in secondary metabolites that possess health benefits, including antioxidative, antibacterial, antiviral, anticancer, and anti-inflammatory properties, and also facilitate in the proper function of the cardiovascular system

(Kalac 2009). The crude proteins content of mushrooms varies from species to species and ranges from 15.2 g/100 g in *L. edodes* to 80.93 g/100 g dried weight in *A. bisporus*. Several mushroom-derived compounds including ceramides; lentinan; exopolysaccharide; schizophyllan; p-coumaric acid; omega-3, omega-6, and omega-9 fatty acids; 2-hydroxytyrosol; carotenoids; resveratrol; betulin; and trametenolic acid are now used in cosmetic industries (Hyde et al. 2010; Seok and Boo 2015; Taofiq et al. 2016). Most of these substances have been described to have anti-tyrosinase, anti-collagenase, anti-elastase, and anti-hyaluronidase effects, which make mushrooms as attractive candidates in cosmeceuticals and nutricosmetic industries (Taofiq et al. 2016). Polysaccharides mainly β-glucans in mushrooms are the most extensively studied bioactive compounds and are known for their immunomodulatory effects (Table 6.1).

Table 6.1 Various compounds reported from mushrooms and their activities

Compounds	Mushroom species	Activities	References
Alkaloids, polyphenols, flavonoids, saponins, p-coumaric acid, tannins	*Agaricus bisporus*, *Ganoderma lucidum*, *Pleurotus florida*, *Phellinus rimosus*, *Leucopaxillus giganteus*, *Sparassis crispa*, *Lentinus edodes*	Free radicals scavenging, inhibition of lipid peroxidation, skin renewal	Ganeshpurkar et al. (2010), Kirar et al. (2017), and Taofiq et al. (2016)
Sterols, triterpenoids	*Ganoderma colossum*, *Lepista nuda*, *Naematoloma sublateritium*, *Panellus serotinus*, *Scleroderma citrinum*, *Tricholoma matsutake*, *Naematoloma fasciculare*, *Lentinus edodes*	Proliferation of neural stem cells, antiaging properties, decrease blood pressure	Gill et al. (2016), Yaoita et al. (2001), and Kim et al. (2013)
Carbohydrates, proteins, amino acids	*Agaricus bisporus*, *Pleurotus ostreatus*, *Termitomyces eurhizus*, *Volvariella volvacea*, *Coprinus comatus*, *Sparassis crispa*, *Lentinula edodes*	Antimicrobial, antiviral, anti-inflammatory effects	Echigo et al. (2012) and Muta (2006)
Benzoquinone derivatives, lignan (flavonoids)	*Antrodia camphorata*, *Pleurotus eryngii*, *Ganoderma lucidum*, *Coriolus versicolor*	Anti-inflammatory effects	Lin et al. (2017) and Walker et al. (2000)
Marasmic acid	*Marasmius androsaceus*, *Marasmius oreades*	Analgesic, partial ganglionic blocking mediated, sedative effects	Mizuno et al. (1995)

6.1.2 Types of Stem Cells and Their Applications

Stem cells can be identified as undifferentiated cells which are able to proliferate, self-renew, and differentiate into specified cell types and eventually into organs (Biehl and Russell 2009). Stem cells are classified into four types based on their source: (1) embryonic stem cells (ESCs) which are derived from human blastocysts during embryogenesis, (2) germinal stem cells, (3) embryonic carcinomas, and (4) adult stem cells which are derived from adult bone marrow (Fig. 6.1) (Thomson 1998). The latter is further divided into hematopoietic stem cells (HSCs), mesenchymal stem cells (MSCs), and tissue-specific progenitor stem cells (TSPSCs) (Alison and Islam 2009).

The cell cycle period in somatic stem cells is comparatively longer, which is more than 16 h, whereas in ESCs the period is nearly half, which is around 8–10 h. ESCs follow a shorter cell cycle period than somatic/adult stem cells because of a considerably reduced G1 (3 h) phase and an elongated S phase (Mens and Ghanbari 2018). This difference in cell cycle duration makes ESCs as quickly dividing cells because somatic stem cells halt in a quiescent state for some time. Based on their differentiation efficiency, stem cells are categorized into (1) unipotent, (2) multipotent, (3) pluripotent, and (4) totipotent (Mahla 2016). Stem cells are the base for all tissue and organ system (heart, liver, kidney, neurons, and muscles) in our body. They play essential and diverse roles in the initiation and the progression of diseases and tissue repair processes (Mahla 2016).

In recent years, stem cell therapy has become a fascinating area of biomedical research, and its use in regenerative medicine is on increase (Sylvester and Longaker 2004). Pre-treatment or priming of MSCs before transplantation could considerably enhance the immunosuppressive properties of MSC-based treatments (Yang et al. 2016). Many chemical compounds (Fig. 6.2) are identified to prime/influence different types of stem cells (e.g., tetrandrine (Yang et al. 2016), isoquinoline sulfonamide (Hwang et al. 2008), ar-turmerone (Hucklenbroich et al. 2014), sulforaphane, and sorafenib (Rausch et al. 2010)).

Fig. 6.1 Classification of stem cells based on their sources

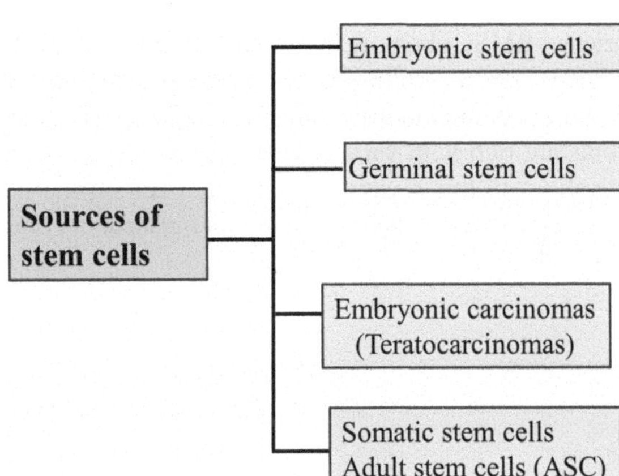

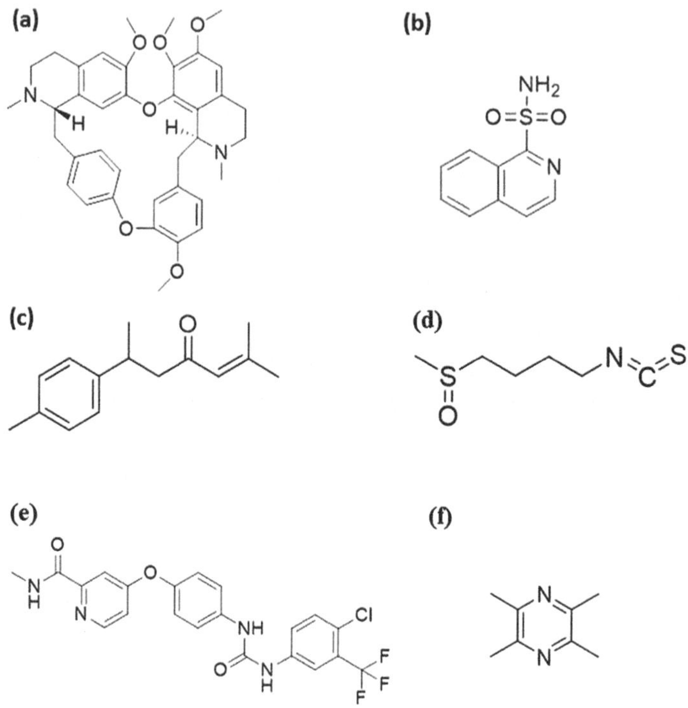

Fig. 6.2 Chemical structures of compounds that display potential effects in stem cell application. (**a**) Tetrandrine, (**b**) isoquinoline sulfonamide, (**c**) ar-turmerone, (**d**) sulforaphane, (**e**) sorafenib, and (**f**) tetramethylpyrazine

Tetramethylpyrazine (TMP), a bioactive compound isolated from *Rhizoma chuanxiong*, was proved to be effective in the treatment of cerebrovascular, bone degeneration, and cardiovascular diseases (Table 6.2) (Guo et al. 2016). TMP also showed to trigger bone marrow-derived mesenchymal stem cells (BMSCs) migration and upregulation of CXCR4. Pre-treatment with TMP prior to transplantation enriched BMSCs homing in ischemic brain (Li et al. 2017). TMP was identified and isolated from the culture of *Pleurotus geesteranus*. It is fascinating to know that mushroom compounds play a definite role in the activation of stem cells. As mushrooms are rich with various chemical components, they could influence stem cell activation. In this review, we have a focus on the medicinal properties of mushrooms and special emphasis given on mushroom compounds that influence various stem cells (Fig. 6.3).

Table 6.2 Protective effects of tetramethylpyrazine (TMP) on various tissues

Effects on stem cells	Mechanism	References
Maintenance of hematopoietic stem cells	Creating an anti-inflammatory and angiogenic environment in bone marrow treatment	Gao et al. (2018)
Protection of bone marrow-derived mesenchymal stem cells from H2O2-induced apoptosis	Regulating the PI3K/Akt and ERK1/2 signaling pathways in the treatment of ischemic stroke	Fang et al. (2017)
Protection of bone marrow-derived mesenchymal stem cells from glucocorticoid-induced apoptosis	Promoting autophagy in an AMP-activated protein kinase (AMPK) and mammalian target of rapamycin (mTOR) in glucocorticoid-induced osteoporosis	Wang et al. (2017)
Differentiation of human umbilical cord-derived mesenchymal stem cells	Differentiated into neuron-like cells by expressing neuronal markers like neuron-specific enolase (NSE), neurofilament protein (NF-H)	Nan et al. (2016)
Neural stem cells proliferation and differentiation	Promotes differentiation by affecting the MAPK signaling pathway under hypoxic condition in neurons	Tian et al. (2010)
Production of cardiac mesenchymal stem cell-derived exosomes	Promotes exosome secretion from C-MSCs via a GTPase-dependent pathway	Ruan et al. (2018)
Migration of neural precursor cells	Activating the phosphatidylinositol 3-kinase/AKT pathway in ischemic stroke	Kong et al. (2016)

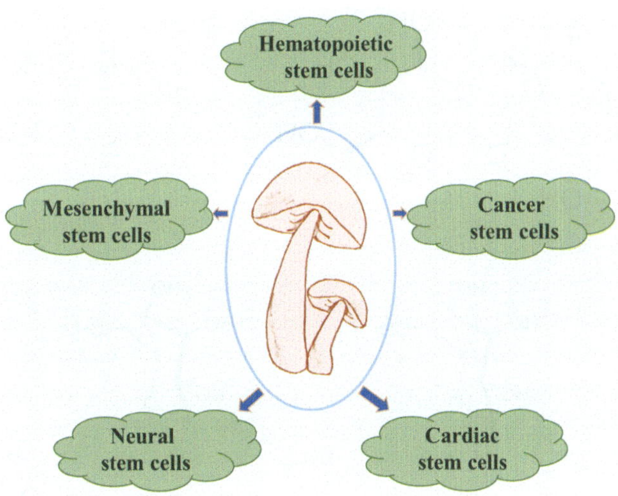

Fig. 6.3 Beneficial effects of mushrooms on various stem cells

6.2 Anti-inflammatory Properties of Mushrooms on Hematopoietic Stem Cells

Inflammation is a complex and protective response of the immune system against pathogens and damaged cells. Immune cells, blood vessels, and molecular mediators are involved in the inflammation process (Dennis and Norris 2015). Immune cells mostly derived from HSCs. HSCs are capable of self-renewing and multilineage differentiation that carry out the replenishment of all blood cells over the lifetime of an organism (Abramson et al. 1977). Acute inflammation persists for a shorter period of time as a protective response, whereas chronic inflammation persists longer and leads to the progression of many diseases including cancer (Okin and Medzhitov 2012).

Mushrooms contain beneficial polysaccharides like trehalose (Fig. 6.4a), which protects cells from protein denaturation. In dried fruiting bodies of *Agaricus bisporus*, around 1–3% of trehalose is found. Trehalose suppresses the expression of cyclooxygenase-2 (COX-2) and nitric oxide synthase which are involved in inflammatory signaling pathways (Echigo et al. 2012). Mushrooms belong to the division, *Basidiomycota*, which is found with β-glucans, which might control the expression of both pro- and anti-inflammatory cytokines (Muta 2006). β-glucans are known as biological response modifiers since they exhibit a wide range of action on various immune responses (Friedman 2016). Lentinan (Fig. 6.4b), the most commonly used β-glucan from shiitake (*Lentinula edodes*) (Fig. 6.5a), is reported to trigger

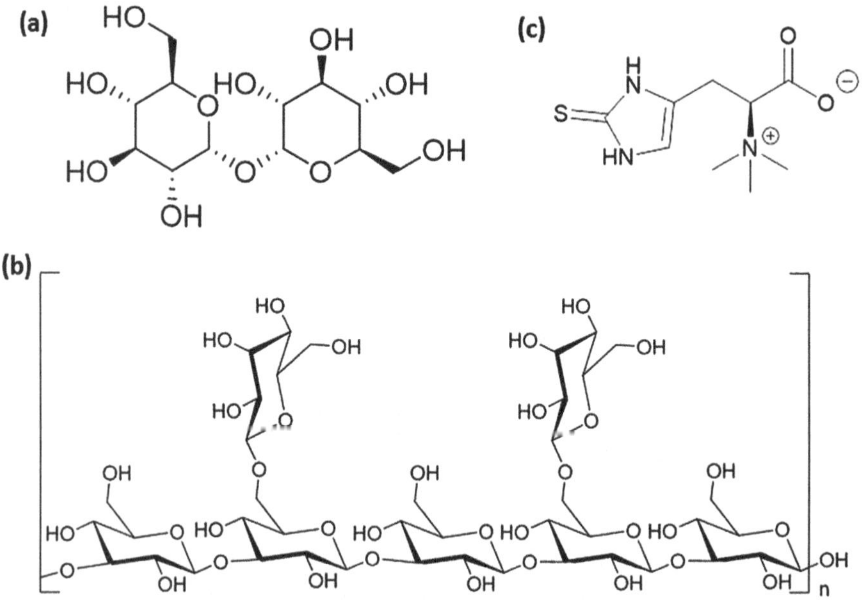

Fig. 6.4 Chemical Structures of compounds anti-inflammatory responses. (**a**) trehalose, (**b**) lentinan, and (**c**) ergothioneine

Fig. 6.5 Mushrooms with potential therapeutic effects on stem cells
*Attributions
Lairich Rig (https://commons.wikimedia.org/wiki/File:Turkeytail_Fungus_(Trametes_versicolor)_-_geograph.org.uk_-_1073528.jpg), Pilát (J.E. Lange) (https://commons.wikimedia.org/wiki/File:2011-06-25_Agaricus_bisporus_3_70149.jpg), Camgroscki (https://commons.wikimedia.org/wiki/File:Image005.jpg)

hematopoietic stem cells, macrophages, and natural killer cells (Akramiene et al. 2007). Many other mushroom species with β-glucan have also been reported for their immunological activities. Mushrooms like *Trametes versicolor*, *Inonotus obliquus*, and *Agaricus bisporus* were found to have anti-inflammatory properties (Ito et al. 2004; Ma et al. 2013; Smiderle et al. 2013). The amino acid composition plays an important role in the anti-inflammatory effects of mushrooms. In oyster mushroom (*Pleurotus ostreatus*), anti-inflammatory property was determined by the amino acids, leucine, isoleucine, tyrosine, and phenylalanine (Jedinak et al. 2011). Ergothioneine (Fig. 6.4c), histidine, and lectins are also shown to have anti-inflammatory effects. Many bio-elements like zinc, copper, iron, and selenium are abundantly present in the fruiting bodies of some common mushrooms like *Boletus edulis*, *Agaricus bisporus*, *Cantharellus cibarius*, and *Imleria badia*. These bio-elements or micronutrients play a significant role as a cofactor in the anti-inflammatory role of mushrooms (Muszynska et al. 2015).

Selenium is one of the most essential elements of the immune system. Selenium facilitates in the scavenging of free radicals along with superoxide dismutase (SOD) or glutathione peroxidase. Selenium further regulates the expression of leucocytes and cytokines (Maseko et al. 2014). Presence of many phenolic compounds makes mushrooms as effective antioxidants since phenolic compounds are known to chelate elements like Fe and Cu that trigger the production of ROS and suppression of free radicals (Ferreira et al. 2009). Presence of many chemical compounds in mushrooms activates lymphocytes and acts as immuno-enhancers, on the other hand, suppress pro-inflammatory factors. The balance between the activation of immune cells and the inhibition of inflammation mediators has been maintained by the chemicals present in mushrooms. Therefore, mushroom extracts could be used to produce drugs against inflammatory diseases like endotoxemia or sepsis. Most of the immune cells are derived from undifferentiated hematopoietic stem cells in the bone marrow and liver. The abovementioned compounds from mushrooms could efficiently activate these hematopoietic stem cells by which mushrooms could act as anti-inflammation mediator.

6.3 Effects of Mushroom Compounds on Cancer Stem Cells

Cancer is one the most common causes of death worldwide. Cells present in a tumor that are capable of self-renewal are called as cancer stem cells (CSCs). These CSCs later differentiate into heterogeneous lineages of cancer cells (Hu and Fu 2012). It was in 2003; the first solid CSCs were documented from breast tumors (Al-Hajj et al. 2003). In a short span of time, CSCs from various human malignant tumors including cancers of the blood, brain, bone, skin, liver, lung, bladder, ovary, prostate, colon, pancreas, and head and neck have been reported (Clevers 2011; Magee et al. 2012; Pardal et al. 2003). CSCs are controlled by specific signaling pathways. Upregulation of Oct4, Nanog, c-Myc, Klf4, and Sox2 was reported in CSCs (Takahashi and Yamanaka 2006). CSCs perform essential roles in tumor progression, metastasis, and drug resistance. Thus, a drug which could target both cancer stem cells and tumor cells would be helpful in cancer survival or in the treatment of cancer. Many mushrooms including *Phellinus linteus* (Lu et al. 2009), *Pleurotus ostreatus* (Lavi et al. 2006), *Agaricus blazei* (Delmanto et al. 2001), *Trametes versicolor or Coriolus versicolor* (Hsieh and Wu 2001), *Grifola frondosa* (Masuda et al. 2009), *Ganoderma lucidum* (Pillai et al. 2010), *Hericium erinaceus* (Wang et al. 2004), and *Inonotus obliquus* (Hu et al. 2009) are reported to have anticancer properties.

Various specific bioactive compounds from mushrooms have also shown to modulate the proliferation and differentiation of stem cells (Table 6.3). Recently antroquinonol (ANQ), a derivative from *Antrodia camphorate*, was found to have inhibitory effects on colon cancer stem cells (Lin et al. 2017). ANQ treatment effectively downregulates stemness-associated gene expressions, which are involved in

Table 6.3 Influence of specific bioactive compounds from mushrooms on stem cells

Stem cells	Bioactive compounds	Mushrooms	References
Cancer stem cells			
(i) Downregulating prostate cancer stem cells by targeting Pten/β--catenin pathway	Polysaccharopeptides (PSP)	*Coriolus versicolor*	Luk et al. (2011)
(ii) Inhibiting colon cancer stem cells by suppressing PI3K/AKT	Antroquinonol (ANQ)	*Antrodia camphorate*	Lin et al. (2017)
Neural stem cells			
(i) Improvement of neural progenitor cells by increasing stromal cell-derived factor 1	Tetramethylpyrazine (TMP)	*Ganoderma lucidum*	Kong et al. (2016)
(ii) Migration of bone marrow-derived mesenchymal stem cells in the ischemic brain by upregulation of CXCR4	Tetramethylpyrazine (TMP)	*Pleurotus geesteranus*	Li et al. (2017)
(iii) Triggering proliferation of neural stem cells	Spirolingzhine A	*Ganoderma lucidum*	Yan et al. (2015)
(iv) Stimulating neuronal differentiation	Ergothioneine	*Pleurotus cornucopiae*	Nakamichi et al. (2016)

CSC production. Simultaneously, ANQ targets and suppresses the PI3K/AKT mechanism and inhibits the viability of colon cancer stem cells. In breast tumor and in C6 glioma cells also ANQ treatment showed a substantial reduction in tumor migration and invasion without affecting normal cells (Lee et al. 2015; Thiyagarajan et al. 2015). Natural compounds like quercetin and myricetin (Walker et al. 2000) extracted from *Agaricus blazei* and kaempferol (Lee et al. 2010) from *Ganoderma lucidum* are also proved to have inhibitory effects on neoplastic transformation by targeting PI3K like ANQ. ANQ, a promising anticancer drug, is presently under phase II clinical trial (Lin et al. 2017). Polysaccharopeptides (PSP) are rich in Turkey tail mushroom (Fig. 6.5b) (*Coriolus versicolor* or Yun-Zhi). PSP extract from *C. versicolor* was found to downregulate prostate CSCs in vitro and also inhibit tumor formation in vivo (Luk et al. 2011). Phosphatase and tensin homolog (Pten) is commonly expressed all over the body and functions as a tumor suppressor by the action of its phosphatase. Knockdown of Pten leads to the initiation of hyperplastic tumor in a mouse and also triggers the Akt/β-catenin pathway which is responsible for the enhancement of mammary stem cells (Korkaya et al. 2009). PSP extracts downregulate the Pten/AKT/β-catenin pathway, by which "stemness" of prostate CSCs might have lost and abolished the tumor initiation (Luk et al. 2011). It is interesting to know the chemopreventive effects of oral administration of PSP against prostate CSCs. All these findings confirm that mushrooms are packed with essential compounds that can be used to treat many types of cancers by targeting cancer stem cells.

6.4 Mushroom Metabolites on Neural Stem Cells

Aging increases the risk of neurodegenerative diseases (NDs) like dementia, Parkinson's disease, Huntington's disease, multiple sclerosis, and Alzheimer's disease. In general, NDs are described by total loss or complete dysfunction of neurons in the central nervous system (Isacson 2003). Earlier, it was believed that no somatic stem cells were present in the brain. However, in 2000, Gage has found out the existence of neural stem cells (NSCs) in the adult brain of all mammals including humans (Gage 2000). NSCs could be efficiently used in the cell-mediated treatment for NDs (Lukaszewicz et al. 2010). Proliferation and differentiation efficiency are the deciding factors for NSCs to be considered for regenerative therapy. Efficiency of regenerative stem cell therapies can be potentially enhanced by specific compounds that could activate the proliferation of NSCs.

Ganoderma lucidum (Fig. 6.5c) was found to suppress the activation of microglial by which it might inhibit neurodegeneration and/or regain the original function of the brain (Zhang et al. 2011). At least ten different compounds isolated from *G. lucidum* were shown to stimulate the proliferation of dentate gyrus NSCs in adult mice. Spirolingzhine-A, a meroterpenoids extracted from *G. lucidum*, activated the proliferation more efficiently than other compounds. 10 μM of spirolingzhine-A triggered the proliferation of NSCs which is almost equal to the proliferative rate of forskolin (10 μM), a commonly used drug for proliferation of NSCs (Yan et al. 2015). In a study by Wong and coworkers, *Hericium erinaceus* was found to accelerate the restoration of infected/injured neurons particularly in the initial stages of recovery (Wong et al. 2011). Ergothioneine extracted from *Pleurotus cornucopiae* was found to transport across the blood-brain barrier and stimulate neuronal differentiation and mitigate the signs of depression in mice (Nakamichi et al. 2016). These findings have set a base for the development of drugs from mushrooms to prevent and to treat NDs. Activation of neural stem cells using compounds extracted from mushrooms seems to be an attractive area worth investigating and could be focused by future researchers. .

6.5 Effects of Mushroom Compounds on Cardiac Stem Cells

Human heart cells stop proliferating and show restricted self-renewing ability soon after birth. It was Bergmann and coworkers who reported that cardiomyocytes are also able to renew. However, the renewal rate is only 1% in a healthy adult (< 20 years), which decreases upon aging (Bergmann et al. 2009). Cardiac stem cells are unique to the cardiac phenotype and have been considered as a reasonable candidate for restoring failing hearts (Yu et al. 2017). Restoration of cardiac tissue could be accomplished by the transplantation of cardiomyocytes, ESCs-derived cardiovascular progenitors (Shiba et al. 2012). Specifically induced pluripotent stem cells (iPSCs) and ESCs are evolving candidates in the treatment of myocardial infarction. Researches on animals and clinical experiments had proved that the stem cell-based treatments are advantageous for restoration and renewal of cardiac cells (Yu et al. 2017).

Mushrooms are back in the attention with a focus on their antiaging benefits, which could trigger many immune cells. Polysaccharides present in *Agaricus brasiliensis* have shown significant potential as natural antioxidants and antiaging agents. Polysaccharides, triterpenes, phenols, flavonoids, nucleobases, coumarin, and ergosterol are the major compounds present in *Ganoderma lucidum*, which make the mushroom medicinally important with antiaging properties (Gill et al. 2016; Russell and Paterson 2006). Oxidative stress is known to induce aging and aging-related pathologies. By reducing oxidative stress, mushrooms act as antiaging agents. *G. lucidum* was found to contain flavonoids, which scavenge superoxides and ultimately reduce oxidative stress in cardiomyocytes (Kirar et al. 2017). *G. lucidum* extracts have shown remarkable cytoprotective and cardioprotective properties on H9c2 cardiomyoblasts. This protective effect could be due to the activation of cardiac stem cells, which normally decline in the aging heart. Mushrooms could recover hearts by minimizing oxidative stress-induced aging at the same time by triggering cardiac stem cells.

6.6 Influence of Mushrooms on Mesenchymal Stem Cells in Tissue Engineering

Tissue engineering-dependent applications require a cell scaffold, a cell-protein-scaffold design prepared by biomaterials such as extracellular matrix (ECM) or hydrogel, cell patch- and 3D printing-based tissue engineering. The scaffold is much needed to improve the survival of cells through cell-cell contacts/cell-scaffold associations (Heyward et al. 2004).Tissue engineering approaches for the generation of new efficient bone tissue using stem cell need to mimic the microenvironment of the original tissue being substituted. Several mechanical factors influence the fate of the stem cells which include tensile strain, shear stress, and pressure (Tuin et al. 2016). Evidences have proved that bone marrow-derived mesenchymal stem cells (MSCs) could increase osteogenesis and new bone formation (Zuk et al. 2002). Human adipose-derived stem cells are multipotent in nature, which might differentiate into adipocytes, osteocytes, and chondrocytes (Gimble and Guilak 2003; Guilak et al. 2004; Zuk et al. 2001). Cardiac patches produced with stem cells (MSCs, cardiac progenitor cells, and endothelial progenitor cells) showed a multi-layered scaffold. Stem cells produce low levels of major histocompatibility complex (MHC) receptors and discharge chemokines that recruit endothelial as well as immune cells which facilitate tissue acceptance at the site of grafting (Petersdorf et al. 2007). Furthermore, several biomaterials, for example, chitosan, silk, and alginate, were reported to be used as 3D scaffolds for MSC culture (Wei et al. 2008).

Researches have used chitosan from mushrooms as a module in calcium-dependent cement in the formation of bone substitutes in tissue engineering (Hirano and Noishiki 1985; Jayakumar et al. 2007). Decellularized and chitin-rich fruiting bodies of *Agaricus bisporus* (Fig. 6.5d) have been used as the 3D scaffold for the differentiation of MSCs into osteocytes (Balasundari et al. 2012). When mushrooms are used as scaffolds, the proliferation was found to be enhanced. This could be because of the presence of nanofibers in mushrooms, which are uniform in size with the width of 10–20 nm and also long fiber length which could offer an excellent 3D network for adhesion and proliferation of cells. This is an interesting and emerging field of research, since only very few research groups are concentrating on this promising area.

6.7 Conclusion

Several mushroom species are small pharmaceutical workshops making a large number of novel compounds with incredible biological functions. Exploring various compounds present in mushrooms is a wide open field in regenerative medicine.

Understanding the benefits of mushroom compounds and the way they influence various stem cells would definitely be an attractive area for future research. Priming stem cells with natural products would be more advantageous than using chemically synthesized compounds.

References

Abramson S, Miller RG, Phillips RA (1977) The identification in adult bone marrow of pluripotent and restricted stem cells of the myeloid and lymphoid systems. J Exp Med 145:1567–1579

Akramiene D, Kondrotas A, Didziapetriene J, Kevelaitis E (2007) Effects of beta-glucans on the immune system. Medicina (Kaunas, Lithuania) 43:597–606

Al-Hajj M, Wicha MS, Benito-Hernandez A, Morrison SJ, Clarke MF (2003) Prospective identification of tumorigenic breast cancer cells. (vol 100, pg 3983, 2003). Proc Nat Acad Sci USA 100:6890–6890. https://doi.org/10.1073/pnas.1131491100

Alison MR, Islam S (2009) Attributes of adult stem cells. J Pathol 217:144–160. https://doi.org/10.1002/path.2498

Balasundari R, Bishi DK, Mathapati S, Naser SB, Cherian KM, Guhathakurta S (2012) Nanocoated botanical scaffold in salvage for human tissue regeneration. J Biomat Tissue Eng 2:330–335. https://doi.org/10.1166/jbt.2012.1058

Bergmann O et al (2009) Evidence for cardiomyocyte renewal in humans. Science 324:98–102. https://doi.org/10.1126/science.1164680

Biehl JK, Russell B (2009) Introduction to stem cell therapy. J Cardiovasc Nurs 24:98–103. https://doi.org/10.1097/JCN.0b013e318197a6a5

Clevers H (2011) The cancer stem cell: Premises, promises and challenges. Nat Med 17:313–319. https://doi.org/10.1038/nm.2304

Delmanto RD, de Lima PLA, Sugui MM, da Eira AF, Salvadori DMF, Speit G, Ribeiro LR (2001) Antimutagenic effect of *Agaricus blazei Murrill* mushroom on the genotoxicity induced by cyclophosphamide. Mutat Res Genet Toxicol Environ Mutagen 496:15–21 doi: https://doi.org/10.1016/S1383-5718(01)00228-5.

Dennis EA, Norris PC (2015) Eicosanoid storm in infection and inflammation. (vol 15, pg 511, 2015). Nat Rev Immunol 15:511–523. https://doi.org/10.1038/nri3928

Echigo R et al (2012) Trehalose treatment suppresses inflammation, oxidative stress, and vasospasm induced by experimental subarachnoid hemorrhage. J Transl Med 10:80. https://doi.org/10.1186/1479-5876-10-80

Fang Y, Chu L, Li L, Wang J, Yang Y, Gu J, Zhang J (2017) Tetramethylpyrazine protects bone marrow-derived mesenchymal stem cells against hydrogen peroxide-induced apoptosis through PI3K/Akt and ERK1/2 pathways. Biol Pharm Bull 40:2146–2152. https://doi.org/10.1248/bpb.b17-00524

Ferreira ICFR, Barros L, Abreu RMV (2009) Antioxidants in wild mushrooms. Curr Med Chem 16:1543–1560. https://doi.org/10.2174/092986709787909587

Friedman M (2016) Mushroom polysaccharides: chemistry and antiobesity, antidiabetes, anticancer, and antibiotic properties in cells, rodents, and humans. Food 5:80. https://doi.org/10.3390/foods5040080

Gage FH (2000) Mammalian neural stem cells. Science 287:1433–1438. https://doi.org/10.1126/science.287.5457.1433

Ganeshpurkar A, Rai G, Jain AP (2010) Medicinal mushrooms: towards a new horizon. Pharmacogn Rev 4:127–135. https://doi.org/10.4103/0973-7847.70904

Gao B et al (2018) Local delivery of tetramethylpyrazine eliminates the senescent phenotype of bone marrow mesenchymal stromal cells and creates an anti-inflammatory and angiogenic environment in aging mice. Aging Cell 17:e12741. https://doi.org/10.1111/acel.12741

Gill BS, Sharma P, Navgeet KS (2016) Chemical composition and antiproliferative, antioxidant, and proapoptotic effects of fruiting body extracts of the lingzhi or reishi medicinal mushroom, *Ganoderma lucidum* (Agaricomycetes), from India. Int J Med Mushrooms 18:599–607. https://doi.org/10.1615/IntJMedMushrooms.v18.i7.50

Gimble JM, Guilak F (2003) Adipose-derived adult stem cells: isolation, characterization, and differentiation potential. Cytotherapy 5:362–369. https://doi.org/10.1080/14653240310003026

Guilak F, Awad HA, Fermor B, Leddy HA, Gimble JM (2004) Adipose-derived adult stem cells for cartilage tissue engineering. Biorheology 41:389–399

Guo M, Liu Y, Shi DZ (2016) Cardiovascular actions and therapeutic potential of tetramethylpyrazine (active component isolated from *Rhizoma Chuanxiong*): Roles and mechanisms. Biomed Res Int 2016:2430329. https://doi.org/10.1155/2016/2430329

Heyward CY, Qidwai H, Loveland AT, Campbell PG, Doctor JS (2004) Evaluating calcium phosphate-based biomaterials as delivery vehicles for human adult mesenchymal stem cells for use in bone regenerative medicine. Dev Biol 271:628–628

Hirano S, Noishiki Y (1985) The blood compatibility of chitosan and n-acylchitosans. J Biomed Mater Res 19:413–417. https://doi.org/10.1002/jbm.820190406

Hsieh TC, Wu JM (2001) Cell growth and gene modulatory activities of Yunzhi (*Windsor Wunxi*) from mushroom Trametes versicolor in androgen-dependent and androgen-insensitive human prostate cancer cells. Int J Oncol 18:81–88

Hu Y, Fu L (2012) Targeting cancer stem cells: a new therapy to cure cancer patients. Am J Cancer Res 2:340–356

Hu HH, Zhang ZY, Lei ZF, Yang YN, Sugiura N (2009) Comparative study of antioxidant activity and antiproliferative effect of hot water and ethanol extracts from the mushroom *Inonotus obliquus*. J Biosci Bioeng 107:42–48. https://doi.org/10.1016/j.jbiosc.2008.09.004

Hucklenbroich J, Klein R, Neumaier B, Graf R, Fink GR, Schroeter M, Rueger MA (2014) Aromatic-turmerone induces neural stem cell proliferation in vitro and in vivo. Stem Cell Res Ther 5:100. https://doi.org/10.1186/scrt500

Hwang KC et al (2008) Chemicals that modulate stem cell differentiation. Proc Nat Acad Sci USA 105:7467–7471. https://doi.org/10.1073/pnas.0802825105

Hyde KD, Bahkali AH, Moslem MA (2010) Fungi-an unusual source for cosmetics. Fungal Divers 43:1–9. https://doi.org/10.1007/s13225-010-0043-3

Isacson O (2003) The production and use of cells as therapeutic agents in neurodegenerative diseases. Lancet Neurol 2:417–424. https://doi.org/10.1016/S1474-4422(03)00437-X

Ito K et al (2004) Long-term effect of 5-fluorouracil enhanced by intermittent administration of polysaccharide K after curative resection of colon cancer – a randomized controlled trial for 7-year follow-up. Int J Color Dis 19:157–164. https://doi.org/10.1007/s00384-003-0532-x

Jayakumar R, Nwe N, Tokura S, Tamura H (2007) Sulfated chitin and chitosan as novel biomaterials. Int J Biol Macromol 40:175–181. https://doi.org/10.1016/j.ijbiomac.2006.06.021

Jedinak A, Dudhgaonkar S, Wu QL, Simon J, Sliva D (2011) Anti-inflammatory activity of edible oyster mushroom is mediated through the inhibition of NF-kappa B and AP-1 signaling. Nutr J 10:52. https://doi.org/10.1186/1475-2891-10-52

Kalac P (2009) Chemical composition and nutritional value of European species of wild growing mushrooms: a review. Food Chem 113:9–16. https://doi.org/10.1016/j.foodchem.2008.07.077

Kalac P (2013) A review of chemical composition and nutritional value of wild-growing and cultivated mushrooms. J Sci Food Agric 93:209–218. https://doi.org/10.1002/jsfa.5960

Kim KH, Moon E, Choi SU, Kim SY, Lee KR (2013) Lanostane triterpenoids from the mushroom *Naematoloma fasciculare*. J Nat Prod 76:845–851. https://doi.org/10.1021/np300801x

Kirar V, Nehra S, Mishra J, Rakhee SD, Misra K (2017) Lingzhi or Reishi medicinal mushroom, *Ganoderma lucidum* (Agaricomycetes), as a cardioprotectant in an oxygen-deficient environment. Intl J Med Mushrooms 19:1009–1021. https://doi.org/10.1615/IntJMedMushrooms.2017024584

Kong X et al (2016) Tetramethylpyrazine promotes migration of neural precursor cells via activating the phosphatidylinositol 3-kinase pathway. Mol Neurobiol 53:6526–6539. https://doi.org/10.1007/s12035-015-9551-1

Korkaya H et al (2009) Regulation of mammary stem/progenitor cells by PTEN/Akt/beta-catenin signaling. PLoS Biol 7:e1000121. https://doi.org/10.1371/journal.pbio.1000121

Lavi I, Friesem D, Geresh S, Hadar Y, Schwartz B (2006) An aqueous polysaccharide extract from the edible mushroom *Pleurotus ostreatus* induces anti-proliferative and pro-apoptotic effects on HT-29 colon cancer cells. Cancer Lett 244:61–70. https://doi.org/10.1016/j.canlet.2005.12.007

Lee KM, Lee DE, Seo SK, Hwang MK, Heo YS, Lee KW, Lee HJ (2010) Phosphatidylinositol 3-kinase, a novel target molecule for the inhibitory effects of kaempferol on neoplastic cell transformation. Carcinogenesis 31:1338–1343. https://doi.org/10.1093/carcin/bgq102

Lee WT, Lee TH, Cheng CH, Chen KC, Chen YC, Lin CW (2015) Antroquinonol from *Antrodia Camphorata* suppresses breast tumor migration/invasion through inhibiting ERK-AP-1- and AKT-NF-kappaB-dependent MMP-9 and epithelial-mesenchymal transition expressions. Food Chem Toxicol 78:33–41. https://doi.org/10.1016/j.fct.2015.01.012

Li L et al (2017) Preconditioning of bone marrow-derived mesenchymal stromal cells by tetramethylpyrazine enhances cell migration and improves functional recovery after focal cerebral ischemia in rats. Stem Cell Res Ther 8:112. https://doi.org/10.1186/s13287-017-0565-7

Lin HC et al (2017) Antroquinonol, a ubiquinone derivative from the mushroom antrodia camphorata, inhibits colon cancer stem cell-like properties: Insights into the molecular mechanism and inhibitory targets. J Agric Food Chem 65:51–59. https://doi.org/10.1021/acs.jafc.6b04101

Lu TL et al (2009) Hispolon from *Phellinus linteus* has antiproliferative effects via MDM2-recruited ERK1/2 activity in breast and bladder cancer cells. Food Chem Toxicol 47:2013–2021. https://doi.org/10.1016/j.fct.2009.05.023

Luk SU et al (2011) Chemopreventive effect of psp through targeting of prostate cancer stem cell-like population. PLoS One 6:e19804. https://doi.org/10.1371/journal.pone.0019804

Lukaszewicz AI, McMillan MK, Kahn M (2010) Small molecules and stem cells. Potency and lineage commitment: the new quest for the fountain of youth. J Med Chem 53:3439–3453. https://doi.org/10.1021/jm901361d

Lull C, Wichers HJ, Savelkoul HF (2005) Antiinflammatory and immunomodulating properties of fungal metabolites. Mediat Inflamm 2005:63–80. https://doi.org/10.1155/MI.2005.63

Ma L, Chen H, Dong P, Lu X (2013) Anti inflammatory and anticancer activities of extracts and compounds from the mushroom *Inonotus obliquus*. Food Chem 139:503–508. https://doi.org/10.1016/j.foodchem.2013.01.030

Magee JA, Piskounova E, Morrison SJ (2012) Cancer stem cells: impact, heterogeneity, and uncertainty. Cancer Cell 21:283–296. https://doi.org/10.1016/j.ccr.2012.03.003

Mahla RS (2016) Stem cells applications in regenerative medicine and disease therapeutics. Intl J Cell Biol 2016:6940283. https://doi.org/10.1155/2016/6940283

Manzi P, Gambelli L, Marconi S, Vivanti V, Pizzoferrato L (1999) Nutrients in edible mushrooms: an inter-species comparative study. Food Chem 65:477–482. https://doi.org/10.1016/S0308-8146(98)00212-X

Maseko T, Howell K, Dunshea FR, Ng K (2014) Selenium-enriched *Agaricus bisporus* increases expression and activity of glutathione peroxidase-1 and expression of glutathione peroxidase-2 in rat colon. Food Chem 146:327–333. https://doi.org/10.1016/j.foodchem.2013.09.074

Masuda Y, Inoue M, Miyata A, Mizuno S, Nanba H (2009) Maitake beta-glucan enhances therapeutic effect and reduces myelosuppression and nephrotoxicity of cisplatin in mice. Int Immunopharmacol 9:620–626. https://doi.org/10.1016/j.intimp.2009.02.005

Mattila P, Suonpaa K, Piironen V (2000) Functional properties of edible mushrooms. Nutrition 16:694–696. https://doi.org/10.1016/S0899-9007(00)00341-5

Mattila P, Salo-Vaananen P, Konko K, Aro H, Jalava T (2002) Basic composition and amino acid contents of mushrooms cultivated in Finland. J Agric Food Chem 50:6419–6422. https://doi.org/10.1021/jf020608m

Mens MMJ, Ghanbari M (2018) Cell cycle regulation of stem cells by MicroRNAs. Stem Cell Rev 14:309–322. https://doi.org/10.1007/s12015-018-9808-y

Mizuno T, Sakai T, Chihara G (1995) Health foods and medicinal usages of mushrooms. Food Rev Int 11:69–81. https://doi.org/10.1080/87559129509541020

Molitoris HP (1994) Mushrooms in medicine. Folia Microbiologica (Praha) 39:91–98

Money NP (2002) Mushroom stem cells. BioEssays 24:949–952. https://doi.org/10.1002/bies.10160

Muszynska B, Krakowska A, Sukowska-Ziaja K, Opoka W, Reczynski W, Bas B (2015) In vitro cultures and fruiting bodies of culinary-medicinal *Agaricus bisporus* (white button mushroom) as a source of selected biologically-active elements. J Food Sci Technol Mysore 52:7337–7344. https://doi.org/10.1007/s13197-015-1830-3

Muta T (2006) Molecular basis for invertebrate innate immune recognition of (1 -> 3)-beta-D-glucan as a pathogen-associated molecular pattern. Curr Pharm Des 12:4155–4161. https://doi.org/10.2174/138161206778743529

Nakamichi N et al (2016) Food-derived hydrophilic antioxidant ergothioneine is distributed to the brain and exerts antidepressant effect in mice. Brain Behavior Cog Neurosci Perspect 6:e00477. https://doi.org/10.1002/brb3.477

Nan CR et al (2016) Tetramethylpyrazine induces differentiation of human umbilical cord-derived mesenchymal stem cells into neuron-like cells in vitro. Int J Oncol 48:2287–2294. https://doi.org/10.3892/ijo.2016.3449

Okin D, Medzhitov R (2012) Evolution of inflammatory diseases. Curr Biol 22:R733–R740. https://doi.org/10.1016/j.cub.2012.07.029

Pardal R, Clarke MF, Morrison SJ (2003) Applying the principles of stem-cell biology to cancer. Nat Rev Cancer 3:895–902. https://doi.org/10.1038/nrc1232

Petersdorf EW, Malkki M, Gooley TA, Martin PJ, Guo Z (2007) MHC haplotype matching for unrelated hematopoietic cell transplantation. PLoS Med 4:e8. https://doi.org/10.1371/journal.pmed.0040008

Pillai TG, Nair CKK, Janardhanan KK (2010) Enhancement of repair of radiation induced DNA strand breaks in human cells by *Ganoderma* mushroom polysaccharides. Food Chem 119:1040–1043. https://doi.org/10.1016/j.foodchem.2009.08.013

Rausch V et al (2010) Synergistic activity of sorafenib and sulforaphane abolishes pancreatic cancer stem cell characteristics. Cancer Res 70:5004–5013. https://doi.org/10.1158/0008-5472.CAN-10-0066

Ruan XF et al (2018) Suxiao Jiuxin pill promotes exosome secretion from mouse cardiac mesenchymal stem cells in vitro. Acta Pharmacol Sin 39:569–578. https://doi.org/10.1038/aps.2018.19

Russell R, Paterson M (2006) *Ganoderma* – a therapeutic fungal biofactory. Phytochemistry 67:1985–2001. https://doi.org/10.1016/j.phytochem.2006.07.004

Seok JK, Boo YC (2015) p-Coumaric acid attenuates uvb-induced release of stratifin from keratinocytes and indirectly regulates matrix metalloproteinase 1 release from fibroblasts. Korean J Physiol Pharmacol 19:241–247. https://doi.org/10.4196/kjpp.2015.19.3.241

Shiba Y et al (2012) Human ES-cell-derived cardiomyocytes electrically couple and suppress arrhythmias in injured hearts. Nature 489:322–325. https://doi.org/10.1038/nature11317

Smiderle FR, Alquini G, Tadra-Sfeir MZ, Iacomini M, Wichers HJ, Van Griensven LJLD (2013) *Agaricus bisporus* and *Agaricus brasiliensis* (1 -> 6)-beta-D-glucans show immunostimulatory activity on human THP-1 derived macrophages. Carbohydr Polym 94:91–99. https://doi.org/10.1016/j.carbpol.2012.12.073

Sylvester KG, Longaker MT (2004) Stem cells – review and update. Arch Surg Chicago 139:93–99. https://doi.org/10.1001/archsurg.139.1.93

Takahashi K, Yamanaka S (2006) Induction of pluripotent stem cells from mouse embryonic and adult fibroblast cultures by defined factors. Cell 126:663–676. https://doi.org/10.1016/j.cell.2006.07.024

Taofiq O, Gonzalez-Paramas AM, Martins A, Barreiro MF, Ferreira ICFR (2016) Mushrooms extracts and compounds in cosmetics, cosmeceuticals and nutricosmetics – a review. Ind Crop Prod 90:38–48. https://doi.org/10.1016/j.indcrop.2016.06.012

Thiyagarajan V, Tsai MJ, Weng CF (2015) Antroquinonol targets FAK-signaling pathway suppressed cell migration, invasion, and tumor growth of c6 glioma. PLoS One 10:e0141285. https://doi.org/10.1371/journal.pone.0141285

Thomson JA (1998) Embryonic stem cell lines derived from human blastocysts (vol 282, pg 1147, 1998). Science 282:1827–1827

Tian Y, Liu Y, Chen X, Zhang H, Shi Q, Zhang J, Yang P (2010) Tetramethylpyrazine promotes proliferation and differentiation of neural stem cells from rat brain in hypoxic condition via mitogen-activated protein kinases pathway in vitro. Neurosci Lett 474:26–31. https://doi.org/10.1016/j.neulet.2010.02.066

Tuin SA, Pourdeyhimi B, Loboa EG (2016) Fabrication of novel high surface area mushroom gilled fibers and their effects on human adipose derived stem cells under pulsatile fluid flow for tissue engineering applications. Acta Biomater 36:220–230. https://doi.org/10.1016/j.actbio.2016.03.025

Wachtel-Galor S, Tomlinson B, Benzie IFF (2004) *Ganoderma lucidum* ('Lingzhi'), a Chinese medicinal mushroom: biomarker responses in a controlled human supplementation study. Br J Nutr 91:263–269. https://doi.org/10.1079/Bjn20041039

Walker EH, Pacold ME, Perisic O, Stephens L, Hawkins PT, Wymann MP, Williams RL (2000) Structural determinants of phosphoinositide 3-kinase inhibition by wortmannin, LY294002, quercetin, myricetin, and staurosporine. Mol Cell 6:909–919. https://doi.org/10.1016/S1097-2765(00)00088-5

Wang ZJ, Luo DH, Liang ZY (2004) Structure of polysaccharides from the fruiting body of Hericium erinaceus. Pers Carbohyd Polym 57:241–247. https://doi.org/10.1016/j.carbpol.2004.04.018

Wang L et al (2017) Tetramethylpyrazine protects against glucocorticoid-induced apoptosis by promoting autophagy in mesenchymal stem cells and improves bone mass in glucocorticoid-induced osteoporosis rats. Stem Cells Dev 26:419–430. https://doi.org/10.1089/scd.2016.0233

Wei HJ et al (2008) Bioengineered cardiac patch constructed from multilayered mesenchymal stem cells for myocardial repair. Biomaterials 29:3547–3556. https://doi.org/10.1016/j.biomaterials.2008.05.009

Wong KH, Naidu M, David P, Abdulla MA, Abdullah N, Kuppusamy UR, Sabaratnam V (2011) Peripheral nerve regeneration following crush injury to rat peroneal nerve by aqueous extract of medicinal mushroom *Hericium erinaceus* (Bull.: Fr) Pers. (Aphyllophoromycetideae). Evid Based Complement Alternat Med 2011:580752. https://doi.org/10.1093/ecam/neq062

Yan YM et al (2015) Metabolites from the mushroom *Ganoderma lingzhi* as stimulators of neural stem cell proliferation. Phytochemistry 114:155–162. https://doi.org/10.1016/j.phytochem.2015.03.013

Yang ZJ et al (2016) Tetrandrine identified in a small molecule screen to activate mesenchymal stem cells for enhanced immunomodulation. Sci Reports-Uk 6:30263. https://doi.org/10.1038/srep30263

Yaoita Y, Matsuki K, Iijima T, Nakano S, Kakuda R, Machida K, Kikuchi M (2001) New sterols and triterpenoids from four edible mushrooms. Chem Pharm Bull:49, 589–594. https://doi.org/10.1248/cpb.49.589

Yu H, Lu K, Zhu J, Wang J (2017) Stem cell therapy for ischemic heart diseases. Br Med Bull 121:135–154. https://doi.org/10.1093/bmb/ldw059

Zhang RP, Xu SL, Cai YN, Zhou M, Zuo XH, Chan P (2011) *Ganoderma lucidum* protects dopaminergic neuron degeneration through inhibition of microglial activation. Evid Based Complement Alternat Med 2011:156810. https://doi.org/10.1093/ecam/nep075

Zuk PA et al (2001) Multilineage cells from human adipose tissue: implications for cell-based therapies. Tissue Eng 7:211–228. https://doi.org/10.1089/107632701300062859

Zuk PA et al (2002) Human adipose tissue is a source of multipotent stem cells. Mol Biol Cell 13:4279–4295. https://doi.org/10.1091/mbc.E02-02-0105

Aqueous and Ethanolic Extracts of Medicinal Mushroom *Trametes versicolor* Interact with DNA: A Novel Genoactive Effect Contributing to Its Antiproliferative Activity in Cancer Cells

7

Tze-Chen Hsieh, Hsiao Hsiang Chao, Yang Chu,
Barbara B. Doonan, and Joseph M. Wu

Contents

T.-C. Hsieh · H. H. Chao · B. B. Doonan · J. M. Wu (✉)
Department of Biochemistry and Molecular Biology, New York Medical College, Valhalla, NY, USA
e-mail: Joseph_Wu@nymc.edu

Y. Chu
Department of Biochemistry and Molecular Biology, New York Medical College, Valhalla, NY, USA

Department of Pharmacy, The First Affiliated Hospital of China Medical University, Shenyang, China

D. C. Agrawal, M. Dhanasekaran (eds.), *Medicinal Mushrooms*,
https://doi.org/10.1007/978-981-13-6382-5_7

Abstract

Trametes versicolor, a medicinal mushroom also known as Yunzhi, has been widely consumed in China as a dietary supplement by individuals diagnosed with various chronic diseases including cancer. Laboratory studies have shown that aqueous and ethanolic extracts of *Trametes versicolor* enhance immune functions and also inhibit the growth of various cancer cells both in vitro and in vivo. However, knowledge of the mechanism by which *Trametes versicolor* exerts its chemotherapeutic action remains incomplete. Efforts to identify its bioactive ingredients have resulted in the isolation of a heterogeneous family of polysaccharopeptides (PSPs). In this study, we investigated whether aqueous and ethanolic extracts of *Trametes versicolor* elicit its antitumor effects by interacting with and changing structure and function of DNA and this novel genoactivity might contribute to suppression of proliferation of human lung A549 cancer cells. Using pBR322 and pSJ3 supercoiled plasmid DNA as substrates, we showed that incubation of DNA with both aqueous and ethanolic extracts of *Trametes versicolor* causes relaxation of double-stranded supercoiled DNA in a concentration- and time-dependent manner. Comparatively, aqueous extracts showed higher conversion rates of supercoiled to relaxed DNA than ethanolic extracts. Aqueous extracts also significantly inhibited A549 lung cancer cell proliferation, downregulated proliferating cell nuclear antigen (PCNA) expression, as well as disrupted cell cycle progression via suppression of levels of cyclins D1 and B. This suggests that immunomodulatory and adjunctive supplementary chemopreventive potential underlying the overall health beneficial effects of *Trametes versicolor* may involve a novel genoactive activity with potential in altering DNA structure and function.

Keywords

Trametes versicolor · Plasmid DNA relaxation · A549 cells, Genoactive · Antiproliferative

Abbreviations

ATCC	American Type Culture Conditions
CPT	Camptothecin
ECL	Enhanced chemiluminescence
FBS	Fetal bovine serum
GMP	Good Manufacturing Practice

MTT	3-(4,5-dimethylthiazol-2-yl)-2,5-diphenyl tetrasodium bromide
PBS	Phosphate-buffered saline
PSPs	Polysaccharopeptides isolated from the medicinal mushroom *Trametes versicolor*
UV-visible	Ultraviolet-visible
YZ	Yunzhi

7.1 Introduction

For millennia, mushrooms have been consumed by humankind both as a culinary delicacy and as a medicinal product to treat a variety of human diseases. Increasingly, there is recognition and acceptance that ingredients derived from mushroom and mushroom products exhibit activity to prevent and/or treat chronic human disorders, attributed to their plethora of bioactivities ranging from antioxidant, anticancer, prebiotic, immunomodulating, anti-inflammatory, cardiovascular, antimicrobial, and antidiabetic effects (Patel and Goyal 2012; Chang and Wasser 2012). In particular, mycelial extracts of *Trametes versicolor*, also known as Yunzhi, have attracted significant interest for their broad health-promoting activities. *Trametes versicolor* is an edible mushroom belonging to species of Basidiomycetes class of fungi. In North America, *Trametes versicolor* is referred to as "turkey tail" mushroom, a nickname symbolic of its fan-shaped wavy edges and varying colored concentric zone appearance (Fig. 7.1a). The fungi are geographically distributed throughout the world and are commonly found on decayed tree trunks, branches, and stumps. The medicinal value of *Trametes versicolor* has been recorded for millennia in the Shen Non Compendium Medica (Wasser and Weis 1999; Borchers et al. 2004), and its therapeutic potential has been confirmed in recent studies (Kidd 2000). Initially, a major focus of research was on its bioactivity to improve immunological function (Borchers et al. 1999; Chu et al. 2002). More recently, attention has been directed to

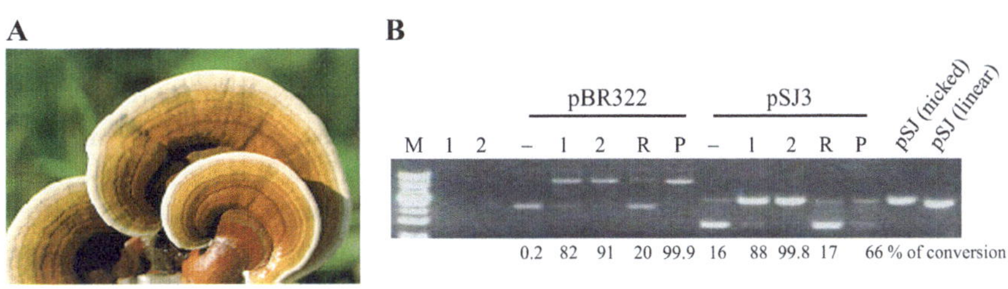

Fig. 7.1 (**a**) An image of *Trametes versicolor*. (Courtesy of Professor Yang Qing-Yao who pioneered the study of Yunzhi mushroom). (**b**) Analysis of DNA conversion/cleavage by *Trametes versicolor* extracts, resveratrol, or piceatannol. DNA conversion pattern of pBR322 or pSJ3 plasmid DNA by *Trametes versicolor* extracts (identified as 1, water extracts; identified as 2, 70% ethanol extracts), resveratrol (R), or piceatannol (P) at 37 °C for 20 h

antitumor effects of Yunzhi which appear to act by mechanisms only partially overlapping its immunomodulatory activities. The bioefficacy of Yunzhi presumably is attributed to its family of heterogeneous polysaccharopeptide complexes. Previously, we demonstrated that aqueous and ethanolic extracts of *Trametes versicolor* inhibited prostate cancer cell proliferation and cell cycle progression as well as induced apoptosis in human leukemic cells (Hsieh and Wu 2001; Hsieh et al. 2006). Extracts of *Trametes versicolor* alone or in combination with extracts of Danshen showed differential growth inhibitory effects in breast cancer cells (Hsieh and Wu 2006). Importantly, studies from this and other laboratories consistently reported that PSP is selectively active in cancer cells while displaying minimal effects in normal cells (Hsieh et al. 2002; Wan et al. 2008; Wan et al. 2010). Mechanistically, bioactive compounds, like PSP, present in medicinal mushrooms, exert a multitude of effects ranging from antioxidant, free radical scavenging, and signal amplification and/or disruptive activities; collectively, these affect gene expression. In a recent FDA-approved, NIH-supported clinical trial, patients with advanced prostate cancer were tested for efficacy of *Trametes versicolor* extracts in augmenting docetaxel-induced tumor suppression and antitumor immune response as had been demonstrated in an immunocompetent murine model of human prostate cancer (Wenner et al. 2012).

The discovery in 1953 by Watson and Crick of double-stranded DNA led to a flurry of research on DNA-small molecule interaction, with respect to diverse binding modes such as groove targeting and DNA intercalation, as well as the perturbation of DNA structure and stability, in turn affecting DNA-protein complex formation and DNA functional changes. Thus, caffeine was found to bind to DNA, which in turn modulated the DNA-binding properties of selective DNA-damaging agents (Traganos et al. 1991). Since PSP derived from *Trametes versicolor* has been shown to potentiate anticancer effects of DNA-damaging agents, notably, camptothecin (CPT), doxorubicin, and etoposide, in human leukemia cells and breast cancer cells (Wan et al. 2010; Hui et al. 2005; Wan et al. 2008), it is of interest to determine whether PSP or mushroom extracts can directly interact with DNA which in turn sensitize responsive cells to DNA-damaging agents. To explore and test this overlooked aspect of PSP, viz., whether *Trametes versicolor* extracts (*aqueous* and *ethanolic*) might interact with DNA, we used the unwinding of supercoiled plasmid DNAs, respectively, pBR322 and pSJ3, as an assay. The results were compared with resveratrol and its metabolite piceatannol which previously have been demonstrated to show DNA modulation activity using pBR322 DNA (Fukuhara and Miyata 1998; Fukuhara et al. 2006; Li et al. 2012). We showed that *Trametes versicolor* aqueous and ethanolic extracts can unwind double-stranded pBR322 and pSJ3 DNA in a time- and concentration-dependent manner, comparable to supercoiled plasmid DNA unwinding effects elicited by piceatannol. Moreover, aqueous extracts of *Trametes versicolor* showed much higher DNA conversion from supercoiled to relaxed form compared to ethanolic extracts. In addition, A549 cell proliferation as measured by colony formation was significantly suppressed by *Trametes versicolor* extracts; the inhibitory effects of *Trametes versicolor* extracts were markedly diminished after a 24 h incubation with plasmid DNA. Comparatively, no effect on A549 cell colony formation ability occurred when tested using plasmid DNA (pBR322 and pSJ3) alone. These results suggest that *Trametes versicolor* extracts can interact with DNA, possibly inducing a change in the topology/function of DNA that contributes to suppression of A549 cell proliferation. Moreover, an aqueous extract of *Trametes versicolor* alone was shown to inhibit A549 cell proliferation, downregulate proliferating cell nuclear antigen (PCNA) expression, and disrupt cell cycle progression as evidenced by suppression of the level of expression of cyclins D1 and B. These findings suggest that immunomodulatory and adjunctive supplementary chemopreventive potential underlying the overall health beneficial effects of *Trametes versicolor* may involve a novel DNA-binding genoactive activity with efficacy in altering DNA structure and function.

7.2 Materials and Methods

7.2.1 Plasmid DNA

Plasmid pBR322 DNA or pSJ3 DNA were prepared using published procedures (Maniatis et al. 1982). Transformation in competent DH5α cells, selection of cells, and isolation of plasmid DNA were performed according to a standard protocol (Maniatis et al. 1982) or procedures provided by the manufacturers. Plasmid DNA identity was authenticated by sequencing using a commercial source (GENEWIZ, South Plainfield, NJ, USA). Resveratrol and piceatannol were purchased from Selleckchem (Houston, TX, USA). All the other chemicals were of analytical grade. Polyphenols were dissolved in DMSO as a 50–100 mM stock solution and stored at −80 °C in the dark. Polyphenols were diluted with water prior to use, and the diluted solution was used immediately.

7.2.2 Source of *Trametes versicolor*, Yunzhi (YZ)

The essence of mushroom Yunzhi (hereinafter referred as YZ) capsules containing 100% *Trametes versicolor* mycelial extract was used in the present studies. The capsules (Batch # 3YZ11482) were supplied by The Hong Kong Health Care Centre Ltd. (Hong Kong, China). The mushroom products were produced from deep-layer cultivated mycelia of Cov-1 according to Good Manufacturing Practice (GMP) standards. Quality control of YZ ensures lack of heavy metal and microorganism contaminations, based on tests performed by government-approved laboratories in Hong Kong.

7.2.3 Preparation of Aqueous and Ethanolic Extracts of YZ

To prepare extracts of YZ, the content of each capsule (400 mg) was suspended in 2 ml of water or 70% ethanol, followed by intermittent mixing and vortexing for 60 min at room temperature with a centrifugation step to remove insoluble particulates. The clear supernatant was sterilized by passing through a 0.22 μm filter, and aliquots were stored at 4 °C. The extract stock used was based on the volume as indicated.

7.2.4 DNA Unwinding Assay

Plasmid DNAs (9.6 ng pBR322 DNA or 10 ng pSJ3 DNA) were incubated with aqueous or ethanolic extracts of YZ or resveratrol or piceatannol. Dose, temperature, and time of incubation were indicated in the figure legends. The reaction was terminated by adding loading buffer (6X) (New England Biolabs Inc., Ipswich, MA, USA). Aliquots were loaded onto a 1% agarose gel, and gel electrophoresis was performed in 1X TBE buffer (pH 8.3); DNA bands were visualized by ethidium bromide staining and photographed using ultraviolet-visible (UV-visible) transillumination. DNA strand conversion/breakage was determined by the conversion of supercoiled pBR322 or pSJ3 plasmid DNAs to either open circular or linear forms of DNA.

7.2.5 Cell Culture

The lung carcinoma A549 cell line was purchased from American Type Culture Conditions (ATCC, Rockville, MD). Cells were maintained in Eagle's minimum essential medium supplemented with 2 mM glutamine and Earle's BSS adjusted to contain 1.5 g/l sodium bicarbonate, 0.1 mM non-essential amino acids, 1 mM sodium pyruvate and additionally supplemented with 0.01 mg/ml bovine insulin and 10% fetal bovine serum (FBS), penicillin (100 U/ml), and streptomycin (100 μg/ml); cells were cultured in a humidified atmosphere of 5% CO_2 in air at 37 °C (Lee et al. 2016). Cells were seeded at a density of $5x10^4$ cells/ml and were passaged by washing the monolayers with phosphate-buffered saline (PBS) followed by a brief incubation with 0.25% trypsin/EDTA.

7.2.6 Effects of YZ on Cell Proliferation and Colony Formation

To determine effects of YZ on cell proliferation, A549 cells were seeded at 5×10^4 cells/ml, 100 μl/well in 96-well plates, and allowed to attach overnight. Various concentrations of aqueous or ethanolic extracts of YZ were added and incubated for 24 h. Cell proliferation on individual wells in the plates was determined by MTT (3-(4,5-dimethylthiazol-2-yl)-2,5-diphenyl tetrasodium bromide) assay (Promega Corp., Madison, Wisconsin), and changes of optical density were measured at 570 nm using an ELISA reader (Thermo Fisher Scientific, Waltham, MA). The effect of YZ on cell viability was assessed as the percentage of viable cells compared with the vehicle-treated control cells, which were arbitrarily assigned a viability of 100%. The assay was performed in triplicate for each concentration of aqueous or ethanolic extracts of YZ added.

Colony formation assay (clonogenicity) was performed as described previously with some modifications (Hsieh et al. 2010). Cells (2000 cells/ml, 2 ml/well) were added to 6-well tissue culture plates containing varying concentrations of water or ethanolic YZ extracts, followed by an additional 8-day incubation to allow colonies to form. Colonies were fixed and stained with 1.25% crystal violet, followed by extensive washing to remove the excess dye, and imaged by an HP scanner (Hsieh et al. 2010).

7.2.7 Preparation of Cell Extracts and Immunoblot Analysis

Cells were collected by centrifugation and were lysed in ice-cold RIPA buffer (50 mM Tris, pH 7.4, 150 mM NaCl, 1 mM EDTA, 1% Triton X-100, 1% deoxycholate, 0.1% SDS, 1 mM dithiothreitol, and 10 μl/ml protease inhibitor cocktail). The extracts were centrifuged, and the clear supernatants were stored in aliquots at −80 °C for further analysis. The protein content of cell lysates was determined by the Coomassie (Bradford) protein assay kit (Pierce, Rockford, IL) using bovine serum albumin as the standard. The aliquots of lysates (20 μg of protein) were boiled with sample buffer for 5 min and resolved by 10% SDS-PAGE. The proteins were transferred to a nitrocellulose membrane and blocked in TBST buffer (10 mM Tris, pH 7.5, 100 mM NaCl, and 0.05% Tween 20) containing 3% nonfat dried milk, overnight at 4 °C. The blots were incubated with various primary antibodies, followed by incubation for 1 h with appropriate secondary antibodies conjugated to horseradish peroxidase in TBST. Actin expression was used as loading control. The primary antibodies, anti-PCNA, anti-cyclin D1, anti-cyclin B, and anti-tubulin, and the secondary antibodies were purchased from Santa Cruz Biotechnology (Santa Cruz, CA). The intensity of the specific immunoreactive bands was detected by enhanced chemiluminescence (ECL), using the manufacturer's protocol (Kirkegaard & Perry Laboratories, Inc. Gaithersburg, MD), and quantified by densitometry and normalized against loading control actin, as previously described (Hsieh et al. 2010; Hsieh et al. 2002; Hsieh and Wu 2001).

7.3 Results

7.3.1 Evidence that Aqueous and Ethanolic Extracts of *Trametes versicolor* (YZ, Also PSP) Interact with Supercoiled Plasmid DNA, Similar to Grape-Derived Polyphenol Piceatannol

Anticancer activities of YZ are supported by regulation of cell proliferation, disruption of cell cycle progression, induction of apoptosis, and elicitation of the immune responses. These and additional effects attributed to YZ implicate the ability of this family of heterogeneous polysaccharopeptides to control the expression of genes selectively in cancer cells. Since cancer cells are known to have global changes in chromatin structure, it is conceivable that the anticarcinogenic activity of YZ may involve its direct interaction with DNA and the chromatin structure. To test this novel genoactive property of YZ, the ability of *Trametes versicolor* extracts (aqueous and ethanolic) to interact/modulate DNA was assayed using unwinding assay of double-stranded plasmid DNA. The results were compared with polyphenols resveratrol and piceatannol, known to display DNA-binding/DNA-interacting activities. Plasmid-based DNA relaxation/breakage assay revealed the strand relaxation of pBR322 and pSJ3 from supercoiled DNA (Form I) to open circular DNA (Form II) by 20 h incubation at 37 °C with aqueous or ethanolic extracts of *Trametes versicolor* (Fig. 7.1b). A similar Form I to II conversion pattern was observed following incubation with piceatannol and not resveratrol (Fig. 7.1b).

Lane 1: marker. Lanes 2–3 (without the addition of plasmid DNA): lane 2, 10 μg of water extracts added, identified as 1; lane 3, 70% ethanol extracts added, identified as 2. Lanes 4–8: lane 4, pBR322 DNA added alone; lane 5, 10 μg of water extracts added, identified as 1; lane 6, 70% ethanol extracts added, identified as 2; lanes 7 and 8, 100 μM of resveratrol (R) (lane 7) and piceatannol (P) (lane 8) incubated with pBR322 DNA. Lanes 9–13: lane 9, pSJ3 DNA added alone; lane 10, 10 μg of water extracts added, identified as 1; lane 11, 70% ethanol extracts added, identified as 2; lanes 12 and 13, 100 μM of resveratrol (R) (lane 12) and piceatannol (P) (lane 13) incubated with pSJ3 DNA. Lanes 14–15: nickase or EcoR1 treated pSJ3 to generate pSJ3 nicked, or linear DNA were included as standard

7.3.2 Effects on pSJ3 DNA Strand Conversion Induced by *Trametes versicolor* Extracts

Different biological activities were previously observed in cells exposed to aqueous vs. ethanolic extracts of *Trametes versicolor* (Hsieh et al. 2006; Hsieh and Wu 2006; Hsieh et al. 2002; Hsieh and Wu 2001, 2013). Thus, we examined whether a differential effects of DNA relaxation/breakage also exist between these two extracts. Time-dependent cleavages of pSJ3 were investigated which showed that both aqueous and ethanolic extracts of YZ converted pSJ3 as early as 5 min of incubation (Fig. 7.2). With prolonged incubation, aqueous extracts showed a much higher percentage of DNA conversion rate from Form I to Form II compared to 70% ethanolic extracts (Fig. 7.2).

7.3.3 The Effects of A549 Cancer Cells in Response to *Trametes versicolor* Extracts Complex with pBR322 or pSJ3 DNA Plasmids

A previous study showed a dose-dependent growth inhibition by liposome-plasmid DNA complex in human ovarian cancer cells (Hofland and Huang 1995), raising the question whether a similar mechanism might operate where plasmid DNA forms a complex with *Trametes versicolor* extracts as a molecular antecedent for control of tumor cell proliferation. To test this possibility, the effects on colony formation by

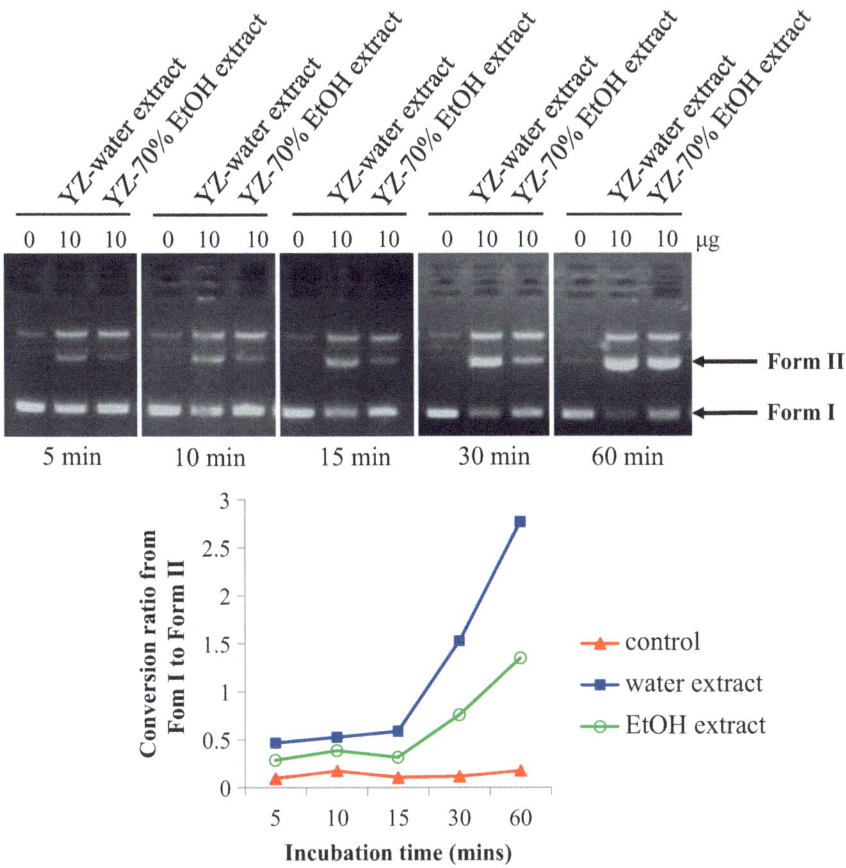

Fig. 7.2 Time-dependent DNA conversion/cleavage of pSJ3 supercoiled DNA by *Trametes versicolor* extracts (water extracts vs. 70% ethanol extracts). DNA conversion pattern of pSJ3 by 10 µg of aqueous extracts or 70% ethanol extracts, at 37 °C for 5, 10, 15, 30, and 60 min. DNA conversion (%) of pSJ3 from supercoiled (Form I) to relaxed form (Form II) following different treatments was calculated

Trametes versicolor extracts in complex with either pBR322 or pSJ3 DNA were investigated using A549 human lung cancer cells. There are no detectable effects on A549 colony formation ability by either plasmid DNA (pBR322 and pSJ3) alone (Fig. 7.3).

Interesting and differential effects were observed on A549 cell growth. The cells exposed to plasmid DNA alone or together with aqueous or ethanolic extracts of YZ for 0 or 24 h showed no effect on A549 proliferation with either plasmid DNA alone, whereas either aqueous or ethanolic extracts of YZ potently suppressed A549 proliferation at the dose used. The inhibitory effects of either YZ extracts were not attenuated by the addition of pBR322 or pSJ3 plasmid DNA at 0 h. At 24 h both aqueous and ethanolic extracts of YZ lose some inhibitory effect on A549 cell proliferation, with certain of the inhibitory effects partially countered by incubation of either extract with pBR322 or pSJ3 plasmid DNA, i.e., the inhibitory effects of

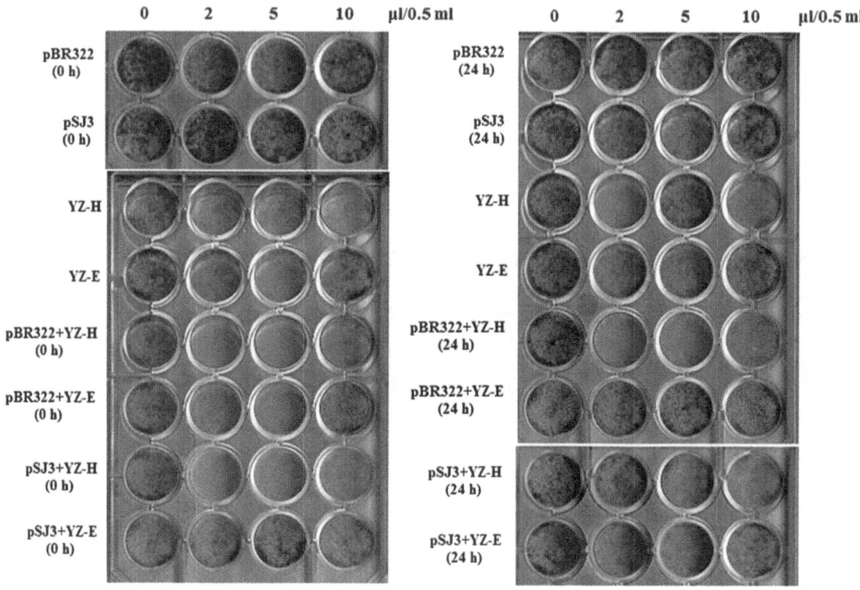

Fig. 7.3 The growth effects of A549 cancer cells in response to pBR322 or pSJ3 DNA plasmids alone, *Trametes versicolor* extracts (YZ-H or YZ-E) alone, or extracts complexed with pBR322 or pSJ3 DNA plasmids. Cells were treated with plasmid alone, varying concentrations of aqueous and ethanolic extracts of YZ (0, 2, 5, and 10 μl/0.5 ml), or plasmid with YZ extracts for 0 or 24 h as indicated. The effects on A549 cancer cell growth by *Trametes versicolor* extracts complexed with either pBR322 or pSJ3 DNA were investigated by colony formation assay as described in "Materials and Methods"

Trametes versicolor extracts were markedly reduced after 24 h incubation with plasmid DNA (Fig. 7.3). Taken together, these results suggest that the inhibitory growth attenuation by the *Trametes versicolor* extracts-plasmid DNA complex appeared to be independent of the DNA sequence since similar results were obtained when using different plasmids such as pBR322 or pSJ3, whereas the proliferation inhibitory counter activities were dependent on the type of *Trametes versicolor* extracts as well as on time of incubation.

7.3.4 Effects of Aqueous and Ethanolic Extracts of *Trametes versicolor* on Cell Growth and Cell Cycle Regulatory Protein Expression in A549 Cells

Trametes versicolor-induced A549 cell growth suppression was investigated using the MTT assay. The results showed that aqueous extracts *of Trametes versicolor* alone inhibit A549 cell growth (Fig. 7.4a), whereas no growth inhibition was observed using ethanolic extracts (Fig. 7.4b). To obtain additional information on the antiproliferative effects by aqueous extracts *of Trametes versicolor*, A549 cells were treated with 0, 4, 10, and 20 μl/ml of *Trametes versicolor* extracts for 3 days, and changes in selected cell cycle regulatory protein expression were investigated by Western blot

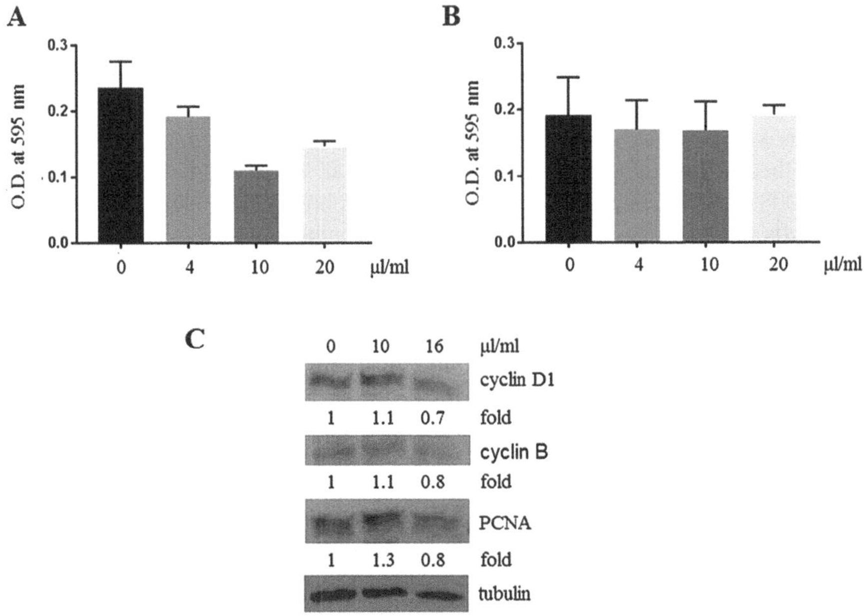

Fig. 7.4 (**a**–**c**) Effects of aqueous and ethanolic extracts of YZ on cell growth and cell cycle regulatory proteins PCNA and cyclins D1 and B expression in A549 cells. Cells were treated with varying concentrations of YZ (0, 4, 10, and 20 μl/ml). (**a**) Effects of aqueous extracts of YZ on cell numbers at day 3 were determined by MTT assay. (**b**) Effects of ethanolic extracts of YZ on cell numbers at day 3 were determined by MTT assay. (**c**) Effects of aqueous extracts (0, 10, and 16 μl/ml) of YZ on protein expressions. Cells were treated and total protein extracts prepared as described in "Materials and Methods." The effects on protein expression were determined by Western blot analysis. Actin was used as a loading control to determine changes in expression by treatment. The intensities of the specific immunoreactive bands were quantified by densitometry and expressed as a fold difference against tubulin

analysis. Decreased levels of expression of PCNA and cyclins D1/B were found, which indicated that aqueous extracts of YZ caused suppression of A549 cell proliferation by inducing the disruption of cell cycle progression (Fig. 7.4c).

7.4 Discussion

Yunzhi has been extensively studied. Ample evidence shows that its general health beneficial effects and chemopreventive potential may lie in its antioxidant, free radical scavenging, and signal modulatory activities, as well as the ability to enhance immunological profiles and induce apoptosis (Kidd 2000; Ooi and Liu 2000; Yang et al. 2005; Wan et al. 2010; Cui and Chisti 2003; Hsieh et al. 2002; Hsieh et al. 2006). Its pharmaco-active constituents are recognized and attributed to the presence of polysaccharides, PSPs, glycoproteins, terpenoids, and other secondary fungal metabolites. Polysaccharides and polysaccharopeptides are typically found in the aqueous extract, whereas components in the ethanolic extract include terpenoids and fungisterol and β-sitosterol (Yokoyama et al. 1975). Previous study demonstrated that PSP inhibits human leukemia HL-60 cell proliferation and blocks cell progression through restriction of both S and G_2 phases (Hui et al. 2005; Wan et al. 2010) as well as sensitizes leukemia and breast cancer cells to apoptosis induced by camptothecin, doxorubicin, or etoposide (Wan et al. 2010; Yang et al. 2005; Wan et al. 2008). To date, how PSP induces apoptosis or how pretreatment of PSP sensitizes cells to DNA-damaging agents remains incompletely understood. Questions that have never been asked include those addressed in this chapter, namely, do aqueous and ethanolic extracts of YZ have any effects on the functions of DNA? Our results provide the first evidence that both aqueous and ethanolic extracts of *Trametes versicolor* relax supercoiled plasmid DNA (Fig. 7.1) and that aqueous extract of YZ is more potent than ethanol extract in inducing conversion of supercoiled to relaxed DNA (Fig. 7.2). These findings add a novel genoactive activity to the armamentarium effects of *Trametes versicolor* extracts, i.e., changing the properties and function of plasmid DNA.

Plasmid DNA is the most commonly used vector for transformation and/or transfection assays. The topology and the size of the plasmid DNA vector could influence the efficiency of transformation and/or transfection. Supercoiled plasmid DNA is the most efficient for both assays. Thus, treatment of *Trametes versicolor* extracts resulted in DNA relaxation (Fig. 7.1) which could affect bacterial transformation and transfection efficacy. Importantly, the latter approach is a core technique for carrying out a number of molecular biology experiments involving recombinant gene expression and disruption of specific gene expression by RNA interference. It is noteworthy that transfection efficiencies are influenced by cell number/density and highest transfection efficiencies are obtained when cells are actively dividing. In our studies, we found that *Trametes versicolor* extracts potently suppress A549 growth (Fig. 7.3 and Fig. 7.4) which may suggest that the use of PSP adversely affects transfection experiments using cultured mammalian cells. Although extracts of *Trametes versicolor* may have a negative impact on transfection efficiencies, this effect is counteracted by the ability to inhibit cell growth and downregulate PCNA expression as well as disruption of cell cycle progression via downregulation of cyclins. These attributes are well aligned with its potential chemopreventive contributions to the overall health benefits of *Trametes versicolor*, thus lending further support that extracts derived from this medicinal mushroom may have adjunctive supplementary as well as clinical efficacies.

7.5 Conclusions

Extracts of cultured mycelia of the commonly found Basidiomycetes mushroom, *Trametes versicolor*, have long been used in many traditional Asian formulas to prevent or treat various types of human diseases. The mechanistic framework frequently used to explain the diverse effects of *Trametes versicolor* has been immunomodulation. In this study, we investigated whether aqueous and ethanolic extracts of *Trametes versicolor* elicit its antitumor effects by interacting with and changing structure and function of DNA and this novel genoactive activity might contribute to suppression of proliferation of human lung A549 cancer cells. Using pBR322 and pSJ3 supercoiled plasmid DNA as substrates, we showed that incubation of DNA with both aqueous and ethanolic extracts of *Trametes versicolor* causes relaxation of double-stranded supercoiled DNA in a concentration- and time-dependent manner. Comparatively, aqueous extracts showed higher conversion rates of supercoiled to relaxed plasmid DNA than ethanolic extracts. However, aqueous extracts were more active in inhibiting A549 lung cancer cell proliferation and in suppressing the level of expression of cell cycle regulatory proteins. Thus the overall health beneficial and therapeutic effects of *Trametes versicolor* may involve a novel DNA-binding genoactive activity with efficacy in altering DNA structure and function. This effect may have important implications for these processes especially during DNA replication, transcription, repair, and maintenance, thus expanding the biological effects and therapeutic impacts of mushroom extracts.

7.6 Acknowledgments

Research in this article was supported in parts by a gift from the Hong Kong Healthcare Center Ltd. We dedicate this article to the fond memory of Professor Yang Qing-Yao who pioneered the study of Yunzhi. We acknowledge with thanks the unwavering support of the CHOU family, especially Ms. Vivien Chou and Ms. Jeannie Chou, for mechanistic studies of medicinal mushrooms in the prevention and management of chronic diseases in humans.

References

Borchers AT, Keen CL, Gershwin ME (2004) Mushrooms, tumors, and immunity: an update. Exp Biol Med (Maywood) 229(5):393–406

Borchers AT, Stern JS, Hackman RM, Keen CL, Gershwin ME (1999) Mushrooms, tumors, and immunity. Proc Soc Exp Biol Med 221(4):281–293

Chang ST, Wasser SP (2012) The role of culinary-medicinal mushrooms on human welfare with a pyramid model for human health. Int J Med Mushrooms 14(2):95–134

Chu KK, Ho SS, Chow AH (2002) Coriolus versicolor: a medicinal mushroom with promising immunotherapeutic values. J Clin Pharmacol 42(9):976–984

Cui J, Chisti Y (2003) Polysaccharopeptides of Coriolus versicolor: physiological activity, uses, and production. Biotechnol Adv 21(2):109–122

Fukuhara K, Miyata N (1998) Resveratrol as a new type of DNA-cleaving agent. Bioorg Med Chem Lett 8(22):3187–3192

Fukuhara K, Nagakawa M, Nakanishi I, Ohkubo K, Imai K, Urano S et al (2006) Structural basis for DNA-cleaving activity of resveratrol in the presence of Cu(II). Bioorg Med Chem 14(5):1437–1443. https://doi.org/10.1016/j.bmc.2005.09.070

Hofland H, Huang L (1995) Inhibition of human ovarian carcinoma cell proliferation by liposome-plasmid DNA complex. Biochem Biophys Res Commun 207(2):492–496. https://doi.org/10.1006/bbrc.1995.1215

Hsieh TC, Elangovan S, Wu JM (2010) Differential suppression of proliferation in MCF-7 and MDA-MB-231 breast cancer cells exposed to alpha-, gamma- and delta-tocotrienols is accompanied by altered expression of oxidative stress modulatory enzymes. Anticancer Res 30(10):4169–4176

Hsieh TC, Kunicki J, Darzynkiewicz Z, Wu JM (2002) Effects of extracts of Coriolus versicolor (I'm-Yunity) on cell-cycle progression and expression of interleukins-1 beta,-6, and -8 in promyelocytic HL-60 leukemic cells and mitogenically stimulated and nonstimulated human lymphocytes. J Altern Complement Med 8(5):591–602

Hsieh TC, Wu JM (2001) Cell growth and gene modulatory activities of Yunzhi (Windsor Wunxi) from mushroom Trametes versicolor in androgen-dependent and androgen-insensitive human prostate cancer cells. Int J Oncol 18(1):81–88

Hsieh TC, Wu JM (2006) Differential control of growth, cell cycle progression, and gene expression in human estrogen receptor positive MCF-7 breast cancer cells by extracts derived from polysaccharopeptide I'm-Yunity and Danshen and their combination. Int J Oncol 29(5):1215–1222

Hsieh TC, Wu JM (2013) Regulation of cell cycle transition and induction of apoptosis in HL-60 leukemia cells by the combination of *Coriolus versicolor* and *Ganoderma lucidum*. Int J Mol Med 32(1):251–257. https://doi.org/10.3892/ijmm.2013.1378

Hsieh TC, Wu P, Park S, Wu JM (2006) Induction of cell cycle changes and modulation of apoptogenic/anti-apoptotic and extracellular signaling regulatory protein expression by water extracts of I'm-Yunity (PSP). BMC Complement Altern Med 6:30

Hui KP, Sit WH, Wan JM (2005) Induction of S phase cell arrest and caspase activation by polysaccharide peptide isolated from *Coriolus versicolor* enhanced the cell cycle dependent activity and apoptotic cell death of doxorubicin and etoposide, but not cytarabine in HL-60 cells. Oncol Rep 14(1):145–155

Kidd PM (2000) The use of mushroom glucans and proteoglycans in cancer treatment. Altern Med Rev 5(1):4–27

Lee YS, Doonan BB, Wu JM, Hsieh TC (2016) Combined metformin and resveratrol confers protection against UVC-induced DNA damage in A549 lung cancer cells via modulation of cell cycle checkpoints and DNA repair. Oncol Rep 35(6):3735–3741. https://doi.org/10.3892/or.2016.4740

Li Z, Yang X, Dong S, Li X (2012) DNA breakage induced by piceatannol and copper(II): mechanism and anticancer properties. Oncol Lett 3(5):1087–1094. https://doi.org/10.3892/ol.2012.597

Maniatis T, Fritsch EF, Sambrook J (1982) Molecular cloning: a laboratory manual. Cold Spring Harbor Laboratory, Cold Spring Harbor

Ooi VE, Liu F (2000) Immunomodulation and anti-cancer activity of polysaccharide-protein complexes. Curr Med Chem 7(7):715–729

Patel S, Goyal A (2012) Recent developments in mushrooms as anti-cancer therapeutics: a review. 3 Biotech 2(1):1–15. https://doi.org/10.1007/s13205-011-0036-2

Traganos F, Kapuscinski J, Darzynkiewicz Z (1991) Caffeine modulates the effects of DNA-intercalating drugs in vitro: a flow cytometric and spectrophotometric analysis of caffeine interaction with novantrone, doxorubicin, ellipticine, and the doxorubicin analogue AD198. Cancer Res 51(14):3682–3689

Wan JM, Sit WH, Louie JC (2008) Polysaccharopeptide enhances the anticancer activity of doxorubicin and etoposide on human breast cancer cells ZR 75 30. Int J Oncol 32(3):689–699

Wan JM, Sit WH, Yang X, Jiang P, Wong LL (2010) Polysaccharopeptides derived from Coriolus versicolor potentiate the S-phase specific cytotoxicity of Camptothecin (CPT) on human leukemia HL-60 cells. Chin Med 5:16

Wasser SP, Weis AL (1999) Therapeutic effects of substances occurring in higher Basidiomycetes mushrooms: a modern perspective. Crit Rev Immunol 19(1):65–96

Wenner CA, Martzen MR, Lu H, Verneris MR, Wang H, Slaton JW (2012) Polysaccharide-K augments docetaxel-induced tumor suppression and antitumor immune response in an immunocompetent murine model of human prostate cancer. Int J Oncol 40(4):905–913. https://doi.org/10.3892/ijo.2011.1292

Yang X, Sit WH, Chan DK, Wan JM (2005) The cell death process of the anticancer agent polysaccharide-peptide (PSP) in human promyelocytic leukemic HL-60 cells. Oncol Rep 13(6):1201–1210

Yokoyama A, Natori S, Aoshima K (1975) Distribution of tetracyclic triterpenoids of lanostane group and sterols in the higher fungi especially of the Polyporaceae and related families. Phytochemistry 14(2):487–497. https://doi.org/10.1016/0031-9422(75)85115-6

Role of Mushrooms in Neurodegenerative Diseases

8

Wooseok Lee, Ayaka Fujihashi, Manoj Govindarajulu, Sindhu Ramesh, Jack Deruiter, Mohammed Majrashi, Mohammed Almaghrabi, Rishi M. Nadar, Timothy Moore, Dinesh Chandra Agrawal, and Muralikrishnan Dhanasekaran

Contents

W. Lee · A. Fujihashi · M. Govindarajulu · S. Ramesh · J. Deruiter · R. M. Nadar
T. Moore · M. Dhanasekaran (✉)
Department of Drug Discovery and Development, Harrison School of Pharmacy, Auburn University, Auburn, AL, USA
e-mail: dhanamu@auburn.edu

M. Majrashi
Department of Drug Discovery and Development, Harrison School of Pharmacy, Auburn University, Auburn, AL, USA

Department of Pharmacology, Faculty of Medicine, University of Jeddah, Jeddah, Kingdom of Saudi Arabia

M. Almaghrabi
Department of Pharmaceutical Chemistry, College of Pharmacy, Taibah University, Almadinah Almunawwarah, Kingdom of Saudi Arabia

D. C. Agrawal (✉)
Department of Applied Chemistry, Chaoyang University of Technology, Taichung, Taiwan
e-mail: dcagrawal@gm.cyut.edu.tw

D. C. Agrawal, M. Dhanasekaran (eds.), *Medicinal Mushrooms*,
https://doi.org/10.1007/978-981-13-6382-5_8

Abstract

Mushrooms have extensively been used not only as a dietary intake but also for the treatment of various central nervous system (CNS) and peripheral nervous system (PNS) disorders. At its early stages, accumulated evidence has suggested that culinary-medicinal mushrooms may play a significant role in the prevention of many age-associated neurodegenerative disorders, such as Alzheimer's and Parkinson's diseases. Therefore, further research and efforts have been devoted to a search for more mushroom species that may improve memory and cognitive functions and, in addition, prevent the progression of dementia and neurodegeneration. Such mushrooms include *Hericium erinaceus*, *Ganoderma lucidum*, *Lignosus rhinocerotis*, *Pleurotus giganteus*, *Sarcodon scabrosus*, *Antrodia camphorata*, *Paxillus panuoides*, *Mycoleptodonoides aitchisonii*, and several other species. This review focuses on the various abovementioned neuroprotective, culinary-medicinal mushrooms and the bioactive secondary metabolites isolated from them. The mushrooms' extracts from basidiocarps/mycelia or isolated compounds have been known to decrease neurotoxicity through various neuroprotective molecular mechanisms such as anti-acetylcholinesterase activity, neurite outgrowth stimulation (neuritogenic), and nerve growth factor (NGF) synthesis (neurotrophic), enhancing mitochondrial functions and reducing endoplasmic reticulum (ER) stress, in addition to antioxidant and anti-inflammatory effects. Therefore, mushrooms can be considered as useful therapeutic agents in the prevention, management, and/or treatment of neurodegenerative diseases.

Keywords

Antioxidant · Culinary mushroom · Neuritogenic · Neurodegeneration · Neuroprotection · Neurotoxicity · Neurotrophic

Abbreviations

AchE	Acetylcholinesterase
AD	Alzheimer's disease
α-KGDH	α-Ketoglutarate dehydrogenase
ALS	Amyotrophic lateral sclerosis
Aβ	Amyloid-β
BBB	Blood-brain barrier

Ca^{2+}	Calcium
CNS	Central nervous system
DNA	Deoxyribonucleic acid
EGL	Ethanol extract of *Ganoderma lucidum*
ER	Endoplasmic reticulum
HD	Huntington's disease
HO-1	Heme oxygenase-1
Hsp70	Heat Shock Protein 70
ICV	Intracerebroventricular
IL-1β	Interleukin 1β
LPE	Lysophosphatidylethanolamine
LPS	Lipopolysaccharide
LXA4	Lipoxin A4
NF-κB	Nuclear factor-kappa B
NGF	Nerve growth factor
NO	Nitric oxide
PD	Parkinson's disease
PDH	Pyruvate dehydrogenase
PGE2	Prostaglandin E2
PNS	Peripheral nervous system
ROS	Reactive oxygen species
SDH	Succinate dehydrogenase
TLR	Toll-like receptor
TNF-α	Tumor necrosis factor-alpha
TRX	Thioredoxin

8.1 Introduction

Neurodegenerative diseases are characterized by progressive and irreversible neuronal cell death. Traditionally, they are defined as disorders with selective loss of neurons and distinct involvement of functional systems defining clinical presentation. Some of the most well-known neurodegenerative diseases include Alzheimer's disease (AD), Parkinson's disease (PD), Huntington's disease (HD), and amyotrophic lateral sclerosis (ALS) (Skovronsky et al. 2006). There are various molecular mechanisms that contribute to neurodegenerative diseases, and most current therapeutic neuroprotective approaches seek to target these mechanisms. Oxidative stress has been implicated in the progression of neurodegenerative diseases due to the generation of toxic reactive oxygen species (ROS) that causes neuronal cell death (Barnham et al. 2004). Some of the therapeutic options against oxidative stress include antioxidants to either reduce the generation of ROS or scavenge existing ROS before any damage can be done or target metal-protein interactions that lead to oxidative stress (Barnham et al. 2004). Mitochondrial dysfunction has also been suggested to play a central role in neurodegenerative diseases since mitochondria serve as regulators of cell death (M. T. Lin and Beal 2006). Additionally, disruption of mitochondrial functions induces glutamate-induced neuronal excitotoxicity leading to neurotoxicity. Mutations in mitochondrial DNA and oxidative stress are the greatest risk factors for neurodegenerative diseases, and therapeutic approaches either target the DNA or protect it from mutations or neurotoxicity caused by toxic ROS (M. T. Lin and Beal 2006).

The connection between neuroinflammation and neurodegenerative diseases has been studied as a potential therapeutic target. Mutation of proteins plays a crucial role in the pathogenesis of many neurodegenerative diseases, such as the accumulation of amyloid beta (A*β*) that is often seen in the pathology of AD (Murphy and LeVine 2010). Proteins often found in the inflammatory response have been linked with the pathology of AD, and therapeutic approaches often target the specific proteins or utilize anti-inflammatory properties (McGeer and McGeer 1995). Exposure to neurotoxicants can require anywhere from weeks to years to produce neuronal dysfunction and neuronal cell death that results in neurological alterations. The consequence of such neurotoxicant exposure may result in neurodegeneration, a broadly used term to describe the loss of structure or function of neurons (Przedborski et al. 2003). Although there are hundreds of diseases that could be described as neurodegenerative diseases, many are rare. However, a few are relatively common, such as Alzheimer's disease (AD), Parkinson's disease (PD), and Huntington's disease (HD) affecting millions of people around the world. As the search for alternative therapeutic targets progresses, some researchers sought out the potential of various culinary and medicinal mushrooms, and this chapter will discuss some of the mushrooms currently being studied for their benefits to neuronal health. Mushrooms that predominantly affect the nervous system produce an immediate response following ingestion. In addition to the medicinal benefits of mushrooms, this chapter will also explore the detrimental effects of mushrooms, for the results of those studies can also be applied in the search for potential therapeutic targets for neurodegenerative diseases.

8.2 Mushrooms with Neuroprotective Effects Against Neurodegeneration

The estimated number of mushroom species is 140,000, although only 10% (~14,000) are known (Wasser 2002). Of the species of mushroom identified, some are currently being studied for their benefits to neuronal health specifically, as a potential therapeutic approach to neurodegenerative diseases. In the subsequent sections, we will discuss the role of the following mushrooms and their molecular mechanisms on neurodegenerative diseases: *Hericium erinaceus*, *Ganoderma lucidum*, *Lignosus rhinocerus*, *Pleurotus giganteus*, *Sarcodon scabrosus*, *Paxillus panuoides*, *Antrodia camphorata*, *Mycoleptodonoides aitchisonii*, and other mushrooms such as *Ganoderma neo-japonicum*, *Dictyophora indusiata*, *Tremella fuciformis*, *Inonotus obliquus*, *Coriolus versicolor*, and *Termitomyces albuminosus*.

8.2.1 *Hericium erinaceus*

Hericium erinaceus is a well-known culinary and medicinal mushroom in Japan and China (Table 8.1; Figs. 8.1 and 8.2a, b). *H. erinaceus*, also known as "Yamabushitake" in Japan, "Houtou" in China, or "Lion's mane" in Western countries, is a mushroom that typically grows on old or dead broadleaf trees (Mori et al. 2008; Wang et al. 2004). *H. erinaceus* is a mushroom that is commonly consumed for culinary purposes in countries like Japan and China, but it is also consumed as a supplementary medicine. It is widely studied for its beneficial effects on neuronal health. A double-blind study conducted by Mori et al. showed that *H. erinaceus* improved mild cognitive impairment. *H. erinaceus* is being investigated for its therapeutic potential as an inducer of neuronal differentiation and for its neuroprotective properties (Mori et al. 2009). To study the molecular mechanism of their effects on neuronal health, extracts from the fruiting bodies and the mycelium were studied, for they contain a large quantity of bioactive components (Trovato Salinaro et al. 2018). Some of these bioactive components include hericenones from the

Table 8.1 Important features of *Hericium erinaceus*

See Fig. 8.1	
Important compounds: See Fig. 8.2	
Mechanisms of action	Increase mitochondrial functions Decrease neuroinflammation Decrease stress Increased expression of cytoprotective proteins Increase lipoxin Neurotrophic
Indication and other aspects	Culinary use Supplementary medicine Mild cognitive impairment CNS disorders: anxiety, depression, multiple sclerosis, Parkinson's disease PNS disorders: gastric ulcers, diabetes mellitus, cancer, hyperlipidemia Obesity

Fig. 8.1 *Hericium erinaceus* (Wikimedia commons)

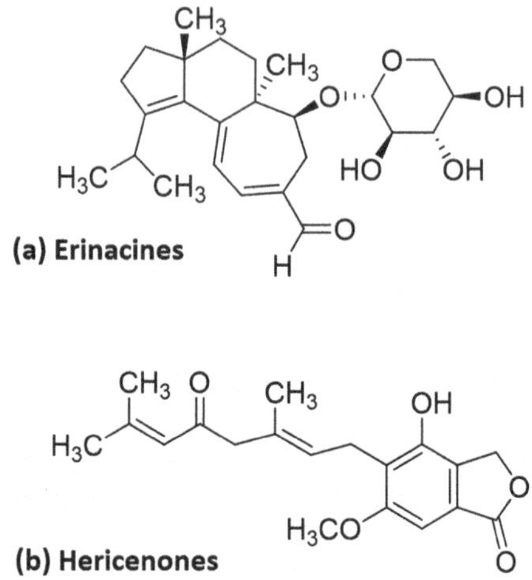

Fig. 8.2 (**a**) Erinacines. (**b**) Hericenones

fruiting bodies and erinacines from the mycelium (Sabaratnam et al. 2013) (Table 8.1). The extracts of *H. erinaceus* have also been used to study the mechanisms through which they contribute to the induction of nerve differentiation. One study found that neurotrophic factors such as nerve growth factor (NGF) have been found to play an important role in differentiation, survival, and maintenance of neuronal cells, but polypeptides such as NGF are not an effective method of therapy due to their inability to cross the blood-brain barrier (BBB) (Chia Wei 2015). In the same study, researchers noted that using smaller molecules that cross the BBB and enhance endogenous neurotrophic factors are predicted to be more effective (Chia Wei 2015). Such molecules are likely to be present in *H. erinaceus*, as studies have found that compounds found in the mushroom stimulate NGF synthesis in cultured astrocytes via activation of JNK pathway (Mori et al. 2008). As to what specific compound is responsible, another study has found erinacines A, B, and C to be strong stimulators of NGF synthesis and not hericenones (Kawagishi et al. 1994; Mori et al. 2008).

Stresses on the endoplasmic reticulum (ER) and the mitochondria have also been found to contribute to neuronal death, and compounds that protect the neuron from stress-induced cell death have been proven to be effective in treating neurodegenerative diseases. One study found that mitochondrial stress is induced by oxidative burden and is linked to neuroinflammation, which causes harmful deleterious effects (Trovato Salinaro et al. 2018). The same study found that treatment with *H. erinaceus* has been found to significantly increase lipoxin A4 (LXA4), a protein known to have anti-inflammatory properties in most areas of the brain, along with increased expression of cytoprotective proteins such as heme oxygenase-1 (HO-1), heat shock protein 70 (Hsp70), and thioredoxin (TRX) (Trovato Salinaro et al. 2018). ER stress, however, is known to be caused by sustained Ca^{2+} depletion, and a study by Ueda et al. has found that *H. erinaceus* contained compounds that protect the neuron from ER stress, which reduces stress-induced neuronal cell death (Ueda et al. 2008). Commonly, *Hericium erinaceus* is indicated in several CNS (anxiety, depression, multiple sclerosis, Parkinson's disease) and PNS (such as gastric ulcers, diabetes mellitus, cancer, and hyperlipidemia) disorders. Interestingly, *H. erinaceus* has also been used to reduce obesity.

8.2.2 *Ganoderma lucidum*

Ganoderma lucidum, or Lingzhi, is an oriental medicinal mushroom that has a long history of promoting health and longevity in countries like China, Japan, and other Asian countries (Wachtel-Galor et al. 2011) (Table 8.2; Figs. 8.3 and 8.4). As a widely accepted medicinal mushroom, there are several research studies on the potential benefits *G. lucidum* for various cancers (breast and prostate) and neurodegenerative disorders, along with the molecular mechanisms of mentioned effects. So far, *G. lucidum* has been noted for its neuroprotective properties. In addition, *Ganoderma lucidum* can serve as a complimentary adjunct medicine for the treatment of chronic fatigue syndrome and liver damage. Most studies used extracts of *G. lucidum* to investigate the molecular mechanism of their neuroprotective effects.

Table 8.2 Important features of *Ganoderma lucidum*

See Fig. 8.3	
Important compounds: See Fig. 8.4	
Mechanisms of action	Decrease neuroinflammation Reduce stress Increase mitochondrial function Antioxidant Increase cholinergic neurotransmission by inhibiting acetylcholinesterase Modulate immune system (Toll-like receptors)
Indication and other aspects	Promoting health and longevity Cancer (breast and prostate) Chronic fatigue syndrome Liver damage Dementia

Fig. 8.3 *Ganoderma lucidum* (Wikimedia commons)

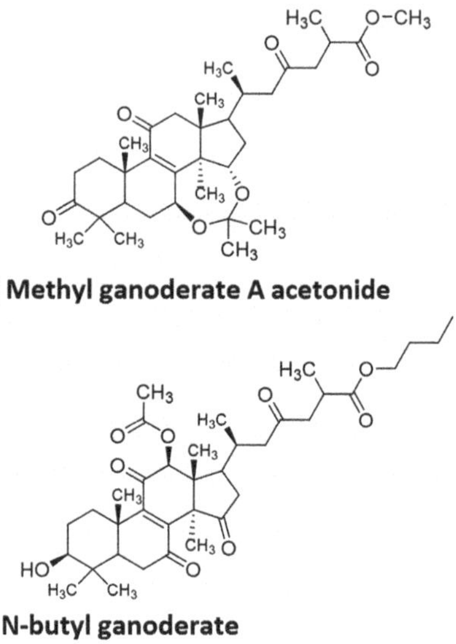

Fig. 8.4 Ganoderic acids. (**a**) Methyl ganoderate A acetonide. (**b**) N-butyl ganoderate

A study done by Zhang and co-workers found that extracts of *G. lucidum* significantly inhibited the production of microglia-derived proinflammatory and cytotoxic factors such as nitric oxide (NO), tumor necrosis factor-α (TNF-α), and interleukin 1β (IL-1β) in a dose-dependent manner, as well as downregulate TNF-α and IL-1β expressions on the mRNA level in lipopolysaccharide (LPS)-stimulated microglia (Zhang et al. 2011). This greatly reduces the stress on the neuron which reduces the risk of stress-induced cell death, a major contributor to neurodegenerative disorders.

Other studies were more specific and researched the effects of *G. lucidum* ethanol extract (EGL) on neuronal health. One study mirrored the findings mentioned earlier, with EGL significantly inhibiting the excessive production of NO, prostaglandin E2 (PGE2), and proinflammatory cytokines such as IL-1β and TNF-α without causing cytotoxicity (Yoon et al. 2013). The same study also found that the inhibition of LPS-stimulated inflammatory responses is associated with the suppression of the nuclear factor-kappa B (NF-κB) and toll-like receptor (TLR) pathways (Yoon et al. 2013). Another study researching the effects of EGL found that EGL significantly enhances the activities of pyruvate dehydrogenase (PDH), α-ketoglutarate dehydrogenase (α-KGDH), succinate dehydrogenase (SDH), and Complex I and II, which is indicative of enhanced function of mitochondria in the rat brain (Ajith et al. 2009). The effects of *G. lucidum* spores on neuronal health were investigated as well. Pre-administration of the spores on the hippocampus significantly reversed the effects of the intracerebroventricular (ICV) injection of streptozotocin (STZ), which is indicative of its ability to alleviate oxidative stress and mitochondrial dysfunction (Zhou et al. 2012). In addition to exhibiting the beneficial effects of *G. lucidum*, some specific chemicals have been isolated from the mushroom to be studied for their potential use as a drug. Two lanostane triterpenes named methyl ganoderate A acetonide and n-butyl ganoderate H were isolated from the fruiting bodies of *G. lucidum*, as they were found to be preferential inhibitors of acetylcholinesterase (AchE) activity (I. Lee et al. 2011). Inhibition of acetylcholinesterase activity is one of the main ways to reduce neuronal cell death and is a candidate for the treatment of neurodegenerative disorders such as Alzheimer's disease.

8.2.3 *Lignosus rhinocerotis*

Lignosus rhinocerotis, or tiger milk mushroom, is often used in Southeast Asia and China as a medicinal mushroom, and its sclerotium has been studied for its potential as an effective functional food or nutraceutical (Yap et al. 2013) (Table 8.3; Fig. 8.5). *L. rhinocerotis* is currently being investigated for its nutrient composition,

Table 8.3 Important features of *Lignosus rhinocerotis*

See Fig. 8.5	
Mechanisms of action	Neurotrophic Neuritogenic Induces neuronal differentiation
Indication and other aspects	Effective functional food or nutraceutical Antioxidant Antiproliferative activity against cancer (human breast and lung cancer carcinoma) Asthma (bronchiole dilator) Cough (suppressant) Infectious disease (bacterial, viral) Liver illness/injury Edema Arthralgia Food poisoning Neurological diseases

Fig. 8.5 *Lignosus rhinocerotis* (Wikimedia commons)

antioxidant properties, and antiproliferative activity against various cancers, such as human breast carcinoma and lung carcinoma (Yap et al. 2013). In addition to its potential as a treatment for various cancers, *L. rhinocerotis* has been investigated for its potential benefits to neuronal health. Similar to the studies investigating *L. rhinocerotis* for its effects on cancer, studies investigating its effects on neuronal health also focus on its sclerotium. *L. rhinocerotis* has also been used to alleviate pulmonary ailments as asthma and chronic cough, as well as general ailments such as food poisoning, joint pain, liver dysfunction, and edema. Infusions of the mushroom have been touted for its effects of enhancing vitality, energy, and attentiveness (Nallathamby et al. 2018). From what is currently known, it is evident that the sclerotium of *L. rhinocerotis* has potential as an inducer of neuronal differentiation. One study reported that *L. rhinocerotis* induced neurite outgrowth, an effect similar to that of nerve growth factor (NGF), and the combination of NGF and *L. rhinocerotis* resulted in additive effects (Eik et al. 2012). Another study investigating the molecular mechanism behind the neuritogenic effect of *L. rhinocerotis* found that the mushroom induced neurite outgrowth via mimicking NGF activity via the MEK/ERK1/2 signaling pathway (Seow et al. 2015).

8.2.4 *Pleurotus giganteus*

Pleurotus giganteus, or "Seri Pagi" in Malaysia, and "Zhudugu" in China, is a culinary mushroom that is gaining popularity for its organoleptic properties (Soytong and Asue 2012) (Table 8.4; Figs. 8.6 and 8.7). Although the mushroom is gaining popularity as a culinary mushroom, some medicinal benefits have been reported. *P. giganteus* can reduce blood glucose, cholesterol, and blood pressure levels in hyperglycemic patients. In addition to alleviating symptoms of diabetes mellitus, *P. giganteus* also has dietary and neurological benefits, for it has a high carbohydrate, fiber, potassium, and phenolic compounds and triterpenoids, making it a prime candidate to be developed as a nutraceutical for neurodegenerative diseases (Khatun

Table 8.4 Important features of *Pleurotus giganteus*

See Fig. 8.6	
Important compound: See Fig. 8.7	
Mechanisms of action	Stimulates neurites and increases neuronal differentiation Neurotrophic Neuritogenic Nutraceutical
Indication and other aspects	Culinary mushroom with organoleptic properties Inflammatory disorders Antioxidant Diabetes mellitus Hypertension Hypercholesteremia Neurological diseases

Fig. 8.6 *Pleurotus giganteus* (Wikimedia commons)

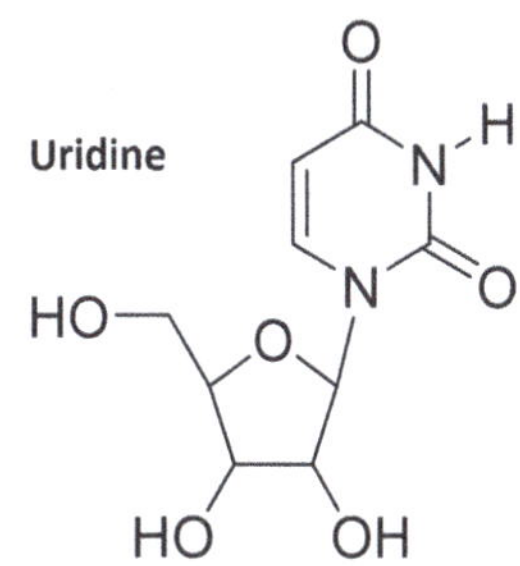

Fig. 8.7 Uridine

et al. 2007). Other medical benefit that was investigated includes its potential in improving neuronal health. The aqueous and ethanolic extracts of *P. giganteus* were used to investigate the effects of the mushroom on neuronal health, and studies have found them to be an effective stimulant of neurites, causing an increase of neuronal differentiation (Phan et al. 2015). Similar to that of *L. rhinocerotis*, *P. giganteus* induce neuronal outgrowth by mimicking NGF activity via the MEK/ERK and PI3K signaling pathways (Phan et al. 2015). In a separate study, it was found that uridine (Fig. 8.7), the key bioactive molecule in *P. giganteus*, is responsible for neurite outgrowth (Chia Wei 2015). In vitro studies of a macrophage cell line treated with ethanol extracts of *P. giganteus* elucidated the mushroom's anti-inflammatory activity—via the suppression of STAT 3 and COX-2 pathways—as well as its ability to scavenge free radicals (Baskaran et al. 2017)

8.2.5 *Sarcodon scabrosus*

Sarcodon scabrosus is a bitter, inedible mushroom that is typically found in coniferous forests of Japan (Shibata et al. 1998) (Table 8.5; Figs. 8.8, 8.9a–g). Although the mushroom is considered to be inedible, there have been many studies investigating the various medicinal benefits of cyathan diterpenoids isolated from the fruiting bodies of *S. scabrosus* called scabronines such as its antibiotic, antimicrobial, anti-inflammatory, and antiproliferative properties (Shi et al. 2011). Different

Table 8.5 Important features of *Sarcodon scabrosus*

See Fig. 8.8	
Important compounds: See Fig. 8.9	
Mechanisms of action	Neurotrophic Neuritogenic
Indication and other aspects	Infectious disease (bacterial) Inflammatory disorders Cancer

Fig. 8.8 *Sarcodon scabrosus* (Wikimedia commons)

scabronines were isolated (Fig. 8.9a–g), and of the various scabronines, some analogues were identified to have neurite outgrowth-promoting activity, similar to that of erinacines found in the fruiting bodies of *H. erinaceus* (Wender et al. 2001).

The molecular mechanism of scabronines is also similar to that of *H. erinaceus*. Since NGF cannot cross the BBB, smaller molecules that can pass the BBB such as erinacines and scabronines stimulate the production of NGF (Wender et al. 2001). Of the various analogues of scabronines, scabronines A and G have been found to be the most potent inducers of NGF synthesis (Ohta et al. 1998; Waters et al. 2005). Other analogues of interest are the sarcodonins, and sarcodonins A and G have been shown to significantly induce neurite outgrowth as well (Shi et al. 2011).

8.2.6 *Paxillus panuoides*

Paxillus panuoides is an inedible mushroom found widely in East Asia and North America on decayed pine trees with reported potential as free radical scavengers (Quang et al. 2003) (Table 8.6, Fig. 8.10). Oxidative stress induced by free radicals has been implicated in the pathology of neurodegenerative diseases, and studies have found leucomentins, the p-terphenyl compounds found in the methanolic

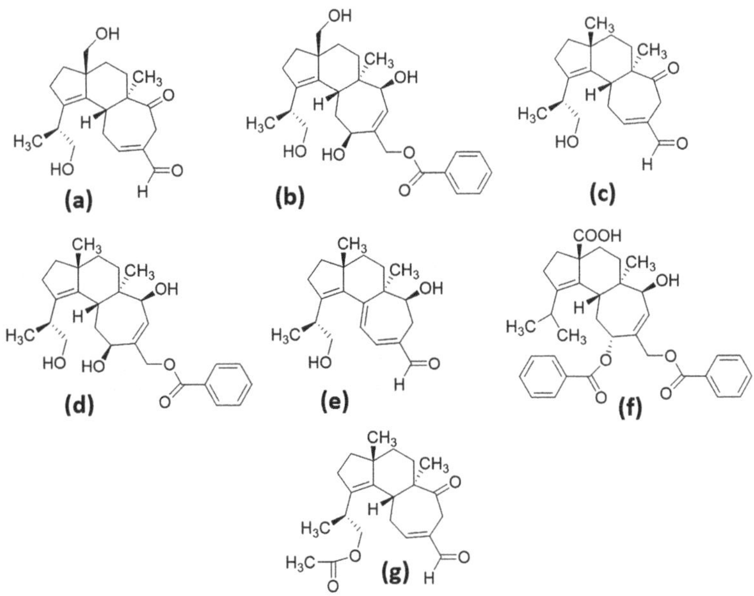

Fig. 8.9 (**a**) Scabronine K. (**b**) Scabronine J. (**c**) Sarcodonin G. (**d**) Scabronine H. (**e**) Sarcodonin A. (**f**) Sarcodonin M. (**g**) 19-O-acetylsarcodonin G

Table 8.6 Important features of *Paxillus panuoides*

See Fig. 8.10	
Important compounds: See Fig. 8.11	
Mechanisms of action	Inhibit lipid peroxidation formation Inhibit hydrogen peroxide neurotoxicity by chelating iron with DNA Scavenge free radicals (hydrogen peroxide and lipid peroxide)
Indication and other aspects	Antioxidant effect Neurological disorders

extract of *P. panuoides*, to be effective inhibitors of lipid peroxidation in rat liver microsomes (Yun et al. 2000). Currently, antioxidants are considered to be potential therapeutic target when treating neurodegenerative diseases, since they would protect from stress-induced cell death. Therefore, leucomentins (Fig. 8.11a–d) have been investigated for their effect on neuronal health. One study has found that leucomentins are capable of neuroprotective activity, and their molecular mechanism was studied (Lee et al. 2003). The study by Lee et al. found that the leucomentins were potent inhibitor of lipid peroxidation and H_2O_2 neurotoxicity but did not have any role in scavenging reactive oxygen species (ROS) (Lee et al. 2003). Instead,

Fig. 8.10 *Paxillus panuoides* (Wikimedia commons)

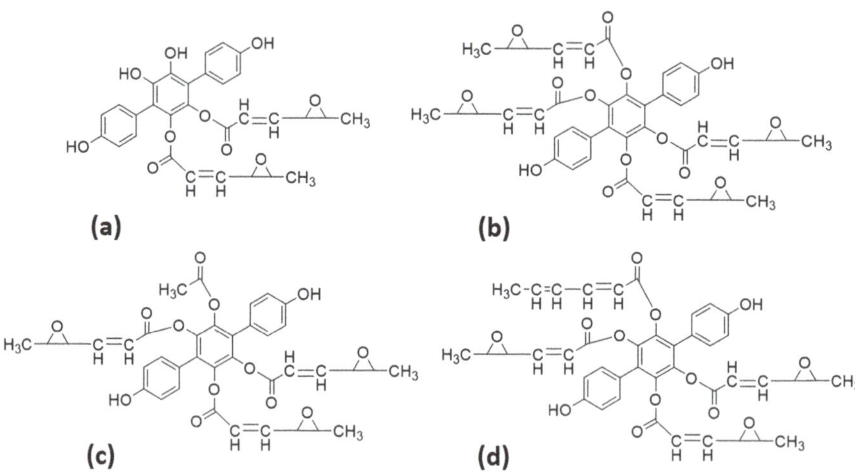

Fig. 8.11 (**a–d**) Leucomentins. (**a**) Leucomentin-1. (**b**) Leomentin-2. (**c**) Leucomentin-3. (**d**) Leucomentin-4. (Lee et al. 2003)

they found that they functioned by chelating iron when DNA was present with H_2O_2, thus preventing damage to the DNA (Lee et al. 2003).

8.2.7 *Antrodia camphorata*

Antrodia camphorata is a fungal parasite on the inner cavity of the Ball camphor tree that originates from Taiwan (Geethangili and Tzeng 2011) (Table 8.7; Figs. 8.12 and 8.13a–h). *A. camphorata* is also known as "Niu-chang-chih," "Chang-Chih," "Niu-chang-ku," or "Chang-ku" in Taiwan and is often used medicinally (Geethangili and Tzeng 2011). It is typically used as a treatment for drug intoxications, diarrhea, abnormal pains, hypertension, itchy skin, and liver cancer, though it has been found

Table 8.7 Important features of *Antrodia camphorata*

See Fig. 8.12	
Important compounds: See Fig. 8.13	
Mechanisms of action	Stimulates immune system Reduces inflammation Reduces oxidative stress Scavenges free radicals Suppresses hyperphosphorylated tau protein expression Prevents neuronal apoptosis
Indication and other aspects	Diarrhea Hypertension Itchy skin (topical use) Cancer (liver cancer) Steatosis (fatty liver) Infectious disease (hepatitis B virus)

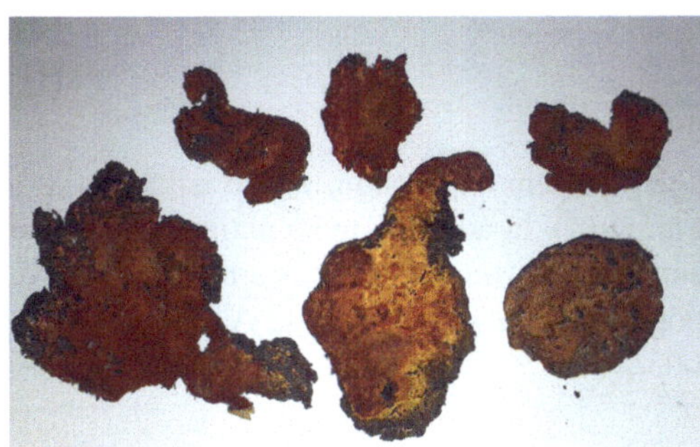

Fig. 8.12 *Antrodia camphorata* (Wikimedia commons)

to have immunostimulatory and anti-inflammatory effects (Chien et al. 2008). In addition, *A. camphorata* has also been found to exert effective protection against chronic chemical-induced hepatic injury via its antioxidant and free radical scavenging activities (Hsiao et al. 2003). Specifically, the fruiting body extracts of *A. camphorata* have been shown to induce apoptosis in hepatoma cells, inhibit cytokine production, inhibit hepatitis B virus replication, and protect against steatosis-induced acute hepatotoxicity, making *A. camphorata* a prime candidate for the treatment of liver cancer, cytokine-induced liver injury, hepatitis B, and fatty liver disease (Ao et al. 2009). *A. camphorata* has been investigated for possible benefits to neuronal health as well. As mentioned earlier, *A. camphorata* is known for its antioxidant and anti-inflammatory effects, both of which can be effective in the treatment of neurodegenerative disorders. One study found that *A. camphorata* possessed strong antioxidant and anti-inflammatory abilities for inhibiting neurotoxicity in an AD animal model, in addition to suppressing hyperphosphorylated tau

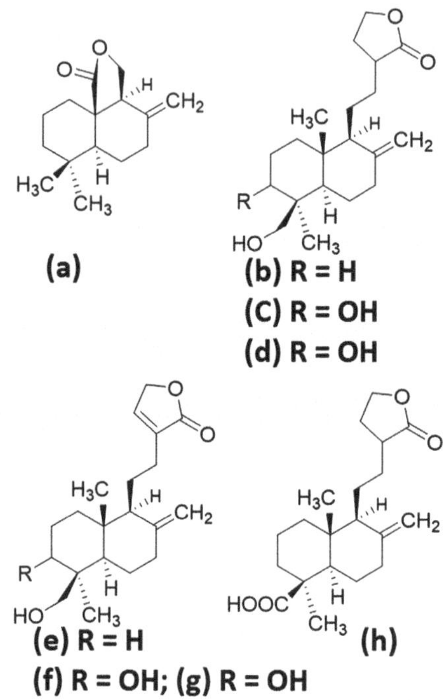

Fig. 8.13 (**a**–**h**) Terpenoids of *Antrodia camphorata*. (**a**) Antrocin. (**b**) 19-Hydroxylabda-8(17)-en-16,15-olide. (**c**) 3β,19-Dihydroxylabda-8(17),11E-dien-16,15-olide. (**d**) 13-epi-3β,19-Dihydroxylabda-8(17),11E-dien-16,15-olide. (**e**) 19-Hydroxylabda-8(17),13-dien-16,15-olide. (**f**) 14-Deoxy-11,12-didehydroandrographolide. (**g**) 14-Deoxyandrographolide. (**h**) Pinusolidic acid

(*p*-tau) protein expression, which is known as an important AD risk factor (Wang et al. 2012). In addition to its antioxidant and anti-inflammatory effects, one study found that *A. camphorata* has a neuroprotective effect, preventing serum-deprived apoptosis via a protein kinase A (PKA)/CREB-dependent pathway (Huang et al. 2005).

8.2.8 *Mycoleptodonoides aitchisonii*

Mycoleptodonoides aitchisonii, also known as "Bunaharitake" in Japan, is a mushroom cultivated on dead broadleaf trees from summer to fall in Asia that is known for its various health benefits such as improving hypertension (Choi et al. 2011, 2014) (Table 8.8; Fig. 8.14a, b). Fruiting body extracts have also been shown to have a strong antioxidant capacity, as indicated by their high activity of total oxy-radical scavenging capacity. In vitro and in vivo studies have elucidated *M. aitchisonii* extracts' ability to reduce blood glucose levels in hyperglycemic mice, demonstrating its potential as a treatment for diabetes mellitus. Fruiting body extracts also have potential as treatments for hyperlipidemia, for the same study by Choi et al. also exhibited the mushroom's ability to reduce total cholesterol, triglyceride, and LDL-cholesterol levels (Choi et al. 2016). *M. aitchisonii* has also been investigated for its various neuroprotective effects. One study by Choi and co-workers found two novel compounds from the fruiting body of *M. aitchisonii* that protected the neuron from thapsigargin (TG)-induced ER stress (Choi et al. 2014).

Table 8.8 Important features of *Mycoleptodonoides aitchisonii*

Important compounds: See Fig. 8.14	
Mechanisms of action	Increase in NGF synthesis Reduction in ER stress and apoptosis
Indication and other aspects	Reduces hypertension Immunostimulant Anti-asthmatic Antioxidant Treats diabetes mellitus (reduce blood glucose) Treats hyperlipidemia Reduces cholesterol

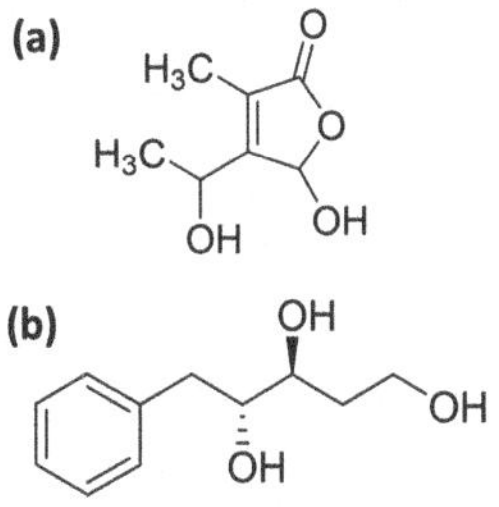

Fig. 8.14 Novel compounds isolated from *Mycoleptodonoides aitchisonii*. (**a**) 5-Hydroxy-4-(1-hydroxyethyl)-3-methylfuran-2(5H)-one. (**b**) 5-Phenylpentane-1,3,4-triol

TG is an inhibitor of the Ca^{2+} ATPases in the endoplasmic reticulum and induces ER stress by disrupting the homeostatic balance of Ca^{2+} in the ER. Compounds found in *M. aitchisonii* aid in retaining the Ca^{2+} balance in the ER, thus protecting the neuron from ER stress-induced cell death (Choi et al. 2014). Another effect that *M. aitchisonii* has on neuronal health is the enhancement of NGF synthesis. Okuyama et al. found that the aqueous extract of *M. aitchisonii* exhibited a significant enhancement of NGF synthesis in the cerebral cortex and hippocampus when administered to newborn rats. They also found that the extracts of the mushroom enhanced synthesis of NGF and catecholamine metabolites (Okuyama et al. 2004).

A summary of neuroprotective mechanisms of various mushrooms in neurodegenerative diseases has been represented in Fig. 8.15. And a review of mushrooms and their active ingredients, neuroprotective mechanisms, and potential therapeutic applications is presented in Table 8.9.

8.2.9 Other Mushrooms with Potential Effects on Neurodegenerative Disorders

Currently, there are many mushrooms that are being investigated for their health benefits, often due to their long history of medicinal use. However, these mushrooms have yet to be studied extensively, unlike the mushrooms listed above that have had several studies investigating their benefits and the molecular mechanisms.

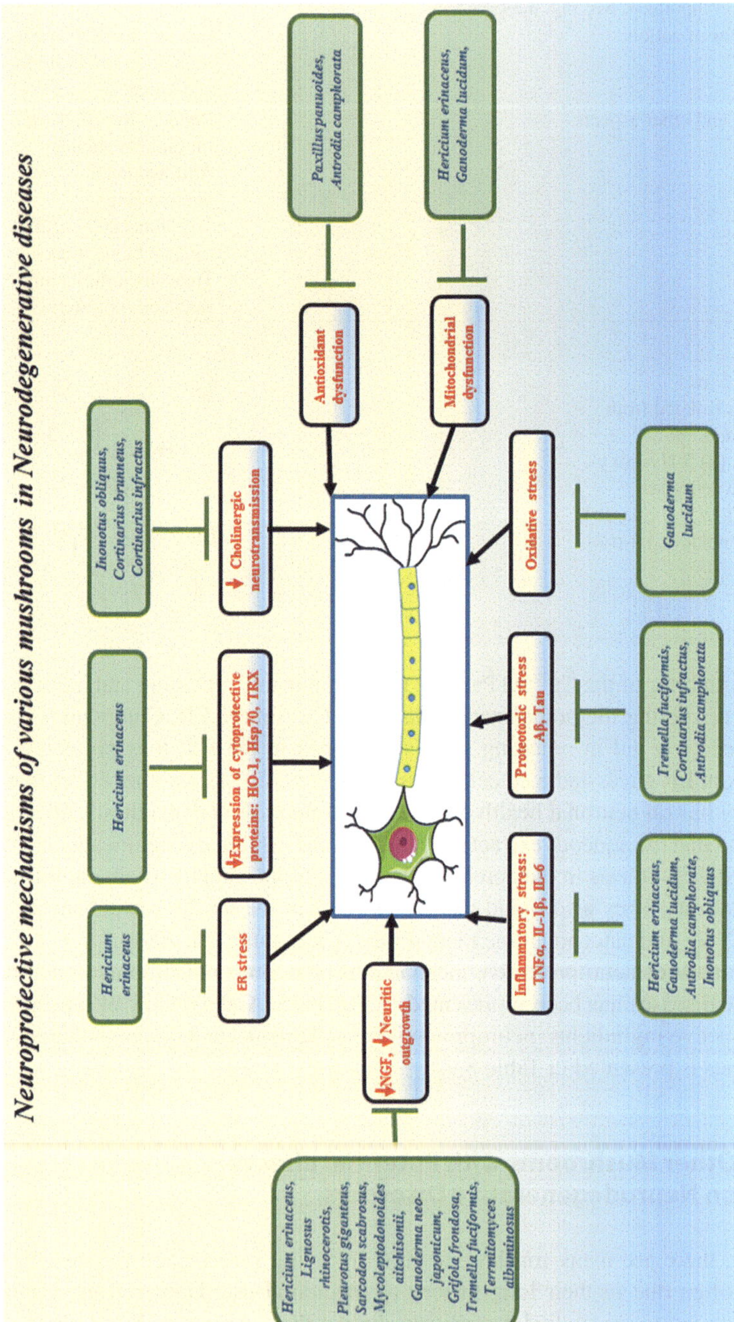

Fig. 8.15 A graphical representation of neuroprotective concepts of various mushrooms

Table 8.9 Review of mushrooms with beneficial effects to reduce neurodegeneration

Mushroom species	Active ingredients and neuroprotective mechanism	Potential therapeutic application
Hericium erinaceus	(a) Erinacines, hericenones (i) Induce NGF (neurotrophic) (b) Lipoxin A4 (i) Decrease endoplasmic reticulum (ER) stress (ii) Increase mitochondrial functions (iii) Decrease neuroinflammation (iv) Increased expression of cytoprotective proteins: heme oxygenase-1 (HO-1), heat shock protein 70 (Hsp70), and thioredoxin (TRX)	Prevention of age-related neurodegenerative disorders such as AD, HD, PD, ALS
Ganoderma lucidum	(a) Ganoderic acids (methyl ganoderate A acetonide and n-butyl ganoderate-triterpenes) (i) Decrease neuroinflammation by inhibiting the production of NO, TNF-α, IL-1β, NF-κB (ii) Suppression of TLR pathway (iii) Increase mitochondrial functions: increase the activity of PDH, α-KGDH, SDH, Complex I, Complex II (iv) Alleviate oxidative stress (v) Inhibit AchE activity	Prevention of AD
Lignosus rhinocerotis	(i) Neurotrophic (mimics NGF) (ii) Neuritogenic: MEK/ERK1/2 pathway	Prevents the neuronal cell death induced by neurodegenerative disorders
Pleurotus giganteus	(a) Uridine (i) Neurotrophic (mimics NGF) (ii) Neuritogenic: MEK/ERK1/2 and PI3K pathway	Prevents the neuronal cell death induced by neurodegenerative disorders
Sarcodon scabrosus	(a) Scabronine and sarcodonin (i) Neurotrophic (mimics NGF) (ii) Neuritogenic	Prevents the neuronal cell death induced by neurodegenerative disorders
Paxillus panuoides	(a) Leucomentins (i) Antioxidant activity	Prevention of AD-like neurodegenerative disorders
Antrodia camphorata	(a) Terpenoids (i) Antioxidant (ii) Anti-inflammatory (iii) Antiapoptotic (iv) Decreases *p*-tau	Prevention of AD
Mycoleptodonoides aitchisonii	(a) Terpenoids (i) Reduces ER stress (ii) Increases NGF synthesis	Prevents the neuronal cell death induced by neurodegenerative disorders

(continued)

Table 8.9 (continued)

Mushroom species	Active ingredients and neuroprotective mechanism	Potential therapeutic application
Other mushrooms with potential effects on neurodegenerative disorders		
Ganoderma neo-japonicum	(i) Antioxidant activity (ii) Increases NGF synthesis	Prevents the neuronal cell death induced by neurodegenerative disorders
Grifola frondosa	(a) Lysophosphatidylethanolamine (i) Neuritogenic effect	Prevents the neuronal cell death induced by neurodegenerative disorders
Cortinarius brunneus	(a) β-Carboline (i) Inhibits AChE activity	Prevention of AD-like neurodegenerative disorders
Cortinarius infractus	(a) Infractopicrin (i) Reduces AChE activity (ii) Decreases A*β*-aggregation	Prevention of AD
Dictyophora indusiata	(a) Dictyoquinazols (i) Reduces excitatory neuronal cell death	Prevention of AD-like neurodegenerative disorders
Tremella fuciformis	(i) Neuritogenic (ii) Decreases *β*-amyloid neuronal cytotoxicity	Prevention of AD
Inonotus obliquus	(a) Pepsin (i) Anti-inflammatory (ii) Reduces AChE activity	Prevention of AD-like neurodegenerative disorders
Termitomyces albuminosus	(a) Cerebrosides, termitomycesphins (i) Neuritogenic	Prevents the neuronal cell death induced by neurodegenerative disorders

8.2.9.1 *Ganoderma neo-japonicum*

Ganoderma neo-japonicum is a mushroom that is noted for its minor free radical scavenging and antihepatotoxic activity (Lin et al. 1995). Similar to *Ganoderma lucidum* which belongs to the same genus, *G. neo-japonicum* also has been investigated for its benefits on neuronal health. Though there haven't been as many studies investigating this mushroom as compared to *G. lucidum*. Despite its free radical scavenging activity, *G. neo-japonicum* has been noted more for its NGF-like bioactive compounds that stimulate neurite growth (Sabaratnam et al. 2013). *Grifola frondosa*, also known as "Maitake," is a culinary mushroom that is viewed as a health food in China and Japan (Ling-Sing Seow et al. 2013). One study found that *G. frondosa* contains an active substance called lysophosphatidylethanolamine (LPE) that induces the activation of mitogen-activated protein kinase (MAPK), which induces neuronal differentiation (Nishina et al. 2006). Two mushrooms of the *Cortinarius* genus are also being studied for their acetylcholinesterase (AChE)-inhibiting abilities. *β*-Carboline alkaloids found in the fruiting bodies of *Cortinarius brunneus*, called brunneins, were studied, and brunnein-A exhibited very low AChE-inhibiting activity and no cytotoxicity (Teichert et al. 2007). Two alkaloids isolated from the extracts of the fruiting bodies of *Cortinarius infractus*, infractopicrin and 10-hydroxy-infractopicrin, were investigated for their potential benefits to neuronal health, and it was found that they show AChE-inhibiting ability, in addition to an inhibitory effect on self-aggregation of Aβ-peptides, an AD pathology-related target (Geissler et al. 2010).

8.2.9.2 *Dictyophora indusiata*

Dictyophora indusiata, also known as veiled lady mushroom or bamboo mushroom, is both a culinary and medicinal mushroom found in Asian countries (Journal et al. 2015). These mushrooms contain neuroprotective compounds, dictyoquinazols A, B, and C, that can reduce excitatory neurotoxin-induced cell death (Lee et al. 2002).

8.2.9.3 *Tremella fuciformis*

Tremella fuciformis is a common culinary and medicinal mushroom found in China that also has reported benefits to neuronal health, including neuroprotective and neurotrophic effects (Park et al. 2007). Extracts of *T. fuciformis* were found to significantly reduce neuronal cytotoxicity caused by *β*-amyloid peptides and increase neurite outgrowth by increasing glucose uptake (Park et al. 2007; Park et al. 2012). In the study by Park et al., *T. fuciformis* also improved cognitive function by regulating the cAMP-responsive element-binding protein (CREB) pathway (Park et al. 2012).

8.2.9.4 *Inonotus obliquus*

Inonotus obliquus, also known as "Chaga," is a medicinal mushroom known for its antitumor and diuretic properties that has also been studied for its benefits to neuronal health, such as its antioxidant and AChE-inhibiting properties (Giridharan et al. 2011). Another study further investigating its antioxidant properties found that the pepsin extract from this mushroom exhibited a protective effect on H_2O_2-induced DNA damage and a reduction of ROS generation (Kim et al. 2011).

8.2.9.5 *Coriolus versicolor*

Coriolus versicolor, or "Yun Zhi," as it is commonly known in China, is a medicinal mushroom that is currently being studied as a potential treatment for AD and other neurodegenerative diseases (Trovato et al. 2016). Similar to the effect of *H. erinaceus*, *C. versicolor* has been found to increase LXA4, which is a compound known for its anti-inflammatory properties (Trovato et al. 2016). The anti-inflammatory property of LXA4 is potentially useful as a treatment for any neurodegenerative disorder associated with neuroinflammation.

8.2.9.6 *Termitomyces albuminosus*

Termitomyces albuminosus is known as "Jizong" in Chinese. Six cerebrosides, termitomycesphins A–D, G, and H, have been found in the extracts of the mushroom (Qi et al. 2000; Qu et al. 2012). All six of these compounds have been found to have a neuritogenic effect, which can be useful in the treatment of neurodegenerative diseases (Qi et al. 2000; Qu et al. 2012).

8.3 Mushrooms That Increase the Risk of Neurotoxicity

Amanita muscaria, *A. panthirina*, and other mushrooms in the *Psilocybe* genus are among the more neuroactive mushrooms (Lima et al. 2012). Hallucinations and euphoria are the primary symptoms of acute mushroom encephalopathy. Most toxic mushroom exposures occur due to intentional recreational use, although accidental ingestion does occasionally occur. The mechanism of neurotoxicity of *A. muscaria and A. panthirina* is attributable to the presence of ibotenic acid (Jo et al. 2014). Ibotenic acid and its metabolite muscimol (isoxazoles) imitate the activity of the excitatory neurotransmitter glutamate (Zinkand et al. 1992). Therefore, the resulting symptoms of neural excitation are observed such as agitation, ataxia, hallucinations, cognitive impairment, and mental status changes. Frank psychosis and seizures are also known to occur. The activity of the *Psilocybe* genus of mushrooms is caused by the chemical psilocybin and its more potent metabolite, psilocin. Both are indole alkylamines with chemical structures that mimic the neurotransmitter serotonin. They bind to serotonin receptors throughout the brain and produce a lysergic acid diethylamide (LSD)-like effect. Symptoms include euphoria, visual illusions, memory deficits, vivid hallucinations, and reckless behavior. Anxiety, agitation, and decreased mental status also frequently occur. Psilocybin is the most abundant tryptamine in *Psilocybe* mushrooms and has been present in concentrations ranging from 0.36% in *P. stuntzii* to 0.98% in *P. semilanceata*. However, after ingestion, it is rapidly dephosphorylated by the enzyme alkaline phosphatase in the intestine. Thus, it is the metabolite psilocin which is thought to be responsible for hallucinations and psychological effects (Badham 1984). Psilocin is the next most abundant compound, ranging from 0.12% in *P. stuntzii* to 0.60% in *Psilocybe cubensis*. Its bioavailability was found to be around 50% in mice and distributes uniformly in most body tissues, though higher concentrations are found in the liver and adrenals (Benjamin 1979). In rats, psilocin concentrates in specific areas of the brain: the neocortex, the hippocampus (involved in learning and memory), and the thalamus (sensory processing) (Sticht and Käferstein 2000). About 20% of psilocin is excreted unaltered in the urine, with the remainder excreted as polar conjugate metabolites such as glucuronides. It has been estimated that less than 4% of psilocin is degraded by monoamine oxidase, the enzyme that degrades endogenous monoamines like serotonin.

The most important short-term aspect of psilocybin intoxication is the unpredictable time course and intensity of the symptoms. Psilocybin mushroom ingestion results in hallucinatory symptoms that begin as early as 10 min post ingestion and typically last anywhere from 4 to 12 h. The following are also common symptoms reported during a typical intoxication: dizziness, giddiness, nausea, weakness, muscle aches, shivering, anxiety, restlessness, and abdominal pain. Hallucinogenic and physiological effects include visual effects including brightening and distortion of colors, afterimages, visual patterns, wavelike motion of surfaces, and altered faces; increased body temperature, facial flushing, tachycardia (increased heart rate), dilation of pupils, and sweating; feelings of unreality and depersonalization, dreaminess, and panic feelings; impaired judgment of distances and incoordination;

impaired judgment of time; and also, a schizophrenoid state of double conception of both slightly altered real-world events and hallucinatory effects. There is no specific antidote to *Psilocybe* intoxication, although a clinical report of reversal of confirmed psilocybin intoxication with physostigmine is interesting and deserves further follow-up study. Management of *Psilocybe* intoxication consists mostly of emotional support and reassurance during panic episodes and monitoring of vital signs. However, in cases of long-term adverse reactions, tranquilizers such as Valium and antipsychotics such as Thorazine have been used. Likewise, in cases where the exact species of mushroom cannot be confirmed, gastric lavage or treatment with activated charcoal has been recommended (Mcdonald 1980).

8.4 Conclusion

Aging of the neurons is an important focal factor in the development of age-related neurodegenerative diseases. Diminutions in the levels of nerve growth factor (NGF) lead to major declines in neuronal function and activity. Culinary-medicinal mushrooms, thought to alleviate this deficiency, will be reaching a plateau in the market of alternative and preventive medicine. In the search for neuroactive and neuroprotective compounds that mimic and imitate the NGF activity for the prevention of neurodegenerative diseases, the potential medicinal values of these culinary and medicinal mushrooms attract intense interest especially those medicinal mushrooms possessing neuritogenic and neuroprotective activity without cytotoxic effects.

References

Ajith TA, Sudheesh NP, Roshny D, Abishek G, Janardhanan KK (2009) Effect of *Ganoderma lucidum* on the activities of mitochondrial dehydrogenases and complex I and II of electron transport chain in the brain of aged rats. Exp Gerontol 44:219–223. https://doi.org/10.1016/J.EXGER.2008.11.002

Ao ZH, Xu ZH, Lu ZM, Xu HY, Zhang XM, Dou WF (2009) Niuchangchih (*Antrodia camphorata*) and its potential in treating liver diseases. J Ethnopharmacol 121:194–212. https://doi.org/10.1016/j.jep.2008.10.039

Badham ER (1984) Ethnobotany of psilocybin mushrooms, especially *Psilocybe cubensis*. J Ethnopharmacol 10:249–254. https://doi.org/10.1016/0378-8741(84)90007-2

Barnham KJ, Masters CL, Bush AI (2004) Neurodegenerative diseases and oxidative stress. Nat Rev Drug Discov 3:205–214. https://doi.org/10.1038/nrd1330

Baskaran A, Chua KH, Sabaratnam V, Ravishankar Ram M, Kuppusamy UR (2017) *Pleurotus giganteus* (Berk. Karun & Hyde), the giant oyster mushroom inhibits NO production in LPS/ H_2O_2 stimulated RAW 264.7 cells via STAT 3 and COX-2 pathways. BMC Complement Altern Med 17:40. https://doi.org/10.1186/s12906-016-1546-6

Benjamin C (1979) Persistent psychiatric symptoms after eating psilocybin mushrooms. Br Med J 1:1319–1320

Chia Wei P (2015) Neurite outgrowth stimulatory activity of an edible mushroom *Pleurotus giganteus* in differentiating neuroblastoma-2a cells. Thesis, University of Malaya, Kuala Lumpur

Chien SC, Chen ML, Kuo HT, Tsai YC, Lin BF, Kuo YH (2008) Anti-inflammatory activities of new succinic and maleic derivatives from the fruiting body of *Antrodia camphorata*. J Agric Food Chem 56:7017–7022. https://doi.org/10.1021/jf801171x

Choi D, Lee JH, Kim YS, Na M-S, Choi OY, Lee HD, Lee MK, Cha WS (2011) A study of mycelial growth and exopolysaccharide production from a submerged culture of *Mycoleptodonoides aitchisonii* in an air-lift bioreactor. Korean J Chem Eng 28:1427–1432. https://doi.org/10.1007/s11814-011-0109-2

Choi JH, Suzuki T, Okumura H, Noguchi K, Kondo M, Nagai K, Hirai H, Kawagishi H (2014) Endoplasmic reticulum stress suppressive compounds from the edible mushroom *Mycoleptodonoides aitchisonii*. J Nat Prod 77:1729–1733. https://doi.org/10.1021/np500075m

Choi MH, Wu Y, Oh DS, Kim SK, Shin HJ (2016) Antioxidant and antidiabetic activities of mycelial and fruit-body extracts from *Mycoleptodonoides aitchisonii*. Biotechnol Bioprocess Eng 21:355–363. https://doi.org/10.1007/s12257-016-0235-0

Eik LF, Naidu M, David P, Wong KH, Tan YS, Sabaratnam V (2012) *Lignosus rhinocerus* (Cooke) Ryvarden: a medicinal mushroom that stimulates neurite outgrowth in PC-12 cells. Evid Based Complement Alternat Med 2012:320308. https://doi.org/10.1155/2012/320308

Geethangili M, Tzeng YM (2011) Review of pharmacological effects of *Antrodia camphorata* and its bioactive compounds. Evid Based Complement Alternat Med 2011:212641. https://doi.org/10.1093/ecam/nep108

Geissler T, Brandt W, Porzel A, Schlenzig D, Kehlen A, Wessjohann L, Arnold N (2010) Acetylcholinesterase inhibitors from the toadstool *Cortinarius infractus*. Bioorg Med Chem 18:2173–2177. https://doi.org/10.1016/J.BMC.2010.01.074

Giridharan VV, Thandavarayan RA, Konishi T (2011) Amelioration of scopolamine induced cognitive dysfunction and oxidative stress by *Inonotus obliquus*– a medicinal mushroom. Food Funct 2:320. https://doi.org/10.1039/c1fo10037h

Hsiao G, Shen MY, Lin KH, Lan MH, Wu LY, Chou DS, Lin CH, Su CH, Sheu JR (2003) Antioxidative and hepatoprotective effects of *Antrodia camphorata* extract. J Agric Food Chem. https://doi.org/10.1021/JF021159T

Huang NK, Cheng JJ, Lai WL, Lu MK (2005) *Antrodia camphorata* prevents rat pheochromocytoma cells from serum deprivation-induced apoptosis. FEMS Microbiol Lett 244:213–219. https://doi.org/10.1016/j.femsle.2005.01.048

Jo WS, Hossain MA, Park SC (2014) Toxicological profiles of poisonous, edible, and medicinal mushrooms. Mycobiology 42:215–220. https://doi.org/10.5941/MYCO.2014.42.3.215

Journal O, Xu T, Beelman RB (2015) The bioactive compounds in medicinal mushrooms have potential protective effects against neurodegenerative diseases. Adv Food Technol Nutr Sci 1:62–65. https://doi.org/10.17140/AFTNSOJ-1-110

Kawagishi H, Shimada A, Shirai R, Okamoto K, Ojima F, Sakamoto H, Ishiguro Y, Furukawa S (1994) Erinacines A, B and C, strong stimulators of nerve growth factor (NGF)-synthesis, from the mycelia of *Hericium erinaceus*. Tetrahedron Lett 35:1569–1572. https://doi.org/10.1016/S0040-4039(00)76760-8

Khatun K, Mahtab H, Khanam PA, Sayeed MA, Khan KA (2007) Oyster mushroom reduced blood glucose and cholesterol in diabetic subjects. Mymensingh Med J 16:94–99

Kim EK, Lee SJ, Hwang JW, Kim CG, Choi DK, Lim BO, Kang H, Moon SH, Jeon BT, Park PJ (2011) *In Vitro* investigation on antioxidative effect of *Inonotus obliquus* extracts against oxidative stress on PC12 cells. J Korean Soc Appl Biol Chem 54:112–117. https://doi.org/10.3839/jksabc.2011.016

Lee IK, Yun BS, Han G, Cho DH, Kim YH, Yoo ID (2002) Dictyoquinazols A, B, and C, new neuroprotective compounds from the mushroom *Dictyophora indusiata*. J Nat Prod 65:1769–1772. https://doi.org/10.1021/np020163w

Lee IK, Yun BS, Kim JP, Ryoo IJ, Kim YH, Yoo ID (2003) Neuroprotective activity of p-terphenyl leucomentins from the mushroom *Paxillus panuoides*. Biosci Biotechnol Biochem 67:1813–1816. https://doi.org/10.1271/bbb.67.1813

Lee I, Ahn B, Choi J, Hattori M, Min B, Bae K (2011) Selective cholinesterase inhibition by lanostane triterpenes from fruiting bodies of *Ganoderma lucidum*. Bioorg Med Chem Lett 21:6603–6607. https://doi.org/10.1016/J.BMCL.2011.04.042

Lima ADL, Costa Fortes R, Carvalho Garbi Novaes MR, Percário S (2012) Poisonous mushrooms: a review of the most common intoxications. Nutr Hosp 27:402–408. https://doi.org/10.1590/S0212-16112012000200009

Lin MT, Beal MF (2006) Mitochondrial dysfunction and oxidative stress in neurodegenerative diseases. Nature 443:787–795. https://doi.org/10.1038/nature05292

Lin JM, Lin CC, Chen MF, Ujiie T, Takada A (1995) Radical scavenger and antihepatotoxic activity of *Ganoderma formosanum*, *Ganoderma lucidum* and *Ganoderma neo-japonicum*. J Ethnopharmacol 47:33–41

Ling Sing Seow S, Naidu M, David P, Wong KH, Sabaratnam V (2013) Potentiation of neuritogenic activity of medicinal mushrooms in rat pheochromocytoma cells. BMC Complement Altern Med 13:157. https://doi.org/10.1186/1472-6882-13-157

Mcdonald A (1980) Mushrooms and madness: hallucinogenic mushrooms and some psychopharmacological implications. Can J Psychiatr 25:586–594. https://doi.org/10.1177/070674378002500709

McGeer PL, McGeer EG (1995) The inflammatory response system of brain: implications for therapy of Alzheimer and other neurodegenerative diseases. Brain Res Brain Res Rev 21:195–218

Mori K, Obara Y, Hirota M, Azumi Y, Kinugasa S, Inatomi S, Nakahata N (2008) Nerve growth factor-inducing activity of *Hericium erinaceus* in 1321N1 human astrocytoma cells. Biol Pharm Bull 31:1727–1732

Mori K, Inatomi S, Ouchi K, Azumi Y, Tuchida T (2009) Improving effects of the mushroom Yamabushitake (*Hericium erinaceus*) on mild cognitive impairment: a double-blind placebo-controlled clinical trial. Phyther Res 23:367–372. https://doi.org/10.1002/ptr.2634

Murphy MP, LeVine H (2010) Alzheimer's disease and the amyloid-β peptide. J Alzheimers Dis 19:311–323. https://doi.org/10.3233/JAD-2010-1221

Nallathamby N, Phan CW, Seow SLS, Baskaran A, Lakshmanan H, Abd Malek SN, Sabaratnam V (2018) A status review of the bioactive activities of tiger milk mushroom *Lignosus rhinocerotis* (Cooke) Ryvarden. Front Pharmacol 8:998. https://doi.org/10.3389/fphar.2017.00998

Nishina A, Kimura H, Sekiguchi A, Fukumoto R, Nakajima S, Furukawa S (2006) Lysophosphatidylethanolamine in *Grifola frondosa* as a neurotrophic activator via activation of MAPK. J Lipid Res 47:1434–1443. https://doi.org/10.1194/jlr.M600045-JLR200

Ohta T, Kita T, Kobayashi N, Obara Y, Nakahata N, Ohizumi Y, Takaya Y, Oshima Y (1998) Scabronine A, a novel diterpenoid having potent inductive activity of the nerve growth factor synthesis, isolated from the mushroom, *Sarcodon scabrosus*. Tetrahedron Lett 39:6229–6232. https://doi.org/10.1016/S0040-4039(98)01282-9

Okuyama S, Lam NV, Hatakeyama T, Terashima T, Yamagata K, Yokogoshi H (2004) *Mycoleptodonoides aitchisonii* affects brain nerve growth factor concentration in newborn rats. Nutr Neurosci 7:341–349. https://doi.org/10.1080/10284150400020490

Park KJ, Lee SY, Kim HS, Yamazaki M, Chiba K, Ha H-C (2007) The neuroprotective and neurotrophic effects of *Tremella fuciformis* in PC12h cells. Mycobiology 35:11–15. https://doi.org/10.4489/MYCO.2007.35.1.011

Park HJ, Shim HS, Ahn YH, Kim KS, Park KJ, Choi WK, Ha HC, Il KJ, Kim TS, Yeo IH, Kim JS, Shim I (2012) *Tremella fuciformis* enhances the neurite outgrowth of PC12 cells and restores trimethyltin-induced impairment of memory in rats via activation of CREB transcription and cholinergic systems. Behav Brain Res 229:82–90. https://doi.org/10.1016/J.BBR.2011.11.017

Phan CW, David P, Naidu M, Wong KH, Sabaratnam V (2015) Therapeutic potential of culinary-medicinal mushrooms for the management of neurodegenerative diseases: diversity, metabolite, and mechanism. Crit Rev Biotechnol 35:355–368. https://doi.org/10.3109/07388551.2014.887649

Przedborski S, Vila M, Jackson Lewis V (2003) Neurodegeneration: what is it and where are we? J Clin Invest 111:3–10. https://doi.org/10.1172/JCI17522

Qi J, Ojika M, Sakagami Y (2000) Termitomycesphins A–D, novel neuritogenic cerebrosides from the edible chinese mushroom *Termitomyces albuminosus*. Tetrahedron 56:5835–5841. https://doi.org/10.1016/S0040-4020(00)00548-2

Qu Y, Sun K, Gao L, Sakagami Y, Kawagishi H, Ojika M, Qi J (2012) Termitomycesphins G and H, additional cerebrosides from the edible Chinese mushroom *Termitomyces albuminosus*. Biosci Biotechnol Biochem 76:791–793. https://doi.org/10.1271/bbb.110918

Quang DN, Hashimoto T, Nukada M, Yamamoto I, Tanaka M, Asakawa Y (2003) Curtisians E-H: four p-terphenyl derivatives from the inedible mushroom *Paxillus curtisii*. Phytochemistry 64:649–654

Sabaratnam V, Kah-Hui W, Naidu M, David PR (2013) Neuronal health – can culinary and medicinal mushrooms help? J Tradit Complement Med 3:62–68. https://doi.org/10.4103/2225-4110.106549

Seow SLS, Eik LF, Naidu M, David P, Wong KH, Sabaratnam V (2015) *Lignosus rhinocerotis* (Cooke) Ryvarden mimics the neuritogenic activity of nerve growth factor via MEK/ERK1/2 signaling pathway in PC-12 cells. Sci Rep 5:16349. https://doi.org/10.1038/srep16349

Shi XW, Liu L, Gao JM, Zhang AL (2011) Cyathane diterpenes from Chinese mushroom *Sarcodon scabrosus* and their neurite outgrowth-promoting activity. Eur J Med Chem 46:3112–3117. https://doi.org/10.1016/J.EJMECH.2011.04.006

Shibata H, Irie A, Morita Y (1998) New antibacterial diterpenoids from the *Sarcodon scabrosus* fungus. Biosci Biotechnol Biochem 62:2450–2452. https://doi.org/10.1271/bbb.62.2450

Skovronsky DM, Lee VMY, Trojanowski JQ (2006) Neurodegenerative diseases: new concepts of pathogenesis and their therapeutic implications. Annu Rev Pathol Mech Dis 1:151–170. https://doi.org/10.1146/annurev.pathol.1.110304.100113

Soytong K, Asue T (2012) Study on physiological and cultural requirements of *Pleurotus giganteus*. Int J Agric Technol 10:923–930

Sticht G, Käferstein H (2000) Detection of psilocin in body fluids. Forensic Sci Int 113:403–407. https://doi.org/10.1016/S0379-0738(00)00213-9

Teichert A, Schmidt J, Porzel A, Arnold N, Wessjohann L (2007) Brunneins A–C, β-carboline alkaloids from *Cortinarius brunneus*. J Nat Prod 70:1529–1531. https://doi.org/10.1021/np070259w

Trovato Salinaro A, Pennisi M, Di Paola R, Scuto M, Crupi R, Cambria MT, Ontario ML, Tomasello M, Uva M, Maiolino L, Calabrese EJ, Cuzzocrea S, Calabrese V (2018) Neuroinflammation and neurohormesis in the pathogenesis of Alzheimer's disease and Alzheimer-linked pathologies: modulation by nutritional mushrooms. Immun Ageing 15:8. https://doi.org/10.1186/s12979-017-0108-1

Trovato A, Siracusa R, Di Paola R, Scuto M, Fronte V, Koverech G, Luca M, Serra A, Toscano MA, Petralia A, Cuzzocrea S, Calabrese V (2016) Redox modulation of cellular stress response and lipoxin A4 expression by *Coriolus versicolor* in rat brain: relevance to Alzheimer's disease pathogenesis. Neurotoxicology 53:350–358. https://doi.org/10.1016/J.NEURO.2015.09.012

Ueda K, Tsujimori M, Kodani S, Chiba A, Kubo M, Masuno K, Sekiya A, Nagai K, Kawagishi H (2008) An endoplasmic reticulum (ER) stress-suppressive compound and its analogues from the mushroom *Hericium erinaceum*. Bioorg Med Chem 16:9467–9470. https://doi.org/10.1016/J.BMC.2008.09.044

Wachtel Galor S, Yuen J, Buswell JA, Benzie IFF (2011) *Ganoderma lucidum* (Lingzhi or Reishi): a medicinal mushroom. Oxid Stress Dis 28:175–200

Wang Z, Luo D, Liang Z (2004) Structure of polysaccharides from the fruiting body of *Hericium erinaceus* Pers. Carbohydr Polym 57:241–247. https://doi.org/10.1016/J.CARBPOL.2004.04.018

Wang LC, Wang SF, Wang JJ, Tsai TY, Lin CH, Pan TM, Lee CL (2012) *In vitro* and *in vivo* comparisons of the effects of the fruiting body and mycelium of *Antrodia camphorata* against amyloid β-protein-induced neurotoxicity and memory impairment. Appl Microbiol Biotechnol 94:1505–1519. https://doi.org/10.1007/s00253-012-3941-3

Wasser S (2002) Medicinal mushrooms as a source of antitumor and immunomodulating polysaccharides. Appl Microbiol Biotechnol 60:258–274. https://doi.org/10.1007/s00253-002-1076-7

Waters SP, Tian Y, Li YM, Danishefsky SJ (2005) Total synthesis of (−)-scabronine G, an inducer of neurotrophic factor production. J Am Chem Soc 127:13514–13515. https://doi.org/10.1021/ja055220x

Wender PA, Bi FC, Brodney MA, Gosselin F (2001) Asymmetric synthesis of the tricyclic core of NGF-inducing cyathane diterpenes via a transition-metal-catalyzed [5 + 2] cycloaddition. Org Lett 3:2105–2108

Yap YH, Tan N, Fung S, Aziz AA, Tan C, Ng S (2013) Nutrient composition, antioxidant properties, and anti-proliferative activity of *Lignosus rhinocerus* Cooke sclerotium. J Sci Food Agric 93:2945–2952. https://doi.org/10.1002/jsfa.6121

Yoon HM, Jang KJ, Han MS, Jeong JW, Kim GY, Lee JH, Choi YH (2013) *Ganoderma lucidum* ethanol extract inhibits the inflammatory response by suppressing the NF-κB and toll-like receptor pathways in lipopolysaccharide-stimulated BV2 microglial cells. Exp Ther Med 5:957–963. https://doi.org/10.3892/etm.2013.895

Yun BS, Lee IK, Kim JP, Yoo ID (2000) Leucomentin-5 and -6, two new leucomentin derivatives from the mushroom *Paxillus panuoides*. J Antibiot (Tokyo) 53:711–713

Zhang R, Xu S, Cai Y, Zhou M, Zuo X, Chan P (2011) *Ganoderma lucidum* protects dopaminergic neuron degeneration through inhibition of microglial activation. Evid Based Complement Alternat Med 2011:156810. https://doi.org/10.1093/ecam/nep075

Zhou Y, Qu Z, Zeng Y, Lin Y, Li Y, Chung P, Wong R, Hägg U (2012) Neuroprotective effect of preadministration with *Ganoderma lucidum* spore on rat hippocampus. Exp Toxicol Pathol 64:673–680. https://doi.org/10.1016/J.ETP.2010.12.011

Zinkand WC, Moore WC, Thompson C, Salama AI, Patel J (1992) Ibotenic acid mediates neurotoxicity and phosphoinositide hydrolysis by independent receptor mechanisms. Mol Chem Neuropathol 16:1–10. https://doi.org/10.1007/BF03159956

Medicinal Mushrooms as Novel Sources for New Antiparasitic Drug Development

9

Daniel A. Abugri, Joseph A. Ayariga, Boniface J. Tiimob, Clement G. Yedjou, Frank Mrema, and William H. Witola

Contents

D. A. Abugri (✉) · B. J. Tiimob
Department of Chemistry and Department of Biology, Laboratory of Ethnomedicine, Parasitology and Drug Discovery, Laboratory of General Mycology and Medical Mycology, Tuskegee University, Tuskegee, AL, USA
e-mail: dabugri@tuskegee.edu

J. A. Ayariga
Department of Microbiology, Alabama State University, Montgomery, AL, USA

C. G. Yedjou
Natural Chemotherapeutics Research Laboratory, RCMI Center for Environmental Health, Jackson State University, Jackson, MS, USA

F. Mrema
Department of Agriculture and Applied Sciences, Alcorn State University, Lorman, MS, USA

W. H. Witola
Department of Pathobiology, College of Veterinary Medicine, University of Illinois, Urbana-Champaign, IL, USA

D. C. Agrawal, M. Dhanasekaran (eds.), *Medicinal Mushrooms*,
https://doi.org/10.1007/978-981-13-6382-5_9

Abstract

An estimate of 60 million to 3.4 billion people are infected with one or more of the globally neglected diseases, with the most vulnerable persons being HIV-AIDS and cancer patients; organ, tissue, and blood recipients; pregnant women and their fetus; children; and immunosuppressive individuals. Some of these diseases are either zoonotic or non-zoonotic in nature. These diseases continue to adversely impact public health, veterinary, and socioeconomic issues worldwide. Several medications are available in the market for the treatment of these parasitic infections. However, these drugs have serious clinical, geographical, and socioeconomic limitations such as high toxicity, parasite-drug resistance, partial absorption by infected sites due to poor solubility, and inability to transverse across the blood-brain barrier and are highly expensive. This chapter captures reports on antiprotozoal properties of medicinal mushroom extracts and secondary metabolites starting from 1990 to 2018 and proposes their possible mechanism of action(s). Here, we have focused on the antiparasitic activity of selected mushrooms against selected parasites (*Plasmodium* spp., *Trypanosome* spp., *Leishmania* spp., *Schistosoma* spp., *Haemonchus contortus*, *Ditylenchus dipsaci*, *Heterodera glycines*, and *Pheretima posthuma*). We observed that the minimum inhibitory concentration for 50% of parasites (IC_{50s}) of most extracts from the common mushrooms and their isolated compounds tested were within the ranges (0.06–54 μg/mL) often reported for common antiparasitic drugs in vitro. The extracts and isolated compounds were found to be nontoxic to host cells in vitro or in vivo, but these extracts and their secondary metabolites targeted cell membrane and lysing cells, caused cell arrest, disrupted mitochondrion function and ribonucleic acid, damaged DNA, disrupted phospholipids, caused auto-oxidation degradation and metal chelation, and induced oxidative stress.

Keywords

Anti-*Leishmania* · Anti-*Plasmodium* · Anti-*Schistosoma* · Anti-*Trypanosoma* · Anti-*Haemonchus* · Mushrooms

Abbreviations

ACT	Artemisinin-based combination therapy
BC	Bioactive compounds
HIV-AIDS	Human immunodeficiency virus (HIV)-acquired immune deficiency syndrome (AIDS)
IC_{50}	Minimum inhibitory concentration that kills 50% of parasites
MM	Medicinal mushrooms
MOA	Mechanism of action
WHO	World Health Organization
WMR	World Malaria Report

9.1 Introduction

Neglected parasitic diseases continue to threaten over 60 million to 3.4 billion people globally, burdening the tropics and subtropical countries the most (Andrews et al. 2014; WHO 2013). These diseases cause high morbidity, mortality, public health and veterinary issues, poverty, and socioeconomic burdens worldwide. These issues and their impact are very well felt by the developing countries (Carpio et al. 2016). The most common parasitic diseases known to cause a huge risk to the global population are malaria, leishmaniasis, toxoplasmosis, schistosomiasis, Chagas diseases (American trypanosomiasis), and trypanosomiasis (Africa sleeping sickness).

Currently, there are no known effective and safe vaccines against these diseases, except some available drugs. The few drugs used to treat these parasitic diseases are constrained by clinical, geographical, and socioeconomic issues (WHO 2018a, b). Several reports show that some of these drugs have serious adverse side effects such as high toxicity, teratogenicity, hypersensitivity, skin rashes, Stevens-Johnson syndromes, vomiting, poor absorption and solubility, and parasite resistance (WHO 2018a, b; Amato et al. 2017; Sutherland et al. 2017; Ariey et al. 2014; Abdo et al. 2003; Wolday et al. 1999; Paredes et al. 2003; Witola et al. 2016; Kaplan 2004; Kotze et al. 2014; Kaplan and Vidyashankar 2012). They are also not available in certain nations and very costly where they are available.

To overcome these limitations in the treatment of malaria, schistosomiasis, leishmaniasis, Chagas, and trypanosomiasis, among other parasitic diseases, natural products from plants, algae, and fungi origins are being explored for bioactive compounds, which upon further research could be developed into alternative drugs (Powers et al. 2017; Abugri et al. 2017; Abugri et al. 2018; Basnet et al. 2017; Abugri et al. 2016; Lindequist et al. 2005).

Medicinal mushrooms are part of the macrofungi kingdom, and are known for their bioactive properties, especially antimicrobial activity (Lindequist et al. 2005; Basnet et al. 2017). This chapter presents a review on medicinal mushroom extracts and their known isolated bioactive compounds that were tested against *Plasmodium* spp., *Schistosoma* spp., *H. contortus*, *Trypanosoma* spp., and *Leishmania* spp. growth in vitro or in vivo and their possible mechanism of action(s).

9.2 Anti-*Plasmodium* Properties of Medicinal Mushrooms and Their Mechanism of Action

Malaria is one of the neglected diseases caused by a unicellular eukaryotic protozoan (*Plasmodium* spp.) parasite. The parasite is transmitted through the bite of female *Anopheles* sp. mosquitoes (Phillips et al. 2017). The disease is prevalent in tropical and subtropical countries of the world and has been reported in North America because of global human migrations. Malaria is the leading cause of illness and deaths in these tropical and subtropical countries, mostly affecting children from 0 to 5 years old, pregnant women and their fetuses, the aged, human immunodeficiency virus (HIV)-acquired immune deficiency syndrome (AIDS) and cancer patients, and organ and tissue transplant and blood transfusion individuals whose immune systems have been compromised (World Health Organization's World Malaria Report 2013). The economic cost associated with malaria and its related threats in 2016 was estimated to be about 2.7 billion USD per year (WHO 2017). In the endemic areas, over 3.2 billion people who reside there are at greater risk of getting malaria, and about 445,000 deaths are recorded annually (WHO WMR 2018).

Several medications are available in these countries for management of the disease. However, parasite-drug-resistant genes have emerged against most mono- and combination therapies against malaria (Amato et al. 2017; Ariey et al. 2014; Lu et al. 2017; Sutherland et al. 2017). Furthermore, these drugs have been known to be toxic when administered to pregnant women and their fetus; cause dizziness, skin rashes, and weakness in immunocompromised individuals coupled with their nonavailability in certain nations; and are highly expensive nature of the medications (Penny et al. 2016; Gutman et al. 2017).

Natural products contain a huge diversity of new robust chemical compounds, which have shown promising results and could replace the noneffective drugs for the treatment of malaria. These natural products could be derived from plants, bacteria, fungus, and algae. Several secondary metabolites from plants and fungi such as terpenoids, polyketides, peroxides, flavonoids, flavonoid glycosides, alkaloids, peptides, terpenes, sesquiterpene, steroids, limonoids, lanostances, polysaccharides, lectin, halenaquinones, schisanlactones, bioactive fatty acids, and proteins have been found to have antiprotozoal activity (Annang et al. 2018; Basnet et al. 2017; Baby et al. 2015; Ferreira et al. 2015; Alves et al. 2012; Nogueira and Lopes 2011). Even though it is known that higher fungus extracts or their secondary metabolites have antiparasitic effects, few have been studied on their anti-*Plasmodium* activity (Table 9.1). Here, we have compiled studies from 1990 to 2018 regarding medicinal mushrooms that have been reported to have antiplasmodial activity and decipher their mechanism of action and the potentials for developing an effective and safe drug that will be more than artemisinin-based combination therapy (ACT) drugs currently used globally.

Sesterterpenes and triterpenes have been reported from Ghanaian mushrooms (*Pleurotus ostreatus*; *Scleroderma areolatum*) to have effective inhibition of *Plasmodium falciparum* 3D7 growth with IC_{50s} between 5.04 and 13.65 μM (Annang et al. 2018). Also, a medicinal mushroom (*Ganoderma lucidum*) was extracted, and bioactive compounds such as lanostances, ganodermalactone F, schisanlactone B, and colossolactone E were isolated from it (Adams et al. 2010). Both extract and compounds were found to have effective inhibitory activity against *Plasmodium falciparum* with IC_{50s} = 6.0–20 μM (Adams et al. 2010). The same mushroom species was extracted and challenged with an animal model of *Plasmodium* (*P. berghei*) in vivo by Oluba et al. (2017).

In another study, using ethyl acetate, methanol, ethanol extracts, and their isolated compounds from *Ganoderma lucidum* such as schisanlactone, ganodermalactone, and colossolactone E were found to be effective in inhibiting *Plasmodium falciparum* in vitro with submicromolar levels IC_{50s} of 6.0 to 10 μM (Lakornwong et al. 2014).

Table 9.1 In vitro and in vivo anti-*Plasmodium falciparum* activity of medicinal mushrooms

Parasite	Compounds/extracts	Type of mushrooms	In vitro	In vivo	References
Plasmodium falciparum	Ethyl acetate/ lanostanes, methanol extracts, schisanlactone B(32), ganodermalactone F(24), colossolactone E(6)	*Ganoderma lucidum*	Yes		Adams et al. (2010) and Lakornwong et al. (2014)
P. falciparum	Extracts	*Phellinus gilvus*, *Polyporus sulphureus*, *Pycnoporus annosus*, *Polyporus pinicola*, *Fomes fomentarius*, *Coprinopsis cinerea*, *Trametes versicolor*	Yes		Ustun et al. (2011)
Plasmodium falciparum	Cordypyridones A and B, naphthoquinones	*Cordyceps nipponica*	Yes		Isaka et al. (2001), Kittakoop et al. (1999), and Wongsa et al. (2005)
Plasmodium berghei	Ethanol and hot water extracts	Milky mushroom (*Calocybe indica*)		Yes	Omoya and Oyedirin (2018)
Plasmodium berghei	Extracts	*Ganoderma lucidum*		Yes	Oluba et al. (2017)
Plasmodium berghei	Ethanol and water extracts	*Pleurotus ostreatus*		Yes	Omoya et al. (2017)
Plasmodium falciparum	Extracts/scalarene sesterterpenes 1 and 2, triterpenes 3 and 4	*Pleurotus ostreatus*, *Scleroderma areolatum*	Yes		Annang et al. (2018)
Plasmodium falciparum	Extracts, 3,4-seco-7-norlanostane triterpenes, ganoboninones A–F, ganoboninketals A–C	*Ganoderma boninense*	Yes		Ma et al. (2015)
Plasmodium falciparum	Extracts	*Ganoderma lucidum* and *Terfezia pfeilii*	Yes		Kadhila-Muandingi (2010)
Plasmodium falciparum	Triterpene lactones	*Ganoderma* sp. *KM01*	Yes		
Plasmodium falciparum	Bioxanthracenes	*Cordyceps pseudomilitaris*	Yes		Jaturapat et al. (2001) and Isaka et al. (2007)
Plasmodium falciparum	n-hexane extracts, sterols	*Pleurotus ostreatus (Jacq. ex. Fr)*	Yes		Ademola and Odeniran (2017)
Plasmodium berghei	Crude extracts	*Ganoderma lucidum*		Yes	

Compounds such as ganoboninketals A, B, and C were isolated from *Ganoderma boninense* and tested on *Plasmodium falciparum* in vitro. The authors found that these compounds were highly effective in inhibiting parasite growth with IC_{50s} values of 4.0, 7.9, and 1.7 μM, respectively. Ustun et al. (2011) also reported that medicinal mushrooms such as *Phellinus gilvus*, *Polyporus sulphureus*, *Pycnoporus annosus*, *Polyporus pinicola*, *Fomes fomentarius*, *Coprinopsis cinerea*, and *Trametes versicolor* extracts possess remarkable antiplasmodial properties with IC_{50s} values 1.39 to 2.73 μg/mL.

In another study, Isaka et al. (2001) extracted the mushroom *Cordyceps nipponica* and found compounds such as cordypyridones A and B and naphthoquinones. When these compounds were purified and tested against *Plasmodium falciparum* growth in vitro, their IC_{50s} were found to be 0.066 and 0.037 ug/ml (Isaka et al. 2001). Other studies by Isaka et al. (2007, 2012) found ES-242 derivatives and sterostrein A that were obtained from the *Cordyceps pseudomilitaris* and *Stereum ostrea* BCC 22955, respectively, to have high anti-*Plasmodium falciparum* activity with IC_{50} values of 3.3 μM and 2.3 μg/mL, respectively. Interestingly, mushrooms known to be poisonous have also been reported to contain bioactive compounds such as aristolane dimeric sesquiterpene and aurisins A and K (Kanokmedhakul et al. 2012). According to Kanokmedhakul et al. (2012), aurisin A and aurisin K were very active against *Plasmodium falciparum* in vitro with IC_{50} values of 0.61 to 20 μg/mL. It has been reported that compounds such as azaphilones named longirostrerones A–D (1–4) and three sterols (ergosterol palmitate, ergosterol, and ergosterol peroxide) which were isolated from fungi *Chaetomium longirostre* had high antiplasmodial activity in vitro (IC_{50s} = 0.62–3.73 μM) (Panthama et al. 2011).

In 2009, Khumkomkhet and co-workers extracted four new depsidones, mollicellins K–N (1–4), and six known depsidones, mollicellins B (5), C (6), E (7), F (8), H (9), and J (10), from the mushroom *Chaetomium brasiliense*. These compounds were tested in vitro to have high anti-*Plasmodium falciparum* activity with IC_{50} values ranging from 1.2 μg/mL to 9.1 μg/mL (Khumkomkhet et al. 2009). In summary, these compounds and extracts from the few mushrooms studied were promising (Table 9.1). However, most of these compounds and the extracts have not been tested in vivo and in clinical trials to ascertain their efficacy and safety. Also, important to the current studies was that several of these compounds have not been combined with current medications to test their synergistic activity, safety, and efficacy in vitro and in vivo.

9.3 Antileishmania Properties of Medicinal Mushrooms

Leishmaniosis is another protozoan disease caused by the *Leishmania* genus which is transmitted through a bite of infected female phlebotomine sandflies and is known to exist in over 98 countries globally (Alvar et al. 2012; Gupta et al. 2007). Leishmaniasis is grouped under three types: visceral leishmaniasis (kala-azar), cutaneous leishmaniasis (oriental sore), and mucocutaneous leishmaniasis (espundia). These forms of the parasitic disease have been reported to cause the greater number of mortality and morbidity (Ashford et al. 1992). This parasitic disease puts over 350 million people at risk of leishmaniosis infection (Andrews et al. 2014; WHO 2013; Gupta et al. 2007). Estimates of over two million people are prone to new infections with the disease, with high morbidity and mortality rates annually (Alvar et al. 2012). Out of this number, an estimated of 20,000–30,000 deaths occur yearly (WHO 2018a, b). The most affected countries are Brazil, East Africa, Southeast Asia, Afghanistan, Algeria, Colombia, Iran, Syria, Kenya, Somalia, Sudan, South Sudan, India, Ethiopia, and Peru (WHO 2018a, b). Presently, there are no vaccines for controlling leishmania infection. However, there are few chemotherapeutics agents such as pentavalent antimonial compounds (e.g., sodium stibogluconate and N-methylglucamine antimoniate), pentamidine, and paromomycin (Al-Qahtani et al. 2009). Even though these compounds seem to manage the acute stages of infections, there are clinical and socioeconomic limitations to their utility in developing countries. Some of the major challenges with these drugs are high cost; intolerable in HIV-AIDS and cancer patients, pregnant women and their fetus, and immunosuppressive individuals (Paredes et al. 2003; Wolday et al. 1999); and drug resistance in some of the strains (Andrews et al. 2014; Abdo et al. 2003). These necessitated the search for alternative bioactive compounds that could be developed into new chemotherapeutics for the treatment of *Leishmania*. Several natural products have been isolated from higher fungi (medicinal mushrooms) and shown to be anti-*Leishmania* amastigote and promastigote (Table 9.2; Mallick et al. 2014; Adams et al. 2010). The compounds that were attributed to the antiparasitic activity originated from lanostanes, which was isolated from the medicinal mushroom *Ganoderma lucidum* (Adams et al. 2010).

In another study, reported that alcoholic extract of *Pleurotus ostreatus* inhibited and caused apoptosis in *Leishmania* promastigotes with IC_{50} = 160 μg/mL.

Also, Mallick et al. (2014) have reported the selective inhibitory activity of *Leishmania* by six mushrooms (*Astraeus hygrometricus*, *Tricholoma giganteum*, *Russula laurocerasi*, *Russula albonigra*, *Termitomyces eurhizus*, and *Russula delica*) used by the Santal tribe in West Bengal in India. The ethanol extracts, polysaccharide fractions, polyphenolic fractions, and water-soluble extracts of these fungi were potent against *Leishmania donovani* promastigotes and amastigote-like form of the parasite. The IC_{50s} for *Astraeus hygrometricus*, *Russula laurocerasi*, *Russula albonigra*, *Termitomyces eurhizus*, *Tricholoma giganteum*, and *Russula delica* against intracellular amastigotes ranged from 90.9 μg/mL to 152 μg/mL (Mallick et al. 2014).

In addition, *Agaricus blazei* Murill water extracts have been reported to inhibit several species of *Leishmania* (*L. amazonensis*, *L. Chagas*, and *L. major*) with minimum inhibitory concentrations (IC_{50s}) of 115.4, 112.3, and 108.4 μg/mL for amastigote-like and 67.5, 65.8, and 56.8 μg/mL for promastigotes, respectively. These extracts did not express any cytotoxic effects on murine macrophages (Valadares et al. 2011). Other medicinal mushrooms have also been reported to have effective inhibitory activity against *Leishmania* parasites at lower concentrations (Inchausti et al. 1997). For instance, 17 extracts and 7 secondary metabolites

Table 9.2 In vitro and in vivo antileishmania activity of medicinal mushrooms

Name of parasite tested	Compounds/extracts	Type of mushrooms	In vitro	In vivo	References
Leishmania	Extracts/polysaccharides/ astrakurkurone	*Astraeus hygrometricus*	Yes	Yes	Mallick et al. (2014)
		Tricholoma giganteum	Yes	Yes	Lai et al. (2012)
	Fractions and extracts	*Agaricus blazei*	Yes		Valadares et al. (2012)
Leishmania amazonensis	Extracts/5-alpha,8-alpha-epidioxy-22E-ergosta-6,22 dien-3beta-ol; 5alpha-ergost-7, 22-dien-3beta-ol; 3beta-hydroxylanostan-8, 24-diene-21-oic acid (trametonolic acid)	*Trametes versicolor*	Yes		Leliebre-Lara et al. (2016)
Leishmania donovani	Extracts	*Phellinus gilvus, Polyporus sulphureus, Pycnoporus annosus, Polyporus pinicola, Fomes fomentarius, Coprinopsis cinerea, Trametes versicolor*	Yes		Ustun et al. (2011)
Leishmania major	Methanolic extracts	*Pleurotus ostreatus*	Yes		Ramezani et al. (2017)

isolated from the basidiomycete's family of mushrooms revealed active compounds such as striatin A, striatin B, and podoscyphic acid with IC_{50s} of 10, 5, and 100 μg/mL, respectively. In vivo testing of striatin A was found to decrease parasite load of 17.6%, while striatin B was not active (Inchausti et al. 1997).

9.4 Antitrypanosomal Properties of Medicinal Mushrooms

Trypanosoma sp. is the causative agent of Chagas (American trypanosomiasis) and trypanosomiasis (African sleeping sickness) diseases, which remain a major public health and socioeconomic problem, especially in Latin America and sub-Saharan Africa (Pérez-Molina and Molina 2018). *Trypanosoma* infections occur in several ways, for example, through organ or tissue transplant and blood transfusion; ingestion of contaminated waters, drink, and food; and accidental injection (Herwaldt 2001). The infection rates range between 0 and 100% depending on the mode of transmission (Pérez-Molina and Molina 2018). For instance, *Trypanosoma* infection through blood has been estimated to range from 10 to 25% (Bern et al. 2008;

Cancino-Faure et al. 2015), while people who received organ transplant (kidney receipt) ranges between 0 and 19% (Riarte et al. 1999; Cicora et al. 2014; Huprikar et al. 2013), and liver recipients (0–29%) and heart recipient can be as high as 75% to 100% (Huprikar et al. 2013; Kun et al. 2009). It has been estimated that Chagas infections and death among the young working ages cause some nations to loss over 752,000 working days and affect their economy of about 1.2 billion USD due to loss of productivity (WHO 2010). The estimated annual global burden of the disease is $627.46 million in health-care costs and $806,170 disability-adjusted life-years, and about 10% of this burden affect non-endemic countries (Lee et al. 2013).

Chronic infection of *Trypanosoma*, when eliminated, will result in cardiomyopathy, arrhythmias, and megaviscera and may result in polyneuropathy and stroke.

There are few therapeutics such as posaconazole, benznidazole, nifurtimox, allopurinol, and triazoles available for the management of acute infections. However, these medications have serious drawbacks such as ineffectiveness and teratogenicity and cause chromosomal aberrations in children (Pérez-Molina and Molina 2017). Thus, recent research in this parasitic drug development has been directed toward discovery of new plant- and non-plant-based natural product drugs to augment the current drugs (De Silva et al. 2013; Nogueira and Lopes 2011).

Fungi secondary metabolites have been reported to have antimicrobial activity against other pathogens including parasites (De Silva et al. 2013; Ferreira et al. 2010). Here, the study reviewed different papers published about different mushroom extracts, their secondary metabolites anti-*Trypanosoma* spp. activity, and their proposed mechanism of action (Table 9.3).

Annang et al. (2018) isolated four compounds (scalarene sesterterpenes (1, 2) and triterpenes (3, 4)) from *Pleurotus ostreatus* and *Scleroderma areolatum*. These compounds were found to have IC_{50s} = 5.34, 6.78, 5.04, and 13.65 μg/mL against *Trypanosoma* spp. All compounds were found to be non-cytotoxic to HepG2 cells that were used to proliferate the parasites for the in vitro assay. A compound ergosterol isolated from *Pleurotus salmoneostramineus* was found to have anti-*Trypanosoma cruzi* activity against the trypomastigote stage with an IC_{50} = 51.3 (46.0–57.0 μg/mL) (Alexandre et al. 2017). Similarly, the authors found that the compound had activity against the amastigote stage but with a high concentration of $IC_{50} < 100$ μg/mL. The compound showed nontoxic effects on mammalian cells at concentrations greater than 200 μg/mL (Alexandre et al. 2017). Also, Ramos-Ligonio et al. (2012) demonstrated that a secondary metabolite ergosterol peroxide extracted from *Pleurotus ostreatus* had a significant anti-*T. cruzi* activity with effective IC_{50} = 6.74 μg/mL. Interestingly, the compound showed non-cytotoxic effect and did not lyse the erythrocytes of host cells at concentrations higher than 1600 μg/mL. Similarly, Cota et al. (2008) reported that hypnophilin and panepoxydone that were isolated from *Lentinus strigosus* mushroom had antiplasmodium IC_{50s} of 0.8 and 38.9, respectively. When a medicinal mushroom (*Lentinus strigosus)* was extracted and hypnophilin and panepoxydone isolated and tested in vitro on *T. cruzi*. They gave remarkable results with EC_{50} and EC_{90} ranging from 0.66 to 5.93 μg/mL and 0.84 to 7.54 μg/mL (Souza-Fagundes et al. 2010). The hypnophilin was found

Table 9.3 Anti-*Trypanosoma* activity of selected medicinal mushrooms

Parasite tested	Compounds/ extracts	Type of mushrooms	In vitro	In vivo	References
Trypanosoma cruzi	Extracts/ ergosterol	*Pleurotus salmoneostramineus*	Yes		Alexandre et al. (2017)
Trypanosome cruzi	5-Alpha, 8-alpha-epidioxy-22E-ergosta-6,22-dien-3beta-ol	*Pleurotus ostreatus*	Yes		Ramos-Ligonio et al. (2012)
Trypanosoma cruzi	Extracts, striatin A, striatin B, and podoscyphic acid	*Panaeolus*, *Mycena*, *Propolis*, *Fomitopsis*, *Phellinus*, *Polyporus*, *Oudemansiella*, *Hyphodontia*, *Galerina*, *Zucoagaricus*, *Agrocybe*, *Cyathus*, *Trametes*, *Porling*, *Pleurocybella*, *Pleurotus*, *Hirneola* spp.	Yes	Yes	Inchausti et al. (1997)
Trypanosoma cruzi	Hypnophilin and panepoxydone	*Lentinus strigosus*	Yes		Cota et al. (2008)
Trypanosoma cruzi	Extracts/ scalarene sesterterpenes 1 and 2, triterpenes 3 and 4	*Pleurotus ostreatus*, *Scleroderma areolatum*	Yes		Annang et al. (2018)
Trypanosoma brucei	Extracts	*Phellinus gilvus*, *Polyporus sulphureus*, *Pycnoporus annosus*, *Polyporus pinicola*, *Fomes fomentarius*, *Coprinopsis cinerea*, *Trametes versicolor*	Yes		Ustun et al. (2011)
Trypanosome brucei brucei	Extracts, alkaloids, saponins, terpenoids, and cardiac glycosides	*Cantharellus cibarius*	Yes	Yes	Abedo et al. (2015)
Trypanosoma congolense	Aqueous extract, triterpenes, and phenolics	*Pleurotus sajor-caju*		Yes	Ademola and Odeniran (2017)

to be a good prototype for further anti-*T. cruzi* drug development. In the same mushroom, Ramos-Ligonio et al. (2012) discovered that 5α,8α-epidioxy-22E-ergosta-6,22-dien-3β-ol exhibited high anti-*T. cruzi* activity with $IC_{50} = 6.74$ μg/mL.

Furthermore, when Inchausti et al. (1997) extracted 17 mushrooms (*Panaeolus*, *Mycena*, *Propolis*, *Fomitopsis*, *Phellinus*, *Polyporus*, *Oudemansiella*, *Hyphodontia*, *Galerina*, *Zucoagaricus*, *Agrocybe*, *Cyathus*, *Trametes*, *Porling*, *Pleurocybella*, *Pleurotus*, *Hirneola* spp.) from the basidiomycetes, their isolated seven compounds (aleurodiscal, striatin A, striatin B, mniopetal E, oosporein, podoscyphic acid, naematolin) showed effective inhibition of *T. cruzi*. The IC_{50s} ranged from 18 to 79 μg/mL (Inchausti et al. 1997).

Other strains such as *Trypanosoma brucei*, *Trypanosoma brucei brucei*, and *Trypanosoma congolense* have been reported in the literature to be susceptible to medicinal mushroom extracts and secondary metabolites (Table 9.1). Again, *Pleurotus sajor-caju* was extracted (phenolic and triterpenes) and tested in vivo using mice model infected with *Trypanosoma congolense*. This study revealed an ED_{50} of 221.5 mg/kg (Ademola and Odeniran 2017). It was found that mice administered with 250 mg/kg of the extracts had survival time over 60 days than the current drug diminazene aceturate (Ademola and Odeniran 2017).

Although these extracts and few secondary metabolites have shown antitrypanosomal activity in either in vitro or in vivo assay, further studies into over 100,000 species of mushrooms identified still remain unknown regarding their antitrypanosomal activity.

9.5 Antinematode Activity of Medicinal Mushrooms

9.5.1 Anti-*Haemonchus contortus* Activity of Medicinal Mushrooms

Haemonchus contortus is a parasitic nematode of high economic and veterinary importance. The parasite infection in ruminants such as cattle, sheep, and goats can cause a huge economic burden to the livestock industry (Wang et al. 2017; Cantacessi et al. 2012). Ruminants infected with this parasitic disease present clinical symptoms such as anemia, weight loss, abortion, and even death (Wang et al. 2017). Presently, the treatment of *Haemonchus contortus*-infected ruminants is by drugs, due to lack of vaccines for this condition (Witola et al. 2016; Wang et al. 2017). The most common anthelmintic use for treatment is benzimidazole and ivermectin. However, these drugs have been challenged with drug resistance, require repeated dosage, are toxic, and are expensive (Witola et al. 2016; Zhao et al. 2010; Kaplan 2004; Kotze et al. 2014; Kaplan and Vidyashankar 2012). Therefore, there is an overwhelming need to develop new anthelmintic that will be safe, effective, and cost-effective against these deadly gastrointestinal nematodes.

Several drugs are currently available in the market as effective antibiotics, which were originally isolated from high fungi (Basnet et al. 2017; De Silva et al. 2013). Therefore, several secondary metabolites derived from higher mushrooms have a promising future for the development of anthelmintic drugs. This section concentrates on current trend of medicinal mushrooms and their inherent bioactive compound inhibitory activity against *H. contortus* in vitro and in vivo (Table 9.4).

Agaricus blazei was reported to reduce *H. contortus* parasite load in infected mice by 28.6 to 54.2% (Almeida Bastos et al. 2016). Interestingly, the mushroom extracts had no hematological effect even at higher doses. In another study by Vieira

Table 9.4 Anti-*H. contortus* activity of medicinal mushrooms

Name of parasite tested	Compounds/extracts	Type of mushrooms	In vitro	In vivo	References
Haemonchus contortus	Extracts, fatty acids (pentadecanoic, hexadecanoic, octadecadienoic, and octadecanoic acids), terpene (β-sitosterol)	*Pleurotus djamor*	Yes		Pineda-Alegría et al. (2017)
Haemonchus contortus	Mushroom formulations	*Agaricus blazei*		Yes	Almeida Bastos et al. (2016)
Haemonchus contortus	Aqueous and ethanol extracts	*Paecilomyces lilacinus* *Trichoderma longibrachiatum*	Yes		Vieira et al. (2017)

et al. (2017), an assessment of both aqueous and ethanolic extracts of *Paecilomyces lilacinus* and *Trichoderma longibrachiatum* against *H. contortus* egg hatching revealed that all extracts possessed efficacy ranging from 79.05% to 100% in a dose-dependent manner.

Heim et al. (2015) reported that fungi lectins obtained from *Coprinopsis cinerea*, *Aleuria aurantia*, *Marasmius oreades*, and *Laccaria bicolor* inhibited more than 95% of *H. contortus* larva development. Also, an in vitro study using *Pleurotus djamor* extracts dominated with fatty acids (pentadecanoic, hexadecanoic, octadecadienoic, and octadecanoic acid) and terpene showed egg-hatching inhibition with percent E1 = 100, E2 = 38.7 and E3 = 5.5 μg/mL (Pineda-Alegría et al. 2017). In the same experiment, it was reported that the extracts and the bioactive compounds contributed to larval mortality after 72 h were 90.6 and 100%. These findings are limited in scope and will require large number screening of species from similar classes of mushrooms to decipher new bioactive compounds for the treatment of nematodes.

9.5.2 Anti-*Schistosoma* Activity of Medicinal Mushrooms

Schistosoma mansoni, *Schistosoma haematobium*, and *Schistosoma japonicum* are the causative agents of the disease schistosomiasis. Schistosomiasis is one of the globally neglected diseases that affects over 207 million people worldwide (Carpio et al. 2016). The disease had been reported to cause disorders in the central nervous system (Ferrari and Moreira 2011). Schistosomiasis infection could be asymptomatic in individuals with a competent immune system. However, on immunocompromised persons, clinical symptoms such as fever, urticarial rashes, encephalopathy, cough pulmonary infiltrates (Carpio et al. 2016). In addition, people with cerebral schistosomiasis do present signs such as a headache, seizures, papilledema, visual difficulties, and oral disorders (Ferrari and Moreira 2011). Currently, there are no vaccines against the disease; however, there are few drugs that could be used to manage the disease, for example, praziquantel. With this unmet medical challenge, there is still a serious need for new effective and safe antischistosomiasis drugs to combat the disease. Mushrooms have been featured as effective sources for new antibiotics and antiparasitic drugs (Basnet et al. 2017; De Silva et al. 2013; Lindequist et al. 2005). Medicinal mushrooms are known to contain polysaccharides, phenolic, flavonoids, alkaloids, terpene, and lectin (Chen et al. 2007; Fatima et al. 2017) as bioactive combatants of diseases. Here, the literature of medicinal mushroom extracts and their bioactive compounds tested on *Schistosoma* spp. and their possible mechanism of action is reviewed.

Antrodia camphorata is a medicinal mushroom that has been used in folk medicine for treatment of viral infection and cancer in Taiwan (Chen et al. 2007). In an in vivo study, using polysaccharide extracts, this mushroom effectively inhibited *S. mansoni* in infected mice model with the worm number, reducing from 120 to 15 at week 6 (Chen et al. 2007). However, studies using mushrooms are few in this parasitic model and therefore require more future studies using different mushroom extracts.

Other parasites such as *Ditylenchus dipsaci*, *Heterodera glycines*, and *Pheretima posthuma* have also been reported in the literature to be susceptible to lectins and methanol extracts of *Ganoderma lucidum* and *Pleurotus highking* (Zhao et al. 2009; Haque et al. 2015). However, further studies are needed to identify the bioactive compounds within the few mushrooms studied and in the characterization of other species of these mushrooms against the parasites.

9.6 Possible Mechanism of Action of Extracts and Secondary Metabolites

Terpenes have been reported to possess high affinity for binding ergosterol when binding to the ergosterol that controls signaling and cell membrane integrity results in the destabilization of pathogen cell membranes (Miron et al. 2014). Additionally, sesquiterpene lactone inhibits glucose uptake in pathogens (Kar 2007). Alkaloids such as quinine, Emeline, and quinidine have been reported to have effective antiprotozoal, especially antimalarial (Wright and Phillipson 1990; Hazra et al. 1987, 1995; Lopes et al. 1978). For instance, some of the alkaloids inhibit DNA, RNA, and lipid synthesis (Wight and Phillipson 1990). This antiparasitic activity observed in mushroom extracts that were rich with quinolone-based compounds corroborated with several studies of quinolone-based compound properties such as metal ion (iron, magnesium) chelation, damage to DNA gyrase, DNA, and RNA (Kadri et al. 2014; Shen et al. 1999). Further, naphthoquinones have been reported to disrupt apicoplast function and thus cause the disappearance of the apicoplast in apicomplexan parasites (i.e., *Toxoplasma gondii*) (Kadri et al. 2014).

Free fatty acids have also been reported to inhibit microbes such as bacteria and fungi (Desbois and Smith 2010). Therefore, it is important to state that some of the mushroom extracts that were rich with fatty acids with C14, C16, and C18 could have been contributed to the high antiparasitic activity observed with some of the parasitic models tested. Desbois and Smith (2010) reported that C14, C16, and C18 free fatty acids are highly potent than the lower C10 and C5 fatty acids. This fatty acid in their free forms solubilizes the membrane or causes over secretions of lipid bilayers internally including its proteins, which turns to impair nutrient uptake, inhibition of enzyme binding, and production of peroxidation products, which becomes toxic for the microorganism survival. The most common fatty acids reported to have antiprotozoal activity include C18:0, C18:1n9, C18:2n6, and C18:3n3 (Desbois and Smith 2010; Rohrer et al. 1986).

The flavonoids and glucoside moiety found in some mushroom extracts might have high antiprotozoal activity. This might be attributed to their abilities to form complexes with extracellular and soluble cell-bound proteins in the cell membranes, which could cause cell plasma membrane perturbations (Moyo et al. 2012; Enwa et al. 2013). The presence of the hydroxyl groups and the sugar moieties enables flavonoids to chemically bind to proteins, enzymes, and ions. This causes deprivation of important nutrients and essential metals required by the parasite for survival, invasion, growth, egress, metabolism, and motility. Other studies published have listed some of the common phenolic acids and flavonoids found in these mushrooms to inhibit protein kinases activity (Lolli et al. 2012; Rao et al. 2010; Agullo et al. 1997). Importantly, protein kinases, especially calcium-dependent protein kinases (CDPKs), have been reported to control protozoal (e.g., *Toxoplasma gondii* and *Plasmodium falciparum*) survival, attachment, replication, motility, and egress in host cells (Johnson et al. 2012; Ojo et al. 2010; Billker et al. 2009; Siden-Kiamos et al. 2006; Lourido et al. 2012). Therefore, the studies reporting phenolic and flavonoids found in medicinal mushrooms to inhibit parasite growth might be targeting the CDPKs. In other studies, polyphenols have been tested to have a high binding affinity for dihydrofolate reductase (DHFR) enzymes, inhibiting DNA gyrase activity (Godstime et al. 2014; Enwa et al. 2013). Some flavanones and phenols also disrupt the mitochondrial and electron transport chain activity of microorganisms, causing calcium stress and damage to the cytoplasmic membranes (Hwang et al. 2012; Rao et al. 2010; Sung et al. 2007; Ito et al. 2007; Darvishi et al. 2013; Ben et al. 2006; Khan and Ahmad 2011; De Vita et al. 2014; Yun et al. 2015; de Oliveira et al. 2013; Sitheeque et al. 2009; Lee and Lee 2014, 2015).

Saponins are classes of organic compounds that are ubiquitously found in plants and higher fungi. These compounds are known to interact with the cell membrane and elicit cell membrane morphological changes, which eventually results in cell lysing and death (Enwa et al. 2013; Moyo et al. 2012).

Ergosterol is known to target parasite plasma membranes and destabilizes the plasma membrane integrity, hence allowing the influx of foreign chemicals, which results in cell lysing, mitochondria dysfunction, and eventually parasite death (Alexandre et al. 2017). A summary of the possible mechanism of actions of mushroom extracts and their isolated compounds is presented in Fig. 9.1.

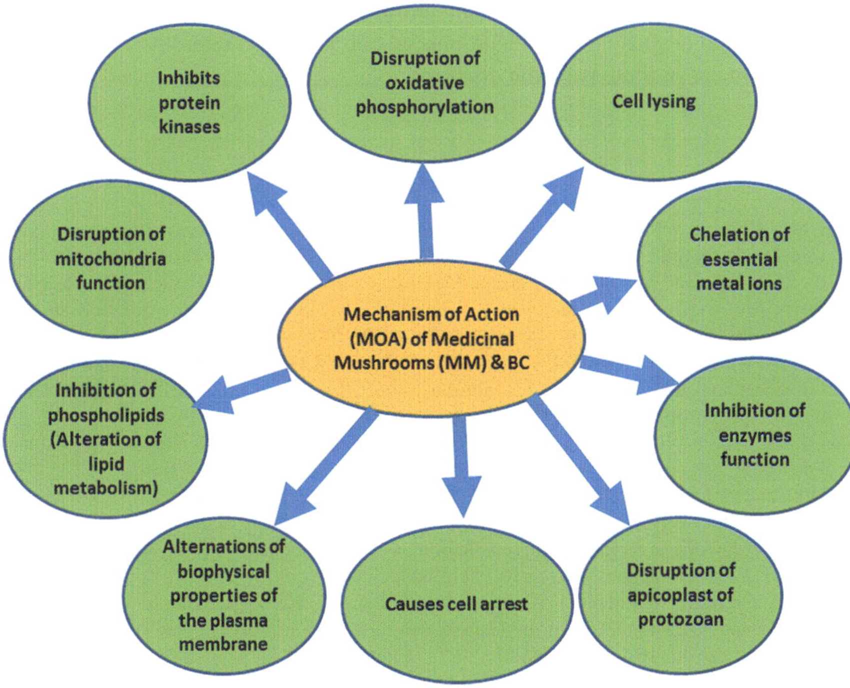

Fig. 9.1 Proposed mechanism of action (MOA) of some medicinal mushroom (MM) extracts and isolated bioactive compounds (BC) against parasite survival

9.7 Future of New Antimicrobials from Medicinal Mushrooms

Out of over 100,000 mushroom species identified on earth, only a fraction of these have been studied to decipher their biological properties. For instance, medicinal mushrooms known to have antiparasitic effects that have been partially investigated in vitro and in vivo come from *Ganoderma* spp., *Cordyceps* spp., *Lentinus* spp., *Trichoderma* spp., *Pleurotus* spp., *Scleroderma* spp., *Hyphodontia* spp., *Fomes* spp., *Trametes* spp., *Antrodia* spp., *Agaricus* spp., *Terfezia* spp., *Astraeus* spp., and *Paecilomyces* spp.

With this number, only a few have been investigated against the numerous zoonotic and non-zoonotic parasites. Most of the studies have been carried out in vitro with less in vivo testing using extracts and secondary metabolites isolated from these higher fungi. It is, therefore, important that future research should be carried out on the majority of the 100,000 species known to decipher their antiparasitic activity and mechanism of action and identify the lead compounds for medicinal chemistry optimization for effective and safe antiparasitic drug development.

From the literature, no evidence exists about these mushroom inhibitory effects on most food and waterborne parasites (e.g., *Cryptosporidium* spp. and *Toxoplasma gondii*).

Future studies should also be considered looking at animal models and human clinical trials to evaluate these mushrooms as nutraceuticals or antiparasitic agents for the treatment of these parasitic diseases.

References

Abdo MG, El Amin WM, Khalil EAG, Mukhtar MM (2003) Antimony-resistant *Leishmania donovani* in eastern Sudan: incidence and in vitro correlation. East Mediterr Health J 9(4):837–843

Abedo AJ, Shettima F, Abdullahi R, Mazadu M, Hussaini M, Muhammed H, Ogar MU, Tasie CP (2015) Anti trypanosomal activity of *Cantharellus cibarius* on *Trypanosoma brucei brucei*. J Pharm Bioallied Sci 10(5):41–45. https://doi.org/10.9790/3008-10514145

Abugri DA, Witola WH, Jaynes JM, Toufic N (2016) *In vitro* activity of *Sorghum bicolor* extracts, 3-deoxyanthocyanidins, against *Toxoplasma gondii*. Exp Parasitol 164:12–19. https://doi.org/10.1016/j.exppara.2016.02.001

Abugri DA, Witola WH, Jaynes JM (2017) *In vitro* antagonistic and indifferent activity of combination of 3-deoxyanthocyanidins against *Toxoplasma gondii*. Parasitol Res 116(12):3387–3400. https://doi.org/10.1007/s00436-017-5661-1

Abugri DA, Witola WH, Russell AE, Troy RM (2018) *In vitro* activity of the interaction between taxifolin (dihydroquercetin) and pyrimethamine against *Toxoplasma gondii*. Chem Biol Drug Des 91(1):194–201. https://doi.org/10.1111/cbdd.13070

Adams M, Christen M, Plitzko I, Zimmermann S, Brun R, Kaiser M, Hamburger M (2010) Antiplasmodial lanostanes from the *Ganoderma lucidum* mushroom. J Nat Prod 73(5):897–900. https://doi.org/10.1021/np100031c

Ademola IO, Odeniran PO (2017) Novel trypanocide from an extract of *Pleurotus sajor-caju* against *Trypanosoma congolense*. Pharm Biol 55(1):132–138. https://doi.org/10.1080/13880209.2016.1230878

Agullo G, Gamet-Payrastre L, Manenti S, Viala C, Rémésy C, Chap H, Payrastre B (1997) Relationship between flavonoid structure and inhibition of phosphatidylinositol 3-kinase: a comparison with tyrosine kinase and protein kinase C inhibition. Biochem Pharmacol 53(11):1649–1657. https://doi.org/10.1016/S0006-2952(97)82453-7

Alexandre TR, Lima ML, Galuppo MK, Mesquita JT, do Nascimento MA, dos Santos AL, Sartorelli P, Pimenta DC, Tempone AG (2017) Ergosterol isolated from the basidiomycete *Pleurotus salmoneostramineus* affects *Trypanosoma cruzi* plasma membrane and mitochondria. J Venomous Anim Toxins Incl Trop Dis 23(1):30. https://doi.org/10.1186/s40409-017-0120-0

Almeida Bastos G, Maia Soares AC, Vieira TM, de Souza Cândido RC, Morais-Costa F, de Oliveira VV, Faria Oliveira NJ, Robson Duarte E (2016) Blood parameters of sheep with high infection of *Haemonchus contortus* and treated with "mushroom of the sun"(*Agaricus blazei*). Semina-Cienc Agrar 37(2):807–817. https://doi.org/10.5433/1679-0359.2016v37n2p807

Al-Qahtani A, Siddiqui YM, Bekhit AA, El-Sayed OA, Aboul-Enein HY, Al-Ahdal MN (2009) Inhibition of growth of *Leishmania donovani* promastigotes by newly synthesized 1, 3, 4-thiadiazole analogs. Saudi Pharm J 17(3):227–232. https://doi.org/10.1016/j.jsps.2009.08.005

Alvar J, Velez ID, Bern C, Herrero M, Desjeux P, Cano J, Jannin J, den Boer M, WHO Leishmaniasis Control Team (2012) Leishmaniasis worldwide and global estimates of its incidence. PLoS One 7(5):e35671. https://doi.org/10.1371/journal.pone.0035671

Alves MJ, Ferreira IC, Dias J, Teixeira V, Martins A, Pintado M (2012) A review on antimicrobial activity of mushroom (Basidiomycetes) extracts and isolated compounds. Planta Med 78:1707–1718. https://doi.org/10.1055/s-0032-1315370

Amato R, Lim P, Miotto O, Amaratunga C, Dek D, Pearson RD, Almagro-Garcia J, Neal AT, Sreng S, Suon S, Drury E (2017) Genetic markers associated with dihydroartemisinin–piperaquine failure in *Plasmodium falciparum* malaria in Cambodia: a genotype–phenotype association study. Lancet Infect Dis 17(2):164–173. https://doi.org/10.1016/S1473-3099(16)30409-1

Andrews KT, Fisher G, Skinner-Adams TS (2014) Drug repurposing and human parasitic protozoan diseases. Int J Parasitol Drugs Drug Resist 4(2):95–111. https://doi.org/10.1016/j.ijpddr.2014.02.002

Annang F, Pérez-Victoria I, Appiah T, Pérez-Moreno G, Domingo E, Martín J, Mackenzie T, Ruiz-Pérez L, González-Pacanowska D, Genilloud O, Vicente F (2018) Antiprotozoan sesterterpenes and triterpenes isolated from two Ghanaian mushrooms. Fitoterapia 127:341–348. https://doi.org/10.1016/j.fitote.2018.03.016

Ariey F, Witkowsk B, Amaratunga C, Beghain J, Langlois AC, Khim N, Kim S, Duru V, Bouchier C, Ma L, Lim P (2014) A molecular marker of artemisinin-resistant *Plasmodium falciparum* malaria. Nature 505(7481):50. https://doi.org/10.1038/nature12876

Ashford RW, Seaman J, Schorscher J, Pratlong F (1992) Epidemic visceral leishmaniasis in southern Sudan: identity and systematic position of the parasites from patients and vectors. Trans R Soc Trop Med Hyg 86(4):379–380. https://doi.org/10.1016/0035-9203(92)90229-6

Baby S, Johnson AJ, Govindan B (2015) Secondary metabolites from *Ganoderma*. Phytochemistry 114:66–101. https://doi.org/10.1016/j.phytochem.2015.03.010

Basnet BB, Liu L, Bao L, Liu H (2017) Current and future perspective on antimicrobial and antiparasitic activities of *Ganoderma* sp.: an update. Mycology 8(2):111–124. https://doi.org/10.1080/21501203.2017.1324529

Ben Arfa A, Combes S, Preziosi-Belloy L, Gontard N, Chalier P (2006) Antimicrobial activity of carvacrol related to its chemical structure. Lett Appl Microbiol 43(2):149–154. https://doi.org/10.1111/j.1472-765X.2006.01938.x

Bern, C., Montgomery, S. P., Katz, L., Caglioti, S., & Stramer, S. L. (2008). Chagas disease and the US blood supply. Current opinion in infectious diseases, 21(5), 476-482

Billker O, Lourido S, Sibley LD (2009) Calcium-dependent signaling and kinases in apicomplexan parasites. Cell Host Microbe 5(6):612–622. https://doi.org/10.1016/j.chom.2009.05.017

Cancino-Faure, B., Fisa, R., Riera, C., Bula, I., Girona-Llobera, E., & Jimenez-Marco, T. (2015). Evidence of meaningful levels of Trypanosoma cruzi in platelet concentrates from seropositive blood donors. Transfusion, 55(6), 1249-1255

Cantacessi C, Campbell BE, Gasser RB (2012) Key strongylid nematodes of animals—Impact of next-generation transcriptomics on systems biology and biotechnology. Biotechnol Adv 30(3):469–488. https://doi.org/10.1016/j.biotechadv.2011.08.016

Carpio A, Romo ML, Parkhouse RME, Short B, Dua T (2016) Parasitic diseases of the central nervous system: lessons for clinicians and policy makers. Expert Rev Neurother 16(4):401–414. https://doi.org/10.1586/14737175.2016.1155454

Chen KC, Peng CC, Peng RY, Su CH, Chiang HS, Yan JH, Hsieh-Li HM (2007) Unique formosan mushroom *Antrodia camphorata* differentially inhibits androgen-responsive LNCaP and-independent PC-3 prostate cancer cells. Nutr Cancer 57(1):111–121. https://doi.org/10.1080/01635580701268360

Choi H, Lee DG (2015) Lycopene induces apoptosis in *Candida albicans* through reactive oxygen species production and mitochondrial dysfunction. Biochimie 115:108–115. https://doi.org/10.1016/j.biochi.2015.05.009

Cicora F, Escurra V, Silguero S, González IM, Roberti JE (2014) Use of kidneys from *Trypanosoma Cruzi*–infected donors in naive transplant recipients without prophylactic therapy: the experience in a high-risk area. Transplantation 97(1):e3–e4. https://doi.org/10.1097/01.TP.0000437673.86339.82

Cota BB, Rosa LH, Fagundes EMS, Martins-Filho OA, Correa-Oliveira R, Romanha AJ, Rosa CA, Zani CL (2008) A potent trypanocidal component from the fungus *Lentinus strigosus* inhibits trypanothione reductase and modulates PBMC proliferation. Mem Inst Oswaldo Cruz 103(3):263–270. https://doi.org/10.1590/S0074-02762008000300007

Darvishi E, Omidi M, Bushehri AAS, Golshani A, Smith ML (2013) The antifungal eugenol perturbs dual aromatic and branched-chain amino acid permeases in the cytoplasmic membrane of yeast. PLoS One 8(10):e76028. https://doi.org/10.1371/journal.pone.0076028

de Oliveira PF, Mendes JM, de Oliveira LE (2013) Investigation on mechanism of antifungal activity of eugenol against *Trichophyton rubrum*. Med Mycol 51(5):507–513. https://doi.org/10.3109/13693786.2012.742966

De Silva DD, Rapior S, Sudarman E, Stadler M, Xu J, Alias SA, Hyde KD (2013) Bioactive metabolites from macrofungi: ethnopharmacology, biological activities and chemistry. Fungal Divers 62(1):1–40. https://doi.org/10.1007/s13225-013-0265-2

De Vita D, Friggeri L, D'Auria FD, Pandolfi F, Piccoli F, Panella S, Palamara AT, Simonetti G, Scipione L, Di Santo R, Costi R (2014) Activity of caffeic acid derivatives against *Candida albicans* biofilm. Bioorg Med Chem Lett 24(6):1502–1505. https://doi.org/10.1016/j.bmcl.2014.02.005

Desbois AP, Smith VJ (2010) Antibacterial free fatty acids: activities, mechanisms of action and biotechnological potential. Appl Microbiol Biotechnol 85(6):1629–1642. https://doi.org/10.1007/s00253-009-2355-3

Enwa, F. O., Omojate, C. G., & Adonu, C. (2013). A review on the phytochemical profile and the antibacterial susceptibility pattern of some clinical isolates to the ethanolic leaves extract of *Moringa oleifera* Lam (Moringaceae). International Journal of Advanced Research, 1(5), 226-238

Ferrari TCA, Moreira PRR (2011) Neuroschistosomiasis: clinical symptoms and pathogenesis. Lancet Neurol 10(9):853–864. https://doi.org/10.1016/S1474-4422(11)70170-3

Ferreira CFRI, Vaz JA, Vasconcelos MH, Martins A (2010) Compounds from wild mushrooms with antitumor potential. Curr Med Chem Anticancer Agents 10(5):424–436

Ferreira IC, Heleno SA, Reis FS, Stojkovic D, Queiroz MJR, Vasconcelos MH, Sokovic M (2015) Chemical features of *Ganoderma* polysaccharides with antioxidant, antitumor and antimicrobial activities. Phytochemistry 114:38–55. https://doi.org/10.1016/j.phytochem.2014.10.011

Godstime OC, Enwa Felix O, Jewo Augustina O, Eze Christopher O (2014) Mechanisms of antimicrobial actions of phytochemicals against enteric pathogens – a review. J Pharm Chem Biol Sci 2(2):77–85

Gupta S, Dube A, Vyas SP (2007) Antileishmanial efficacy of amphotericin B bearing emulsomes against experimental visceral leishmaniasis. J Drug Target 15(6):437–444. https://doi.org/10.1080/10611860701453836

Gutman J, Kovacs S, Dorsey G, Stergachis A, ter Kuile FO (2017) Safety, tolerability, and efficacy of repeated doses of dihydroartemisinin-piperaquine for prevention and treatment of malaria: a systematic review and meta-analysis. Lancet Infect Dis 17(2):184–193. https://doi.org/10.1016/S1473-3099(16)30378-4

Haque MA, Sarker AK, Paul RK, Khan SS, Islam MAU (2015) Screening for antiparasitic activity of crude extracts of *Pleurotus highking*, a Bangladeshi edible mushroom. Bangladesh Pharm J 18(1):38–41. https://doi.org/10.3329/bpj.v18i1.23512

Hazra B, Saha AK, Ray R, Roy DK, Sur P, Banerjee A (1987) Antiprotozoal activity of diospyrin towards *Leishmania donovani* promastigotes *in vitro*. Trans R Soc Trop Med Hyg 81(5):738–741. https://doi.org/10.1016/0035-9203(87)90013-7

Hazra B, Ghosh R, Banerjee A, Kirby GC, Warhurst DC, Phillipson JD (1995) *In vitro* antiplasmodial effects of diospyrin, a plant-derived naphthoquinoid, and a novel series of derivatives. Phytother Res 9(1):72–74. https://doi.org/10.1002/ptr.2650090118

Heim C, Hertzberg H, Butschi A, Bleuler-Martinez S, Aebi M, Deplazes P, Künzle M, Štefanić S (2015) Inhibition of *Haemonchus contortus* larval development by fungal lectins. Parasit Vectors 8(1):425. https://doi.org/10.1186/s13071-015-1032-x

Herwaldt BL (2001) Laboratory-acquired parasitic infections from accidental exposures. Clin Microbiol Rev 14(4):659–688. https://doi.org/10.1128/CMR.14.3.659-688.2001

Huprikar S, Bosserman E, Patel G, Moore A, Pinney S, Anyanwu A, Neofytos D, Ketterer D, Striker R, Silveira F, Qvarnstrom Y (2013) Donor-derived *Trypanosoma cruzi* infection in solid organ recipients in the United States, 2001–2011. Am J Transplant 13(9):2418–2425. https://doi.org/10.1111/ajt.12340

Hwang IS, Lee J, Jin HG, Woo ER, Lee DG (2012) Amentoflavone stimulates mitochondrial dysfunction and induces apoptotic cell death in *Candida albicans*. Mycopathologia 173(4):207–218. https://doi.org/10.1007/s11046-011-9503-x

Inchausti A, Yaluff G, Rojas de Arias A, Torres S, Ferreira ME, Nakayama H, Schini A, Lorenzen K, Anke T, Fournet A (1997) Leishmanicidal and trypanocidal activity of extracts and secondary metabolites from basidiomycetes. Phytother Res 11:193–197. https://doi.org/10.1002/(SICI)1099-1573(199705)11:3<193::AID-PTR68>3.0.CO;2-R

Isaka M, Kongsaeree P, Thebtaranonth Y (2001) Bioxanthracenes from the insect pathogenic fungus *Cordyceps pseudomilitaris* BCC 1620. J Antibiot 54(1):36–43. https://doi.org/10.7164/antibiotics.54.29

Isaka M, Srisanoh U, Lartpornmatulee N, Boonruangprapa T (2007) ES-242 derivatives and cycloheptapeptides from *Cordyceps* sp. strains BCC 16173 and BCC 16176. J Nat Prod 70(10):1601–1604. https://doi.org/10.1021/np070357h

Isaka M, Srisanoh U, Sappan M, Supothina S, Boonpratuang T (2012) Sterostreins F–O, illudalanes and norilludalanes from cultures of the Basidiomycete *Stereum ostrea* BCC 22955. Phytochemistry 79:116–120. https://doi.org/10.1016/j.phytochem.2012.04.009

Ito SI, Ihara T, Tamura H, Tanaka S, Ikeda T, Kajihara H, Dissanayake C, Abdel-Motaal FF, El-Sayed MA (2007) α-Tomatine, the major saponin in tomato, induces programmed cell death mediated by reactive oxygen species in the fungal pathogen *Fusarium oxysporum*. FEBS Lett 581(17):3217–3222. https://doi.org/10.1016/j.febslet.2007.06.010

Jaturapat A, Isaka M, Hywel-Jones NL, Lertwerawat Y, Kamchonwongpaisan S, Kirtikara K, Tanticharoen M, Thebtaranonth Y (2001) Bioxanthracenes from the insect pathogenic fungus *Cordyceps pseudomilitaris* BCC 1620. J Antibiot 54(1):29–35. https://doi.org/10.7164/antibiotics.54.29

Johnson SM, Murphy RC, Geiger JA, DeRocher AE, Zhang Z, Ojo KK, Larson ET, Perera BGK, Dale EJ, He P, Reid MC (2012) Development of *Toxoplasma gondii* calcium-dependent protein kinase 1 (Tg CDPK1) inhibitors with potent anti-toxoplasma activity. J Med Chem 55(5):2416–2426. https://doi.org/10.1021/jm201713h

Kadhila-Muandingi NP (2010) The distribution, genetic diversity and uses of Ganoderma mushrooms in Oshana and Ohangwena regions of Northern Namibia. Doctoral dissertation

Kadri D, Crater AK, Lee H, Solomon VR, Ananvoranich S (2014) The potential of quinoline derivatives for the treatment of *Toxoplasma gondii* infection. Exp Parasitol 145:135–144. https://doi.org/10.1016/j.exppara.2014.08.008

Kanokmedhakul S, Lekphrom R, Kanokmedhakul K, Hahnvajanawong C, Bua-Art S, Saksirirat W, Prabpai S, Kongsaeree P (2012) Cytotoxic sesquiterpenes from luminescent mushroom *Neonothopanus nambi*. Tetrahedron 68(39):8261–8266. https://doi.org/10.1016/j.tet.2012.07.057

Kaplan RM (2004) Drug resistance in nematodes of veterinary importance: a status report. Trends Parasitol 20(10):477–481. https://doi.org/10.1016/j.pt.2004.08.001

Kaplan RM, Vidyashankar AN (2012) An inconvenient truth: global worming and anthelmintic resistance. Vet Parasitol 186(1–2):70–78. https://doi.org/10.1016/j.vetpar.2011.11.048

Kar A (2007) Pharmaocgnosy and pharmacobiotechnology. New Age International Limited Publishers, New Delhi, p 332

Khan MSA, Ahmad I (2011) Antibiofilm activity of certain phytocompounds and their synergy with fluconazole against *Candida albicans* biofilms. J Antimicrob Chemother 67(3):618–621. https://doi.org/10.1093/jac/dkr512

Khumkomkhet P, Kanokmedhakul S, Kanokmedhakul K, Hahnvajanawong C, Soytong K (2009) Antimalarial and cytotoxic depsidones from the fungus *Chaetomium brasiliense*. J Nat Prod 72(8):1487–1491. https://doi.org/10.1021/np9003189

Kittakoop P, Punya J, Kongsaeree P, Lertwerawat Y, Jintasirikul A, Tanticharoen M, Thebtaranonth Y (1999) Bioactive naphthoquinones from *Cordyceps unilateralis*. Phytochemistry 52(3):453–457. https://doi.org/10.1016/S0031-9422(99)00272-1

Kotze AC, Hunt PW, Skuce P, von Samson-Himmelstjerna G, Martin RJ, Sager H, Krücken J, Hodgkinson J, Lespine A, Jex AR, Gilleard JS (2014) Recent advances in candidate-gene and

whole-genome approaches to the discovery of anthelmintic resistance markers and the description of drug/receptor interactions. Int J Parasitol Drugs Drug Resist 4(3):164–184. https://doi.org/10.1016/j.ijpddr.2014.07.007
Kui-zheng C, Xiao-ye Y, Xiao-liang W, Cheng H, Gui-hua MA, Feng-bao Y, Yan-bin F, Xu-bin Z, De-xin R, Zhen-lian L, Yu-ling Z (2007) Investigation on resistance of gastrointestinal nematodes in sheep and goats to anthelmintics in Ningxia, China. Vet Sci China 6:009
Kun H, Moore A, Mascola L, Frank S, Gena L, Kubak B, Radhakrishna S, Leiby D, Herron R, Mone T, Hunter R (2009) Transmission of *Trypanosoma cruzi* by heart transplantation. Clin Infect Dis 48(11):1534–1540. https://doi.org/10.1086/598931
Lai TK, Biswas G, Chatterjee S, Dutta A, Pal C, Banerji J, Bhuvanesh N, Reibenspies JH, Acharya K (2012) Leishmanicidal and anticandidal activity of constituents of Indian edible mushroom *Astraeus hygrometricus*. Chem Biodivers 9(8):1517–1524. https://doi.org/10.1002/cbdv.201100272
Lakornwong W, Kanokmedhakul K, Kanokmedhakul S, Kongsaeree P, Prabpai S, Sibounnavong P, Soytong K (2014) Triterpene lactones from cultures of Ganoderma sp. KM01. J Nat Prod 77(7):1545–1553. https://doi.org/10.1021/np400846k
Lee W, Lee DG (2014) An antifungal mechanism of curcumin lies in membrane-targeted action within *Candida albicans*. IUBMB Life 66(11):780–785. https://doi.org/10.1002/iub.1326
Lee H, Lee DG (2015) Mode of action of bioactive phytochemicals, plant secondary metabolites, possessing antimicrobial properties. The battle against microbial pathogens: basic science, technological advances and educational programs. FORMATEX, Spain, pp 185–192
Lee BY, Bacon KM, Bottazzi ME, Hotez PJ (2013) Global economic burden of Chagas disease: a computational simulation model. Lancet Infect Dis 13(4):342–348. https://doi.org/10.1016/S1473-3099(13)70002-1
Leliebre-Lara V, Monzote Fidalgo L, Pferschy-Wenzig EM, Kunert O, Nogueiras Lima C, Bauer R (2016) *In Vitro* antileishmanial activity of sterols from *Trametes versicolor* (Bres. Rivarden). Molecules 21(8):1045. https://doi.org/10.3390/molecules21081045
Lindequist U, Niedermeyer TH, Jülich WD (2005) The pharmacological potential of mushrooms. Evid-based Complementary Alternat Med 2(3):285–299. https://doi.org/10.1093/ecam/neh107
Lolli G, Cozza G, Mazzorana M, Tibaldi E, Cesaro L, Donella-Deana A, Meggio F, Venerando A, Franchin C, Sarno S, Battistutta R (2012) Inhibition of protein kinase CK2 by flavonoids and tyrphostins. A structural insight. Biochemistry 51(31):6097–6107. https://doi.org/10.1021/bi300531c
Lopes JN, Cruz FS, Docampo R, Vasconcellos ME, Sampaio MCR, Pinto AV, Gilbert B (1978) *In vitro* and *in vivo* evaluation of the toxicity of 1, 4-naphthoquinone and 1, 2-naphthoquinone derivatives against *Trypanosoma cruzi*. Ann Trop Med Parasitol 72(6):523–531. https://doi.org/10.1080/00034983.1978.11719356
Lourido S, Tang K, Sibley LD (2012) Distinct signalling pathways control *Toxoplasma* egress and host-cell invasion. EMBO J 31(24):4524–4534. https://doi.org/10.1038/emboj.2012.299
Lu F, Culleton R, Zhang M, Ramaprasad A, Von Seidlein L, Zhou H, Zhu G, Tang J, Liu Y, Wang W, Cao Y (2017) Emergence of indigenous artemisinin-resistant *Plasmodium falciparum* in Africa. N Engl J Med 376(10):991–993. https://doi.org/10.1056/NEJMc1612765
Ma K, Li L, Bao L, He L, Sun C, Zhou B, Si S, Liu H (2015) Six new 3, 4-seco-27-norlanostane triterpenes from the medicinal mushroom *Ganoderma boninense* and their antiplasmodial activity and agonistic activity to LXRβ. Tetrahedron 71(12):1808–1814. https://doi.org/10.1016/j.tet.2015.02.002
Mallick S, Dutta A, Dey S, Ghosh J, Mukherjee D, Sultana SS, Mandal S, Paloi S, Khatua S, Acharya K, Pal C (2014) Selective inhibition of *Leishmania donovani* by active extracts of wild mushrooms used by the tribal population of India: an *in vitro* exploration for new leads against parasitic protozoans. Exp Parasitol 138:9–17. https://doi.org/10.1016/j.exppara.2014.01.002
Miron D, Battisti F, Silva FK, Lana AD, Pippi B, Casanova B, Gnoatto S, Fuentefria A, Mayorga P, Schapoval EE (2014) Antifungal activity and mechanism of action of monoterpenes against dermatophytes and yeasts. Rev Bras Farmacognosia 24(6):660–667. https://doi.org/10.1016/j.bjp.2014.10.014

Moyo B, Masika PJ, Muchenje V (2012) Antimicrobial activities of *Moringa oleifera* lam leaf extracts. Afr J Biotechnol 11(11):2797–2802. https://doi.org/10.5897/AJB10.686

Nogueira CR, Lopes LM (2011) Antiplasmodial natural products. Molecules 16(3):2146–2190. https://doi.org/10.3390/molecules16032146

Ojo KK, Larson ET, Keyloun KR, Castaneda LJ, DeRocher AE, Inampudi KK, Kim JE, Arakaki TL, Murphy RC, Zhang L, Napuli AJ (2010) *Toxoplasma gondii* calcium-dependent protein kinase 1 is a target for selective kinase inhibitors. Nat Struct Mol Biol 17(5):602. https://doi.org/10.1038/nsmb.1818

Oluba OM, Josiah SJ, Adebisi KE, Ojeaburu SI, Onyeneke EC (2017) Antiplasmodial and antioxidant activities of chloroform extract of *Ganoderma lucidum* fruit body in *Plasmodium berghei*-infected mice. Orient Pharm Exp Med 17(4):389–395. https://doi.org/10.1007/s13596-017-0288-4

Omojate Godstime C, Enwa Felix O, Jewo Augustina O, Eze Christopher O (2014) Mechanisms of antimicrobial actions of phytochemicals against enteric pathogens – a review. J Pharm Chem Biol Sci 2(2):77–85

Omoya FO, Oyedirin O (2018) *In vivo* prophylactic evaluation of the antiplasmodial property of ethanol and hot water extracts of milky mushroom (*Calocybe indica*). J Adv Microbiol 8(2):2456–7116. https://doi.org/10.9734/JAMB/2018/39198

Omoya OF, Momoh AO, Adegoke OA (2017) *In Vivo* evaluation of the antiplasmodial property of ethanol and hot water extracts of *Pleurotus ostreatus* mushroom. Sci J Public Health 5(3):178–185. https://doi.org/10.11648/j.sjph.20170503.14

Panthama N, Kanokmedhakul S, Kanokmedhakul K, Soytong K (2011) Cytotoxic and antimalarial azaphilones from *Chaetomium longirostre*. J Nat Prod 74(11):2395–2399. https://doi.org/10.1021/np2004903

Paredes R, Munoz J, Diaz I, Domingo P, Gurgui M, Clotet B (2003) Leishmaniasis in HIV infection. Postgrad Med 49(1):39. https://doi.org/10.4103/0022-3859.929

Penny MA, Verity R, Bever CA, Sauboin C, Galactionova K, Flasche S, White MT, Wenger EA, Van de Velde N, Pemberton-Ross P, Griffin JT (2016) Public health impact and cost-effectiveness of the RTS, S/AS01 malaria vaccine: a systematic comparison of predictions from four mathematical models. Lancet 387(10016):367–375. https://doi.org/10.1016/S0140-6736(15)00725-4

Pérez-Molina JA, Molina I (2017) Chagas disease. Lancet 91(10115):82–94. https://doi.org/10.1016/S0140-6736(17)31612-4

Pérez-Molina, J. A., & Molina, I. (2018). Chagas disease. The Lancet, 391(10115), 82-94

Phillips MA, Burrows JN, Manyando C, van Huijsduijnen RH, Van Voorhis WC, Wells TNC (2017) Malaria. Nat Rev Dis Primers 3:17050. https://doi.org/10.1038/nrdp.2017.51

Pineda-Alegría JA, Sánchez-Vázquez JE, González-Cortazar M, Zamilpa A, López-Arellano ME, Cuevas-Padilla EJ, Mendoza-de-Gives P, Aguilar-Marcelino L (2017) The edible mushroom *Pleurotus djamor* produces metabolites with lethal activity against the parasitic nematode *Haemonchus contortus*. J Med Food 20(12):1184–1192. https://doi.org/10.1089/jmf.2017.0031

Powers JL, Zhang X, Kim CY, Abugri DA, Witola WH (2017) Activity of green algae extracts against *Toxoplasma gondii*. Med Aromat Plants (Los Angels) 6(293):2167–0412. https://doi.org/10.4172/2167-0412.1000293

Ramezani P, Hejazi SH, Narimani M, Soleimanifard S (2017) *In vitro* antileishmanial activity and apoptosis induction of *Pleurotus ostreatus* alcoholic extract on *Leishmania major*. Res J Pharmacognosy (RJP) 4(3):51–58

Ramos-Ligonio A, López-Monteon A, Trigos Á (2012) Trypanocidal activity of ergosterol peroxide from *Pleurotus ostreatus*. Phytother Res 26(6):938–943. https://doi.org/10.1002/ptr.3653

Rao A, Zhang Y, Muend S, Rao R (2010) Mechanism of antifungal activity of terpenoid phenols resembles calcium stress and inhibition of the TOR pathway. Antimicrob Agents Chemother 54(12):5062–5069. https://doi.org/10.1128/AAC.01050-10

Riarte A, Luna C, Sabatiello R, Sinagra A, Schiavelli R, Rissio AD, Maiolo E, Garcia MM, Jacob N, Pattin M, Lauricella M (1999) Chagas' disease in patients with kidney transplants: 7 years of experience, 1989–1996. Clin Infect Dis 29(3):561–567. https://doi.org/10.1086/598634

Rohrer LUCIA, Winterhalter KH, Eckert J, Köhler P (1986) Killing of *Giardia lamblia* by human milk is mediated by unsaturated fatty acids. Antimicrob Agents Chemother 30(2):254–257. https://doi.org/10.1128/AAC.30.2.254

Shen AY, Chen CP, Roffler S (1999) A chelating agent possessing cytotoxicity and antimicrobial activity: 7-morpholinomethyl-8-hydroxyquinoline. Life Sci 64(9):813–825. https://doi.org/10.1016/S0024-3205(98)00623-7

Siden-Kiamos I, Ecker A, Nybäck S, Louis C, Sinden RE, Billker O (2006) *Plasmodium berghei* calcium-dependent protein kinase 3 is required for ookinete gliding motility and mosquito midgut invasion. Mol Microbiol 60(6):1355–1363. https://doi.org/10.1111/j.1365-2958.2006.05189.x

Sitheeque MAM, Panagoda GJ, Yau J, Amarakoon AMT, Udagama URN, Samaranayake LP (2009) Antifungal activity of black tea polyphenols (catechins and theaflavins) against *Candida* species. Chemotherapy 55(3):189–196. https://doi.org/10.1159/000216836

Souza-Fagundes EM, Cota BB, Rosa LH, Romanha AJ, Corrêa-Oliveira R, Rosa CA, Zani CL, Teixeira-Carvalho A, Martins-Filho OA (2010) *In vitro* activity of hypnophilin from *Lentinus strigosus:* a potential prototype for Chagas disease and leishmaniasis chemotherapy. Braz J Med Biol Res 43(11):1054–1061. https://doi.org/10.1590/S0100-879X2010007500108

Sung WS, Lee IS, Lee DG (2007) Damage to the cytoplasmic membrane and cell death caused by lycopene in *Candida albicans*. J Microbiol Biotechnol 17(11):1797–1804

Sutherland CJ, Lansdell P, Sanders M, Muwanguzi J, van Schalkwyk DA, Kaur H, Nolder D, Tucker J, Bennett HM, Otto TD, Berriman M (2017) Pfk13-independent treatment failure in four imported cases of *Plasmodium falciparum* malaria given artemether-lumefantrine in the UK. Antimicrob Agents Chemother 61:e02382. https://doi.org/10.1128/AAC.02382-16

Ustun O, Kaiser M, Tasdemir D (2011) Antiprotozoal and cytotoxic activities of some mushrooms from Turkey. Planta Med 77(12):PL32. https://doi.org/10.1055/s-0031-1282681

Valadares DG, Duarte MC, Oliveira JS, Chávez-Fumagalli MA, Martins VT, Costa LE, Leite JP, Santoro MM, Régis WC, Tavares CA, Coelho EA (2011) Leishmanicidal activity of the *Agaricus blazei* Murill in different Leishmania species. Parasitol Int 60(4):357–363. https://doi.org/10.1016/j.parint.2011.06.001

Vieira TM, Fonseca LD, Bastos GA, de Oliveira VV, Silva MLF, Morais-Costa F, de Paiva Ferreira AV, de Oliveira NJF, Duarte ER (2017) Control of *Haemonchus contortus* in sheep using basidiocarps of *Agaricus blazei* Murril. Vet Res Commun 41(2):99–106. https://doi.org/10.1007/s11259-017-9677-x

Wang C, Li F, Zhang Z, Yang X, Ahmad AA, Li X, Du A, Hu M (2017) Recent research progress in China on *Haemonchus contortus*. Front Microbiol 8:1509. https://doi.org/10.3389/fmicb.2017.01509

Witola WH, Cooks-Fagbodun S, Ordonez AR, Matthews K, Abugri DA, McHugh M (2016) Knockdown of phosphoethanolamine transmethylation enzymes decreases viability of *Haemonchus contortus*. Vet Parasitol 223:1–6. https://doi.org/10.1016/j.vetpar.2016.04.008

Wolday D, Berhe N, Akuffo H, Britton S (1999) Leishmania–HIV interaction: immunopathogenic mechanisms. Parasitol Today 15(5):182–187. https://doi.org/10.1016/S0169-4758(99)01431-3

Wongsa P, Tasanatai K, Watts P, Hywel-Jones N (2005) Isolation and in vitro cultivation of the insect pathogenic fungus *Cordyceps unilateralis*. Mycol Res 109(8):936–940. https://doi.org/10.1017/S0953756205003321

World Health Organization (2013) World malaria report 2013. Accessed 7 June 2018

World Health Organization (WHO) (2010) Control of the leishmaniasis: report of a meeting of the WHO Expert Committee on the Control of Leishmaniosis. Geneva, 22–26 March 2010. WHO Technical Report Series; no. 949

World Health Organization (WHO) (2017) World malaria report 2017. http://www.who.int/malaria/media/world-malaria-report-2017/en/. Accessed 7 May 2018

World Health Organization (WHO) (2018a) Facts sheets of Malaria. http://www.who.int/news-room/fact-sheets/detail/malaria

World Health Organization (WHO) (2018b) Facts sheets of Leishmaniosis. http://www.who.int/news-room/fact-sheets/detail/leishmaniasis

Wright CW, Phillipson JD (1990) Natural products and the development of selective antiprotozoal drugs. Phytother Res 4(4):127–139. https://doi.org/10.1002/ptr.2650040402

Yun J, Lee H, Ko HJ, Woo ER, Lee DG (2015) Fungicidal effect of isoquercitrin via inducing membrane disturbance. BBA Biomembranes 1848(2):695–701. https://doi.org/10.1016/j.bbamem.2014.11.019

Zhao S, Guo YX, Liu QH, Wang HX, Ng TB (2009) Lectins but not antifungal proteins exhibit anti-nematode activity. Environ Toxicol Pharmacol 28(2):265–268. https://doi.org/10.1016/j.etap.2009.05.003

Zhao J, Pu W, Zhan Z, Yue C (2010) Anthelmintic resistance of Trichostrongylidae to benzimidazoles. Zhongguo Shouyi Ke-ji 40(5):528–531

Antiviral Potency of Mushroom Constituents

10

Prabin Pradeep, Vidya Manju, and Mohammad Feraz Ahsan

Contents

Abstract

Mushrooms have various bioactive compounds, which have anticancer, antibacterial, antifungal, antiviral, antioxidant, and anti-inflammatory compounds. The limitation of natural antiviral compounds in the market and the viruses becoming resistant to currently available antivirals warranted the need for new antivirals with fewer side effects. Although many different compounds from various mushrooms (both edible and nonedible) have been isolated and shown to have antiviral effects, successful introduction of a new antiviral to the market requires extensive studies and clinical trials. Antiviral compounds isolated from the

Authors Prabin Pradeep and Vidya Manju have been equally contributed to this chapter.

P. Pradeep · V. Manju · M. F. Ahsan (✉)
Inter University Centre for Biomedical Research & Super Speciality Hospital, Kottayam, Kerala, India

D. C. Agrawal, M. Dhanasekaran (eds.), *Medicinal Mushrooms*,
https://doi.org/10.1007/978-981-13-6382-5_10

mushrooms have shown potential activity against prominent viruses such as human immunodeficiency virus, influenza, herpes simplex virus, hepatitis B and C viruses, etc. State of the art suggests more in-depth exploration of mushrooms for more potent antivirals and other lifesaving drugs.

Keywords
Edible mushrooms · Poisonous mushrooms · Antivirals · *Basidiomycetes* · Natural products

Abbreviations

AZT	Zidovudine
DNA	Deoxyribonucleic acid
EBV	Epstein-Barr virus
EMCV	Encephalomyocarditis virus
FMDV	Foot-and-mouth disease virus
GLPG	*Ganoderma lucidum* proteoglycan
GLTA	*Ganoderma lucidum* triterpenoids Lanosta-7,9(11),24-trien-3-one, 15;26- dihydroxy
HBV	Hepatitis B virus
HCV	Hepatitis C virus
HIV	Human immunodeficiency virus
HSV	Herpes simplex virus
JLS	Water-soluble lignin-rich fraction
KS-2	Extract from culture mycelia of *Lentinus edodes*
LAC	Laccase
NA	Neuraminidase
NDV	Newcastle disease virus
RNA	Ribonucleic acid
RSV	Respiratory syncytial virus
RT	Reverse transcriptase
sAAP	Sulfated *Auricularia auricula* polysaccharides
VSV	Vesicular stomatitis virus
VZV	Varicella zoster virus

10.1 Introduction

10.1.1 Mushrooms

The term mushroom was used for the identification of edible sporophores, while toadstool was generally used for inedible and poisonous sporophores. As there was no scientific distinction between both terms, the term mushroom is applicable for any umbrella-shaped fruiting body-containing fungi. The umbrella-shaped sporophores are mainly found in the family *Agaricaceae*. They bear thin, gill-like structure underside the cap from where the spores are shed.

The *Agarics* and *Boletes* include most known mushrooms. Other groups of fungi considered to be mushrooms are hydnums or hedgehog mushrooms, polypores, shelf fungi or bracket fungi, calvatias, cantharelloid fungi, puffballs, stinkhorns, earthstars, bird's nest fungi, morels, false morels, or lorchels. Other unusual forms, which are not closely related to true mushrooms but are included with them, are the jelly fungi, the ear fungi or Jew's ear, and edible truffle. Mushrooms contain small quantities of amino acids and vitamin B and are also free of cholesterol. The commercially grown common mushroom is a good source of nutrients such as carbohydrate and protein, low in fat with trace amount of mineral salt and vitamins and abundant of water content, making it a healthy intake. Currently more than 10,000 mushrooms are known to exist in the world (Kara Rogers, Encyclopaedia Britannica 1999).

10.1.2 Antivirals

An agent that kills a virus or has the ability to suppress the replication and therefore inhibits its capability to propagate is called as antiviral. The development of antivirals has been lagging far behind compared to that of antibiotics. The main reason is that a virus is just a genetic material, either a deoxyribonucleic acid (DNA) or ribonucleic acid (RNA), maybe with a few enzymes, wrapped within a protein coat, and sometimes an extra lipid envelope. The fact that viruses use host machinery for replicating makes targeting the virus more difficult without harming the cell. Along with that some viruses can stay dormant in the body of the host without replicating, therefore able to avoid the drugs that inhibit replication. It is very difficult to find agents or compounds, which are specific to viruses as they share most of the metabolic processes of the infected host cell (Drugs.com, Medicinenet.com).

In Table 10.1, the generic drugs or formulations currently available as antiviral agents are provided along with their original sources. *AIDS* acquired immunodeficiency syndrome, *CMV* cytomegalovirus, *EBV* Epstein-Barr virus, *HIV* human immunodeficiency virus, *HSV* herpes simplex virus, *RSV* respiratory syncytial virus, *VZV* varicella zoster virus, *HTLV* human T-cell lymphotropic virus.

Table 10.2 describes certain products isolated from natural sources that have been shown to possess antiviral activity. The antiviral elements or the targets are also mentioned. *HBV* hepatitis B virus, *HCV* hepatitis C virus, *HTLV* human T-cell lymphotropic virus and *RSV* respiratory syncytial virus.

Table 10.1 Synthetic antiviral drugs

Generics/active compounds	Target/activity	Source/ References
Acyclovir	Acyclovir acts against HSV mainly. It is also active against other *herpes* viruses such as VZV, CMV, and EBV	De Clercq and Li (2016)
Valacyclovir	Used for the treatment and suppression of genital *herpes* infection	De Clercq and Li (2016)
Famciclovir	Used for the treatment and suppression of genital *herpes* infection	De Clercq and Li (2016)
Ganciclovir	Ganciclovir has potent in vitro activity against all *herpes* viruses, including CMV	De Clercq and Li (2016)
Ribavirin	It has activity against adenoviruses; herpesviruses; CMV; vaccinia; influenza A and B; parainfluenza 1, 2, and 3; measles; mumps; RSV; rhinovirus; Lassa fever; hantavirus disease; and hepatitis C	De Clercq and Li (2016)
Zidovudine (AZT)	AZT is active in vitro against many human retroviruses, including HTLV-I and HIV	De Clercq and Li (2016)
Amantadine	Amantadine is only effective against influenza A, and some naturally occurring strains of influenza A are resistant to it	De Clercq and Li (2016)
Zanamivir	Zanamivir and oseltamivir are licensed for the treatment of influenza A and B infections	De Clercq and Li (2016)

10.2 Methods

The search for relevant articles was performed using Boolean operators in PubMed with Medical Subject Headings with different combinations of keywords like "mushroom and antiviral," "mushroom and HIV," "basidiomycetes and antiviral," and "mushroom and virus inhibitors." Other combinations of keywords were also used to cross-check the database for relevant article. Search includes the articles published till December 2017.

Duplicate citations were removed and relevant papers were further selected for full text analysis. Search included all types of studies including research findings, review, and clinical trials, with no language restrictions. Articles in languages other than English were translated and incorporated for the studies.

10.2.1 Inclusion Criteria

1. Studies conducted to prevent viral diseases using mushroom
2. Studies based on antiviral activity of whole mushroom or a compound isolated from mushroom

Table 10.2 Natural antiviral drugs

Products	Sources	Target/ activity	References
Fulvic acid	Cellulose	Enveloped viruses	Van Rensburg et al. (2010), Kotwal (2008), and Kotwal et al. (2005)
Pomegranate polyphenols	Pomegranate	Enveloped viruses	Kotwal (2008), Neurath et al. (2005), and Sundararajan et al. (2010)
Resveratrol+	Red vine leaves	HIV and EBV	Heredia et al. (2000) and De Leo et al. (2012)
Sesquiterpenoids and alkaloids	Roots of *Alangium chinense*	HCV	Zhang et al. (2013)
Longumosides and amide alkaloids	*Piper longum*	HBV	Jiang et al. (2013)
Epigallocatechin lucidone	Green tea	HBV	Xu et al. (2008)
Lectins	*Lindera erythrocarpa*	HCV	Chen et al. (2013)
3-Hydroxy caruilignan	Red and blue green algae	HCV	Takebe et al. (2013) and Wu et al. (2012)
Hypericin	St. John's wort	HCV	Jacobson et al. (2001)
Cimifugin carnosic acid	*Cimicifuga foetida*	RSV	Wang et al. (2012) and Shin et al. (2013)
Polybromocatechol	Rhodophyta	Rhinovirus	Park et al. (2012)
Sesquiterpene coumarins	*Ferula assa-foetida*	Influenza	Lee et al. (2009)
Chalcones	*Glycyrrhiza* leaflets	Influenza	Dao et al. (2011)
Diarylheptanoids	*Alpinia katsumadai*	Influenza	Grienke et al. (2010)
Griffithsin	*Griffithsia* (Red algae)	HIV	Férir et al. (2011)

10.2.2 Exclusion Criteria

1. Studies on mushroom other than antiviral activities
2. Studies without clearly specified results on mushroom

After applying exclusion criteria for the searched articles, the relevant details were analyzed and tabulated. These were categorized into compounds extracted from mushroom showing antiviral activity, into mushroom extracts against specific viruses, and into edible or nonedible mushrooms showing antiviral activity.

10.3 Results

The search in PubMed database using different combinations of keywords resulted in 2620 hits. After eliminating 1561 duplicate hits, 1059 papers were selected for screening. On applying inclusion and exclusion criteria, relevant citations identified were 218. Among the several combinations of keywords used, "mushroom *and* antiviral" resulted in maximum hits.

Other search hits and combinations of keywords resulting in duplications and papers other than antiviral activity and redundant hits were excluded. Search included all types of mushrooms including medicinal mushrooms. Research papers related to antiviral activity of mushroom were selected based on mushroom compounds or extracts showing antiviral activity.

Both edible and nonedible mushrooms for antiviral activity were considered. Certain nonedible mushrooms possess harmful compounds but with antiviral activity. Selected studies were further analyzed and categorized into compounds showing antiviral activity to specific virus or group of viruses. Among different groups of mushrooms, basidiomycetes showed maximum antiviral activity.

10.3.1 Human Immunodeficiency Virus

Several varieties of mushrooms have shown antiviral activity against HIV. Most of the activity is against reverse transcriptase (RT) and protease enzymes and against a few unspecified targets. A low molecular weight laccase (LAC) from *Tricholoma giganteum* inhibited RT activity of HIV-1 (Wang and Ng 2004a, b). LAC from varieties of mushrooms have been found to be inhibitory for the RT of HIV-1, viz., from the dried fruiting body of *Hericium erinaceus* (Wang and Ng 2004a, b), fruiting body of *Pleurotus eryngii* (Wang and Ng 2006b), *Lentinus edodes*, and *Ganoderma lucidum* (Sun et al. 2011; Wang and Ng 2006a, b). Marmorin from mushroom *Hypsizygus marmoreus* is a ribosome inactivation protein, which also showed HIV-1 RT inhibitory activity (Wong et al. 2008).

The HIV-1 RT inhibitory activity has been found in hemagglutinin from *Cordyceps militaris* (Wong et al. 2009); lectin isolated from the toxic mushroom *Inocybe umbrinella* and *Pholiota adiposa*; schizolysin, a hemolysin from *Schizophyllum commune* mushroom; and metalloprotease from *Lepista nuda* (Han et al. 2010; Wu et al. 2011). A novel polysaccharide-peptide complex from *Pleurotus abalonus* and a fraction isolated from *Russula paludosa* inhibited the RT of HIV-1 (Li et al. 2012; Wang et al. 2007). Lectins from *Pleurotus citrinopileatus* and *Russula delica* and a novel ribonuclease from *Hohenbuehelia serotina* have also been shown to inhibit the RT of HIV-1 (Li et al. 2008; Zhao et al. 2010; Zhang et al. 2014a, b).

A water-soluble high molecular weight lignin from the mushroom, *Fuscoporia oblique*, showed inhibitory effect against HIV-1 protease (Ichimura et al. 1998). A ubiquitin-like protease from *Pleurotus ostreatus* and lanostane triterpenes isolated from *Ganoderma colossus* showed dose-dependent inhibition of the HIV-1 protease activity (Wang and Ng 2000; El Dine et al. 2008).

Polycarboxylated water-solubilized lignin from the mycelium of *Lentinus edodes* showed antiviral activity against HIV; it inhibited the HIV antigen expression in HIV-infected murine equivalent of human T4 cells (Suzuki et al. 1990). *Inonotus obliquus* a parasitic birch fungi and *Lentinus edodes* mycelium extracts have shown potent antiviral activity against HIV-1 (Shibnev et al. 2015; Tochikura et al. 1988). Several triterpenoids isolated from the mushroom *Ganoderma lucidum* have shown to have potent anti-HIV-1 activity (Lindequist et al. 2005).

Table 10.3 describes the extracts/compounds from mushroom that have been shown to possess antiviral activity and target enzymes against HIV. LB-1b polysaccharide-peptide complex, SU2 fraction isolated from *Russula paludosa*.

10.3.2 Herpes Simplex Virus

Subfractions of *Inonotus obliquus* have been shown to protect Vero cells against HSV when added before or 1 hour within the addition of virus cultures (Polkovnikova et al. 2014). *Rozites caperata* Peptide 28 (RC28), a novel protein from the mushroom extract in phosphate-buffered saline, showed potent antiviral activity against HSV (Gong et al. 2009). Acidic protein-bound polysaccharide isolated from water-soluble extracts of *Ganoderma lucidum* showed potent antiviral activity against both HSV-1 and HSV-2 (Eo et al. 2000). A proteoglycan *Ganoderma lucidum* proteoglycan (GLPG) inhibited either the entry or the attachment to target cells and successfully inhibited HSV-1 and HSV-2 replication and infection in host cells (Liu et al. 2004). Culture broth of basidiomycete *Macrocystidia cucumis* showed antiviral response against HSV-1 in baby hamster kidney cells (Saboulard et al. 1998).

A novel antiviral protein isolated from *Grifola frondosa* fruiting bodies (*Grifola frondosa* anti-HSV-1 protein, GFAHP) showed potent antiviral activity against HSV-1 (Gu et al. 2007). *Agaricus brasiliensis* mycelial polysaccharide was chemically modified by sulfation and used against HSV-1; it showed effective inhibition in attachment, entry, and cell-to-cell transfer (Cardozo et al. 2013). Illudin S isolated from fruiting bodies of *Omphalotus illudens* showed antiviral activity against HSV (Lehmann et al. 2003). A clinical study where a mixture of *Wisteria floribunda*, *Trapa natans*, *Terminalia chebula*, *Coicis lacryma-jobi*, *Ganoderma lucidum*, and *Elfuinga applanata* extracts is given to 15 HSV patients showed to reduce recurrent symptoms (Hijikata et al. 2007). Several compounds from *Ganoderma pfeifferi* akin to *Ganoderma lucidum* have been shown to have antiviral effect against HSV-1 (Lindequist et al. 2015). Lanostane-type triterpenoid derived from *Scleroderma citrinum* was also shown to have antiviral response against HSV (Kanokmedhakul et al. 2003).

Aqueous extract from the mushroom *Inonotus obliquus* inhibits the entry of HSV into cells by acting on its glycoproteins (Pan et al. 2013). Water-soluble lignin-rich fraction (JLS-S001) of *Lentinus edodes* blocked HSV virus, probably by interfering with the assembly and budding of the virus (Sarkar et al. 1993). *Pleurotus ostreatus*,

Table 10.3 Antiviral mushrooms against HIV

Mushrooms	Extracts/compounds	Virus	Target/activity	References
Lentinus edodes	Polycarboxylated water-solubilized lignin	HIV	Antigen expression	Suzuki et al. (1990)
Fuscoporia oblique	Water-soluble lignin	HIV-1	Protease	Ichimura et al. (1998)
Inonotus obliquus	–	HIV-1	–	Shibnev et al. (2015)
Pleurotus ostreatus	Ubiquitin-like protein	HIV-1	Protease	Wang and Ng (2000)
Tricholoma giganteum	LAC	HIV-1	RT	Wang and Ng (2004a, b)
Hericium erinaceus	LAC	HIV-1	RT	Wang and Ng (2004a, b)
Pleurotus eryngii	LAC	HIV-1	RT	Li and Wang (2006); Wang and Ng (2006b)
Lentinus edodes	LAC	HIV-1	RT	Sun et al. (2011)
Ganoderma lucidum	LAC	HIV-1	RT	Wang et al. (2006)
Ganoderma lucidum	Several triterpenoids	HIV	–	Lindequist et al. (2005)
Ganoderma colossus	Lanostane triterpenes	HIV-1	Protease	El Dine et al. (2008)
Hypsizygus marmoreus	Marmorin	HIV-1	RT	Wong et al. (2008)
Cordyceps militaris	Hemagglutinin	HIV-1	RT	Wong et al. (2009)
Inocybe umbrinella	Lectin	HIV-1	RT	Zhao et al. (2009)
Pholiota adipose	Lectin	HIV-1	RT	Zhang et al. (2009)
Schizophyllum commune	Schizolysin	HIV-1	RT	Han et al. (2010)
Lepista nuda	Metalloprotease	HIV-1	RT	Wu et al. (2011)
Pleurotus abalonus	LB-1b	HIV-1	RT	Li et al. (2012)
Russula paludosa	SU2	HIV-1	RT	Wang et al. (2007)
Pleurotus citrinopileatus	Lectin	HIV-1	RT	Li et al. (2008)
Russula delica	Lectin	HIV-1	RT	Zhao et al. (2010)
Hohenbuehelia serotina	Ribonuclease	HIV-1	RT	Zhang et al. (2014a, b)

Fomes fomentarius, *Auriporia aurea*, and *Trametes versicolor* were promising in blocking HSV in RK-13 cells (Krupodorova et al. 2014). A cloned peptide, RC28, from *Rozites caperata* showed excellent antiviral activity against HSV in Vero cells; the same group also showed antiviral activity in a herpes simplex virus-1 mouse keratitis model (Yan et al. 2015). Sulfated derivative of *Agaricus brasiliensis* fruiting bodies (FR-S) inhibited HSV-1 and HSV-2 and also has synergistic effect with acyclovir (Cardozo et al. 2014). Cardozo et al. showed that sulfated derivative of *Agaricus brasiliensis* mycelia (MI-S), another sulfated derivative from *Agaricus brasiliensis*, also inhibited HSV-1 and HSV-2 in similar way and synergistic with acyclovir (Cardozo et al. 2011). A protein named RC-183 purified from *Rozites caperata* inhibited HSV-1 and HSV-2 (Piraino and Brandt 1999).

Table 10.4 describes the extracts/compounds from mushroom that have been shown to possess antiviral activity and target enzymes against HSV. APBP, acidic protein-bound polysaccharide; GFAHP, *Grifola frondosa* anti-HSV-1 protein; JLS, water-soluble lignin-rich fraction; MI-S sulfated derivative of *Agaricus brasiliensis* mycelia; RC 28, *Rozites caperata* peptide 28.

10.3.3 Influenza Virus

Phellinus linteus extract used as an adjuvant for H5N1 vaccine reassortant prepared by reverse genetics from A/Vietnam/1194/2004 (H5N1) virus and the influenza A virus strain A/Puerto Rico/8/1934 H1N1 (PR8) HA vaccine showed cross-immunity and protection (Ichinohe et al. 2010). Five polyphenols isolated from *Phellinus baumii* through ethanolic extract identified as hispidin, hypholomine B, inoscavin A, davallialactone, and phelligridin D inhibited the neuraminidase (NA) of influenza A H1N1, H5N1, and H3N2, successfully reducing the cytopathic effect in Madin-Darby canine kidney cells (Hwang et al. 2015). *Phellinus igniarius* water extract pre-treated cells were shown to have protected the cells from influenza virus infection (Lee et al. 2013). *Auriporia aurea*, *Flammulina velutipes*, *Fomes fomentarius*, *Ganoderma lucidum*, *Lentinus edodes*, *Lyophyllum shimeji*, *Pleurotus eryngii*, *Pleurotus ostreatus*, *Schizophyllum commune*, and *Trametes versicolor* extracts were tested for inhibition against influenza A H1N1 virus, and all of them inhibited the H1N1 virus in Madin-Darby canine kidney cells with varying titers (Krupodorova et al. 2014).

Two phenolic compounds from *Inonotus hispidus* showed antiviral action against influenza A and B virus (Awadh et al. 2003). Teplyakova et al. studied the antiviral activity of 11 fungal species from Altai Mountains, Russia, and found *Datronia mollis*, *Daedaleopsis confragosa*, *Ischnoderma benzoinum*, *Trametes versicolor*, *Trametes gibbosa*, *Laricifomes officinalis*, *and Lenzites betulina* as promising antivirals against H5N1 and H3N2 (Teplyakova et al. 2012). Extract from culture mycelia of *Lentinus edodes* (KS-2), a peptidomannan isolated from *Lentinus edodes* and tested in mice by intraperitoneal and oral routes, showed increased interferon levels in mice and was also inhibitory for influenza A H2N2 virus (Suzuki et al. 1979). Several compounds from *Ganoderma pfeifferi* were tested and found to have

Table 10.4 Antiviral mushrooms against HSV

Mushrooms	Extracts/ compounds	Virus	Target/activity	References
Inonotus obliquus	–	HSV	–	Polkovnikova et al. (2014)
Rozites caperata	RC28	HSV	–	Gong et al. (2009)
Ganoderma lucidum	APBP	HSV-1 and HSV-2	–	Eo et al. (2000)
Ganoderma lucidum	GLPG	HSV-1 and HSV-2	Entry/attachment	Liu et al. (2004)
Macrocystidia cucumis	–	HSV-1	–	Saboulard et al. (1998)
Grifola frondosa	GFAHP	HSV-1	–	Gu et al. (2007)
Agaricus brasiliensis	Sulfated polysaccharide	HSV-1	Attachment/entry/ cell-to-cell spread	Cardozo et al. (2013)
Omphalotus illudens	Illudin S	HSV-1	–	Lehmann et al. (2003)
Ganoderma pfeifferi	Several compounds	HSV-1	–	Lindequist et al. (2015)
Scleroderma citrinum	Triterpenoid	HSV	–	Kanokmedhakul et al. (2003)
Inonotus obliquus	Aqueous extract	HSV	Entry	Pan et al. (2013)
Lentinus edodes	JLS-S001	HSV	Assembly/budding	Sarkar et al. (1993)
Pleurotus ostreatus	–	HSV	–	Hijikata et al. (2007)
Fomes fomentarius	–	HSV	–	Hijikata et al. (2007)
Auriporia aurea	–	HSV	–	Hijikata et al. (2007)
Trametes versicolor	–	HSV	–	Hijikata et al. (2007)
Rozites caperata	RC28	HSV-1	–	Yan et al. (2015)
Agaricus brasiliensis	FR-S/MI-S	HSV-1 and HSV-2	–	Cardozo et al. (2011, 2014)
Rozites caperata	RC-183	HSV-1 and HSV-2		Piraino and Brandt (1999)

antiviral activity against influenza virus like applanoxidic acid G, lucialdehyde B, lucidadiol, and ergosta-7, 22-diene-3b-ol (Lindequist et al. 2015).

Cryptoporus volvatus tested against influenza A H1N1 and H3N2 demonstrated inhibition and reduction of viral effects (Gao et al. 2014). Sesquiterpenoid isolated from *Phellinus igniarius* inhibited the neuraminidase activity of influenza virus (Song et al. 2014). *Poria cocos* polysaccharide-II (PCP-II) elicited better immune response when administered as an adjuvant along with influenza vaccine (Wu et al. 2016). *Inonotus obliquus* polysaccharides were tested for their antiviral activity against a number of feline viruses and found to be active against feline influenza H3N2 and H5N6 (Tian et al. 2017). Supplementation of C57 black six mice with hexose-related compound from basidiomycetes mushroom increased the survival rate of mice infected with influenza virus (Nogusa et al. 2009). The lectin from *Agrocybe aegerita* was tested as an adjuvant for influenza vaccination and indicated that the IgG level was higher than that of non-adjuvanted vaccine (Ma et al. 2017).

Table 10.5 describes the extracts/compounds from mushroom that have been shown to possess antiviral activity and target enzymes against influenza virus. KS-2, extract from culture mycelia of *Lentinus edodes*; NA neuraminidase; PCP-II, *Poria cocos* polysaccharide-II.

10.3.4 Hepatitis Virus

A clinical study of *Agaricus blazei* Murill extract administration in hepatitis B patients for 12 months showed normalization in liver function (Hsu et al. 2008). D-fraction extracted from *Grifola frondosa* in combination with human interferon alpha-2b showed a better response compared to each of them alone against HBV (Gu et al. 2006). *Ganoderma lucidum* in a submerged cultured broth containing Radix sophorae flavescentis extracted in organic acids were found to have an inhibitory effect on HBV (Li and Zhang 2005). *Poria cocos* polysaccharide-II, a polysaccharide from *Poria cocos*, also elicited stronger immune response when administered as an adjuvant along with HBV vaccine (Wu et al. 2016). Polysaccharides from *Antrodia camphorata* caused an efficient inhibition of HBV antigens and further did not show any cytopathic effect (Lee et al. 2002).

Ganoderic acid from *Ganoderma lucidum* protected mice from liver injury during HBV infection (Li and Wang 2006). *Lentinula edodes* mycelia solid culture extract that was used to study the inhibitory activity against HCV found that mycelia solid culture extract inhibited HCV entry into the cell (Matsuhisa et al. 2015). *Agaricus blazei* Murill extracts used against chronic HCV patients showed insignificant but a slight decrease in viral load (Johnson et al. 2009). LAC purified from *Pleurotus ostreatus* showed inhibitory effect against HCV (El-Fakharany et al. 2010), whereas lectins from *Pleurotus ostreatus* when used as adjuvant in HBV DNA vaccine showed better immune response (Gao et al. 2013).

Table 10.6 describes the extracts/compounds from mushroom that have been shown to possess antiviral activity and target enzymes against hepatitis virus.

Table 10.5 Antiviral mushrooms against influenza

Mushrooms	Extracts/ compounds	Virus	Activity	References
Phellinus linteus	Extract	Influenza	Adjuvant (cross protection)	Ichinohe et al. (2010)
Phellinus baumii	Hispidin	H1N1, H5N1, H3N2	NA	Hwang et al. (2015)
Phellinus baumii	Hypholomine B	H1N1, H5N1, H3N2	NA	Hwang et al. (2015)
Phellinus baumii	Inoscavin A	H1N1, H5N1, H3N2	NA	Hwang et al. (2015)
Phellinus baumii	Davallialactone	H1N1, H5N1, H3N2	NA	Hwang et al. (2015)
Phellinus baumii	Phelligridin D	H1N1, H5N1, H3N2	NA	Hwang et al. (2015)
Phellinus igniarius	Water extract	Influenza virus	–	Lee et al. (2013)
Auriporia aurea	Extract	H1N1	–	Krupodorova et al. (2014)
Flammulina velutipes	Extract	H1N1	–	Krupodorova et al. (2014)
Fomes fomentarius	Extract	H1N1	–	Krupodorova et al. (2014)
Ganoderma lucidum	Extract	H1N1	–	Krupodorova et al. (2014)
Lentinus edodes	Extract	H1N1	–	Krupodorova et al. (2014)
Lyophyllum shimeji	Extract	H1N1	–	Krupodorova et al. (2014)
Pleurotus eryngii	Extract	H1N1	–	Krupodorova et al. (2014)
Pleurotus ostreatus	Extract	H1N1	–	Krupodorova et al. (2014)
Schizophyllum commune	Extract	H1N1	–	Krupodorova et al. (2014)
Trametes versicolor	Extract	H1N1	–	Krupodorova et al. (2014)
Inonotus hispidus	Phenolic extracts	Influenza A and B	–	Lindequist et al. (2005)
Daedaleopsis confragosa	Extract	H5N1 and H3N2	–	Teplyakova et al. (2012)
Datronia mollis	Extract	H5N1 and H3N2	–	Teplyakova et al. (2012)
Ischnoderma benzoinum	Extract	H5N1 and H3N2	–	Teplyakova et al. (2012)
Trametes gibbosa	Extract	H5N1 and H3N2	–	Teplyakova et al. (2012)

(continued)

Table 10.5 (continued)

Mushrooms	Extracts/ compounds	Virus	Activity	References
Trametes versicolor	Extract	H5N1 and H3N2	–	Teplyakova et al. (2012)
Laricifomes officinalis	Extract	H5N1 and H3N2	–	Teplyakova et al. (2012)
Lenzites betulina	Extract	H5N1 and H3N2	–	Teplyakova et al. (2012)
Lentinus edodes	KS-2	H2N2	–	Suzuki et al. (1979)
Ganoderma pfeifferi	Several compounds	Influenza virus	–	Lindequist et al. (2015)
Cryptoporus volvatus	Extract	H1N1 and H3N2	–	Gao et al. (2014)
Phellinus igniarius	Sesquiterpenoid	Influenza virus	NA	Song et al. (2014)
Poria cocos	PCP-II	Influenza virus	Adjuvant	Wu et al. (2016)
Inonotus obliquus	Polysaccharides	Feline H3N2 and H5N6	Viral binding/ absorption	Tian et al. (2017)
Agrocybe aegerita	Lectin	Influenza virus	Adjuvant	Ma et al. (2017)

Table 10.6 Antiviral mushrooms against hepatitis virus

Mushrooms	Extracts/Compound	Virus	Activity	References
Agaricus blazei Murill	Extract	HBV	Supplement	Hsu et al. (2008)
Grifola frondosa	D-fraction	HBV	Combination	Gu et al. (2006)
Ganoderma lucidum	Extract	HBV	–	Li and Zhang (2005)
Poria cocos	PCP-II	HBV	Adjuvant	Wu et al. (2016)
Antrodia camphorate	Polysaccharides	HBV	–	Lee et al. (2002)
Ganoderma lucidum	Ganoderic acid	HBV	–	Li and Wang (2006)
Lentinula edodes	Mycelia solid culture extract	HCV	Entry	Matsuhisa et al. (2015)
Agaricus blazei Murill	Extract	HCV	–	Johnson et al. (2009)
Pleurotus ostreatus	LAC	HCV	–	El-Fakharany et al. (2010)
Pleurotus ostreatus	Lectin	HBV	Adjuvant	Gao et al. (2013)

10.3.5 Other Viruses

Sterols from the edible mushroom *Hypsizygus marmoreus* showed inhibitory property against Epstein-Barr virus (Akihisa et al. 2005). The aqueous ethanol extract and polysaccharide from the mushroom *Agaricus brasiliensis* were found to have antiviral activity against polio virus type 1 (Faccin et al. 2007). Two triterpenoids from *Ganoderma lucidum* exhibited antiviral activity against enterovirus 71 with no cytotoxicity to human rhabdomyosarcoma cells (Zhang et al. 2014a, b). LAC, endopolysaccharides from *Cerrena unicolor*, demonstrated antiviral action against human herpes virus type-1 (HHV-1) and encephalomyocarditis virus (EMCV) (Mizerska-Dudka et al. 2015). JLS-18, a water-soluble lignin from *Lentinula edodes*, exhibited antiviral response against Sendai virus in vitro (Yamamoto et al. 1997). Sulfated *Auricularia auricula* polysaccharides sAAP(1) and sAAP(t) indicated substantial antiviral response against Newcastle disease virus (NDV) (Nguyen et al. 2012).

Agaricus blazei Murill fractions strongly inhibited the Western equine encephalitis virus from causing cytopathic effect in Vero cells (Sorimachi et al. 2001). *Elfvingia applanata* water-soluble compounds work as a very potent antiviral and growth inhibitor against vesicular stomatitis virus (Eo et al. 2001). Sulfated *Tremella* polysaccharide showed antiviral response against NDV infection to chick embryo fibroblast cells (Zhao et al. 2011). Aqueous and ethanol extracts and polysaccharides extracted from *Lentinula edodes* inhibited the replication of poliovirus type 1 (Rincão et al. 2012). A novel heteropolysaccharide from *Grifola frondosa* mycelia inhibited the propagation of enterovirus 71, and capsid protein (VP1) expression and RNA synthesis were also suppressed (Zhao et al. 2016). Administering *Agaricus blazei* Murill extract combined with foot-and-mouth disease virus DNA vaccine that provided significant response for antibodies against foot-and-mouth disease virus (FMDV) along with stronger T-cell response was observed (Chen and Shao 2006). Fve protein isolated from *Flammulina velutipes* co-administered with HPV-16 E7 showed higher production of HPV-16 E7-specific interferon (Ding et al. 2009).

Table 10.7 describes the extracts/compounds from mushroom that have been shown to possess antiviral activity against the aforementioned viruses. HPV, human papillomavirus; FMDV, foot-and-mouth disease virus; NDV, Newcastle disease virus; VSV, vesicular stomatitis virus; EMCV, encephalomyocarditis virus; EBV, Epstein-Barr virus c-EPL *Cerrena unicolor* endopolysaccharides; GLTA, *Ganoderma lucidum* triterpenoids Lanosta-7,9(11),24-trien-3-one,15;26- dihydroxy; GLTB, *Ganoderma lucidum* triterpenoids ganoderic acid Y; HHV-1, human herpes virus type-1; WEE, Western equine encephalomyelitis.

Table 10.7 Antiviral mushrooms against other viruses

Mushrooms	Extracts/compounds	Viruses	Activity	References
Hypsizygus marmoreus	Sterols	EBV	–	Akihisa et al. (2005)
Agaricus brasiliensis	Extract	Polio	–	Faccin et al. (2007)
Ganoderma lucidum	GLTA, GLTB	Enterovirus 71	Adsorption	Zhang et al. (2014a, b)
Cerrena unicolor	LAC, c-EPL	HHV-1, EMCV	–	Mizerska-Dudka et al. (2015)
Lentinula edodes	JLS-18	Sendai virus	–	Yamamoto et al. (1997)
Auricularia auricular	Polysaccharides	NDV	–	Nguyen et al. (2012)
Agaricus blazei	Extract	WEE	–	Sorimachi et al. (2001)
Elfvingia applanata	Water-soluble compounds	VSV	Entry/endocytosis	Eo et al. (2001)
Tremella	Polysaccharide	NDV	–	Zhao et al. (2011)
Lentinula edodes	Aqueous and ethanol extracts and polysaccharides	Polio	Replication	Rincão et al. (2012)
Grifola frondosa	Mycelia extract	Enterovirus 71	Replication, VP1 protein expression, and RNA synthesis	Zhao et al. (2016)
Agaricus blazei	Murill extract	FMDV	Adjuvant	Chen and Shao (2006)
Flammulina velutipes	Fve	HPV-16	Adjuvant	Ding et al. (2009)

10.4 Discussion

Most of the higher *Basidiomycetes* possess biologically active compounds in their fruiting bodies, cultured mycelia, and cultured broth (Wasser 2017). Common antibiotics cannot be utilized as an antiviral, so specific drugs against viruses should be developed (Lindequist et al. 2005). The need for natural antiviral compounds arises from the issues of side effects and development of drug-resistant mutant viruses (De Clercq 1996). Mushrooms having antiviral activity with no side effects and being readily available make it an effective source of foods and therapeutics. Mushrooms are being used since the ancient times in China and Japan as traditional medicines and nutritional source (Oyetayo 2011). Several other countries like Korea and Russia use mushrooms in their modern clinical practices (Wasser 2010). Different compounds extracted from mushrooms include lectins (Hassan et al. 2015), polysaccharides (Friedman 2016), polysaccharopeptides (Ng 1998), enzymes, and other molecular compounds (Lindequist et al. 2005), which possess antiviral activity against HIV, herpes virus, influenza virus, Epstein-Barr virus, coxsackievirus, etc. Medicinal mushrooms having different components each of which showing antiviral action against different viruses make them more interesting for pharmacological exploration. Compounds isolated from fruiting body and culture filtrates of different mushrooms possess a broad spectrum of antiviral activity. Some of the *Basidiomycetes* possess toxins such as amatoxins, phallotoxins, and virotoxins, which show adverse reactions like neurotoxic and psychotropic effects (Ng 1998).

Direct effects of the antiviral compounds include inhibition of viral enzymes, nucleic acid synthesis, or adsorption and uptake of viruses by host cells. Immunostimulatory activity of complex molecules results in indirect antiviral effects.

Antifungal and antibacterial compounds are present in mushrooms which are needed for their survival in the natural environment. Hence these compounds having strong activities will be useful for humans (Lindequist et al. 2005). Other than antiviral activities, mushrooms possess antimicrobial, antitumor, cytostatic, anti-allergic, anti-complement, anti-atherogenic, anti-inflammatory, and other biological activities, many of which are commonly shared by the same compound of mushrooms (Fig. 10.1). Mushrooms contain 19–35% protein, 72–85% poly-unsaturated fatty acids, and all essential amino acids especially lysine and leucine. Fresh

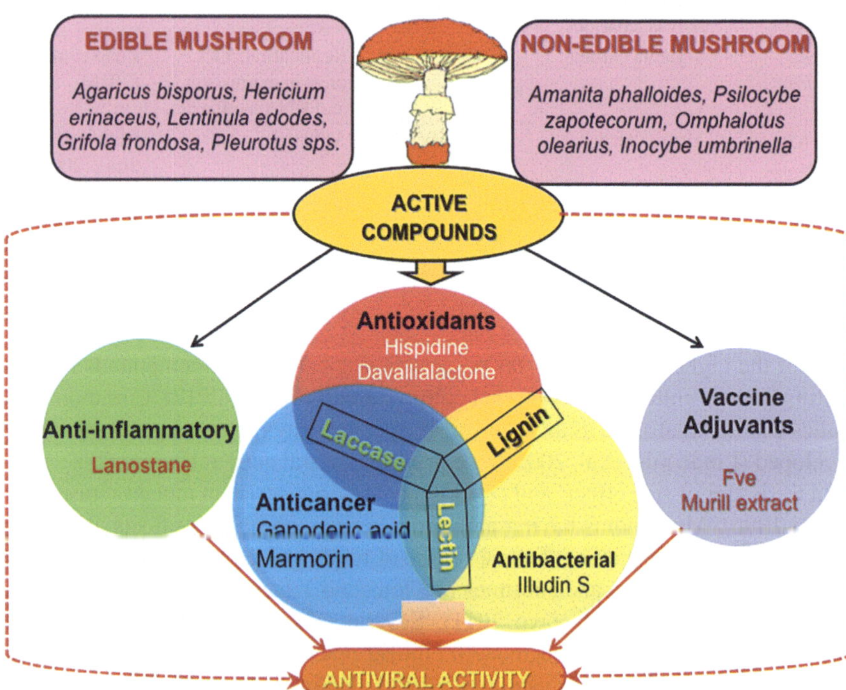

Fig. 10.1 Schematic representation of medicinal values of mushrooms and the antiviral properties. *Fve* protein isolated from *Flammulina velutipes*

mushrooms contain 51–88% carbohydrate and 4–20% fiber. They are also a good source of vitamins (thiamine, riboflavin, niacin, ascorbic acid, and biotin) and minerals (Chang and Buswell 1996). Among the known species, well-studied mushrooms are limited. Molecular mechanisms underlying antiviral compounds from mushrooms are to be investigated. Isolated and extracted compounds can be formulated into tablets, capsules, and teas depending on the source. Usage of mushrooms as dietary supplements should further be encouraged.

10.5 Conclusions

The studies have pointed out many compounds in the mushroom have excellent antiviral potency, but further testing with other viruses could be interesting. The mechanistic interaction of these compounds against viruses is also poorly understood; investigating the mechanism will give a great advantage of compound or combinations to be used as therapeutics, reducing the chances of emergence of resistant viruses. As there is a shortage of antiviral drugs in the market against many viruses, exploration of new antiviral components from mushroom could add to the existing armament of antivirals. The ease of mushroom cultivation could provide an ubiquitous supply of antiviral compounds once large-scale production has been formulated for these compounds. Overall mushrooms are a good source of many antioxidant, anti-inflammatory, and antiviral compounds, inspiring further exploration on the journey of novel antivirals.

References

Akihisa T, Franzblau SG, Tokuda H et al (2005) Antitubercular activity and inhibitory effect on Epstein-Barr virus activation of sterols and polyisoprenepolyols from an edible mushroom, *Hypsizygus marmoreus*. Biol Pharm Bull 28:1117–1119. https://doi.org/10.1248/bpb.28.1117

Awadh ANA, Mothana RA, Lesnau A et al (2003) Antiviral activity of *Inonotus hispidus*. Fitoterapia 74:483–485. https://doi.org/10.1016/S0367-326X(03)00119-9

Cardozo FT, Camelini CM, Mascarello A et al (2011) Antiherpetic activity of a sulfated polysaccharide from *Agaricus brasiliensis* mycelia. Antivir Res 92:108–114. https://doi.org/10.1016/j.antiviral.2011.07.009

Cardozo FT, Larsen IV, Carballo EV et al (2013) In vivo anti-herpes simplex virus activity of a sulfated derivative of *Agaricus brasiliensis* mycelial polysaccharide. Antimicrob Agents Chemother 57:2541–2549. https://doi.org/10.1128/AAC.02250-12

Cardozo FT, Camelini CM, Leal PC et al (2014) Antiherpetic mechanism of a sulfated derivative of *Agaricus brasiliensis* fruiting bodies polysaccharide. Intervirology 57:375–383. https://doi.org/10.1159/000365194

Chang ST, Buswell JA (1996) Mushroom nutriceuticals. World J Microbiol Biotechnol 12:473–476. https://doi.org/10.1007/BF00419460

Chen L, Shao HJ (2006) Extract from *Agaricus blazei* Murill can enhance immune responses elicited by DNA vaccine against foot-and-mouth disease. Vet Immunol Immunopathol 109:177–182. https://doi.org/10.1016/j.vetimm.2005.08.028

Chen WC, Wang SY, Chiu CC et al (2013) Lucidone suppresses hepatitis c virus replication by Nrf2-mediated heme oxygenase-1 induction. Antimicrob Agents Chemother 57:1180–1191. https://doi.org/10.1128/AAC.02053-12

Dao TT, Nguyen PH, Lee HS et al (2011) Chalcones as novel influenza A (H1N1) neuraminidase inhibitors from *Glycyrrhiza inflata*. Bioorg Med Chem Lett 21:294–298. https://doi.org/10.1016/j.bmcl.2010.11.016

De Clercq E (1996) Chemotherapy of viral infections. In: Medical microbiology, 4th edn. University of Texas Medical Branch, Galveston

De Clercq E, Li G (2016) Approved antiviral drugs over the past 50 years. Clin Microbiol Rev 29:695–747. https://doi.org/10.1128/CMR.00102-15

De Leo A, Arena G, Lacanna E et al (2012) Resveratrol inhibits Epstein Barr virus lytic cycle in Burkitt's lymphoma cells by affecting multiple molecular targets. Antivir Res 96:196–202. https://doi.org/10.1016/j.antiviral.2012.09.003

Ding Y, Seow SV, Huang CH et al (2009) Coadministration of the fungal immunomodulatory protein FIP-Fve and a tumour-associated antigen enhanced antitumour immunity. Immunology 128:881–894. https://doi.org/10.1111/j.1365-2567.2009.03099.x

Drugs.com; Available via: https://www.drugs.com/drug-class/antiviral-agents.html

El Dine RS, El Halawany AM, Ma CM, Hattori M (2008) Anti-HIV-1 protease activity of lanostane triterpenes from the Vietnamese mushroom *Ganoderma colossum*. J Nat Prod 71:1022–1026. https://doi.org/10.1021/np8001139

El-Fakharany EM, Haroun BM, Ng TB, Redwan ER (2010) Oyster mushroom laccase inhibits hepatitis C virus entry into peripheral blood cells and hepatoma cells. Protein Pept Lett 17:1031–1039. https://doi.org/10.2174/092986610791498948

Eo SK, Kim YS, Lee CK, Han SS (2000) Possible mode of antiviral activity of acidic protein bound polysaccharide isolated from *Ganoderma lucidum* on herpes simplex viruses. J Ethnopharmacol 72:475–481. https://doi.org/10.1016/S0378-8741(00)00266-X

Eo SK, Kim YS, Oh KW et al (2001) Mode of antiviral activity of water soluble components isolated from *Elfvingia applanata* on vesicular stomatitis virus. Arch Pharm Res 24:74–78

Faccin LC, Benati F, Rincão VP et al (2007) Antiviral activity of aqueous and ethanol extracts and of an isolated polysaccharide from *Agaricus brasiliensis* against poliovirus type 1. Lett Appl Microbiol 45:24–28. https://doi.org/10.1111/j.1472-765X.2007.02153.x

Férir G, Palmer KE, Schols D (2011) Synergistic activity profile of griffithsin in combination with tenofovir, maraviroc and enfuvirtide against HIV-1 clade C. Virology 417:253–258. https://doi.org/10.1016/j.virol.2011.07.004

Friedman M (2016) Mushroom polysaccharides: chemistry and antiobesity, antidiabetes, anticancer, and antibiotic properties in cells, rodents, and humans. Foods 5:80. https://doi.org/10.3390/foods5040080

Gao W, Sun Y, Chen S et al (2013) Mushroom lectin enhanced immunogenicity of HBV DNA vaccine in C57BL/6 and HBsAg-transgenic mice. Vaccine 31:2273–2280. https://doi.org/10.1016/j.vaccine.2013.02.062

Gao L, Sun Y, Si J et al (2014) *Cryptoporus volvatus* extract inhibits influenza virus replication in vitro and in vivo. PLoS One 9:e113604. https://doi.org/10.1371/journal.pone.0113604

Gong M, Piraino F, Yan N et al (2009) Purification, partial characterization and molecular cloning of the novel antiviral protein RC28. Peptides 30:654–659. https://doi.org/10.1016/j.peptides.2008.11.016

Grienke U, Schmidtke M, Kirchmair J et al (2010) Antiviral potential and molecular insight into neuraminidase inhibiting diarylheptanoids from *Alpinia katsumadai*. J Med Chem 53:778–786. https://doi.org/10.1021/jm901440f

Gu CQ, Li JW, Chao FH (2006) Inhibition of hepatitis B virus by D-fraction from *Grifola frondosa*: Synergistic effect of combination with interferon-α in HepG2 2.2.15. Antivir Res 72:162–165. https://doi.org/10.1016/j.antiviral.2006.05.011

Gu CQ, Li JW, Chao F et al (2007) Isolation, identification and function of a novel anti-HSV-1 protein from *Grifola frondosa*. Antivir Res 75:250–257. https://doi.org/10.1016/j.antiviral.2007.03.011

Han CH, Zhang GQ, Wang HX, Ng TB (2010) Schizolysin, a hemolysin from the split gill mushroom *Schizophyllum commune*. FEMS Microbiol Lett 309:115–121. https://doi.org/10.1111/j.1574-6968.2010.02022.x

Hassan MAA, Rouf R, Tiralongo E et al (2015) Mushroom lectins: specificity, structure and bioactivity relevant to human disease. Int J Mol Sci 16:7802–7838. https://doi.org/10.3390/ijms16047802

Heredia A, Davis C, Redfield R (2000) Synergistic inhibition of HIV-1 in activated and resting peripheral blood mononuclear cells, monocyte-derived macrophages, and selected drug-resistant isolates with nucleoside analogues combined with a natural product, resveratrol. J Acquir Immune Defic Syndr 25:246–255. https://doi.org/10.1097/00126334-200011010-00006

Hijikata Y, Yamada S, Yasuhara A (2007) Herbal mixtures containing the mushroom *Ganoderma lucidum* improve recovery time in patients with herpes genitalis and labialis. J Altern Complement Med 13:985–987. https://doi.org/10.1089/acm.2006.6297

Hsu C-H, Hwang K-C, Chiang Y-H, Chou P (2008) The mushroom *Agaricus blazei* Murill extract normalizes liver function in patients with chronic hepatitis B. J Altern Complement Med 14:299–301. https://doi.org/10.1089/acm.2006.6344

Hwang BS, Lee IK, Choi HJ, Yun BS (2015) Anti-influenza activities of polyphenols from the medicinal mushroom *Phellinus baumii*. Bioorg Med Chem Lett 25:3256–3260. https://doi.org/10.1016/j.bmcl.2015.05.081

Ichimura T, Watanabe O, Maruyama S (1998) Inhibition of HIV-1 protease by water-soluble lignin-like substance from an edible mushroom, *Fuscoporia obliqua*. Biosci Biotechnol Biochem 62:575–577. https://doi.org/10.1271/bbb.62.575

Ichinohe T, Ainai A, Nakamura T et al (2010) Induction of cross-protective immunity against influenza A virus H5N1 by an intranasal vaccine with extracts of mushroom mycelia. J Med Virol 82:128–137. https://doi.org/10.1002/jmv.21670

Jacobson JM, Feinman L, Liebes L et al (2001) Pharmacokinetics, safety, and antiviral effects of hypericin, a derivative of St. John's wort plant, in patients with chronic hepatitis C virus infection. Antimicrob Agents Chemother 45:517–524. https://doi.org/10.1128/AAC.45.2.517

Jiang ZY, Liu WF, Zhang XM et al (2013) Anti-HBV active constituents from *Piper longum*. Bioorg Med Chem Lett 23:2123–2127. https://doi.org/10.1016/j.bmcl.2013.01.118

Johnson E, Førland DT, Sætre L et al (2009) Effect of an extract based on the medicinal mushroom *Agaricus blazei* murill on release of cytokines, chemokines and leukocyte growth factors in human blood ex vivo and in vivo. Scand J Immunol 69:242–250. https://doi.org/10.1111/j.1365-3083.2008.02218.x

Kanokmedhakul S, Kanokmedhakul K, Prajuabsuk T et al (2003) A bioactive triterpenoid and vulpinic acid derivatives from the mushroom *Scleroderma citrinum*. Planta Med 69:568–571. https://doi.org/10.1055/s-2003-40639

Kara Rogers, Encyclopaedia Britannica (1999). https://www.britannica.com/science/mushroom

Kotwal GJ (2008) Genetic diversity-independent neutralization of pandemic viruses (e.g. HIV), potentially pandemic (e.g. H5N1 strain of influenza) and carcinogenic (e.g. HBV and HCV) viruses and possible agents of bioterrorism (variola) by enveloped virus neutralizing com. Vaccine 26:3055–3058. https://doi.org/10.1016/j.vaccine.2007.12.008

Kotwal GJ, Kaczmarek JN, Leivers S et al (2005) Anti-HIV, anti-poxvirus, and anti-SARS activity of a nontoxic, acidic plant extract from the *Trifolium* species Secomet-V/anti-Vac suggests that it contains a novel broad-spectrum antiviral. Ann N Y Acad Sci 1056:293–302. https://doi.org/10.1196/annals.1352.014

Krupodorova T, Rybalko S, Barshteyn V (2014) Antiviral activity of Basidiomycete mycelia against influenza type A (serotype H1N1) and herpes simplex virus type 2 in cell culture. Virol Sin 29:284–290. https://doi.org/10.1007/s12250-014-3486-y

Lee IH, Huang R-L, Chen C-T et al (2002) *Antrodia camphorata* polysaccharides exhibit anti-hepatitis B virus effects. FEMS Microbiol Lett 209:63–67. https://doi.org/10.1111/j.1574-6968.2002.tb11110.x

Lee CL, Chiang LC, Cheng LH et al (2009) Influenza A (H1N1) antiviral and cytotoxic agents from *Ferula assa-foetida*. J Nat Prod 72:1568–1572. https://doi.org/10.1021/np900158f

Lee S, Il KJ, Heo J et al (2013) The anti-influenza virus effect of *Phellinus igniarius* extract. J Microbiol 51:676–681. https://doi.org/10.1007/s12275-013-3384-2

Lehmann VKB, Huang A, Ibanez-Calero S et al (2003) Illudin S, the sole antiviral compound in mature fruiting bodies of *Omphalotus illudens*. J Nat Prod 66:1257–1258. https://doi.org/10.1021/np030205w

Li YQ, Wang SF (2006) Anti-hepatitis B activities of ganoderic acid from *Ganoderma lucidum*. Biotechnol Lett 28:837–841. https://doi.org/10.1007/s10529-006-9007-9

Li YQ, Zhang KC (2005) In vitro inhibitory effects on HBsAg and HBeAg secretion of 3 new components produced by *Ganoderma lucidum* in the medium contained *Radix sophorae* flavescentis extract. Wei Sheng Wu Xue Bao 45:643–646

Li YR, Liu QH, Wang HX, Ng TB (2008) A novel lectin with potent antitumor, mitogenic and HIV-1 reverse transcriptase inhibitory activities from the edible mushroom *Pleurotus citrinopileatus*. Biochim Biophys Acta, Gen Subj 1780:51–57. https://doi.org/10.1016/j.bbagen.2007.09.004

Li N, Li L, Fang JC et al (2012) Isolation and identification of a novel polysaccharide–peptide complex with antioxidant, anti-proliferative and hypoglycaemic activities from the abalone mushroom. Biosci Rep 32:221–228. https://doi.org/10.1042/BSR20110012

Lindequist U, Niedermeyer THJ, Jülich WD (2005) The pharmacological potential of mushrooms. Evid Based Complement Alternat Med 2:285–299. https://doi.org/10.1093/ecam/neh107

Lindequist U, Jülich WD, Witt S (2015) *Ganoderma pfeifferi* – A European relative of *Ganoderma lucidum*. Phytochemistry 114:102–108. https://doi.org/10.1016/j.phytochem.2015.02.018

Liu J, Yang F, Ye LB et al (2004) Possible mode of action of antiherpetic activities of a proteoglycan isolated from the mycelia of *Ganoderma lucidum* in vitro. J Ethnopharmacol 95:265–272. https://doi.org/10.1016/j.jep.2004.07.010

Ma L-B, Xu B-Y, Huang M et al (2017) Adjuvant effects mediated by the carbohydrate recognition domain of *Agrocybe aegerita* lectin interacting with avian influenza H9N2 viral surface glycosylated proteins. J Zhejiang Univ Sci B 18:653–661. https://doi.org/10.1631/jzus.B1600106

Matsuhisa K, Yamane S, Okamoto T et al (2015) Anti-HCV effect of *Lentinula edodes* mycelia solid culture extracts and low-molecular-weight lignin. Biochem Biophys Res Commun 462:52–57. https://doi.org/10.1016/j.bbrc.2015.04.104

Medicinenet.com; Available via: https://www.medicinenet.com/script/main/art.asp?articlekey=10211

Mizerska-Dudka M, Jaszek M, Błachowicz A et al (2015) Fungus *Cerrena unicolor* as an effective source of new antiviral, immunomodulatory, and anticancer compounds. Int J Biol Macromol 79:459–468. https://doi.org/10.1016/j.ijbiomac.2015.05.015

Neurath AR, Strick N, Li YY, Debnath AK (2005) *Punica granatum* (pomegranate) juice provides an HIV-1 entry inhibitor and candidate topical microbicide. Ann N Y Acad Sci 4:41. https://doi.org/10.1196/annals.1352.015

Ng TB (1998) A review of research on the protein-bound polysaccharide (polysaccharopeptide, PSP) from the mushroom *Coriolus versicolor* (basidiomycetes: Polyporaceae). Gen Pharmacol 12:473–476. https://doi.org/10.1016/S0306-3623(97)00076-1

Nguyen TL, Chen J, Hu Y et al (2012) In vitro antiviral activity of sulfated *Auricularia auricula* polysaccharides. Carbohydr Polym 90:1254–1258. https://doi.org/10.1016/j.carbpol.2012.06.060

Nogusa S, Gerbino J, Ritz BW (2009) Low-dose supplementation with active hexose correlated compound improves the immune response to acute influenza infection in C57BL/6 mice. Nutr Res 29:139–143. https://doi.org/10.1016/j.nutres.2009.01.005

Oyetayo OV (2011) Medicinal uses of mushrooms in Nigeria: Towards full and sustainable exploitation. Afr J Tradit Complement Altern Med 8:267–274. https://doi.org/10.4314/ajtcam.v8i3.65289

Pan H, Yu X, Li T et al (2013) Aqueous extract from a Chaga medicinal mushroom, *Inonotus obliquus* (higher Basidiomycetes), prevents Herpes simplex virus entry through inhibition of viral-induced membrane fusion. Intl J Med Mushrooms 15:29–38. https://doi.org/10.1615/IntJMedMushr.v15.i1.40

Park SH, Song JH, Kim T et al (2012) Anti-human rhinoviral activity of polybromocatechol compounds isolated from the Rhodophyta, *Neorhodomela aculeate*. Marine Drugs 10:2222–2233. https://doi.org/10.3390/md10102222

Piraino F, Brandt CR (1999) Isolation and partial characterization of an antiviral, RC-183, from the edible mushroom *Rozites caperata*. Antivir Res 43:67–78. https://doi.org/10.1016/S0166-3542(99)00035-2

Polkovnikova MV, Nosik NN, Garaev TM et al (2014) A study of the antiherpetic activity of the Chaga mushroom (*Inonotus obliquus*) extracts in the Vero cells infected with the herpes simplex virus. Vopr Virusol 59:45–48

Rincão VP, Yamamoto KA, Silva Ricardo NMP et al (2012) Polysaccharide and extracts from *Lentinula edodes*: structural features and antiviral activity. Virol J 9:37. https://doi.org/10.1186/1743-422X-9-37

Saboulard D, Gaspar A, Roussel B, Villard J (1998) New antiherpetic nucleoside from a Basidiomycete. *C R Acad Sci III* 321:585–591. https://doi.org/10.1016/S0764-4469(98)80461-7

Sarkar S, Koga J, Whitley RJ, Chatterjee S (1993) Antiviral effect of the extract of culture medium of *Lentinus edodes* mycelia on the replication of herpes simplex virus type 1. Antivir Res 20:293–303. https://doi.org/10.1016/0166-3542(93)90073-R

Shibnev VA, Garaev TM, Finogenova MP et al (2015) Antiviral activity of aqueous extracts of the birch fungus *Inonotus obliquus* on the human immunodeficiency virus. Vopr Virusol 60:35–38

Shin H-B, Choi M-S, Ryu B et al (2013) Antiviral activity of carnosic acid against respiratory syncytial virus. Virol J 10:303. https://doi.org/10.1186/1743-422X-10-303

Song AR, Le SX, Kong C et al (2014) Discovery of a new sesquiterpenoid from *Phellinus ignarius* with antiviral activity against influenza virus. Arch Virol 159:753–760. https://doi.org/10.1007/s00705-013-1857-6

Sorimachi K, Ikehara Y, Maezato G et al (2001) Inhibition by *Agaricus blazei* Murill fractions of cytopathic effect induced by Western Equine Encephalitis (WEE) virus on VERO cells in vitro. Biosci Biotechnol Biochem 65:1645–1647. https://doi.org/10.1271/bbb.65.1645

Sun J, Wang H, Ng TB (2011) Isolation of a laccase with HIV-1 reverse transcriptase inhibitory activity from fresh fruiting bodies of the *Lentinus edodes* (Shiitake mushroom). Indian J Biochem Biophys 48:88–94

Sundararajan A, Ganapathy R, Huan L et al (2010) Influenza virus variation in susceptibility to inactivation by pomegranate polyphenols is determined by envelope glycoproteins. Antivir Res 88:1–9. https://doi.org/10.1016/j.antiviral.2010.06.014

Suzuki F, Suzuki C, Shimomura E et al (1979) Antiviral and interferon-inducing activities of a new peptidomannan, KS-2, extracted from culture mycelia of *Lentinus edodes*. J Antibiot 32:1336–1345. https://doi.org/10.7164/antibiotics.32.1336

Suzuki H, Iiyama K, Yoshida O, Yamazaki S, Yamamoto N, Toda S (1990) Structural characterization of the immunoactive and antiviral water-solubilized lignin in an extract of the culture medium of Lentinus edodes mycelia (LEM). Agric Biol Chem 54:479–487

Takebe Y, Saucedo CJ, Lund G et al (2013) Antiviral lectins from red and blue-green algae show potent in vitro and in vivo activity against Hepatitis C virus. PLoS One 8:e64449. https://doi.org/10.1371/journal.pone.0064449

Teplyakova TV, Psurtseva NV, Kosogova TA et al (2012) Antiviral activity of polyporoid mushrooms (higher Basidiomycetes) from Altai mountains (Russia). Intl J Med Mushrooms 14:37–45. https://doi.org/10.1615/IntJMedMushr.v14.i1.40

Tian J, Hu X, Liu D et al (2017) Identification of *Inonotus obliquus* polysaccharide with broad-spectrum antiviral activity against multi-feline viruses. Int J Biol Macromol 95:160–167. https://doi.org/10.1016/j.ijbiomac.2016.11.054

Tochikura TS, Nakashima H, Ohashi Y, Yamamoto N (1988) Inhibition (in vitro) of replication and of the cytopathic effect of human immunodeficiency virus by an extract of the culture medium of *Lentinus edodes* mycelia. Med Microbiol Immunol 177:235–244. https://doi.org/10.1007/BF00189409

Van Rensburg CEJ, Gandy JJ, Snyman JR (2010) An observational trial: patient profile of users of Secomet V. S Afr Fam Pract 52:165. http://hdl.handle.net/2263/16819

Wang HX, Ng TB (2000) Isolation of a novel ubiquitin-like protein from *Pleurotus ostreatus* mushroom with anti-human immunodeficiency virus, translation-inhibitory, and ribonuclease activities. Biochem Biophys Res Commun 276:587–593. https://doi.org/10.1006/bbrc.2000.3540

Wang HX, Ng TB (2004a) A new laccase from dried fruiting bodies of the monkey head mushroom *Hericium erinaceum*. Biochem Biophys Res Commun 322:17–21. https://doi.org/10.1016/j.bbrc.2004.07.075

Wang HX, Ng TB (2004b) Purification of a novel low-molecular-mass laccase with HIV-1 reverse transcriptase inhibitory activity from the mushroom *Tricholoma giganteum*. Biochem Biophys Res Commun 315:450–454. https://doi.org/10.1016/j.bbrc.2004.01.064

Wang HX, Ng TB (2006a) A laccase from the medicinal mushroom *Ganoderma lucidum*. Appl Microbiol Biotechnol 72:508–513. https://doi.org/10.1016/j.jsv.2005.04.011

Wang HX, Ng TB (2006b) Purification of a laccase from fruiting bodies of the mushroom *Pleurotus eryngii*. Appl Microbiol Biotechnol 69:521–525. https://doi.org/10.1007/s00253-005-0086-7

Wang J, Wang HX, Ng TB (2007) A peptide with HIV-1 reverse transcriptase inhibitory activity from the medicinal mushroom *Russula paludosa*. Peptides 28:560–565. https://doi.org/10.1016/j.peptides.2006.10.004

Wang K-C, Chang J-S, Lin L-T et al (2012) Antiviral effect of cimicifugin from *Cimicifuga foetida* against human respiratory syncytial virus. Am J Chin Med 40:1033–1045. https://doi.org/10.1142/S0192415X12500760

Wasser SP (2010) Medicinal mushroom science: History, current status, future trends, and unsolved problems. Intl J Med Mushrooms 12:1–16. https://doi.org/10.1615/IntJMedMushr.v12.i1.10

Wasser SP (2017) Medicinal mushrooms in human clinical studies. Part I. Anticancer, oncoimmunological, and Immunomodulatory activities: a review. Intl J Med Mushrooms 19:279–317. https://doi.org/10.1615/IntJMedMushrooms.v19.i4.10

Wong JH, Wang HX, Ng TB (2008) Marmorin, a new ribosome inactivating protein with antiproliferative and HIV-1 reverse transcriptase inhibitory activities from the mushroom *Hypsizygus marmoreus*. Appl Microbiol Biotechnol 81:669–674. https://doi.org/10.1007/s00253-008-1639-3

Wong JH, Wang H, Ng TB (2009) A haemagglutinin from the medicinal fungus *Cordyceps militaris*. Biosci Rep 29:321–327. https://doi.org/10.1042/BSR20080153

Wu YY, Wang HX, Ng TB (2011) A novel metalloprotease from the wild Basidiomycete mushroom *Lepista nuda*. J Microbiol Biotechnol 21:256–262. https://doi.org/10.4014/jmb.1010.10060

Wu SF, Lin CK, Chuang YS et al (2012) Anti-hepatitis C virus activity of 3-hydroxy caruilignan C from *Swietenia macrophylla* stems. J Viral Hepat 19:364–370. https://doi.org/10.1111/j.1365-2893.2011.01558.x

Wu Y, Li S, Li H et al (2016) Effect of a polysaccharide from *Poria cocos* on humoral response in mice immunized by H1N1 influenza and HBsAg vaccines. Int J Biol Macromol 91:248–257. https://doi.org/10.1016/j.ijbiomac.2016.05.046

Xu J, Wang J, Deng F et al (2008) Green tea extract and its major component epigallocatechin gallate inhibits hepatitis B virus in vitro. Antivir Res 78:242–249. https://doi.org/10.1016/j.antiviral.2007.11.011

Yamamoto Y, Shirono H, Kono K, Ohashi Y (1997) Immunopotentiating activity of the water-soluble lignin rich fraction prepared from LEM – the extract of the solid culture medium of *Lentinus edodes* mycelia. Biosci Biotechnol Biochem 61:1909–1912. https://doi.org/10.1271/bbb.61.1909

Yan N, He F, Piraino F et al (2015) Antiviral activity of a cloned peptide RC28 isolated from the higher basidiomycetes mushroom *Rozites caperata* in a mouse model of HSV-1 keratitis. Intl J Med Mushrooms 17:819–828. https://doi.org/10.1615/IntJMedMushrooms.v17.i9.20

Zhang GQ, Sun J, Wang HX, Ng TB (2009) A novel lectin with antiproliferative activity from the medicinal mushroom *Pholiota adiposa*. Acta Biochim Pol 56:415–421. doi: 20091798 [pii]

Zhang Y, Liu Y-B, Li Y et al (2013) Sesquiterpenes and alkaloids from the roots of *Alangium chinense*. J Nat Prod 76:1058–1063. https://doi.org/10.1021/np4000747

Zhang R, Zhao L, Wang H, Ng TB (2014a) A novel ribonuclease with antiproliferative activity toward leukemia and lymphoma cells and HIV-1 reverse transcriptase inhibitory activity from the mushroom, *Hohenbuehelia serotina*. Int J Mol Med 33:209–214. https://doi.org/10.3892/ijmm.2013.1553

Zhang W, Tao J, Yang X et al (2014b) Antiviral effects of two *Ganoderma lucidum* triterpenoids against enterovirus 71 infection. Biochem Biophys Res Commun 449:307–312. https://doi.org/10.1016/j.bbrc.2014.05.019

Zhao JK, Wang HX, Ng TB (2009) Purification and characterization of a novel lectin from the toxic wild mushroom *Inocybe umbrinella*. Toxicon 53:360–366. https://doi.org/10.1016/j.toxicon.2008.12.009

Zhao S, Zhao Y, Li S et al (2010) A novel lectin with highly potent antiproliferative and HIV-1 reverse transcriptase inhibitory activities from the edible wild mushroom *Russula delica*. Glycoconj J 27:259–265. https://doi.org/10.1007/s10719-009-9274-5

Zhao X, Hu Y, Wang D et al (2011) Optimization of sulfated modification conditions of *Tremella* polysaccharide and effects of modifiers on cellular infectivity of NDV. Int J Biol Macromol 49:44–49. https://doi.org/10.1016/j.ijbiomac.2011.03.010

Zhao C, Gao L, Wang C et al (2016) Structural characterization and antiviral activity of a novel heteropolysaccharide isolated from *Grifola frondosa* against enterovirus 71. Carbohydr Polym 144:382–389. https://doi.org/10.1016/j.carbpol.2015.12.005

Discovery of Muscarine Leading to the Basic Understanding of Cholinergic Neurotransmission and Various Clinical Interventions

11

Sindhu Ramesh, Mohammed Majrashi, Mohammed Almaghrabi, Manoj Govindarajulu, Maali Fadan, Jack Deruiter, Randall Clark, Vanisree Mulabagal, Dinesh Chandra Agrawal, Timothy Moore, and Muralikrishnan Dhanasekaran

Contents

S. Ramesh · M. Govindarajulu · M. Fadan · J. Deruiter · R. Clark · T. Moore
M. Dhanasekaran (✉)
Department of Drug Discovery and Development, Harrison School of Pharmacy, Auburn University, Auburn, AL, USA
e-mail: dhanamu@auburn.edu

M. Majrashi
Department of Drug Discovery and Development, Harrison School of Pharmacy, Auburn University, Auburn, AL, USA

Department of Pharmacology, Faculty of Medicine, University of Jeddah, Jeddah, Kingdom of Saudi Arabia

D. C. Agrawal, M. Dhanasekaran (eds.), *Medicinal Mushrooms*,
https://doi.org/10.1007/978-981-13-6382-5_11

Abstract

Muscarine was initially detected in mushrooms, and this discovery has led to the treatment of various diseases, and the therapeutic efficacy has been a godsend to the humans. The objective of this book chapter is to address the discovery of muscarine leading to the basic understanding of cholinergic neurotransmission and various clinical interventions. Furthermore, we have also explicated about the discovery of "muscarine" and its important role in delineating the basic concepts in understanding the characterization of central and peripheral cholinergic neurotransmission and its relevant clinical interventions. Muscarinic receptors stimulation or blockade plays an important role in the pathophysiology of a variety of disease states. Consequently, drugs acting on muscarinic receptors have revolutionized the human and animal health industry. Actions of muscarinic agonists or antagonists have led to potential therapeutic outcomes to treat numerous pathological conditions associated with central and peripheral nervous system. However, toxic symptoms or the adverse drug reactions are attributed to excess stimulation of cholinergic receptors by muscarine.

Keywords

Cholinergic neurotransmission · Muscarinic receptors · Mushrooms · Mushroom poisoning · Therapeutic value · Toxicity

Abbreviations

AD	Alzheimer's disease
ANS	Autonomic nervous system
CNS	Central nervous system
COMT	Catechol-O-methyltransferase
COPD	Chronic obstructive pulmonary disease
CVS	Cardiovascular system
CYP	Cytochrome P450
GDP	Guanosine diphosphate
GERD	Gastro esophageal reflux disease
GIT	Gastrointestinal tract

M. Almaghrabi
Department of Pharmaceutical Chemistry, College of Pharmacy, Taibah University, Almadinah Almunawwarah, Kingdom of Saudi Arabia

V. Mulabagal
Department of Civil Engineering, Auburn University, Auburn, AL, USA

D. C. Agrawal (✉)
Department of Applied Chemistry, Chaoyang University of Technology, Taichung, Taiwan
e-mail: dcagrawal@gm.cyut.edu.tw

GPCR	G protein-coupled receptors
GTP	Guanosine-5′-triphosphate
IBS	Irritable bowel syndrome
mAChR	Muscarinic acetylcholine receptors
MAO	Monoamine oxidase
OAB	Overactive bladder

11.1 Introduction

Mushrooms, the fruiting bodies belonging to the higher fungi group, are thought to be a delicacy among various cultural groups over the world. Mushroom poisoning or mushroom toxicity occurs from consumption of various toxins present in it. Interestingly, some mushrooms other than the edible species are poisonous on ingestion. At times, poisonous mushrooms resembling edible mushrooms are consumed unknowingly resulting in toxicity or poisoning. Various toxic syndromes associated with mushroom consumption have been well documented (Deshmukh et al. 2006). Muscarine was first detected in mushrooms particularly Inocybe and Clitocybe species (Saviuc and Danel 2006).

Muscarine, a water-soluble toxin was primarily isolated from the mushroom species *Amanita muscaria*. Historically, the studies of (Riker and Wescoe 1951) on the cardioselective inhibition of muscarinic acetylcholine receptors (mAChR) function by gallamine (M_2 specific antagonist) gave the first indication of the existence of more than one subtype of mAChR. Several years later, Roszkowski (1961) established that the muscarinic agonist 4-(3-chlorophenylcarbamoyloxy) - 2-butynyltrimethylammonium chloride (McN-A343) had a preferential effect on mAChRs located in sympathetic ganglia while having practically no activity on mAChRs in the heart and smooth muscle. Furthermore, Barlow et al. (1976) also provided early functional evidence for differences between the pharmacology of mAChRs located in the ileum (smooth muscle) and those in the atrium (cardiac muscle). However, it was not until the introduction of the muscarinic antagonist, pirenzepine, that could differentiate between various types of mAChRs in both radioligand binding (Hammer et al. 1980) and functional assays (Brown et al. 1980; Hammer and Giachetti 1982). This study formally proposed the existence of at least two distinct mAChR subtypes (Hammer and Giachetti 1982). The subsequent advent of molecular cloning techniques has of course confirmed the existence of five distinct subtypes of mAChR protein, but the pharmacological classification of mAChRs remains of paramount importance in confirming the functional relevance of any expressed gene product, as well as validating novel therapeutic agents that target mAChRs. In recent years, the advent of newer, high-throughput functional screening assays promises to yield novel mAChR agonists with the potential for improved selectivity over existing agents. In addition, the use of a variety of snake toxins, predominantly from the black and green mamba, appear to show quite impressive degrees of selectivity for 1 or 2 mAChR subtypes relative to all others.

Hence, these subtypes have established and provided a more definitive pharmacological classification of mAChR subtypes. However, mAChR pharmacology is still generally characterized by a lack of truly selective agonists, as well as a lack of readily available antagonists that can differentiate any one mAChR subtype to the exclusion of all others.

The muscarinic acetylcholine receptors (AChRs) are named from M_1 to M_5, and they belong to the family of G protein-coupled receptors (GPCR). These GPCR mediate a slow metabolic response via a second messenger cascades. However, there is insignificant activity of muscarine on nicotinic receptors. Muscarine is involved in various autonomic physiological functions including heart rate and force of contraction, contraction of smooth muscles, and the release of neurotransmitters. Mushrooms that contain muscarine are commonly found in yards, parks, and wooded areas throughout the United States, Europe, and Asia. Since muscarine is a quaternary amine, it does not readily cross the blood-brain barrier and does not directly cause CNS effects. Muscarine is not metabolized by cholinesterase and has a longer biologic half-life than the natural ligand, acetylcholine. Muscarine-containing mushrooms predominantly induce cholinergic symptoms, as mentioned above, which occurs within an hour of consumption and lasts for 6–24 h. If the consumed doses are minimal, the cholinergic symptoms can resolve without any treatment or atropine administration (Benjamin 1995). The original interest in *Amanita muscaria* lays in its psychopharmacological actions. But, it is highly doubtful if muscarine contributes at all to the above CNS effect, because of its relative small amount in the CNS and due to the presence of other alkaloids in the mushroom such as ibotenic acid.

11.2 Discovery of Muscarine Leading to the Basic Understanding of Cholinergic Neurotransmission and Various Clinical Interventions

The discovery of "muscarine" played a critical role in the characterization of central and peripheral cholinergic neurotransmission. Muscarinic receptors are characterized through their interaction (binding with intrinsic properties) with muscarine, a water-soluble toxin derived from the mushroom *Amanita muscaria* (Fig. 11.1). With regard to the cholinergic neurotransmission, acetylcholine acts on muscarinic (G protein-coupled) and nicotinic (ion-channeled) receptors to exert its physiological action in the body. The muscarinic receptors are part of a large family of G protein-coupled receptors ("G proteins"), which use an intracellular secondary messenger system involving an increase of intracellular calcium to transmit signals inside cells. Binding of acetylcholine to a muscarinic receptor causes a conformational change in the receptor that is responsible for its association with and activation of an intracellular G protein, the latter converting GTP to GDP in order to become activated and dissociate from the receptor. The activated G protein can then act as an enzyme to catalyze downstream intracellular events.

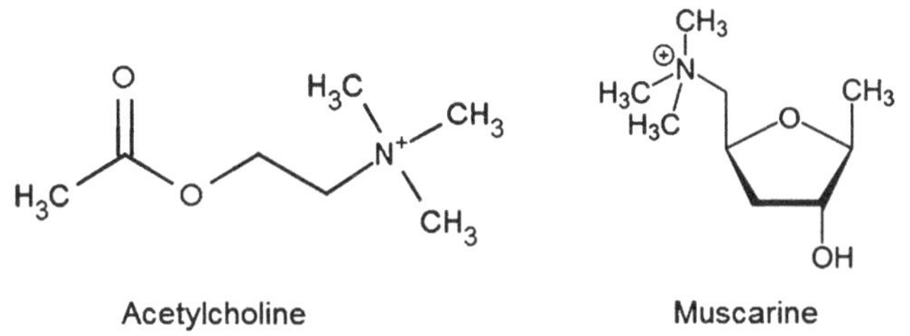

Fig. 11.1 Structure of acetylcholine and muscarine

Table 11.1 Muscarine receptor location and its mode of action

Receptor	Locations	Mode of action
Muscarinic M_1	CNS neurons, sympathetic postganglionic neurons, presynaptic sites	Formation of IP3 and DAG, increased intracellular calcium
Muscarinic M_2	Myocardium, smooth muscle, some presynaptic sites; CNS neurons	Opening of potassium channels, inhibition of adenylyl cyclase
Muscarinic M_3	Exocrine glands, vessels (smooth muscle and endothelium); CNS neurons	Like M1 receptor-ligand binding
Muscarinic M_4	CNS neurons; possibly vagal nerve endings	Like M2 receptor-ligand binding
Muscarinic M_5	Vascular endothelium, especially cerebral vessels; CNS neurons	Like M1 receptor-ligand binding

There are five subtypes of muscarinic receptors (M_1–M_5) based on structural differences, G protein coupling mechanisms, and differences in affinities for various agonist and antagonist ligands (Wess 1996; Abrams et al. 2006; Wess et al. 2007). The M_1, M_3, and M_5 receptors are stimulatory, while M_2 and M_4 are inhibitory (Table 11.1). Excessive stimulation or inhibition of muscarinic receptors is a significant component of the pathophysiology of a variety of disease states. This essential finding has revolutionized the medical therapy and led to potential therapeutic outcomes to treat hundreds of pathological conditions associated with central and peripheral nervous system. In this section, we will review the role of muscarinic receptor-mediated action on each organ, major pathological conditions, and the muscarinic drugs used for clinical interventions.

Figure 11.2 represents the muscarinic neurotransmission in various disorders.

11.2.1 Discovery of Muscarine and Its Relevance in the Etiology and Treatment of CNS Disorders

In the central nervous system, the cholinergic system originates from the Ch4 cell group of the nucleus basalis of Meynert. From these perikarya (cell body), two bundles of cholinergic fibers extend to the cerebral cortex and amygdala and are designated as the medial and lateral cholinergic pathways. These pathways contain muscarinic and nicotinic receptors, and stimulation of these receptors controls

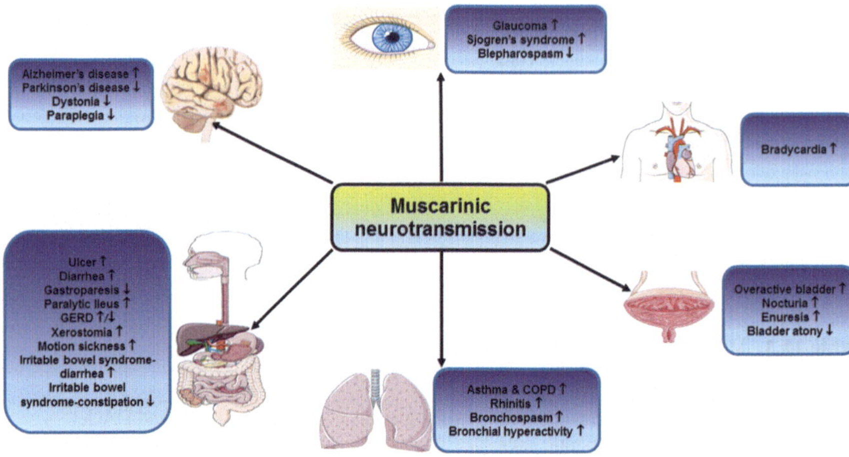

Fig. 11.2 Muscarinic neurotransmission in various disorders

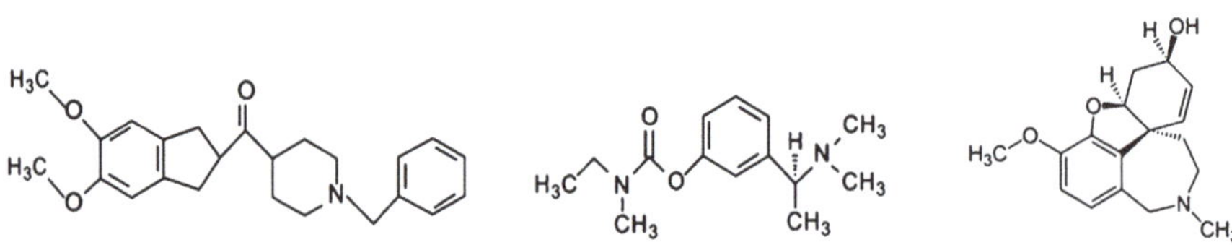

Donepezil Rivastigmine Galantamine

Fig. 11.3 Cholinergic drugs used in the treatment of Alzheimer's disease

different physiological functions. The foremost functions controlled by muscarinic receptor-mediated actions include memory formation and control of movement. Major pathologies associated with the disturbance of muscarinic neurotransmission are dementia (Alzheimer's disease), Parkinson's disease, dystonia, tremor, and spastic paraplegia. Dementia is a neurological disorder that seriously affects a person's ability to carry out daily activities. The most common form of dementia among older people is Alzheimer's disease, which initially involves the neurodegeneration in the parts of the brain that control thought, memory, and language. Indirect-acting cholinomimetics or acetylcholinesterase inhibitors are the first line of therapy currently used to treat Alzheimer's disease. These drugs distribute to the brain and prevent the metabolism of acetylcholine, thereby extending activity. The main acetylcholinesterase inhibitors in use for Alzheimer's include donepezil, rivastigmine, and galantamine (Fig. 11.3).

Parkinson's disease belongs to a group of conditions called motor system disorders, which are the result of a depletion of dopamine in nigrostriatal dopaminergic nerves. The primary symptoms of Parkinson's disease are tremor (trembling in hands, arms, legs, jaw, and face), rigidity (stiffness of the limbs

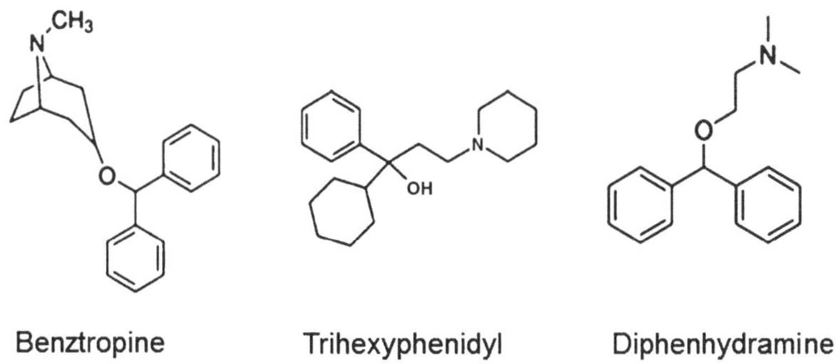

Fig. 11.4 Antimuscarinic drugs

and trunk), bradykinesia (slowness of movement), and postural instability (impaired balance and coordination). Progression of neurodegeneration leads to more pronounced symptoms resulting in difficulties with walking, speaking, swallowing, and chewing, as well as urinary problems, gastrointestinal tract issues (constipation), dermatological pathologies, and sleep disturbances. The primary treatment for Parkinson's is L-DOPA in combination with carbidopa, but dopamine agonists such as pramipexole, ropinirole, and rotigotine are also used as adjunctive therapies, as well as monoamine oxidase (MAO) inhibitor (selegiline, rasagiline) and catechol-O-methyltransferase (COMT) inhibitors (entacapone). The tremor associated with Parkinson's disease can be treated with drugs blocking the effect of muscarinic receptor-mediated neurotransmission. The drugs used for this indication include benztropine, trihexyphenidyl, and diphenhydramine (Fig. 11.4).

Dystonia is the movement disorder due to sustained muscle contractions resulting in twisting and repetitive movements and abnormal postures. Dystonia affects the muscles in the arms, legs, neck, or the entire body. Sustained muscle contraction occurs due excessive stimulation of muscarinic receptors in the brain by acetylcholine. Similar to Parkinson's disease, muscarinic antagonists are used to reduce the symptoms in dystonia.

Spastic paraplegia is a condition that occurs due to paresis of the lower extremities with increased muscle tone and spasmodic contraction of the muscles. However, in paraplegia there is paralysis of the hind limb. The paralysis may be acute in onset as in fracture of a lumbar vertebra, or gradual; it may be spastic or flaccid. In the majority of cases, paraplegia results from disease or injury of the spinal cord that causes interference with nerve paths connecting the brain and the muscles. Muscarinic antagonists such as those described above have been used to reduce symptoms associated with paraplegia. Interestingly, lesion of the brain regions results in excessive laughing and crying. In the Table 11.2, we have summarized the pathological condition associated with increased or decreased muscarinic neurotransmission in the central nervous system and the therapy due to muscarinic receptor activation or blockade.

Table 11.2 List of pathological conditions in the central nervous system associated with altered muscarinic neurotransmission and the muscarinic receptor-mediated therapeutic approach

Pathological conditions in the *central nervous system* associated with altered muscarinic neurotransmission	Drugs affecting muscarinic neurotransmission used in therapy
Dementia: Alzheimer's disease	Cholinomimetics
Parkinson's disease-tremor	Muscarinic antagonist
Dystonia	Muscarinic antagonist
Paraplegia	Muscarinic antagonist
Excessive laughing and crying	Muscarinic antagonist

11.2.2 Discovery of Muscarine and Its Relevance in the Etiology and Treatment of Ophthalmic Disorders

The ophthalmic system is innervated by the cholinergic nerves and adrenergic nerves. Cholinergic muscarinic transmission innervates radial and sphincter muscles of iris, ciliary muscle, and lacrimal glands. Stimulation of muscarinic receptors in the eye controls the contraction of iris muscles (miosis), lacrimal secretion, and aqueous humor outflow. The major medical problems due to the imbalance of the muscarinic-mediated cholinergic neurotransmission in the eye are pinpoint pupil (miosis), decreases in lacrimal gland secretion (xerophthalmia, Sjogren's syndrome), increased intraocular pressure (Glaucoma), and altered smooth muscle contraction leading to strabismus, esotropia, and blepharitis. Blockade of muscarinic receptors and alpha-adrenergic agonist are the most common drugs used as diagnostic agents for ophthalmic examinations. Glaucoma is a group of diseases that damage the optic nerve and result in vision loss and blindness. Open-angle and closed-angle glaucoma are the most common types of glaucoma. Glaucoma occurs due to the increased production and decreased outflow of the aqueous humor. The accumulation of aqueous humor increases the intraocular pressure inside the eyes gradually over time. While beta-blockers and prostaglandins are the primary therapies for glaucoma, cholinomimetics including muscarinic agonists and the cholinesterase inhibitors echothiophate are used to treat open-angle glaucoma (Fig. 11.5). Xerophthalmia (dry) is a very common ocular problem. Tearing and/or dry eye problems are a set of conditions that affect the quality and/or quantity of the tears in the eye. The more modern term is dysfunctional tear syndrome, as that covers the many causes of this very complex problem. Anything that disturbs either the production of tears and/or the quality of the tears leads to dry eye syndrome (dysfunctional tear syndrome). Sjogren's syndrome is an autoimmune disorder where excessive antibody production targets the glands specifically, and the chronic inflammation leads to dry eyes.

Muscarinic agonists such as cevimeline have utility in Sjogren's because of their ability to increase the release of fluid from lacrimal glands and reduce the dryness in the eyes. Blepharospasm is an uncontrolled muscle contraction that leads to forcible closure of the eyelids (abnormal, involuntary blinking or spasm of the eyelids). Blepharospasm is also associated with an abnormal function of the basal ganglia and stress. Antagonists at muscarinic receptors have been shown to relax the muscle

Fig. 11.5 "Cholinomimetics"

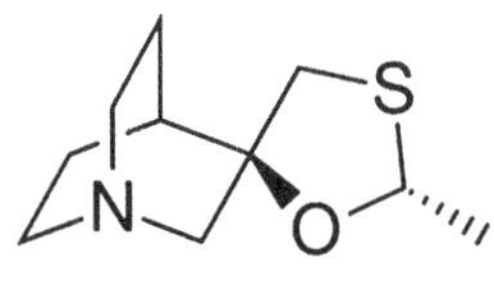

Cevimeline

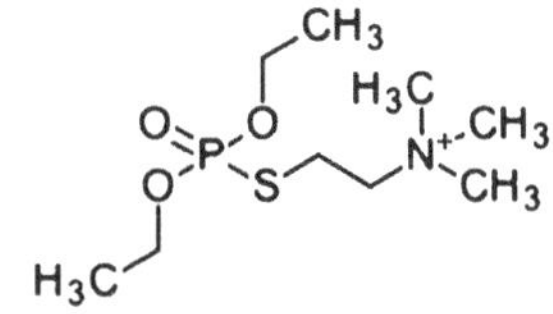

Echothiophate

and reduce the spasm. In the Table 11.3, we have summarized the list of ophthalmic pathologies that can be treated with drugs affecting the muscarinic receptors. In Table 11.3, we have compared the drugs affecting muscarinic receptor used in the treatment of glaucoma. Furthermore, in Table 11.4, we have listed the pathological condition associated with increased or decreased muscarinic neurotransmission in the eye and the therapy due to muscarinic receptor activation or blockade.

11.2.3 Discovery of Muscarine and Its Relevance in the Etiology and Treatment of Respiratory Disorders

The autonomic nervous system (ANS) innervates tracheal and bronchial smooth muscle, bronchial glands, and the blood vessels of the respiratory tract. Respiratory functions controlled by ANS include the relaxation/constriction of tracheal and bronchial smooth muscle and the increase/decrease in bronchial secretions. In the parasympathetic nervous system, muscarinic receptor (M_3) stimulation leads to the constriction of tracheal and bronchial smooth muscle and increased bronchial secretion. The pathologies linked at least in part to the muscarinic receptor in the respiratory tract are asthma, bronchospasm, and chronic obstructive pulmonary disorder (COPD) and rhinitis. Asthma is a reversible chronic inflammatory disease of the respiratory tract, and often the symptoms are temporarily reversed spontaneously or after treatment with drugs. Asthma is distinct from bronchial narrowing resulting from viral infections (acute or chronic bronchitis), destructive disease of the lungs (emphysema, bronchiectasis), or cardiovascular disease. However, COPD is an obstructive disorder in which the volume and/or rate of airflow into the lungs is altered/obstructed. Airflow obstruction in COPD occurs due to chronic bronchitis and/or emphysema and this airflow limitation that is not entirely reversible. The airflow limitation is usually progressive and is associated with an abnormal inflammatory response of the lungs to noxious particles or gases, primarily caused by cigarette smoking. Although COPD affects the lungs, it also produces significant systemic consequences. Muscarinic antagonists have been shown to reduce the pathologies related to the above respiratory tract constrictive disorders. In asthma beta-2 agonists are the primary therapies applied to promote relaxation of bronchial smooth muscle, but muscarinic receptor blockers including ipratropium, tiotropium, and umeclidinium are useful adjunctive agents (Fig. 11.6). For COPD muscarinic antagonists are the frontline therapies.

Table 11.3 Drugs affecting muscarinic receptor used in glaucoma

	Carbachol	Pilocarpine	Cevimeline
	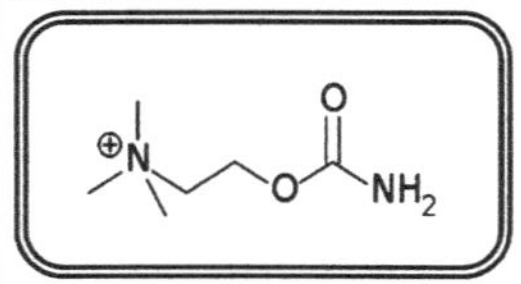	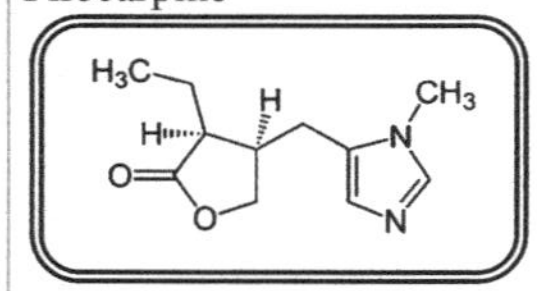	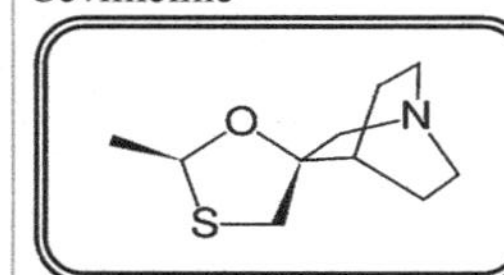
Structure	Carbamate substitution highly polar quaternary ammonium	Natural alkaloid imidazole ring, chiral secondary amine	Synthetic quinuclidine ring tertiary amine
Chemical stability	More stable than ach with regard to esterase degradation	Epimerization and some ester hydrolysis	Not metabolized by esterase (metabolized by CYP)
Receptor specificity	Non-specific (muscarinic, nicotinic)	Muscarinic specificity and no nicotinic	Specific muscarinic
Effect on AchE	Inhibition (due to binding ability)	None	None (not a substrate)
Distribution	Not well distributed	Good distribution profile, significant therapeutic value	Good distribution profile significant therapeutic value
Metabolism	Not metabolized by CYP renal excretion	Not metabolized by CYP renal excretion	CYP metabolism
Therapeutic use	Glaucoma induce miosis	Glaucoma, Sjogren's	Sjogren's

Table 11.4 List of pathological conditions in the ophthalmic system associated with altered muscarinic neurotransmission and the muscarinic receptor-mediated therapeutic approach

Pathological conditions in the *eye* associated with altered muscarinic neurotransmission	Drugs affecting muscarinic neurotransmission used
Drugs to induce miosis	Cholinomimetics
Drugs to induce mydriasis	Muscarinic antagonist
Drugs to induce cycloplegia	Muscarinic antagonist
Glaucoma- therapy	Cholinomimetics
Sjogren's therapy	Cholinomimetics
Blepharospasm	Muscarinic antagonist

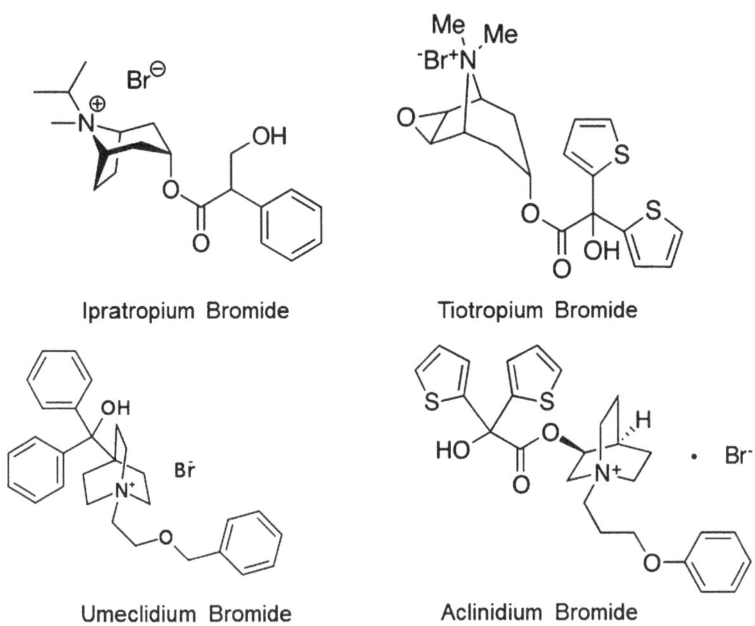

Fig. 11.6 Antimuscarinics used in COPD and asthma

Unlike asthma and COPD, bronchospasm can occur due to excessive exercise or any other acute conditions associated with infection or inflammation. Muscarinic antagonists have been shown to reduce bronchospasm. Interestingly, cholinomimetics have been used in the diagnosis of bronchial hyperactivity. Bronchial hyper-responsiveness is a condition where there is an increase in sensitivity to an extensive variety of airway narrowing stimuli. Bronchial hyper-responsiveness is a composite functional disorder that requires appropriate diagnosis and needs therapy. In asthma and COPD, patients have increased bronchial sensitivity. The understanding of bronchial hyper-responsiveness has clinical implications in the diagnosis and therapy of asthma and COPD.

Table 11.5 List of pathological conditions in the respiratory tract associated with altered muscarinic neurotransmission and the muscarinic receptor-mediated therapeutic approach

Respiratory tract-muscarinic neurotransmission	Drugs affecting muscarinic neurotransmission used
Asthma	Muscarinic antagonist
COPD	Muscarinic antagonist
Rhinitis	Muscarinic antagonist
Bronchospasm	Muscarinic antagonist
Diagnosis of bronchial hyperactivity	Cholinomimetics

Rhinitis occurs predominantly due to the inflammation of the nasal membrane that causes periods of nasal discharge, sneezing, and congestion that persists for at least 1 h/day. It is characterized by one or more of the following symptoms such as nasal congestion, rhinorrhea (anterior and posterior), sneezing, and itching. Rhinitis also affects the eyes and throat. Histamine is the major mediator of rhinitis; however, muscarinic neurotransmission also plays a vital role in the pathology, and muscarinic antagonists have been used to treat rhinitis. Please see the Table 11.5, where the list of respiratory pathologies and the therapeutic approach associated with muscarinic receptor.

11.2.4 Discovery of Muscarine and Its Relevance in the Etiology and Treatment of Cardiovascular Disorders

With regard to the cardiovascular system, muscarinic (M_2) receptors are present in the heart and regulate its function in part. The muscarinic (M_2) receptor is a Gi protein-coupled receptor. Thus, the stimulation of muscarinic (M_2) receptor results in decreased contraction leading to bradycardia, an abnormally slow heart rate of less than 60 beats per minute. Bradycardia can cause dizziness, weakness, lack of energy, or fainting spells. Muscarinic antagonists have been used to treat bradycardia. Cardiac arrest also can be treated with drugs affecting muscarinic receptor. The heart has an internal electrical system that controls the rhythm of the heartbeat, and when pathologies develop affecting this system, abnormal heart rhythms or arrhythmias can occur. During an arrhythmia, the heart can beat too fast, too slow, or it can stop beating. Sudden cardiac arrest occurs when the heart develops an arrhythmia that causes it to stop beating. This is different than a heart attack, where the heart usually continues to beat but blood flow to the heart is blocked. Depending on the condition, cholinomimetic or cholinolytics can be used. Blood vessels are not innervated by the parasympathetic nervous system, but muscarinic receptors are present on the blood vessels, and stimulation of these receptors can cause vasodilation. High blood pressure (hypertension) occurs when the systolic pressure is consistently over 140 mm Hg, or the diastolic blood pressure is consistently over 90 mm Hg. Ganglionic blockers such as mecamylamine have been used to treat hypertension (Fig. 11.7). A list of CVS pathological conditions and the possible therapies using muscarinic receptor as a target is given in Table 11.6.

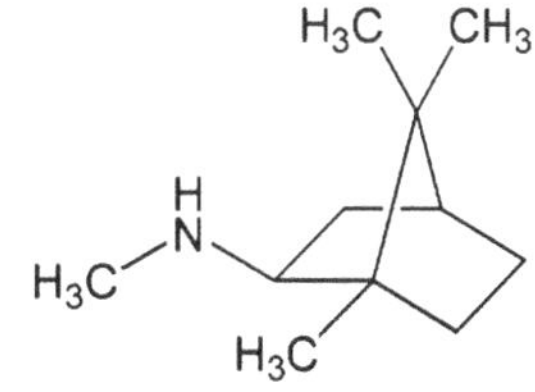

Fig. 11.7 Ganglionic blocker mecamylamine

Table 11.6 List of pathological conditions in the cardiovascular system associated with altered muscarinic neurotransmission and the muscarinic receptor-mediated therapeutic approach

Cardiovascular system-muscarinic neurotransmission	Drugs affecting muscarinic neurotransmission used
Bradycardia	Muscarinic antagonist

11.2.5 Discovery of Muscarine and Its Relevance in the Etiology and Treatment of Gastrointestinal Disorders

The autonomic nervous system innervates the smooth muscle, sphincter muscles, and the glands of the gastrointestinal tract. Therefore the physiological functions controlled by ANS in this GIT include motility and tone, sphincter control, and secretions (glands and gastric acid). Parasympathetic nervous system activation increases the motility and tone (smooth muscle contraction) via acetylcholine-induced stimulation of M_3 receptors. Stimulation of M_3 receptors relaxes the sphincter, while stimulation of M_1 receptors promotes secretion (glands and gastric acid). Peptic ulcer is a sore in the lining of the esophagus, stomach, or duodenum; an ulcer in the stomach is a gastric ulcer, and an ulcer in the duodenum is a duodenal ulcer. Ulcer is usually caused by drugs which enhance acid secretion or by the bacterium *Helicobacter pylori* or a combination of both factors. Therefore, ulcer is treated most commonly by drugs that block acid secretion (proton pump inhibitors, H2-antagonists, bismuth salts, antacids) and antimicrobial drugs such as amoxicillin, tetracyclines, and metronidazole. However, muscarinic antagonists can reduce the acid secretion and have demonstrated some therapeutic efficacy in ulcer.

Diarrheal diseases states are characterized by excessive watery stools occur very frequently. Diarrhea can be acute, usually lasting only a few days, and is typically caused by an infection (bacteria, viruses, or parasites), or it may be chronic, persisting longer than 4 weeks. Chronic diarrhea can indicate a serious disorder, such as ulcerative colitis or Crohn's disease, or a less serious condition, such as irritable bowel syndrome. Muscarine receptor antagonists such as those discussed previously have therapeutic utility in the treatment of diarrhea because of their ability to antagonize the actions of acetylcholine in the GI tract and decrease gastric motility and inhibit secretion.

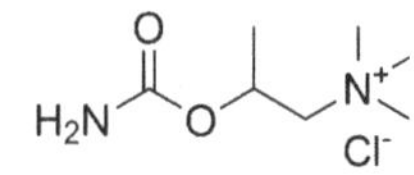

Fig. 11.8 Bethanechol: a muscarinic receptor agonist

Postabdominal distension refers to the sensation of fullness in the abdomen. The main cause of distension is linked to conditions that lead to accumulation of gas in the GIT such as dyspepsia, intestinal infection, severe constipation, peritonitis, entero-paralysis, intestinal obstruction, and severe deficiency in potassium. Muscarinic agonists can be used therapeutically to reduce the distension.

Gastroparesis (delayed gastric emptying) is a disorder in which the stomach takes too long to empty its contents. Food moves slowly or stops moving through the digestive tract. Normally, the stomach contracts to move food down into the small intestine for digestion. The vagus nerve controls the movement of food from the stomach through the digestive tract. Gastroparesis can occur due to the damage of vagus nerve, and the muscles of the stomach and intestines do not work normally. Therapies for gastroparesis include nutritional modifications, medications to stimulate gastric emptying, drugs that reduce vomiting, endoscopic and surgical approaches, and psychological interventions. The muscarinic agonist bethanechol is an approved smooth muscle muscarinic agonist that increases lower esophageal sphincter pressure and evokes fundoantral contractions (Fig. 11.8). Also antiemetic agents of the dopamine antagonist and antimuscarinic pharmacologic classes are used alone or in combination with prokinetic drugs for gastroparesis.

Paralytic ileus (referred as intestinal volvulus, bowel obstruction, intestinal obstruction) involves a partial or complete blockage of the bowel that results in the failure of the intestinal contents to pass through. Paralytic ileus, also called pseudo-obstruction, is one of the major causes of obstruction in infants and children. Similar to the therapy in gastroparesis, muscarinic agonists can improve the patient's conditions in paralytic ileus. Gastroesophageal reflux disease (GERD) occurs when a muscle at the end of the esophagus does not close properly. This allows stomach contents to leak back, or reflux, into the esophagus and irritate it. Esophagus is the tube that carries food from the mouth to the stomach. Patient feels a burning in the chest or throat called heartburn. Sometimes, patient can taste stomach fluid in the back of the mouth. Both muscarinic agonist and antagonist have been used in the therapy for GERD.

Xerostomia (dry mouth, pasties, or cotton mouth) is due to a lack of saliva. Xerostomia can cause difficulty in speech and eating. It also leads to halitosis and a dramatic rise in the number of cavities, as the protective effect of saliva is no longer present, and can make the mucosa of the mouth more vulnerable to infection. Notably, a symptom of methamphetamine abuse usually called "meth mouth" is largely caused by xerostomia. Xerostomia also occurs due to Sjogren's syndrome as discussed above. Muscarinic agonists can increase salivary secretions and reduce the symptoms. Nausea and vomiting are not diseases per se but can be symptoms of many different conditions including morning sickness during pregnancy, infections, migraine headaches, motion sickness, food poisoning, cancer chemotherapy, or

Table 11.7 List of pathological conditions in the gastrointestinal tract associated with altered muscarinic neurotransmission and the muscarinic receptor-mediated therapeutic approach

Pathological conditions in the *gastrointestinal tract* associated with altered muscarinic neurotransmission	Drugs affecting muscarinic neurotransmission used in therapy
Ulcer	Muscarinic antagonist
Diarrhea	Muscarinic antagonist
Postabdominal distension	Muscarinic agonist
Gastroparesis	Muscarinic agonist
Paralytic ileus	Muscarinic antagonist
Gastroesophageal reflux disease	Muscarinic agonist/antagonist
Xerostomia	Muscarinic antagonist
Nausea and vomiting	Muscarinic antagonist
Motion sickness	Muscarinic antagonist
Irritable bowel syndrome (IBS-D)-diarrhea	Muscarinic antagonist
Irritable bowel syndrome (IBS-C)-constipation	Muscarinic agonist

other medicines. Muscarinic antagonists along with dopamine antagonists and setron drugs are used to reduce nausea and vomiting. Motion sickness (airsickness, carsickness, and seasickness) is a common problem in people traveling by car, train, airplanes, and especially boats. Motion sickness can start suddenly, with a queasy feeling and cold sweats. It can then lead to nausea, dizziness, and vomiting. Muscarinic antagonists such as scopolamine and others discussed above are widely used to reduce the symptoms in motion sickness. Irritable bowel syndrome (IBS) is a problem that affects the large intestine. It can cause abdominal cramping, bloating, and a change in bowel habits. Some people with the disorder have constipation and some have diarrhea. Based on the symptoms, diarrhea or constipation, muscarinic antagonists or muscarinic agonists are used, respectively. In Table 11.7, we have summarized the muscarinic receptor-related pathologies in the GIT and the possible muscarinic receptor-based therapy.

11.2.6 Discovery of Muscarine and Its Relevance in the Etiology and Treatment of Bladder Disorders

The detrusor, trigone, and sphincter muscle in the urinary tract are controlled by the autonomic nervous system. Activation of the parasympathetic nervous system contracts detrusor muscle through acetylcholine-mediated M_3 receptor activation, and M_3 activation also relaxes trigone and sphincter muscle. Therefore acetylcholine's action result on muscarinic receptors causes detrusor muscle to contract and the bladder to empty, and when there is overstimulation of this system, uncontrollable spasm and contraction lead to overactive bladder (OAB). The major clinical features of OAB include urinary frequency, urinary urgency, incontinence, and accidental loss of urine that occurs after the strong and sudden urge to urinate. Uncontrollable detrusor muscle contraction also leads to nocturia (waking up one or more times

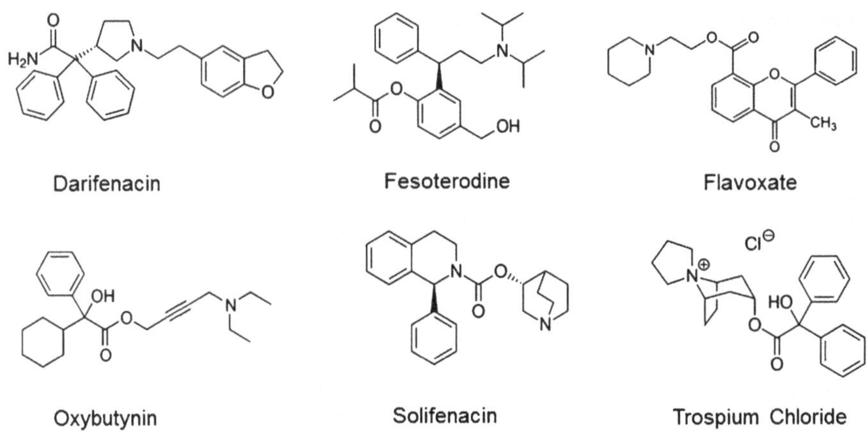

Fig. 11.9 Antimuscarinic drugs used in overactive bladder (OAB)

Fig. 11.10 Cholinomimetics used to treat bladder atony

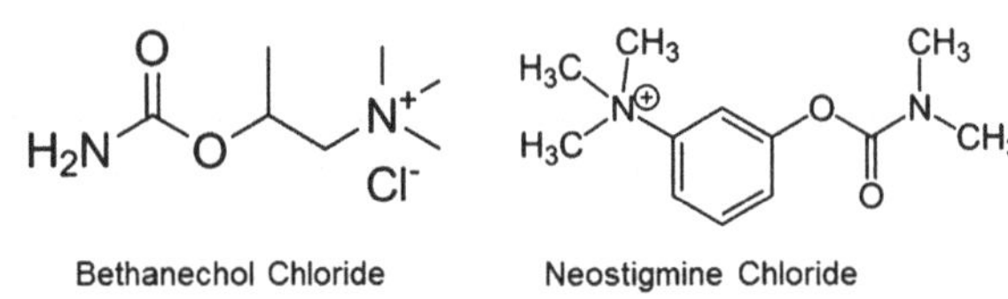

Table 11.8 List of pathological conditions in the urinary bladder associated with altered muscarinic neurotransmission and the muscarinic receptor-mediated therapeutic approach

Pathological conditions in the *urinary bladder* associated with altered muscarinic neurotransmission	Drugs affecting muscarinic neurotransmission used in therapy
Overactive bladder	Muscarinic antagonist
Nocturia	Muscarinic antagonist
Enuresis	Muscarinic antagonist
Atony of the bladder	Muscarinic agonist

during the night to urinate). Muscarinic antagonists including darifenacin, fesoterodine, flavoxate, oxybutynin, solifenacin, and trospium chloride are the gold standards in the treatment of overactive bladder and nocturia (Fig. 11.9). Enuresis (bed wetting) in children also can be treated with muscarinic antagonist.

However, in atony of the bladder, there is lack of tone or tension, and there is muscle relaxation and flaccidity of the bladder. Bladder atony leads to inability to empty the bladder completely, and it is usually a disorder of the voiding phase of micturition. In atony of the bladder, cholinomimetics such as the muscarinic agonist bethanachol and the acetylcholinesterase inhibitor neostigmine are used in therapy (Fig. 11.10). Please see Table 11.8 for the list of bladder pathologies and therapies associated with muscarinic receptor.

Muscarine initially helped the scientists with the characterization of the cholinergic nervous system. The representation and understanding of the physiology of the cholinergic system led to the diagnosis and treatment of various central and peripheral nervous system disorders. In the next section, we discuss about the different types of edible mushrooms, their mechanisms of action, and its indication in various central and peripheral pathological conditions. We have looked into disease states associated with each organ and the efficacy of mushrooms in curing various disease states.

11.3 Conclusion

A combination of various signs and symptoms can occur with muscarinic poisoning from mushrooms. Manifestations may vary among individuals who consumed mushrooms. Accurate diagnosis depends on clinical suspicion and recognition of muscarinic manifestations, particularly diaphoresis, salivation, bladder cramping, diarrhea, and difficulty with visual accommodation. However, confusion can occur if mushroom poisoning is attributed to a suspected species rather than to the toxin suggested by signs and symptoms. Muscarinic toxicity due to cholinergic medications necessitates an adjustment in drug dosage. Extreme stimulation or inhibition of muscarinic receptors is an important component of the pathophysiology of a variety of disease states. Recent advances in molecular and structural biology have led to a better understanding of the structure, function, and regulation of mAChRs and their associated signaling pathways. This critical discovery has revolutionized the medical therapy and led to potential therapeutic outcomes to treat hundreds of pathological conditions associated with central and peripheral nervous system.

References

Abrams P, Andersson KE, Buccafusco JJ, Chapple C, Groat WC, Fryer AD, Kay G, Laties A, Nathanson NM, Pasricha PJ, Wein AJ (2006) Muscarinic receptors: Their distribution and function in body systems, and the implications for treating overactive bladder. Br J Pharmacol 148:565–578. https://doi.org/10.1038/sj.bjp.0706780

Barlow RB, Berry KJ, Glenton PAM, Nikolaou NM, Soh KS (1976) A comparison of affinity constants for muscarine-sensitive acetylcholine receptors in Guinea-pig atrial pacemaker cells at 29°c and in ileum at 29°C and 37°C. Br J Pharmacol 58:613–620. https://doi.org/10.1111/j.1476-5381.1976.tb08631.x

Benjamin DR (1995) Mushrooms: Poisons and panaceas: A handbook for naturalists, mycologists, and physicians, 1st edn. W.H. Freeman & Company, New York

Brown DA, Fatherazi S, Garthwaite J, White RD (1980) Muscarinic receptors in rat sympathetic ganglia. Br J Pharmacol 70:577–592. https://doi.org/10.1111/j.1476-5381.1980.tb09777.x

Deshmukh SK, Verekar SA, Natarajan K (2006) Poisonous and hallucinogenic mushrooms of India. Intl J Med Mushrooms 8:251–262. https://doi.org/10.1615/IntJMedMushr.v8.i3.70

Hammer R, Giachetti A (1982) Muscarinic receptor subtypes: M1 and M2 biochemical and functional characterization. Life Sci 31:2991–2998

Hammer R, Berrie CP, Birdsall NJ, Burgen AS, Hulme EC (1980) Pirenzepine distinguishes between different subclasses of muscarinic receptors. Nature 283:90–92

Riker WF, Wescoe WC (1951) The pharmacology of flaxedil, with observations on certain analogs. Ann N Y Acad Sci 54:373–394. https://doi.org/10.1111/j.1749-6632.1951.tb39932.x
Roszkowski AP (1961) An unusual type of sympathetic ganglionic stimulant. J Pharmacol Exp Ther 132:156–170
Saviuc P, Danel V (2006) New syndromes in mushroom poisoning. Toxicol Rev 25:199–209
Wess J (1996) Molecular biology of muscarinic acetylcholine receptors. Crit Rev Neurobiol 10:69–99
Wess J, Eglen RM, Gautam D (2007) Muscarinic acetylcholine receptors: Mutant mice provide new insights for drug development. Nat Rev Drug Discov 6:721–733. https://doi.org/10.1038/nrd2379

Current Research on Medicinal Mushrooms in Italy

12

Giuseppe Venturella, Paola Saporita, and Maria Letizia Gargano

Contents

Abstract

Historical data and current research in Italy reveal the permanent interest of people and scientists toward the importance of fungi as functional food and medicine. This chapter reports the main studies on medicinal mushrooms carried out by Italian researchers on the antibacterial activity, the characterization of β-glucans content, antitumor activity, Alzheimer's disease, characterization of bioactive compounds, antimicrobial activity, dietary supplementation, chemical

Authors Giuseppe Venturella, Paola Saporita and Maria Letizia Gargano have been equally contributed to this chapter.

G. Venturella (✉) · P. Saporita
Department of Agricultural, Food and Forest Sciences, University of Palermo, Palermo, Italy
e-mail: giuseppe.venturella@unipa.it

M. L. Gargano
Department of Earth and Marine Sciences, University of Palermo, Palermo, Italy

D. C. Agrawal, M. Dhanasekaran (eds.), *Medicinal Mushrooms*,
https://doi.org/10.1007/978-981-13-6382-5_12

contents, and animal health. Currently, in vitro experiments are prevalent with respect to clinical trials on human and animals.

Keywords
Medicinal mushrooms · Health benefits · Pharmacological activities · Mediterranean area · Italy

Abbreviations

FMG	Fungus myceliated grains
HPV	Human papillomavirus
LXA4	Lipoxin A4
MBC	Minimal bactericidal concentrations
MIC	Minimum inhibitory concentrations
MOP	Mycotherapic plus homeopathic (thymic hormones) and a probiotics
PCA	Androgen-independent human prostate cancer
SAF	Agricultural and Forest Sciences

12.1 Introduction

Since the first and second century AD, knowledge about medicinal mushrooms was widespread in Italy.

Pedanius Dioscorides reported the medicinal properties of an *Agaricus*, later identified as *Fomitopsis officinalis* (Vill.) Bondartsev & Singer. At that time, the "agaricin," an organic acid extracted from the fungus *F. officinalis*, was already used against tuberculosis sweating.

An ethnomycological investigation carried out by Bernardino da Ucria (1789) highlighted the use of *Bovista* and *Langermannia* species as curative for amputations, hemorrhages, hemorrhoids, ulcers, and excoriations.

In the "Commentario della farmacopea italiana" (Commentary of the Italian Pharmacopoeia), by Guareschi (1897), the "agaricin" is reported as antidiaphoretic in a patient affected by tuberculosis and arthritis.

Polypores were also part of Ötzi's (the Similaun Man) first aid kit, two pieces of birch fungus [*Fomitopsis betulina* (Bull.) B.K. Cui, M.L. Han & Y.C. Dai] strung on leather thongs. The fungus, which includes both anti-inflammatory and antibacterial compounds, was highly valued for its medicinal properties (Capasso 1998). The powder of the birch fungus has been used as food by Otzi (Pelkonen et al. 2017) to eliminate the intestinal parasites (*Trichuris trichiura* L.).

The Italian Renaissance Scholar, Ermolao Barbaro il Giovane (Bernardi 1851), highlights the use of *Fomes fomentarius* (L.) Fr. to produce a cotton-like mass with hemostatic action.

An Italian researcher (Tiberio 1895) reported the antibacterial power of some extracts of mold. His studies anticipated the discovery of the drug penicillin by Alexander Fleming 35 years later (Bucci and Galli 2011).

Despite these interesting historical data, for a long time, however, the Italian research on medicinal mushrooms has not been in step with that carried out in the East and in other European countries.

The publication of the first volume of the International Journal of Medicinal Mushrooms in 1999 has made an important contribution to the increase in interest also in Italy toward medicinal mushrooms by many researchers.

12.2 Studies on Medicinal Mushrooms in Italy

Medicinal mushrooms offer important health benefits and exhibit a broad spectrum of pharmacological activities (Gargano et al. 2017).

The studies on medicinal mushrooms carried out by Italian researchers are mainly oriented toward the antibacterial activity, the characterization of β-glucans content, antitumor activity, Alzheimer's disease, characterization of bioactive compounds, antimicrobial activity, dietary supplementation, chemical contents, and animal health. Currently, in vitro experiments are prevalent if compared with clinical trials on human and animals (Table 12.1).

12.2.1 Antibacterial Activity

Smania et al. (1999) isolated three sterols from *Ganoderma applanatum* (Pers.) Pat.: 5α-ergost-7en-3β-ol, 5α-ergost-7,22-dien-3β-ol, and 5,8-epidioxy-5α,8α-ergost-6,22-dien-3β-ol, as well as a novel lanostanoid. The antibacterial activity of these compounds was determined by minimal inhibitory concentration (MIC) and minimal bactericidal concentration (MBC). The antimicrobial activity was tested on *Bacillus cereus* MIP 96016; *Corynebacterium diphtheriae* MIP 96048, *E. coli* ATCC 25922, *Pseudomonas aeruginosa* ATCC 27853, *Staphylococcus aureus* ATCC 25923, *S. saprophyticus* MIP 97018, and *Streptococcus pyogenes* MIP 83111. The Gram-positives were more sensitive (MICs of 0.003 to 2.0 mg/ml; MBCs of 0.06 to 4.0 mg/ml) than the Gram-negatives (MICs of 1.0 to 4.0 mg/ml; MBCs of 2.0 to >4.0 mg/ml).

Schillaci et al. (2013) highlighted the effects of some extracts of *Pleurotus* species against a group of bacterial reference strains of medical relevance: *S. aureus* ATCC 25923, *S. epidermidis* RP62A, *P. aeruginosa* ATCC 15442, and *Escherichia coli* ATCC10536.

Schillaci et al. (2017) and Cusimano et al. (2017) with the aim to obtain novel agents against bacteria of clinical importance focused on the edible desert truffles *Tirmania pinoyi* (Maire) Malençon, *Terfezia claveryi*, Chatin, and *Picoa juniperi* Vittad. as sources of new antimicrobial agents. They investigated in vitro antibacterial activity of acid-soluble protein extracts (aqueous extracts) of the

Table 12.1 Active compounds, medicinal properties, and applications of medicinal mushrooms

Taxa	Active compounds	Medicinal properties	Applications		References
			Human	Animals	
Agaricus blazei	Agaritine	Immunostimulant, prevention and treatment of atherosclerosis, hepatitis, hyperlipidemia, diabetes, dermatitis, cancer	In vitro, in vivo	–	Firenzuoli et al. (2008)
Auricularia auricula-judae	β-glucans	Prevention of arterial hypertension	In vivo	–	Ardigò (2017)
Cortinarius caperatus	β-(1 → 6)-D-glucan	Antioxidant	In vitro	–	Zacchigna et al. (2017)
Cyclocybe aegerita	Proteins, carbohydrates, lipids	Antiradical	In vitro	–	Landi et al. (2017)
	β-glucans	Inhibition ochratoxigenic microfungi	In vitro	–	Ricelli et al. (2002)
Cordyceps sinensis	β-glucans	Prevention of the side effects of adjuvant chemotherapy (breast cancer)	In vivo	–	Crosta et al. (2017)
Fomitopsis pinicola	β-glucans	Antifungal, antitumor	In vitro, in vivo	–	Angelini et al. (2017)
Ganoderma applanatum	5α-ergost-7en-3β-ol, 5α-ergost-7,22-dien-3β-ol, and 5,8-epidioxy-5α,8α-ergost-6,22-dien-3β-ol, applanoxidic acid	Antibacterial, antimicrobial	In vitro	–	Smania et al. (1999)
Ganoderma lucidum	β-glucans	Antioxidant, improve hormone production	In vivo	–	Oliviero (2017)
	Phenol	Antioxidant	In vitro	–	Saltarelli et al. (2017)
	β-glucans	Immunostimulant	In vitro	–	Ferrari et al. (2017)
	β-glucans	Antioxidant	In vivo	–	Rossi et al. (2014)
	β-glucans	Virologic and clinical efficacy	In vivo	–	Scaglione et al. (2017)
	β-glucans	Prevention of the side effects of adjuvant chemotherapy (breast cancer)	In vivo	–	Crosta et al. (2017)

Grifola frondosa	β-glucans	Antibacterial	In vitro	–	Saporita et al. (2017)
	β-glucans	Antioxidant, improve hormone production	In vivo	–	Oliviero (2017)
	β-glucans	Virologic and clinical efficacy	In vivo	–	Scaglione et al. (2017)
	β-glucans	Prevention of the side effects of adjuvant chemotherapy (breast cancer)	In vivo	–	Crosta et al. (2017)
Hericium erinaceus	β-glucans	Neuroprotective	–	In vivo	Trovato et al. (2016b, 2017)
	β-glucans	Anti-inflammatory and antibacterial, the cytotoxic effect on cancer cells	In vivo	–	Cazzavillan and Gregori (2017)
	Polysaccharides	Sleep and mood disorders	In vivo	–	Rossi et al. (2017)
Lentinula edodes	β-glucans, vitamin D	Antibacterial, antioxidant	In vitro	–	Parola et al. (2017a, b)
	β-glucans, chitosan	Antiproliferative	In vitro	–	Manzi and Pizzoferrato (2000)
	β-glucans, lentinan	Antioxidant, inhibition aflatoxin production	In vitro	–	Fanelli et al. (2000)
	β-glucans	Inhibition ochratoxigenic microfungi	In vitro	–	Ricelli et al. (2002)
	β-glucans	Inhibition of aflatoxin	In vitro	–	Reverberi et al. (2005, 2011)
	β-glucans	Antioxidant, improve hormone production	In vivo	–	Oliviero (2017)
	β-glucans	Antioxidant	In vitro	–	Tolaini et al. 2010
	β-glucans	Virologic and clinical efficacy	In vivo	–	Scaglione et al. (2017)
	β-glucans	Prevention of the side effects of adjuvant chemotherapy (breast cancer)	In vivo	–	Crosta et al. (2017)

(continued)

Table 12.1 (continued)

Taxa	Active compounds	Medicinal properties	Applications		References
			Human	Animals	
Lentinula edodes (fungus myceliated grains)	β-glucans	Antioxidant, reduction of intestinal parasite infection	–	In vivo	Bonanno et al. (2017)
Leucopaxillus gentianeus	Cucurbitacin B, cucurbitacin D, 16-deoxycucurbitacin B, leucopaxillones, 18-deoxyleucopaxilloneA	Antiproliferative	In vitro	–	Clericuzio et al. (2006)
Ophiocordyceps sinensis	β-glucans	Antioxidant	In vivo	–	Rossi et al. (2014)
	β-glucans	Antioxidant, improve hormone production	In vivo	–	Oliviero (2017)
	β-glucans	Virologic and clinical efficacy	In vivo	–	Scaglione et al. (2017)
Pholiota spumosa	Putrescine-1,4-dicinnamide	Antitumor (prostate cancer)	In vitro	–	Russo et al. (2007)
Picoa juniperi	Peptides	Antibacterial, antimicrobial	In vitro	–	Schillaci et al. (2017)
Pleurotus eryngii var. *elaeoselini*	β-glucans	Antibacterial	In vitro	–	Schillaci et al. (2013)
Pleurotus eryngii var. *eryngii*	β-glucans, chitosan	Antiproliferative	In vitro	–	Manzi and Pizzoferrato (2000)
	β-glucans	Inhibition ochratoxigenic microfungi	In vitro	–	Ricelli et al. (2002)
	β-glucans	Antioxidant, antimicrobial and anti-inflammatory	In vitro	–	Salviato et al. (2017)
	β-glucans	Antibacterial	In vitro	–	Schillaci et al. (2013)
	β-glucans	Antitumor (colon cancer)	–	In vivo	Marino Gammazza et al. (2017)
	(1 → 6)-α-glucan, (1 → 6)-β-glucan, ergosterol-3-O-β-linoleate	Antioxidant and antitumor	In vitro	–	Cateni et al. (2017)

Pleurotus eryngii var. *ferulae*	β-glucans	Antitumor (colon cancer)	In vitro	–	Fontana et al. (2014)
	β-glucans	Antibacterial	In vitro	–	Schillaci et al. (2013)
Pleurotus nebrodensis	β-glucans	Antibacterial	In vitro	–	Schillaci et al. (2013)
	β-glucans	Antitumor (colon cancer)	In vitro	–	Fontana et al. (2014)
Pleurotus ostreatus	β-glucans, vitamin D	Antibacterial, antioxidant	In vitro	–	Parola et al. (2017a, b)
	β-glucans, chitosan	Antiproliferative	In vitro	–	Manzi and Pizzoferrato (2000)
Pleurotus pulmonarius	β-glucans, chitosan	Antiproliferative	In vitro	–	Manzi and Pizzoferrato (2000)
Polyporus umbellatus	β-glucans	Prevention arterial hypertension	In vivo	–	Ardigò (2017)
Pseudoinonotus dryadeus	Cerevisterol, sphingosine, diacylglycerophospholipids	Antioxidant	In vitro	–	Cateni et al. (2015)
Russula rosea	Sesquiterpenes	Antiproliferative	In vitro	–	Clericuzio et al. 2012
Russula amarissima	Sesquiterpenes	Antiproliferative	In vitro	–	Clericuzio et al. 2012
Terfezia claveryi	Peptides	Antibacterial, antimicrobial	In vitro	–	Schillaci et al. (2017), Cusimano et al. (2017)
Tirmania pinoyi	Peptides	Antibacterial, antimicrobial	In vitro	–	Schillaci et al. (2017), Cusimano et al. (2017)
Trametes versicolor	β-glucans	Antioxidant, inhibition of aflatoxins	In vitro	–	Scarpari et al. (2016)
	β-glucans	Neuroprotective	–	In vivo	Trovato et al. (2016a)

abovementioned three species against the Gram-positive human pathogenic reference strain *S. aureus* ATCC 29213 and the Gram-negative strain *P. aeruginosa* ATCC 15442. The acid-soluble protein extracts of *T. pinoyi* and *T. claveryi* showed minimum inhibitory concentrations of 50 μg/mL against tested pathogens. Besides the same authors identified new antimicrobial peptides as potential substances to fight pathogens such as *S. aureus* and *P. aeruginosa* of great clinical importance, very common in hospital settlement, but also in animal health and in the field of food safety.

Parola et al. (2017a, b) analyzed the crude water extracts of *Lentinula edodes* (shiitake) and *Pleurotus* sp., which showed antibacterial activity against *S. aureus* and *P. aeruginosa*. In particular, they also put in comparison the best performing strain extract with the commercial antibiotic ceftriaxone against *P. aeruginosa*, assessing that 20 mg of crude extract corresponds to 0.2 mg of the pure antibiotic when studied by means of disk diffusion assay.

Saporita et al. (2017) analyzed the cold extracts of a rare white form of maitake [*G. frondosa* (Dicks.) Gray] which is effective in inhibiting the growth of *S. epidermidis* 12228 and *P. aeruginosa* 15442 at the maximum screening concentration of 50% v/v.

12.2.2 Antitumor Activity

As regards the evaluation of antitumor action, there are a number of surveys which report in vitro activities of different fungal species.

Fontana et al. (2014) evaluated in vitro antitumor effects of the cold water extracts of some *Pleurotus* species on human colon cancer cells. The treatment of human colon cancer HCT116 cells with cold water extracts of *Pleurotus eryngii* (DC.) Quél. var. *ferulae* (Lanzi) Sacc. and *Pleurotus nebrodensis* (Inzenga) Quél. resulted in significant inhibition of the viability of tumor cells and promoted apoptosis. The extracts were also able to inhibit cell migration and affected homotypic and heterotypic cell-cell adhesion.

Russo et al. (2007) reported the activity in vitro of putrescine-1,4-dicinnamide, a phenylpropanoid derivative conjugated with polyamine putrescine isolated from *Pholiota spumosa* (Fr.) Singer. In particular, they investigated the response of DU-145 cells, a well-characterized androgen-independent human prostate cancer (PCA) cell line, to this phenylpropanoid derivative. The putrescine-1,4-dicinnamide inhibits the cell growth of cancer cells inducing apoptosis cell death.

Marino Gammazza et al. (2017) indicated a new generation of "biotherapeutics" and evaluated the anticancer effect of *P. eryngii* var. *eryngii* extract in an animal model of ectopically implanted C26 colon carcinoma, widely used as an experimental model of cancer cachexia.

12.2.3 Nutritional Value, Chemical Content, and Antiproliferative Activity of Medicinal Mushroom Extracts

Manzi and Pizzoferrato (2000) characterized the β-glucans content of cultivated edible mushrooms such as *Pleurotus ostreatus* (Jacq.) P. Kumm., *P. pulmonarius* (Fr.) Quél., *P. eryngii* (DC.) Quél., and *Lentinula edodes* (Berk.) Pegler. The investigation shows that β-glucans in mushrooms are distributed both in the soluble and in the insoluble dietary fractions. The physiological effects of mushrooms are due to a cumulative action of different fiber components (soluble β-glucans, chitin/chitosan).

Clericuzio et al. (2006) analyzed basidiomata and mycelium of *Leucopaxillus gentianeus* (Quél.) Kotl. and isolated different compounds (cucurbitacin B, cucurbitacin D, 16-deoxycucurbitacin B, leucopaxillones, and 18-deoxyleucopaxillone). The antiproliferative activity of the isolated triterpenes was determined against the NCI-H460 human tumor cell line, in comparison with the antitumor compound topotecan, a well-known topoisomerase I inhibitor.

2-acylcyclopentene-1,3-dione derivatives were isolated by Gilardoni et al. (2007) from *Hygrophorus chrysodon* (Batsch) Fr. The authors reported interesting antifungal activity of chrysotriones against *Fusarium fujikuroi* Nirenberg [sub: *F. verticillioides* (Sacc.) Nirenberg], one of the common worldwide pathogens of cultivated plants.

Aristolane sesquiterpenes and moderate cell growth inhibitory activity were reported for *Russula rosea* Pers. [sub: *R. lepida* Fr.] and *R. amarissima* Romagn. & E.-J. Gilbert. (Clericuzio et al. 2012).

Cateni et al. (2017) isolated two water-soluble glucans from *P. eryngii* var. *eryngii* extracts. The two fractions PEPS-1 and PEPS-2 were characterized, and a structure of the repeating unit of the glucan was determined as linear (1 → 6)-α-glucan and (1 → 6)-β-glucan, respectively. The methanolic extract of *P. eryngii* var. *eryngii* was also investigated, and a complex mixture of triglycerides together with ergosterol and ergosterol-3-O-β-linoleate were identified.

Saltarelli et al. (2017) showed a higher level of total phenol content in young primordia of *Ganoderma lucidum* (Curtis) P. Karst. strains.

Landi et al. (2017) characterized the nutritional value, antiradical properties and chemical composition of wild edible basidiomata of *Cyclocybe aegerita* (V. Brig.) Vizzini [sub: *Agrocybe aegerita* (Brig.) Sing.].

The role of β-glucans content from *G. lucidum* mixed with medicinal plants in dietary supplements was evaluated by Ferrari et al. (2017). The components of the mixing dietary supplements are able to modulate the immune system increasing the natural defenses of the human body.

12.2.4 Antioxidant Activity

The in vitro antioxidant activity of *Pseudoinonotus dryadeus* (Pers.) T. Wagner & M. Fisch. was tested by Cateni et al. (2015). The extract of the fungus showed free radical-scavenging activity compared with the positive control quercetin. A complex mixture of free fatty acids, cerevisterol, sphingosine, and a complex mixture of diacylglycerophospholipids were also isolated from P. *dryadeus*.

Zacchigna et al. (2017) highlighted the antioxidant activity of polysaccharides in hot water extract of edible mushrooms *Cortinarius caperatus* (Pers.) Fr. In particular, a water-soluble fraction characterized as a β-(1 → 6)-D-glucan was for the first time isolated from *C. caperatus*. This compound showed significant antioxidant activity.

12.2.5 Medicinal Mushrooms against Aflatoxins in Food and Feeds

The effect of bioactive compounds produced by some *Basidiomycetes* on the growth of *Aspergillus* spp. and the synthesis of aflatoxins was investigated by several authors. Fanelli et al. (2000) reported that mycelia of *L. edodes*, incubated on wheat seeds for 20 and 30 days and subsequently inoculated with *Aspergillus parasiticus* Speare, delayed the fungal growth of the toxigenic strain and inhibited aflatoxin production.

Ricelli et al. (2002) analyzed the effects of the extracts of *L. edodes*, *C. aegerita*, and *P. eryngii* in the inhibition of the growth of two ochratoxigenic microfungi *Aspergillus ochraceus* G. Wilh. and *Penicillium verrucosum* Dierckx.

Reverberi et al. (2005) analyzed the inhibition of aflatoxin production by *L. edodes* culture filtrates. *L. edodes* filtrates play a role as external stimulus affecting the antioxidant status in the fungal cell that, in turn, leads to aflatoxin inhibition.

L. edodes was also tested as an agent which enhances the biocontrol activity of *Papiliotrema laurentii* (Kuff.) Xin Zhan Liu, F.Y. Bai, M. Groenew. & Boekhout [sub: *Cryptococcus laurentii* (Kuff.) C.E. Skinner] against *Penicillium expansum* Link contamination and patulin production in apple fruits. In vitro *L. edodes* culture filtrates improved the growth of *P. laurentii* and the activity of its catalase, superoxide dismutase, and glutathione peroxidase, which play a key role in oxidant scavenging. In addition, LF23 also delayed *P. expansum* conidia germination (Tolaini et al. 2010).

In another study, Reverberi et al. (2011) showed that the use of paraquat 0.5 and 1 mM resulted in an enhancement of the expression of the β-glucan synthase gene Lefks1 and a consequent stimulating effect (about 30–35%) on β-glucan production. Moreover, oxidative stress (PQ)-induced polysaccharides showed higher aflatoxin inhibiting capacity in two different strains of *A. parasiticus* in comparison with non-induced polysaccharides.

Scarpari et al. (2016) evaluated the effect of *Trametes versicolor* (L.) Lloyd bioactive compounds in the stimulation of the antioxidant system of *Aspergillus* mycelia. These bioactive compounds act by modulating the fungal antioxidant system, leading to the inhibition of aflatoxins in food and feed.

12.2.6 Case Reports

Firenzuoli et al. (2008) highlighted the role of *Agaricus blazei* Murrill in the prevention and treatment of atherosclerosis, hepatitis, hyperlipidemia, diabetes, dermatitis, and cancer.

Rossi et al. (2014) reported the capacity of protection from oxidative stress of mixed dietary supplements based on *G. lucidum* and *Ophiocordyceps sinensis* (Berk.) G.H. Sung, J.M. Sung, Hywel-Jones & Spatafora [sub: *Cordyceps sinensis* (Berk.) Sacc.] on amateur cyclists involved in cycling races. The results showed that, after 3 months of the administration, the testosterone/cortisol ratio changed in a statistically significant manner. The cyclists were thus protected from nonfunctional overreaching and overtraining syndrome. After 3 months of fungal integration, the data show a greater ability to scavenge free radicals in the serum of cyclists after the race.

Trovato et al. (2016a) highlighted the neuroprotective action of *T. versicolor* [sub: *Coriolus versicolor* (L.) Quél.] administered orally to the rat. They observed the highest induction of lipoxin A4 in the cortex and the hippocampus of brains of rats. The activations of the LXA4 signal and the modulation of the proteins that respond to stress could serve as a potential therapeutic target for Alzheimer's disease-related inflammation and for neurodegenerative damage.

Trovato et al. (2016b) provided evidence of the neuroprotective action of *Hericium erinaceus* (Bull.) Pers. when administered orally to rat. In the brain of rats receiving the fungus, maximum induction of LXA4 was observed in cortex and hippocampus followed by "substantia nigra" striatum and cerebellum.

Trovato et al. (2017) underlined that oxidative stress and altered antioxidant systems had been considered an important factor underlying the pathogenesis of Alzheimer's disease. They confirm the role of lipoxin A4 and inflammasome in the neurodegeneration and the importance of the use of *T. versicolor* against neuroinflammation and mitochondrial dysfunction in the pathogenesis of Alzheimer's disease.

Angelini et al. (2017) evaluated the antioxidant activity and the antiproliferative and pro-apoptotic effect of *Fomitopsis pinicola* (Sw.) P. Karst. in adherent and not adherent human tumor cells, apoptosis, and the genotoxicity. Antifungal activity was tested against cultures of different human pathogenic fungi.

Oliviero (2017) tested a preparation based on shiitake, maitake, reishi, and *Cordyceps mushrooms* during a 6-month therapy. The treatment showed a clear improvement of patient's clinical conditions and a normalization or clear improvement of lab values, even in patients undergoing chemotherapy. The study confirmed that shiitake, maitake, reishi, and *Cordyceps* mushrooms have a protective and radical scavenger action on the endogenous antioxidant system, restore and improve hormone production, drain mesenchyme from metabolic wastes thus balancing pH and having toxic metals chelation effects, act on the intestinal ecosystem, protect the hepatic cell, and improve NO-mediated cell oxygenation.

Scaglione et al. (2017) treated 32 patients, 20 females, and 12 males, with HPV DNA positive test. All patients were treated with "MOP therapy": a combination of a mycotherapic plus homeopathic (thymic Hormones) and probiotics. After 3 months of treatment of 22 patients, 20 females and 2 males (68.75%) had a negative HPV test. MOP therapy showed virologic and clinical efficacy.

Ardigò (2017) highlighted the efficacy of *Auricularia auricula-judae* (Bull.) Quél. and *Polyporus umbellatus* (Pers.) Fr. in restoring high arterial hypertension to normal values. *A. auricula-judae* produces a hypotensive effect due to adenosine which acts through two mechanisms of action: (a) a vasodilator effect leading to dilation of the arterial vessel and (b) an anti-adrenergic effect, which prevents vasoconstriction and the resulting increase in blood pressure due to the action of adrenaline. Ergone in *P. umbellatus* has an effect similar to that of spironolactone, an antialdosteronic diuretic medicine. This compound also regulates aquaporins, the membrane proteins of kidney cells, hence eliminating the excess of sodium and water.

Cazzavillan and Gregori (2017) highlighted the role of *H. erinaceus* as a potential anti-inflammatory agent in gastrointestinal diseases. In particular, the administration of *H. erinaceus* food supplements in patients with abdominal pain, blood in stool, fatigue in remission and life quality improvement resulted in anti-inflammatory effects and a significant reduction of calprotectin in feces.

Rossi et al. (2017) reported the positive effects of *H. erinaceus* supplementation in improving mood and sleep disorders in a group of overweight female subjects treated with hypocaloric diet.

Salviato et al. (2017) indicated *P. eryngii* as a promising natural source of bioactive substances that may help reduce macrophage polarization in subjects affected by metabolic syndrome.

Crosta et al. (2017) tested a preparation based on shiitake, maitake, reishi, and *Cordyceps* as support of traditional therapies in the prevention of the side effects of adjuvant chemotherapy in breast cancer, especially of nausea and asthenia.

12.2.7 Culture Collections

The main culture collections of medicinal mushroom strains are currently kept in the University of Pavia and Palermo. The researchers from Pavia's University (Savino et al. 2014, 2017) selected strains of *H. erinaceus* and stored more than 150 fungal strains including the rare *Ganoderma pfeifferi* Bres., *Fomitopsis officinalis*, *Cellulariella warnieri* (Durieu & Mont.) Zmitr. & Malysheva (sub: *Lenzites warnieri* Durieu & Mont.), and *Perenniporia meridionalis* Decock & Stalpers. In the Herbarium SAF of the Department of Agricultural, Food and Forest Science of the University of Palermo (Gargano 2017; Venturella et al. 2016), a number of medicinal mushroom strains of great commercial and nutraceutical interest are currently available.

12.2.8 Medicinal Mushrooms in Animal Production and Health

Trials on animals are very scarce in Italy, and only a contribution was recently presented during the ninth International Medicinal Mushrooms Conference (IMMC9) by Bonanno et al. (2017). The authors reported the preliminary results of the effects of diets supplemented with fungus myceliated grains (FMG) to lactating ewes in terms of intestinal parasites control, milk production, and cheese oxidative stability. In front of analogous dry matter and nutrient intake, the ewes fed FMG at higher levels showed a reduction of intestinal parasite infection, a tendency to improve milk yield and higher milk casein content. The increase of FMG in the diet was responsible for the production of cheeses with lower secondary lipid oxidation and a higher antioxidant capacity, suggesting a major oxidative stability of cheese fat and an enrichment of cheese in antioxidant compounds induced by the FMG.

12.3 Conclusions

In the last years, the interest on medicinal mushrooms is progressively increasing among the population. Italian scientists, like those of other European countries, are trying to fill the gap of knowledge against Eastern countries on the use of medicinal mushrooms as functional food and/or integrated medicine. The nutraceutical industry in Italy is attracted by fungi and their potential application as a functional food and therapeutic products. The number of private industries producing nutraceuticals based on mushrooms is constantly growing in the last decade. The main problem for the industries is represented by the finding of the dry extracts being currently forced to buy it from abroad with a little guarantee on the origin of the product and on its certification. In light of the increasing number of in vitro experiments and trials on human and domestic and farm animals, it is necessary to create the conditions for the cultivation of different species of medicinal mushrooms on the Italian territory able to guarantee significant quantities of product to be allocated not only in the food sector but also in the nutraceutical sector.

References

Angelini P, Rosignoli P, Bistocchi G, Arcangeli A, Riccioni C, Belfiori B, Rubini A, Tirillini B, Fabiani R, Venanzoni R (2017) *Fomitopsis pinicola*, a promising pharmacological macrofungus. In: Gargano ML, Venturella G (eds.), Book of abstracts: the 9th international medicinal mushrooms conference (IMMC9), Palermo 24–28 September 2017

Ardigò W (2017) *Auricularia auricula-judae* and *Polyporus umbellatus* effectively contrast arterial hypertension: a clinical study. In: Gargano ML, Venturella G (eds) Book of abstracts: the 9th international medicinal mushrooms conference (IMMC9), Palermo 24–28 September 2017

Bernardi J (1851) Ermolao Barbaro o la scienza del pensiero dal secolo decimoquinto a noi., Venezia

Bernardino da Ucria (1789) Horthus Regius Panhormitanus. Tipis Regis Panormi, pp 498

Bonanno, A., Di Grigoli A, Vitale F, Di Miceli G, Todaro M, Alabiso M, Gargano ML, Venturella G, Anike FN, Isikhuemhen OS (2017) Effects of feeding diets supplemented with fungus myceliated grains on some production, health and oxidation traits of dairy ewes. In: Gargano ML, Venturella G (eds) Book of abstracts: the 9th international medicinal mushrooms conference (IMMC9), Palermo 24–28 September 2017

Bucci R, Galli P (2011) Vincenzo Tiberio: a misunderstood researcher. Ita J Pub Health 8:404–406. https://doi.org/10.2427/5688

Capasso L (1998) 5300 years ago, the Iceman used natural laxatives and antibiotics. Lancet 352:1864. https://doi.org/10.1016/S0140-6736(05)79939-6

Cateni F, Zacchigna M, Altieri T, Procida G, Cichelli A (2015) Antioxidant properties of oak bracket mushroom, *Pseudoinonotus dryadeus* (Higher Basidiomycetes): A mycochemical study. Int J Med Mushrooms 17:627–637. https://doi.org/10.1615/IntJMedMushrooms.v17.i7.30

Cateni F, Venturella G, Gargano ML, Zacchigna M, Procida G (2017) Mycochemical investigation of *Pleurotus eryngii* var. *eryngii*. In: Gargano ML, Venturella G (eds) Book of abstracts: the 9th international medicinal mushrooms conference (IMMC9), Palermo 24–28 September 2017

Cazzavillan S, Gregori A (2017) *Hericium erinaceus* in the management of fecal calprotectin. In: Gargano ML, Venturella G (eds) Book of abstracts: the 9th international medicinal mushrooms conference (IMMC9), Palermo 24–28 September 2017

Clericuzio M, Tabasso S, Bianco MA, Pratesi G, Beretta G, Tinelli S, Zunino F, Vidari G (2006) Cucurbitane triterpenes from the fruiting bodies and cultivated mycelia of *Leucopaxillus gentianeus*. J Nat Prod 69:1796–1799. https://doi.org/10.1021/np060213n

Clericuzio M, Cassino C, Corana F, Vidari G (2012) Terpenoids from *Russula lepida* and *R. amarissima* (*Basidiomycota*, *Russulaceae*). Phytochemistry 84:154–159

Crosta L, Galanti D, Catarella MT, Valerio MR (2017) Micetrin to prevent nausea and asthenia induced by chemotherapy in the adjuvant treatment of breast cancer. In: Gargano ML, Venturella G (eds) Book of abstracts: the 9th international medicinal mushrooms conference (IMMC9), Palermo 24–28 September 2017

Cusimano MG, Schillaci D, Cascioferro SM, Arizza V, Chiaramonte M, Inguglia L, Davino S, Saletti R, Cunsolo V, Venturella G, Gargano ML (2017) New antimicrobial peptides from *Tirmania pinoyi* and *Terfezia boudieri* in the struggle against antibiotic resistance. In: Gargano ML, Venturella G (eds) Book of abstracts: the 9th international medicinal mushrooms conference (IMMC9), Palermo 24–28 September 2017

Dioscoride Pedanio (1544) De materia medica III,1

Fanelli C, Tasca V, Ricelli A, Reverberi M, Zjalic S, Finotti E, Fabbri AA (2000) Inhibiting effect of medicinal mushroom *Lentinus edodes* (Berk.) Sing. (*Agaricomycetideae*) on aflatoxin production by *Aspergillus parasiticus* Speare. Int J Med Mushrooms 2(3):1–8. https://doi.org/10.1615/IntJMedMushr.v2.i3.70

Ferrari B, Insolia V, Priori EC, Occhinegro A, Ratto D, Guglielminetti ML, Savino E, Bottone MG, Rossi P (2017) Dietary supplement based on *Ganoderma lucidum* tested. In: Gargano ML, Venturella G (eds) Book of abstracts: the 9th international medicinal mushrooms conference (IMMC9), Palermo 24–28 September 2017

Firenzuoli F, Gori L, Lombardo G (2008) The medicinal mushroom *Agaricus blazei* Murrill: review of literature and pharmaco-toxicological problems. Evid-Based Complement Alt Med 5:3–15. https://doi.org/10.1093/ecam/nem007

Fontana S, Flugy A, Schillaci O, Cannizzaro A, Gargano ML, Saitta A, De Leo G, Venturella G, Alessandro R (2014) In vitro antitumor effects of the cold water extracts of Mediterranean species of genus *Pleurotus* (Higher Basidiomycetes) on human colon cancer cells. Int J Med Mushrooms 16:49–63. https://doi.org/10.1615/IntJMedMushr.v16.i1.50

Gargano ML (2017) The herbarium SAF fungal culture collection as a potential source of nutraceuticals and cultivated mushrooms. In: Gargano ML, Venturella G (eds) Book of abstracts: the 9th international medicinal mushrooms conference (IMMC9), Palermo 24–28 September 2017

Gargano ML, van Griensven LJLD, Isikhuemhen OS, Lindequist U, Venturella G, Wasser SP, Zervakis GI (2017) Medicinal mushrooms: valuable biological resources of high exploitation potential. Plant Biosyst 151(3):548–565. https://doi.org/10.1080/11263504.2017.1301590

Gilardoni G, Clericuzio M, Tosi S, Zanoni G, Vidari G (2007) Antifungal acylcyclopentenediones from fruiting bodies of *Hygrophorus chrysodon*. J Nat Prod 70:137–139. https://doi.org/10.1021/np060512c

Guareschi I (1897) Commentario della farmacopea italiana, 3. UTET, Torino

Landi N, Pacifico S, Ragucci S, Di Giuseppe AMA, Iannuzzi F, Zarrelli A, Piccolella S, Di Maro A (2017) Pioppino mushroom in southern Italy: an undervalued source of nutrients and bioactive compounds. J Sci Food Agric 97(15):5388–5397. https://doi.org/10.1002/jsfa.8428

Manzi P, Pizzoferrato L (2000) Beta-glucans in edible mushrooms. Food Chem 68(3):315–318. https://doi.org/10.1016/S0308-8146(99)00197-1

Marino Gammazza A, Barone R, Gargano ML, Caruso Bavisotto C, Macaluso F, D'Amico D, Trovato E, Rappa F, Campanella C, Di Felice V, Cappello F, Venturella G (2017) Anti-cancer effects of *Pleurotus eryngii* var. *eryngii*: an in vitro and in vivo model focusing on Heat Shock Proteins. In: Gargano ML, Venturella G (eds) Book of abstracts: the 9th international medicinal mushrooms conference (IMMC9), Palermo 24–28 September 2017

Oliviero F (2017) Antioxidant and anti-inflammatory properties of a shiitake-maitake-reishi-Cordyceps-based preparation for the prevention of oxidative stress and autoimmune diseases. In: Gargano ML, Venturella G (eds) Book of abstracts: the 9th international medicinal mushrooms conference (IMMC9), Palermo 24-28 September 2017

Parola S, Chiodaroli L, Orlandi V, Vannin C, Panno L (2017a) *Lentinula edodes* and *Pleurotus ostreatus*: Functional food with antioxidant – antimicrobial activity and an important source of Vitamin D and medicinal compounds. Funct Foods Health Dis 7(10):780–794

Parola S, Chiodaroli L, Orlandi V, Vannini C, Panno L (2017b). *Lentinula edodes* and *Pleurotus ostreatus* as functional food with antioxidant, antimicrobial and medicinal properties. In: Gargano ML, Venturella G (eds) Book of abstracts: The 9th international medicinal mushrooms conference (IMMC9), Palermo 24–28 September 2017

Pelkonen O, Duez P, Vuorela PM, Vuorela H (eds) (2017) Toxicology of herbal products. Springer, Cham, p 478

Reverberi M, Fabbri AA, Zjalic S, Ricelli A, Punelli F, Fanelli C (2005) Antioxidant enzymes stimulation in *Aspergillus parasiticus* by *Lentinula edodes* inhibits aflatoxin production. Appl Microbiol Biotechnol 69(2):207–215. https://doi.org/10.1007/s00253-005-1979-1

Reverberi M, Zjalic S, Ricelli A, Di Meo C, Scarpari M, Fanelli C, Fabbri AA (2011) Mushroom versus fungi: natural compounds from *Lentinula edodes* inhibit aflatoxin biosynthesis by *Aspergillus parasiticus*. World Mycotoxin J 4(3):217–224. https://doi.org/10.3920/WMJ2010.1270

Ricelli A, Fabbri AA, Trionfetti-Nisini P, Reverberi M, Zjalic S, Fanelli C (2002) Inhibiting effect of different edible and medicinal mushrooms on the growth of two Ochratoxigenic microfungi. Int J Med Mushrooms 4(2):173–179. https://doi.org/10.1615/IntJMedMushr.v4.i2.110

Rossi P, Buonocore D, Altobelli E, Brandalise F, Cesaroni V, Iozzi D, Savino E, Marzatico F (2014) Improving training condition assessment in endurance cyclists: Effects of *Ganoderma lucidum* and *Ophiocordyceps sinensis* dietary supplementation. Evid Based Compl Alt:11 pp. https://doi.org/10.1155/2014/979613

Rossi P, Vigna L, Nostrini I, Barberi CE, Tomaino SCM, Conti DM, Savino E (2017) Mood and sleep disorders variations after eight weeks supplementation with *Hericium erinaceus*: A pilot study in 18 overweight women. In: Gargano ML, Venturella G (eds) Book of abstracts: the 9th international medicinal mushrooms conference (IMMC9), Palermo 2--28 September 2017

Russo A, Piovano M, Clericuzio M, Lombardo L, Tabasso S, Chamy MC, Vidari G, Cardile V, Vita-Finzi P, Garbarino JA (2007) Putrescine-1,4-dicinnamide from *Pholiota spumosa*

(Basidiomycetes) inhibits cell growth of human prostate cancer cells. Phytomedicine 14(2–3):185–191. https://doi.org/10.1016/j.phymed.2006.09.010

Saltarelli R, Ceccaroli P, Vallorani L, Leonardi P, Iotti M (2017) Antioxidant activities of two Italian *Ganoderma lucidum* strains at different growth stages. In: Gargano ML, Venturella G (eds) Book of abstracts: the 9th international medicinal mushrooms conference (IMMC9), Palermo 24–28 September 2017

Salviato N, Salvia R, Palla F (2017) Antioxidant, antimicrobial and anti-inflammatory properties of *Pleurotus eryngii* for metabolic syndrome's prevention. In: Gargano ML, Venturella G (eds) Book of abstracts: the 9th international medicinal mushrooms conference (IMMC9), Palermo 24–28 September 2017

Saporita P, Burruano S, Conigliaro G, Cusimano MG, Gargano ML, Giambra S, Padovan F, Palazzolo E, Saiano F, Schillaci D, Venturella G (2017) A white maitake (*Grifola frondosa*): nutritional value and antibacterial preliminary activity test. In: Gargano ML, Venturella G (eds) Book of abstracts: the 9th international medicinal mushrooms conference (IMMC9), Palermo 24-28 September 2017

Savino E, Girometta C, Chinaglia S, Guglirlminetti M, Rodolfi M, Bernicchia A, Perini C, Salerni E, Picco AM (2014) Medicinal mushroom in Italy and their ex situ conservation through culture collection. In: Proceedings of the 8th international conference on mushroom biology and mushroom products (ICMBMP8) 2014.

Savino E, Girometta C, Cesaroni V, Rodolfi M, Perini C, Angelini P, Picco AM (2017) Conservation of biodiversity of *Hericium erinaceus* in Italy. In: Gargano ML, Venturella G (eds) Book of abstracts: The 9th international medicinal mushrooms conference (IMMC9), Palermo 24–28 September 2017

Scaglione G, Alio W, Maiorana A, Prestileo T (2017) The M.O.P. therapy in HPV-related diseases of the female and male genital tract. A prospective observation clinical study. In: Gargano ML, Venturella G (eds) Book of abstracts: the 9th international medicinal mushrooms conference (IMMC9), Palermo 24–28 September 2017

Scarpari M, Parroni A, Zaccaria M, Fattorini L, Bello C, Fabbri AA, Bianchi G, Sala V, Zjalic S, Fanelli C (2016) Trametes versicolor bioactive compounds stimulate *Aspergillus flavus* antioxidant system and inhibit aflatoxin synthesis. Plant Biosyst 150(4):653–659. https://doi.org/10.1080/11263504.2014.981235

Schillaci D, Arizza V, Gargano ML, Venturella G (2013) Antibacterial activity of Mediterranean Oyster mushrooms, species of genus *Pleurotus* (higher Basidiomycetes). Int J Med Mushrooms 15(6):591–594. https://doi.org/10.1615/IntJMedMushr.v15.i6.70

Schillaci D, Cusimano MG, Cascioferro SM, Di Stefano V, Arizza V, Chiaramonte M, Inguglia L, Bawadekji A, Davino S, Gargano ML, Venturella G (2017) Antibacterial activity of desert truffles from Saudi Arabia against *Staphylococcus aureus* and *Pseudomonas aeruginosa*. Int J Med Mushrooms 19(2):121–125. https://doi.org/10.1615/IntJMedMushrooms.v19.i2.30

Smania A Jr, Delie Monache F, de Fatima E, Smania A, Cuneo RS (1999) Activity of steroidal compounds isolated from *Ganoderma applanatum* (Pers.) Pat. (Aphyllophoromycetideae) fruit body. Int J Med Mushrooms 1(4):325–330. https://doi.org/10.1615/IntJMedMushr.v1.i4.40

Tiberio V (1895) Sugli estratti di alcune muffe. Annali d'Igiene Sperimentale 5:91–103

Tolaini V, Zjalic S, Reverberi M, Fanelli C, Fabbri AA, Del Fiore A, De Rossi P, Ricelli A (2010) *Lentinula edodes* enhances the biocontrol activity of *Cryptococcus laurentii* against *Penicillium expansum* contamination and patulin production in apple fruits. Int J Food Microbiol 138(3):243–249. https://doi.org/10.1016/j.ijfoodmicro.2010.01.044

Trovato A, Siracusa R, Di Paola R, Scuto M, Fronte V, Koverech G, Luca M, Serra A, Toscano MA, Petralia A, Cuzzocrea S, Calabrese V (2016a) Redox modulation of cellular stress response and lipoxin A4 expression by *Coriolus versicolor* in rat brain: relevance to Alzheimer's disease pathogenesis. Neurotoxicology 53:350–358. https://doi.org/10.1016/j.neuro.2015.09.012

Trovato A, Siracusa R, Di Paola R, Scuto M, Ontario ML, Bua O, Di Mauro P, Toscano A, CCT P, Maiolino L, Serra A, Cuzzocrea S, Calabrese V (2016b) Redox modulation of cellular stress response and lipoxin A4 expression by *Hericium erinaceus* in rat brain: relevance to Alzheimer's disease pathogenesis. Immun Ageing 13:23. https://doi.org/10.1186/s12979-016-0078-8

Trovato A, Pennisi M, Crupi R, Di Paola R, Alario A, Modafferi S, Di Rosa G, Fernandes T, Signorile A, Maiolino L, Cuzzocrea S, Calabrese V (2017) Neuroinflammation and mitochondrial dysfunction in the pathogenesis of Alzheimer's disease: modulation by *Coriolus Versicolor* (Yun-Zhi) nutritional mushroom. J Neurol Neuromed 2(1):19–28

Venturella G, Zervakis G, Polemis E, Gargano ML (2016) Taxonomic identity, geographic distribution, and commercial exploitation of the culinary-medicinal mushroom *Pleurotus nebrodensis* (Basidiomycetes). Int J Med Mushrooms 18(1):59–65. https://doi.org/10.1615/IntJMedMushrooms.v18.i1.70

Zacchigna M, Altieri T, Beltrame G, Cateni F, Procida G (2017) Mycochemical study of polysaccharides from edible mushroom *Cortinarius caperatus* (Gypsy mushroom). In: Gargano ML, Venturella G (eds) Book of abstracts: the 9th international medicinal mushrooms conference (IMMC9), Palermo 24–28 September 2017

African Medicinal Mushrooms: Source of Biopharmaceuticals for the Treatment of Noncommunicable Diseases – A Review

13

Kenneth Anchang Yongabi

Contents

Abstract

African mushrooms remain underexploited for biopharmaceuticals for application in modern medicine. Yet, amidst the rich use of mushrooms in African folk medicine and the high rates of endemic diseases plaguing the continent, the need for a critical exposé of African mushrooms as biopharmaceuticals and phytoceuticals is imperative and hereby presented. Generally, mushrooms are rich sources of many bioactive compounds that are important to our health servicing.

K. A. Yongabi (✉)
Phytobiotechnology Research Foundation, Bamenda, Cameroon

Ebonyi State University, Abakaliki, Nigeria

Imo State University, Owerri, Nigeria

D. C. Agrawal, M. Dhanasekaran (eds.), *Medicinal Mushrooms*,
https://doi.org/10.1007/978-981-13-6382-5_13

They are a good source of proteins that are important to all body functions. Mushroom proteins are of very high quality and are rich in the most important essential amino acids. They are an excellent source of most B vitamins, and the primary natural source of ergosterol or provitamin D, and also essential minerals such as zinc and selenium. Malnutrition and stunting are largely a health problem in rural Africa. Mushrooms are abundant in nutrients that can potentially stamp out nutritional deficiencies in African children. While many people who eat balanced diets receive all the needed minerals, some get more sodium than they need. Mushrooms have the benefits of low sodium, low cholesterol, and higher potassium and iron than most foods. Chitin which is present in many African mushrooms such as *Pleurotus tuber-regium*, *Termitomyces* spp., *Pleurotus* spp., and *Agaricus* spp. is the primary structural material in mushrooms and has been shown to be of immense value as dietary fiber. It can also be hydrolyzed to glucosamine, which is widely accepted by physicians for treatment of chronic diseases of aging people such as autoimmune disorders and arthritis. Also it is used as a valuable food supplement for the prevention and alleviation of osteoarthritis. β-Glucans widely identified in a number of African mushrooms such as *Ganoderma* spp., *African truffles*, and *Agaricus* spp. are valuable immune regulatory substances that can treat and manage different cancers, diabetes, and other cardiovascular disorders.

Keywords
β-Glucans · Chitosans · Chitin · Dietary fiber · Glucosamine · Medicinal mushroom · Noncommunicable diseases · Osteoarthritis · Selenium

Abbreviations

DNA	Deoxyribonucleic acid
FDA	Food and Drug Administration
HDL	High-density lipoprotein
LDL	Low-density lipoprotein
NCDs	Noncommunicable Diseases

13.1 Introduction

Africa's mushroom taxonomy scores high, but the exploitation of Africa's medicinal mushrooms through actual transformation and its use in modern medicine remains barren for too long. It is no news to restate that mushrooms contain many nutritional and medicinal ingredients and are described as food supplements or nutraceuticals (Wittlif and Airth 1970; Kurtzman 1991; Aoyagi et al. 1993; Chang 1999; Yongabi et al. 2015). In modern drug discovery track, drugs must undergo extensive testing before they are acceptable (Fenton et al. 2000). Many food plants and mushrooms are considered food supplements and are acceptable because they have been tested from other sources, and especially in Africa in folklore uses (Yongabi et al. 2004). Mushrooms contain a number of culinary, flavor, and medicinal components that fit the definition of food supplements (Kurtzman 1993; Yongabi 2014). As such, one species of mushroom may be richer in one of these components than others (Sanchez et al. 2002).

These nutritional components of mushrooms with therapeutic properties can be used to prevent or at least reduce the burden of pain and debilitating disease condition in noncommunicable diseases such as in cancer and other malignancies (Kurtzman 1997; Chioza and Ohga 2014). Unlike synthetic medicines, nutraceuticals benefit from promoting natural body functions and not from selective toxicity (Yusof et al. 2003). Although mushroom bioactive compounds have been to a certain extent analyzed for toxicity, African medicinal mushroom used in folklore medicine needs to be analyzed for toxicity notwithstanding. For example, high doses of niacin and other vitamins from mushrooms are toxic (Windholtz 1983; Yongabi 2014).

Food with a high nutritional value such as mushrooms must be considered a "healthy food." Many reviews have been published on the nutritional value of mushrooms, and it is no intention in this paper to duplicate that effort (Noller 1951; Chang 1972; Presant and Kornfeld 1972; Albershiem 1976; Kurtzman 1993, 1997; Cho et al. 1999; Su et al. 2004; Okhuoya et al. 2010; Chioza and Ohga 2014; Yongabi et al. 2015). The thrust of this review is to page the glaring limitation of the development of Africa's medicinal mushrooms by-products in the management of chronic diseases and to showcase the application of bioactive components of some of these mushrooms from the African mycoflora in the healthcare systems as biopharmaceuticals.

13.2 The Myth that African Mushroom Species Are Poisonous, Yet Drug Leads in Disguise

Mushrooms may contain hemagglutinins which cause malabsorption. Hemagglutinins have also been found in some indigenous African beans such as *Glycine max* and other African black bean varieties (Aoyagi et al. 1993; Kurtzman 1991). It is likely that other mushrooms from the mycoflora of Africa may contain hemagglutinins. There are limited studies on this subject (Yongabi 2014). Studies on medicinal mushrooms in Africa have been limited to taxonomy, ethnofolkloric uses, spatial distribution, and antimicrobial and phytochemistry without a focus on domestication and development of biopharmaceuticals (Yongabi et al. 2004; Mpeketula 2008). It is difficult to ascertain that infectious diseases have found a home perpetually on the continent with now increasing epidemiologic transition to noncommunicable diseases such as cancer, diabetes, and cardiovascular diseases, with a continent that prides more than 25% of the world mushroom biodiversity (Okhuoya et al. 2010; Yongabi 2014). Can these medicinal mushrooms' biodiversity be fully exploited through biotechnology for healthcare in the continent?

The toxicology of African mushrooms still is virgin and leaves a local myth and impression in situ that these mushrooms are generally poisonous. Mushroom literacy is required to diffuse this cloud. All mushrooms that have been studied for thiaminase content have been reported that thiaminase destroys vitamin B1 (Ortiz et al. 1992). Although both hemagglutinin and thiaminase may be considered poison, they offer only a minor danger (Wittliff and Airth 1970; Albershiem 1976; Su et al. 2004). Both of these poisons are proteins, and as long as mushrooms are cooked, their hemagglutinin and thiaminase activity will be destroyed (Kurtzman 1975; Robson 1999; Gallaher et al. 2002). Similarly, these compounds are found in beans and are generally termed as anti-nutritional agents and get destroyed when cooked (Noller 1951; Kogure 1975; Brzezinski et al. 2004). If not properly cooked, such compounds in mushrooms may rather cause gastrointestinal problems. The question remains which mushrooms in Africa have these hemagglutinins and thiaminase, and how can these be best processed for maximal nutritional and medicinal benefits? These are some of the unanswered questions in the field of mushroom biotechnology in Africa. Like many other plants and animals identified across Africa with toxic properties, medicinal and edible mushrooms from the continent do possess some anti-nutritional properties and poisons to a certain level and must be properly reported.

13.3 African Medicinal Mushrooms: Source of Valuable Dietary Fibers

African medicinal mushrooms are good source of valuable dietary fibers. However, these are yet to be transformed for large-scale biopharmaceutical applications in the treatment and management of cancers and osteoarthritis.

The fiber of plants is cellulose, lignin, and hemicelluloses (Noller 1951; Wittliff and Airth 1970; Kurtzman 1993). But the fiber of mushrooms is chitin. Often, chitin is referred as "cellulose-like." Chitin and cellulose do have many chemical and mechanical properties in common, but they are quite different from each other. Chitin is also the material that makes up the horny shells of crabs, lobsters, shrimp, and other arthropods which are currently being used as therapeutic elements in formulations from China such as Tianshi and are heavily marketed in the African continent (Chang 1972, 1993, 1999; FDA 2000; Yongabi 2007) of course; those shells are not eaten, but they are processed into high-price food supplements (Yusof et al. 2003). There are cheap and cultivable dietary fibers from some African mushrooms reported in the literature, but extensive studies to classify these fibers and more so the development of supplements from them to treat cancer, diabetes, and osteoarthritis are still required. Truly, arthropod shells are not food at all, and they certainly do not grow quickly as mushrooms. When chitin is purified, it is always somewhat degraded, so all shellfish chitin that has been studied is degraded. Generally, most mushroom chitin that is eaten would be in or at least nearly in its natural state (Foster and Webber 1960; Albershiem 1976; Austin et al. 1981). For large-scale production, it is easier to grow mushrooms and purify chitin than shellfish chitin for pharmaceutical applications in Africa. If chitin loses its acetyl groups through hydrolysis, it is called chitosan (Furda 1983; Gordon and Williford 1983; Razdan and Pettersson 1994). That hydrolysis is quite easily accomplished under alkaline conditions and should be exploited by African mycoflora. Mushrooms, such as *Ganoderma lucidum* and *Ganoderma applanatum* wastes, have been converted to chitosan and used as a dressing on human wounds (Austin et al. 1981; Katz 2001). In Africa, more than 25% of diabetic patients have ulcers and wounds that are often hard to treat (Yongabi 2014). Mushroom chitosan will be very useful in the treatment of diabetic foot ulcer as well as Buruli ulcer which are classified as a neglected tropical disease in the continent. *Ganoderma* species in Africa (Figure 13.1a, b) are widespread as epiphytes, and these remain to study for the various chitins and lignocellulosic materials for large-scale bioconversion into therapeutic chitosan (FDA 1996, 1998; Katz 2001). Several researchers have studied "water-soluble chitin" as a wound dressing. The biochemical processes of acetylation are the primary difference between chitin and chitosan. Chitosans are water-soluble below pH 6.0, but chitins are not (Cheung 1998; Fenton et al. 2000). At the moment, there is no large-scale production of medicinal chitosan from African mushrooms.

Both chitin and chitosan from shellfish have been studied as dietary fiber (Bano et al. 1981; Windholtz 1983; Braham et al. 2003). Fibers potentially can clean the human digestive tract and thus can serve as a complementary and alternative therapy for colon cancer and other gastrointestinal ulcers (Fenton et al. 2000). Incidents of both colon and other gastrointestinal cancers are increasing in Africa (Yongabi 2014). But these cancers and also heart disease and stroke can be significantly reduced. The studies suggest that chitin and chitosans (Fig. 13.2) are excellent hypocholesterolemics, that is, they reduce cholesterol and thus reduce heart disease and stroke. However, reports differ on which is more effective, chitin or chitosan.

The biotechnologists must be able to design simple and scalable protocols to generate bio-industries for the commercial production of chitosan from African mushrooms such as *Pleurotus tuber-regium*, *Termitomyces titanicus*, *Ganoderma lucidum*, *Agaricus* spp., and *Volvariella* spp. In Nigeria, the STK group has

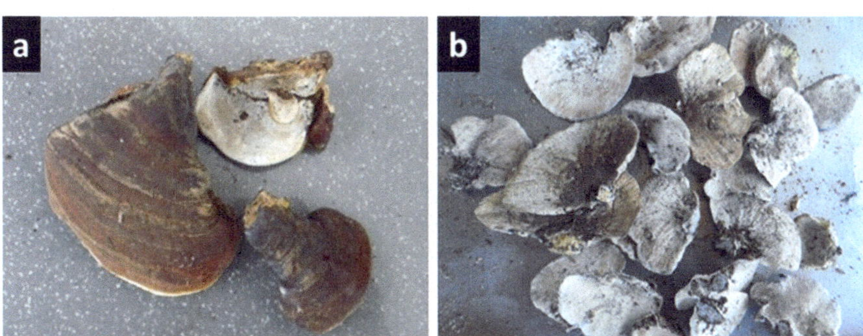

Fig. 13.1 (**a**) *Ganoderma* spp. from Bali, subdivision, Northwest Cameroon (contain chitin). (**b**) *Ganoderma* spp. growing as an epiphyte on a local tree at Bachuo Akagbe, Southwest region of Cameroon (sporophore contain chitin) (both specimen collected by the author)

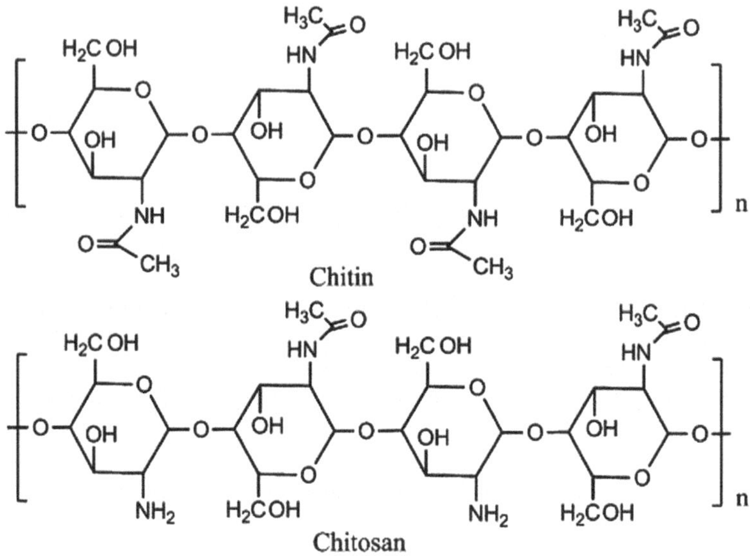

Fig. 13.2 Chemically, chitosan is poly-B-(1->4)-D-glucose, while chitin is poly-B-(1->4)-2-acetamido-2-deoxy-D-glucose

developed simple bioprocess pathways for the conversion of chitin extracted from cultivated *Pleurotus ostreatus* into the production of a supplement for the treatment of prostate cancer (www.stkbiotech.com).

13.4 Bioprocessing of Chitosan and β-Glucans from African Mushrooms for a Large-Scale Application in the Management of Noncommunicable Diseases (NCDs)

Chitosans and to a lesser degree chitin may chelate some undesirable nutrients (McCarty 1998; Mello et al. 2004). Studies, however, have investigated only chitosans. Reports suggest that chitosans capture cholesterol and bile in the intestines and carry them out so that the body cannot absorb or reabsorb (Orth et al. 2002). Bile is required by the intestines, cholesterol and recycled bile are the normal sources of material to supply bile, and so if the liver cannot get bile or cholesterol from the intestines, it uses cholesterol from the blood, thus reducing the amount there. However, further studies are required to explain the mechanism. Chitosan is a polymeric amine, so it might act in the same manner.

The standard dose of poly-allylamine is 3.8 g/day. Experiments show that intake of this dose reduces LDL (dangerous cholesterol) by 15% and increases HDL (acceptable cholesterol) by 3% after 24 weeks after intake. Synthetic drugs developed in the western countries for the treatment of cholesterol are somewhat similar to the mechanism in which chitosan functions in reducing cholesterol. These commercial products are cholestyramine. Dietary fiber supplements are imported into Africa from China, India, the USA, and Europe and constitute a big market for elderly and people in Africa especially those on restricted diets who may have problems with constipation, diabetes, and hypertension (Yusof et al. 2003; Sanchez et al. 2002). Chitosans are also for use in reducing weight for overweight people at risk of cardiovascular diseases. It has been reported that total serum cholesterol decreased by 7%, LDL by 10%, and HDL by 5% (Kurtzman 1991; Ormrod et al. 1998; FDA 2000; Orth et al. 2002).

Studies on mushrooms and mycelium containing chitin as hypocholesterolemics in rats and hamsters found a significant decrease in serum total LDL in those with the mushroom diet (Fenton et al. 2000). It is expected that more experiments with chitin and chitosans would give similar benefits from African medicinal mushrooms. Like cellulose, chitin cannot be digested in the human gastrointestinal tract, but also like cellulose, chemical digestion is possible. The right biotechnological applications and bioprocesses are exigent. However, unlike glucose, the product of the chemical digestion of cellulose, glucosamine (Fig. 13.3), generates more nutritional and therapeutic value than just calories (Wakita 1976; Outila et al. 1999; Yusof et al. 2003).

Similarly, the applications of β-glucans in the treatment of arthritis and other cardiovascular diseases have been reported (FDA 2000; Mattila et al. 2002; Yongabi 2014). Apparently, these β-glucans have been identified from a number of mushrooms in Asia including *Lentinus edodes*, *Cordyceps sinensis*, and *Agaricus* spp., and a number of dietary supplements have been commercially exploited and imported into Africa. However, African truffles and mushrooms (Figure 13.4a, b) containing β-glucans (Fig. 13.5) have yet to be exploited.

13.5 Alternative and Complementary Treatment of Cancer and Osteoarthritis in Africa

Cancer and osteoarthritis are becoming a more common problem than ever before in Africa. The population of sub-Saharan Africa is made up of many aging people whose common health predicament to manage is osteoarthritis particularly in women. Thousands of people get complete hip and joint surgical operation each year. Those operations have exorbitant cost for most Africans and usually are not

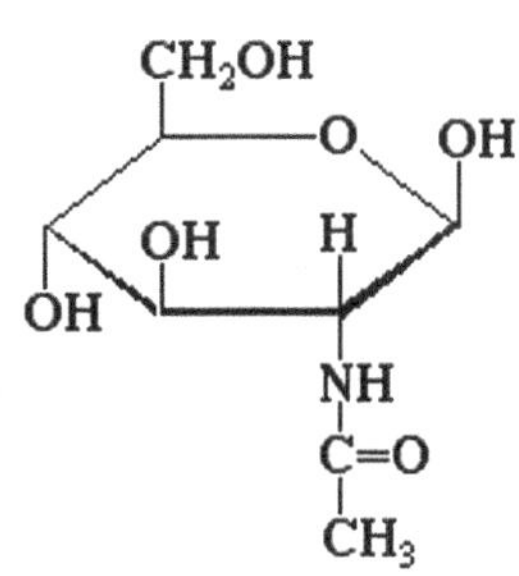

Fig. 13.3 N-Acetyl-D-glucosamine from African *Pleurotus tuber-regium*, *Agaricus bisporus truffles*

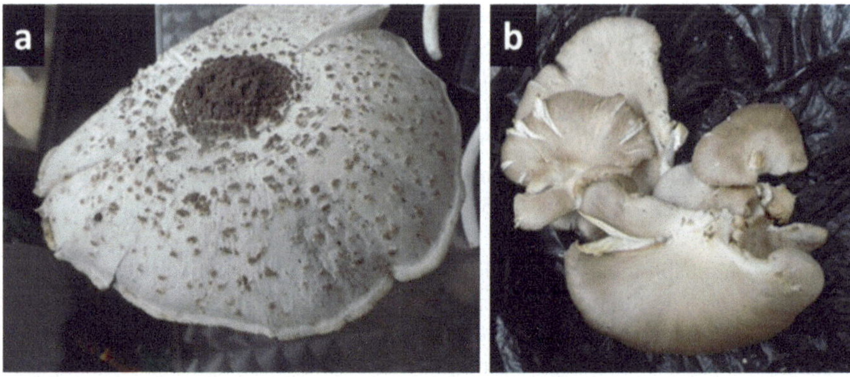

Fig. 13.4 (**a**) Truffle from the central region of Cameroon, (**b**) species of *Pleurotus* containing β-glucans collected from a decaying Avocado tree in Bali, Northwest region of Cameroon (specimens collected by the author)

successful. And the aftermath is fraught with a considerable level of opportunistic diseases due to immune suppression (Mello et al. 2004). These postsurgical diseases are hardly considered by African physicians. Osteoarthritis is the loss of joint cartilage, so that instead of the cartilage at the ends of the bones rubbing as the joint moves, the bones of the joint rub directly on one another (Mattila et al. 2002; Braham et al. 2003). Like chitin, cartilage is primarily acetylglucosamine and a similar sugar acetylgalactosamine. Enzymes normally hydrolyze damaged cartilage so that it can be disposed of by the body. If everything is working well, new cartilage replaces the damaged one. However, often athletes and older people begin to lose cartilage faster than they can replace it. Glucosamine often with chondroitin has become a very common but expensive food supplement found in abundance in mushroom species in Africa (Chioza and Ohga 2014). It is generally recommended in the USA by orthopedic physicians.

Refining stipe and sporophore into glucosamine require complex processing and a large-scale cultivation of these mushrooms. Unfortunately, domestication of African mushroom species and germplasm preservation still remains a challenge for more than half a century. Yet, the dietary fibers chitin and chitosan, as well as glucans, are substances that can treat and prevent osteoarthritis and cancer and can also

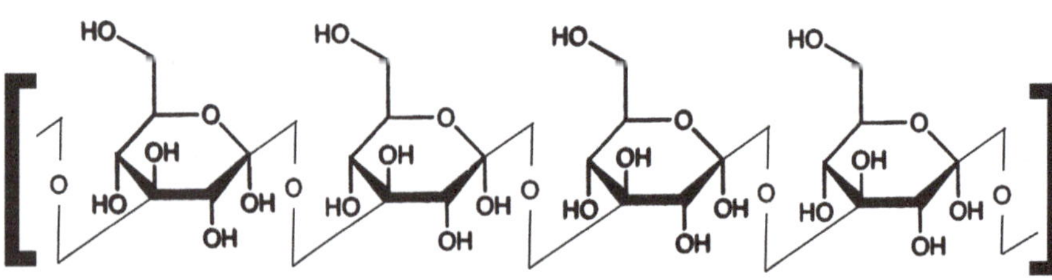

Fig. 13.5 1-6 β-Glucan (*Agaricus blazei*) and identified in Cameroon species of *Pleurotus tuber-regium*

serve as precursors for cancer drug development for Africa's biopharmaceutical industry. The production of dietary fiber needs not be as sophisticated. The appropriate mushroom biotechnological processes to transform Africa's medicinal mushroom potentials are still at bay (Mpeketula 2008; Yongabi 2014).

Another mushroom ingredient unexplored from Africa's medicinal mushroom is glucosamine. Many patients with arthritis and cancer from Africa import glucosamine from western drug markets at exorbitant cost. Mushroom glucosamine is accepted as an over-the-counter food supplement in western nations, but generally as ostentatious drugs on shelves of big pharmacies and shopping malls in Africa (Katz 2001; Yongabi 2014). There is only limited research, and more has to be done to explore and develop the large-scale production of therapeutic glucosamine from African medicinal mushrooms. African mushroom species such as *Ganoderma* spp., *Agaricus* spp., and *Pleurotus* spp. are important sources of glucosamine. With more people exercising regularly, future interest in and the need for glucosamine would seem assured (Mattila et al. 2002). This is also relevant for sports nutrition and medicine.

13.6 Nutrition

β-glucans are glucose biopolymers (complex polysaccharides) special found in yeast and mushrooms. Each type of β-glucan has a unique structure in which glucose is linked together in different ways, giving them different physical characteristics and bioactivities (Bano et al. 1981; Windholtz 1983; Kurtzman 1991; Yusof et al. 2003). There are reports on the nutritional and proximate quantification of a number of African mushroom species (Chang 1999; Su et al. 2004; Okhuoya et al. 2010), but the bioproducts for nutritional use are rare on the market across Africa. Mushroom nutrients are safe for patients with cancer- and arthritis-related disorders.

13.7 Trace Minerals and Vitamins and Their Role in the Therapy of NCDs

A number of minerals in mushrooms are desirable and useful in enhancing the immune system. Sodium in mushrooms is generally low, potassium is high, and iron is high. Phosphorus is also high. Calcium might be expected to be high in *Agaricus* spp. and some other local species. Some minerals are controlled by the organism and others by the substrates on which mushrooms are cultivated. More studies are required to quantitate the actual level of trace elements in African medicinal and edible mushrooms (Kurtzman 1991; Sanchez et al. 2002). Apparently, calcium is controlled by the organism, and most likely sodium is controlled by the substrate on which the mushroom is cultivated or grows on in the wild. However, if the osmotic pressure of the substrate becomes high, it is the growth of the mushrooms that will be reduced by the substrate (Kurtzman 1993; Robson 1999). A quick review of the importance of minerals in the reducing chronic disease burden is worth mentioning. Potassium is important in many enzymatic and muscle functions; calcium is required for the bones and teeth, so a very good support for people with osteoarthritis. Apparently, childbearing and lactation require that the calcium metabolism of women must be quite different from men, and so women require more calcium. Iron is required for blood, liver, and muscle formation, and this is very important in the treatment of leukemia (Yongabi 2014). Once again, women of all ages require more iron than men due to the loss of blood during menstruation.

For one reason, women of childbearing age require more iron since it is normal for them to lose more blood than men. However, it is also true that older women require more iron than men of the similar age (Razdan and Pettersson 1994; FDA 2000; Orth et al. 2002). Phosphorus is also required for bones and teeth, but it is of great importance in all tissues. It is used to transfer energy within the body and for genetic information. The primary function of sodium is in osmotic balance. Analysis for sodium in *Agaricus* has been reported to vary from 3.90 to 14.98 mg/100 g, a factor of just over four times. Many amino acids can be easily formed within the body, but the essential amino acids are needed to build the proteins that make our body functions. Those called the essential amino acids must be in the food we eat; however, some can replace others to some degree. In general, mushrooms supply about half as much of each of the essential amino acids supplied by milk. In most mushrooms, cysteine is low, and methionine is high, compared to milk. Fortunately, methionine can substitute to a large degree for cystine. They are both sulfur-containing. Based on amino acid composition, mushrooms are excellent food. Some authors make a strong differentiation between proteins and free amino acids (Wakita 1976; Yusof et al. 2003). Free amino acids may be important to taste, and large quantities may cause undesirable effects on some people. However, normal levels in foods are metabolized in exactly the same manner as their minus-H2O residues in the protein. Almost all vitamins are found in greater amount in mushrooms (Mello et al. 2004). Thiamine is low in the mushrooms, but niacin is very high. Niacin is the anti-pellagra vitamin, but it is also recommended for controlling blood cholesterol. However, it appears to be effective, only in massive dose 1–5 g/day. Thus one would need to eat about 20 kg of fresh mushrooms every day. Even dose as low as 50 mg of niacin can cause severe flushing, so there are real dangers with its use (Braham et al. 2003; Mello et al. 2004). Apparently, all fungi produce ergosterol which is provitamin D. It becomes vitamin D when it is exposed to ultraviolet radiation (Fenton et al. 2000). Vitamin D is required to support bone formation and necessary for patients with arthritis. That radiation can occur in the mushroom after it is harvested. Sunlight is the normal source of the ultraviolet.

Some doubt has often been expressed about the availability of nutrients found in mushrooms. A preliminary experiment was done by feeding humans with *Cantharellus tubaeformis* with vitamin D2. Others receive the same dose without mushrooms, and a third group received no vitamin D. There was no significant difference between the groups receiving the vitamin, but the group that received none showed less in their serum at the end of the study (Fenton et al. 2000).

13.8 Probiotic Foods from African Mushrooms and Fungi Can Help Patients with Chronic Diseases

Reports show that fungal mycelia are high in B vitamins as compared to cheeses (Kurtzman 1993; Robson 1999). Fermented mushrooms can actually provide a plethora of minerals and useful by-products that can be used as probiotics. However, there is very limited information on these products from African mushrooms, and the accuracy of the available information warrants verification.

13.9 Conclusion and Recommendations

African mushroom contains many ingredients that are healthy for culinary and healthcare applications. These include chitins, chitosans, protein, trace minerals, glucans and glucosamine, vitamin D, etc. These are ingredients that possess properties such as antioxidant, anticancer, and immune enhancing effects. These nutrients are very critical in the treatment and management of noncommunicable diseases such as osteoarthritis, cancer, and other cardiovascular diseases and diabetes. In this article, it is noted that these ingredients are not well researched and exploited in the African mushroom species for the use in the healthcare sector. It is noted that similar products from other continents are imported into Africa. Although African mushrooms contain these ingredients but have not been exploited to address the increasing non-communicable diseases. Like almost all foods, mushrooms also contain materials that can be poisonous, yet all known poisonous materials in commercially grown mushrooms are completely harmless when the mushrooms are prepared in a customary manner. In this chapter, it is recommended that emphasis in the development of biopharmaceuticals for the treatment of noncommunicable diseases in Africa be emphasized through the exploitation of African medicinal mushrooms.

References

Albershiem P (1976) The primary cell wall. In: Bonner J, Varner JE (eds) Plant biochemistry, 3rd edn. Academic, New York, pp 226–274

Aoyagi Y, Kasuga A, Sasaki H, Matuzawa M, Tsutagawa Y, Kawai H (1993) Chemical composition of shitake mushroom cultivated on logs and sawdust substrate beds and their relations to composition of the substrate. Nippon Shokuhin Kogyo Gakkaishi 40:771–775

Austin PR, Brine CJ, Castle JE, Zikakis JP (1981) Chitin: new facets of research. Science 212:749–753

Bano Z, Bhagya S, Srinivasan KS (1981) Essential amino acid composition and proximate analysis of the mushroom *Pleurotus eous* and *P. florida*. Mushroom Newslett Trop 1(3):6–10

Braham R, Dawson B, Goodman C (2003) The effect of glucosamine supplementation on people experiencing regular knee pain. Br J Sports Med 37:45–49

Brzezinski R, LeHoux JG, Kelly A (2004) Clinical studies on the innocuousness of chitosan and its short-chain derivative generated by enzymatic hydrolysis. Asia Pacific J Clin Nutr 13:S96

Chang ST (1972) The Chinese mushroom. The Chinese University of Hong Kong, Hong Kong, p 113
Chang ST (1993) Mushroom biology: the impact on mushroom production and mushroom products. In: Chang et al (eds) Mushroom biology and mushroom products. The Chinese University Press, Hong Kong, pp 3–20
Chang ST (1999) Global impact of edible and medicinal mushrooms on human welfare in the 21st Century: non-green revolution. Int J Med Mushrooms 1:1–7. https://doi.org/10.1615/IntJMedMushrooms.v1.i1.10
Cheung PC (1998) Plasma and hepatic cholesterol levels and fecal neutral sterol excretion are altered in hamsters fed straw mushroom diets. J Nutr 128:1512–1516
Chioza A, Ohga S (2014) Cultivated mushrooms in Malawi: a look at the present situation. Adv Microbiol 4(1):6–1174
Cho YW, Cho YN, Chung SH, Yoo G, Ko SW (1999) Water-soluble chitin as a wound healing accelerator. Biomaterials 20:2139–2145
FDA (1996) http://www.fda.gov/cder/guidance/cholesty.pdf
FDA (1998) http://www.fda.gov/cder/foi/label/1998/209261bl.pdf
FDA (2000) http://www.fda.gov/cder/foi/label/2000/211761bl.pdf
Fenton JI, Chlebek-Brown KA, Peters TL, Carson JP, Orth MW (2000) Glucosamine HCL reduces equine articular cartilage degradation in explant culture. Osteoarthritis Cartilage 8:258–265
Foster AB, Webber JM (1960) Chitin. Adv Carbohydrate Chem 15:371–393
Furda I (1983) Aminopolysaccharides-their potential as dietary fiber. In: Furda I (ed) Unconventional sources of dietary fibers. American Chemical Society, Washington, DC, pp 105–122
Gallaher DD, Gallaher CM, Mahrt GJ, Carr TP, Hollingshead CH, Heslink R Jr, Wise J (2002) A glucomannan and chitosan fiber supplement decreases plasma cholesterol and increases cholesterol excretion in overweight normocholesterolemic humans. J Am College Nutr 21:428–433
Gordon DT, Williford CB (1983) Chitin and chitosan: influence on absorption in rats. In: Furda I (ed) Unconventional sources of dietary fibers. American Chemical Society, Washingon, DC, pp 155–184
Katz DL (2001) A scientific review of the health benefit of oats. The Quaker Oats Company. http://www.quakeroatmeal.com/healthpros/IHP/HealthBenefitsofOats.pdf
Kogure T (1975) On the specificity of mushroom *Pleurotus ostreatus* and *Pleurotus spodoleucus* extracts. Vox Sanguinis 29:221–227
Kurtzman RH Jr (1975) Mushrooms as a source of food proteins. In: Friedman M (ed) Protein nutritional quality of foods and feeds, part 2. Marcel Dekker, New York, pp 305–318
Kurtzman RH Jr (1991) Dolomite upsets the carbon dioxide balance. Mushroom Sci 13:747–751
Kurtzman RH Jr (1993) Analysis, digestibility and the nutritional value of mushrooms. In: Chang et al (eds) Mushroom biology and mushroom products. Chinese University Press, Shatin
Kurtzman RH Jr (1997) Nutrition from mushrooms, understanding and reconciling available data. Mycoscience 40:247–253
Mattila P, Lampi AM, Ronkainen R, Toivo J, Piironen V (2002) Sterols and vitamin D2 contents in some wild and cultivated mushrooms. Food Chem 76:293–298. https://doi.org/10.1016/S0308-8146(01)00275-8
McCarty MF (1998) Enhanced synovial production of hyaluronic acid may explain rapid clinical responses to high-dose glucosamine in osteoarthritis. Med Hypotheses 50(6):507–510
Mello DM, Brian MS, Nielson D et al (2004) Comparison of inhibitory effects of glucosamine and mannosamine on bovine articular cartilage degradation in vitro. Am J Vet Res 65:1440–1445. https://doi.org/10.2460/ajvr.2004.65.1440
Mpeketula PMG (2008) Indigenous mushroom species cultivation, processing and utilization for food security and conservation. National Research Council of Malawi, Conference Proceedings, Malawi, pp 95–105
Noller CR (1951) Chemistry of organic compounds. WB Saunders, Philadelphia, p 885
Okhuoya JA, Akpaja EO, Osemwegie OO, Oghenekaro AO, Ihayaere CA (2010) Nigerian mushrooms: underutilized non-wood forest resources. Environ Manag 14(1):43–54

Ormrod DJ, Holmes CC, Miller TE (1998) Dietary chitosan inhibits hypercholesterolemia and atherogenesis in apolipoprotein E-deficient mouse model of atherosclerosis. Atherosclerosis 138(2):329–334

Orth MW, Peters TL, Hawkins JN (2002) Inhibition of articular cartilage degradation by glucosamine-HCL and chondroitin sulfate. Equine Vet J Suppl 34:224–229

Ortiz R, Sanchez R, Piez A, Montaio LF, Zenteno E (1992) Induction of intestinal malabsorption syndrome in rats fed with *Agaricus bisporus* mushroom lectin. J Agric Food Chem 40:1375–1378

Outila T, Mattila PH, Piironen VI, Lamberg-Allardt CJ (1999) Bioavailability of vitamin D from wild edible mushrooms (*Cantharellus tubaeformis*) as measured with a human bioassay. Am J Clin Nutr 69:95–98

Presant P, Kornfeld S (1972) Characterization of the cell surface receprot for the *Agaricus bisporus* Hemagglutinin. J Biol Chem 247(21):6937–6945

Razdan A, Pettersson R (1994) Effect of chitin and chitosan on nutrient digestibility and plasma lipid concentrations in broiler chickens. Br J Nutr 72:277–288

Robson G (1999) Hyphal cell biology. In: Oliver RP, Schweizer M (eds) Molecular fungal biology. Cambridge University Press, Cambridge, pp p164–p184

Sanchez JE, Huerta G, Montiel E (eds) (2002) Mushroom biology and mushroom products, Proceedings of the 1V international conference. University of Morelos, Cuernavaca, Mexico, p 468

Su CH, Lin BW, Yu SY, Liu SW (2004) Use of *Ganoderma tsugae* for the treatment of human chronic skin ulcers. Mushroom Sci 16:659–662

Wakita S (1976) Thiamine-destruction by mushrooms. Sci Rep Yokohama Natl Univ Sect 2 23:39–70

Windholtz M (1983) The Merck Index, 10th ed. Merck & Co., Rahway, pp 1463–1481

Wittliff JL, Airth RL (1970) Thiaminase I. (Thiamine: base 2-methyl-4aminopyrimidine-5-methenyl transferase, EC 2.5.1.2). Methods Enzymol 18(A):234–238. https://doi.org/10.1016/0076-6879(71)18310-3

Yongabi KA (2007) Ethnomycology of grassland fields and tropical rainforest of Cameroon. Poster presentation at the 'First World Conference on Conservation and Sustainable Use of Wild Fungi', organized by the Regional Government of Andalucia, Spain, 10–16 December, pp 298–299

Yongabi A (2014) Current developments in mushroom biotechnology in sub Saharan Africa. WSMBMP Bull 11:4–13

Yongabi K, Agho M, Martinez-Carrera D (2004) Ethnomycological studies on wild mushrooms in Cameroon, Central Africa. Micologia Aplicada Int 16(2):34–36

Yongabi A, Laura D, Keto M, Suki KKM, Alex D, Francisca NN (2015) Can we exploit and adapt indigenous knowledge and ethnobotanicals for a healthy living in the face of emerging diseases like Ebola in Africa. Am J Clin Exp Med 3:24–28. https://doi.org/10.11648/jcem.s.2015030101.15

Yusof NL, Wee A, Lim LY, Khor E (2003) Flexible chitin films as potential wound-dressing materials: wound model studies. J Biomed Mater: Res A 66(2):224–232

Tiger Milk Mushroom (The *Lignosus* Trinity) in Malaysia: A Medicinal Treasure Trove

14

Shin-Yee Fung and Chon-Seng Tan

Contents

S.-Y. Fung (✉)
Medicinal Mushroom Research Group (MMRG), Department of Molecular Medicine, University of Malaya, Kuala Lumpur, Malaysia

Centre for Natural Products Research and Drug Discovery (CENAR), University of Malaya, Kuala Lumpur, Malaysia

University of Malaya Centre for Proteomics Research (UMCPR), University of Malaya, Kuala Lumpur, Malaysia

C.-S. Tan
Ligno Research Foundation, Balakong Jaya, Selangor, Malaysia

D. C. Agrawal, M. Dhanasekaran (eds.), *Medicinal Mushrooms*,
https://doi.org/10.1007/978-981-13-6382-5_14

Abstract

The Tiger Milk mushroom is traditionally used by aborigines and locals alike for various bodily ailments and to improve general health. In Malaysia, three species of genus *Lignosus* have been identified: *Lignosus rhinocerus*, *Lignosus tigris*, and *Lignosus cameronensis*. Taxonomically, *Lignosus rhinocerus* is distinct from the rest of the *Lignosus* species. Prior to the discovery and identification of *Lignosus tigris* and *Lignosus cameronensis* as two separate and distinct species from *Lignosus rhinocerus*, these were often described and used interchangeably as *Lignosus rhinocerus*. To date, the wild types of these three species are still being referred to as *Lignosus rhinocerus*, as the sclerotia are often collected without intact stipe and cap, which make it hard to differentiate among the three. Scientific investigations to validate the traditional claims recorded by ethnomycological surveys have focused on the investigation of the composition and functional properties of *Lignosus rhinocerus* sclerotium. Chronic, subacute and acute toxicity, genotoxicity, antifertility and teratogenic, nutritive, anticancer, anti-inflammatory, antioxidative activity and inhibition of protein glycation effects have been reported using the cultivated species (TM02) of this mushroom. The genomic, transcriptomic, and proteomic studies on *Lignosus rhinocerus* sclerotium have also been performed. A substantial amount of data has been generated since the inception of the research in 2009 by the Medicinal Mushroom Research Group (MMRG) in the Department of Molecular Medicine, Faculty of Medicine, University of Malaya, Malaysia. Active research is ongoing in our laboratories to elucidate more scientific data of this much sought-after medicinal mushroom.

Keywords

Lignosus · *Rhinocerus* · *Tigris* · *Cameronensis* · Tiger Milk mushroom · Medicinal mushroom

Abbreviations

184B5	Human mammary gland/breast epithelial cells
AGE	Advanced glycation end-product
AOAC	Association of Official Analytical Communities
CAT	Catalase
CML	Nε-(carboxymethyl)lysine
DF	Dietary fiber
MCF-7	Human breast adenocarcinoma cell line
MCM	Mushroom complete medium
MMRG	Medicinal Mushroom Research Group
NDCs	Nondigestible carbohydrates
OECD	Organization for Economic Cooperation and Development

Pent	Pentosidine
SOD	Superoxide dismutase
TM02	*Lignosus rhinocerus* TM02 cultivar
TNF	Tumor necrosis factor

14.1 Introduction

14.1.1 The History of Tiger Milk Mushroom

Mordecai Cubitt Cooke was the first to scientifically document the taxonomy of Tiger Milk mushroom which he located in Penang Island, Malaysia, naming it *Polyporus rhinocerus* (Cooke 1879). Preceding this record was the writings in *The Diary of John Evelyn* (1664) which mentioned that *Lac tygridis* (*tigridis*) was used by local communities to treat diseases that physicians and druggist from the West were not able to find a cure for. Tiger Milk mushroom is known by a variety of local names, *cendawan susu harimau*, *betes kismas*, *hurulingzhi*, and *hijiritake* (Burkill and Haniff 1930; Burkill et al. 1966; Haji Taha 2006; Huang 1999a,b; Yokota 2011), and synonyms, *Polyporus sacer* var. *rhinocerus* and *Microporus rhinocerus* (Imazeki 1952a, b). Its current accepted name is *Lignosus rhinocerus* (Cooke) (Ryvarden 1972). It is interesting to follow the historical path of the gradual popularity of Tiger Milk mushroom from the maritime trading days of the Malay Peninsula. Back in the days when Chinese traders were active in the Malay Peninsula, local medical supplies would be obtained for their long journey, and it is thought that Tiger Milk mushroom would be part of their supply. An example would be the great admiral, diplomat, soldier, and trader Zheng He, who is well-known in Chinese and Muslim history. Zheng He (birth name: Ma He) was born in 1371 in the southern China region of Yunnan to a Hui (a Muslim Chinese ethnic group) family. In China, the family name is said first, followed by the given name. "Ma" is known in China as short for "Muhammad," indicating Zheng He's Muslim heritage. In 1405, the third emperor of the Ming dynasty, Emperor Zhu Di, decided to send out a giant fleet of ships to explore and trade with the rest of the world. He chose Zheng He to lead the massive expedition, where almost 30,000 sailors were involved in each voyage, with Zheng He commanding all of them. Between 1405 and 1433, Zheng He led seven expeditions that sailed to present-day Malaysia, Indonesia, Thailand, India, Sri Lanka, Iran, Oman, Yemen, Saudi Arabia, Somalia, Kenya, and many other countries. Wherever they sailed, they commanded respect of the local people, who offered tributes to the Chinese emperor. Zheng He would sail back to China with exotic goods such as ivory, camels, gold, and medicinals. As the Tiger Milk mushroom has an outward appearance similar to that of *Lingzhi*, an already established medicinal mushroom in China, it was also collected and brought along as a potential medicinal source. Due to this maritime trade link between the Malay

Peninsula and the Chinese traders which valued such mushrooms for its medicinal potential, Tiger Milk mushroom became well-known mainly among the Peranakan culture – a result of Malay/Chinese intermarriage and assimilation. This may very well be the reason why Tiger Milk mushroom is widely known for its medicinal properties (as medicinal mushroom) only among the Malay and Chinese communities which form the foundation of the Peranakan communities. Literature also reports the use of Tiger Milk mushroom as a form of medicine in Singapore and China (where the traders eventually return). Other than Malaysia, it has also been located in Indonesia, Thailand, the Philippines, and Papua New Guinea. It is listed as a mushroom with important economic value (Burkill et al. 1966) in *A Dictionary of the Economic Products of the Malay Peninsula*. Sir Henry Nicholas Ridley, the founder of Malaysian rubber industry, has attempted (albeit unsuccessful) to cultivate this important medicinal mushroom (Ridley 1890). The Tiger Milk mushroom is traditionally used by aborigines and locals alike for ailments such as cough, asthma, fever, sinusitis, nausea and vomiting, indigestion, allergies, breast cancer, chronic hepatitis, gastric ulcer, and food poisoning, for wound healing, and as a general tonic to strengthen the body (Chang and Lee 2004). Malaysian folklore believed that the Tiger Milk mushroom can be found at the spot where a prowling tigress's milk would have dropped to the ground. The gap between mention in the literature is possibly due to its rarity, scarcity (not more than one is found within a 5 km radius), and inconsistencies of the mushroom obtained from the wild which might have hampered scientific investigations to verify its traditional usages. Chang and Lee (2004) noted at time of their writing that no local species in Malaysia have been cultivated successfully enough to have penetrated the local medicinal pharmaceutical market. Vikineswary and Chang (2013) associated the decline of this species with modern development, deterioration of environment, and the advent of modern medicine. The wild-type mushroom collected from the state of Pahang, Malaysia, was reported to contain low (trace) levels of toxic metal contents (Lai et al. 2013) and may be a cause of concern depending on amounts taken.

14.2 The Linkages of Tiger Milk Mushroom in Malaysia with Other *Lignosus* Species

At present, the following eight species of this *Polyporales* macrofungi are recognized and documented, shown in the Table 14.1.

One unique feature one can notice from the taxonomy tree (Fig. 14.1) is that *Lignosus rhinocerus* is distinctly set apart from the rest of the *Lignosus* species, from which we hypothesize that its bioactive properties may possibly be different from species in the same genus. The outlook of the different *Lignosus* species is shown in Fig. 14.2.

The identifying feature of the species lies on the pore size (Fig. 14.3, Table 14.2).

Polyporaceae are a family of bracket fungi with complex macrostructure. The flesh of their fruiting bodies is composed of several kinds of hyphae, from soft to very tough. This family is regarded as one of the most valued medicinal fungi across

Table 14.1 Species of *Lignosus*

No.	Species	Distribution/location
i.	*Lignosus rhinocerus* (synonym: *Polyporus rhinocerus* (Cooke 1879); *Microporus rhinocerus* (Cooke) Imazeki (Imazeki 1952a, b); *Lignosus rhinocerus* (Cooke) (Ryvarden 1972); *Lignosus rhinocerotis* (Cooke) Ryvarden [as *Lignosus rhinocerus*] (Ryvarden 1972)	Only found in the tropical forest of Southeast Asia including Malaysia
ii.	*Lignosus dimiticus* (Ryvarden 1975)	Zaire
iii.	*Lignosus ekombitii* (Douanla-Meli and Langer 2003)	Cameroon
iv.	*Lignosus goetzei* (Henn.) (Ryvarden 1972) (synonym: *Polyporus goetzei*)	Tanzania
v.	*Lignosus sacer* (synonym: *Polyporus sacer* Afzel. ex Fr. (Fries and Nyman 1837); *Leucoporus sacer* (Afzel. ex Fr.) Pat., (Patouillard 1900); *Microporus sacer* (Afzel. ex Fr.) (Imazeki 1952a, b); *Trametes sacra* (Afzel. ex Fr.) (Corner 1989); *Polyporus scleropodius* Lév., (Leveille 1846)	Africa
vi.	*Lignosus hainanensis* (Cui et al. 2011)	Hainan Province, Southern China
vii.	*Lignosus tigris* (Tan et al. 2013)	Malaysia
viii.	*Lignosus cameronensis* (Tan et al. 2013)	Pahang, Malaysia

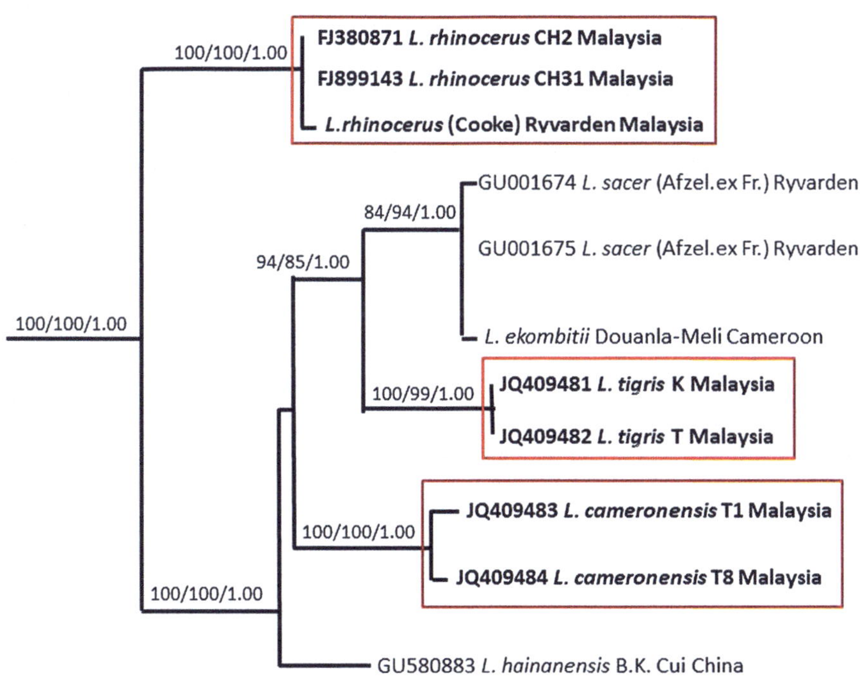

Fig. 14.1 The taxonomy of *Lignosus* sp.

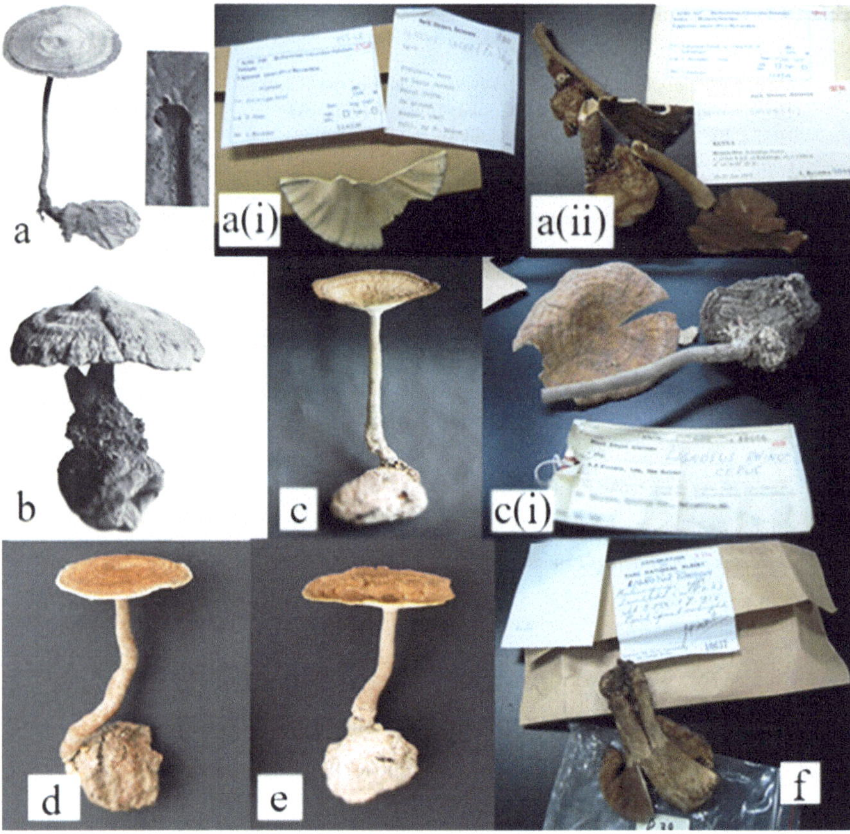

Fig. 14.2 Species of *Polyporales* macrofungi **a**#, **a(i)***, **a(ii)***, and **a(iii)***, *Lignosus sacer*; **b**#, *Lignosus goetzei*; **c** and **c(i)***, *Lignosus rhinocerus*; **d**, *Lignosus cameronensis*; **e**, *Lignosus tigris*; **f***, *Lignosus dimiticus*
#Picture from Index of Mycological Writings Volume III, Lloyd, C.G. 1909–1912
*Specimens provided by (Prof Leif Ryvarden) Herbarium of the University of Oslo, Norway (Pictures with herbarium label)

Southeast Asia. Species in this taxon has a wide range of ethnomycological uses. Traditionally, this mushroom is being used as a tonic, to maintain general health, for immune enhancement, or in the treatment of numerous ailments including cancer, asthma, sinusitis, bronchitis, nausea, and a variety of allergic reactions. It is also used to treat discomfort caused by fright, fever, coughing, vomiting, and cuts (Chang and Lee 2004).

The mushroom's sclerotium is the source of its medicinal value. The most popular member of this taxon, *Lignosus rhinocerus*, was first described some 400 years ago, in June 1664 by John Evelyn. His book entitled *The Diary of John Evelyn* recorded an event which described the giving of a rare, important medicinal product named *Lac tigridis* (tiger's milk) from the Jesuits of Japan and China to their Order at Paris as repository's collection. In his writings, the gift was described as "fungus-like, but

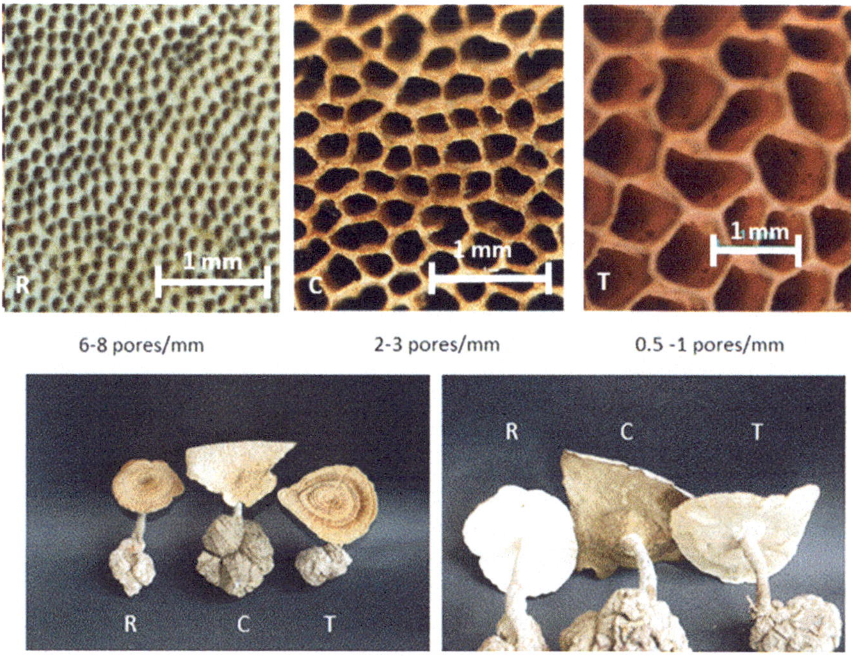

Fig. 14.3 Pore sizes and morphology of three species of genus *Lignosus*. (R) *Lignosus rhinocerus*, (C) *Lignosus cameronensis*, (T) *Lignosus tigris*. They are also differentiated based on the genetic DNA sequences of internal transcribed spacer (ITS) regions

Table 14.2 Comparative size of pore and basidiospore in *Lignosus* species

Species	Pores (per mm)	Basidiospores (μm)
Lignosus cameronensis	2–4	2.4–4.8 × 1.9–3.2
Lignosus dimiticus	6–8	3–4.5 × 2.5–3
Lignosus ekombitii	2–4	8.1–9.3 × 2.5–3.8
Lignosus goetzei	0.5–2	6–9 × 5–8
Lignosus hainanensis	3–4	4.9–6 × 2.2–2.9
Lignosus sacer	1–3	5–7 × 3–4.5
Lignosus rhinocerus	7–8	3–3.5 × 2.5–3
Lignosus tigris	1–2	2.5–5.5 × 1.8–3.6

Source: Tan et al. (2013)

was weighty like metal, yet was a concretion, or coagulation, of some other matter" that Western druggist and physicians could not figure out (Evelyn 1664). This gift was sent along with other medically useful items such as the horn of rhinoceros, which is believed to have healing properties such as to treat fever, rheumatism, gout, hallucinations, typhoid, headaches, carbuncles, vomiting, and food poisoning (http://www.pbs.org/wnet/nature/rhinoceros-rhino-horn-use-fact-vs-fiction/1178/). Perhaps this is the rationale behind the name of this species. Cooke (1879) pioneered the scientific documentation of this fungus and named it as *Polyporus rhinocerus* using a specimen obtained from Penang Island, Malaysia.

Lignosus rhinocerus has been reported to be found in southern regions of China (commonly known as hurulingzhi), Sri Lanka, Thailand, the Philippines, Indonesia, Papua New Guinea, Australia, Vanuatu, and Malaysia (Cui et al. 2011; Huang 1999b; Núñez and Ryvarden 2001).

Lignosus rhinocerus is the most studied species. Since 2009, upon the success of its cultivation, several studies have been conducted generating plenty of data. All the other species are not as well researched and not well understood. Information on complete phylogenetic relationship, evolution, and genetic diversity of the *Lignosus* species is not available.

14.3 Aborigines

This section describes the folk practices and belief system of aborigines (Semai, Temuan, Jakun) on the use of Tiger Milk mushroom as medication and the categories of diseases for which it is used and its outcomes.

Burkill et al. (1966) documented the use of several fungi by local communities for the treatment of a variety of ailments. Among others, he noted that the sclerotium of a *Lignosus* species ("cendawan susu harimau" or "susu rimau" in Malay language or tiger's milk fungus in English) was used to treat congestions and coughs. Chang and Lee (2004) did a preliminary study, and their observations strongly suggest that there is a dearth of information on macrofungal utilization in Malaysia although still used by the Malays, Chinese, and indigenous communities in Malaysia. From our various informal talks with the medicinal practitioners/villagers in various locations where we obtained the wild type of Tiger Milk mushroom, there is a strong belief within the local indigenous communities that the consumption of Tiger Milk mushroom prior to going into the jungle will "enable them to find their way home and reduces the chances of them getting lost" in their day trip into the wild. They would also carry with them the sclerotium of the mushroom, and during their rest, they would take a bite/chew the sclerotium and subsequently continue their journey uneventfully. We hypothesize that the mushroom acts as a tonic, that as they make their journey into the wild, the chewing on the sclerotium during their intermittent rest will give them the energy boost and hence enable them to think clearly and remember their way home. Many men and women that we have met throughout our journey in collecting the mushroom have also mentioned to us that the mushroom is widely used among them whenever they have "swelling" of any kind, for the ladies mainly for "swelling of the breast" (*bengkak payudara*). Some ladies who have stopped breastfeeding experience swelling and use the Tiger Milk mushroom sclerotial powder, made into a paste, and apply it topically. Some villagers from Cameron Highlands, Pahang, Malaysia, also told us that they use this same paste for wound healing. These experiences convey to us that the mushroom has anti-inflammatory properties that could assist in reducing swelling and improvement of wound healings. It may also contain antiproliferative properties, in cases where it is seen to reduce the size of tumors (*bengkak*). The local Chinese communities have also consistently used this sclerotium for lung and respiratory ailments. These random, informal conversations that we have had with the locals of varying ethnicities have encouraged us to further pursue scientifically based research into its bioactivity (Sect. 14.7).

14.4 *Lignosus* Species in Malaysia

This section describes the discovery of three species of genus *Lignosus* (*rhinocerus*, *tigris*, and *cameronensis*) in Malaysia and their morphological features, growth characteristics, and medicinal usage.

Cooke (1879) first described the *rhinocerus* species in Penang Island, Malaysia, followed by Sir Henry Nicholas Ridley, an English botanist and the "father" of Malaya's rubber industry, who described *Pleurotus rhinocerus* as a highly priced, valuable medicine and gave an account of the specimens which he obtained from Pekan, Pahang, Malaysia, and Bukit Mandai, Singapore. He further morphologically compared the fungus with *Pleurotus tuber-regium* and *Wolfiporia cocos* (Ridley 1890). *P. rhinocerus* is found in the Index Fungorum database as *Lignosus rhinocerotis* (Cooke) Ryvarden (an orthographic variant is *Lignosus rhinocerus*) which belongs to *Basidiomycota*, *Agaricomycotina*, *Agaricomycetes*, *Incertae sedis*, *Polyporales*, and *Polyporaceae* (Index Fungorum 2014).

In Malaysia, for the first time, *Lignosus rhinocerus* (MycoBank MB534962) was described in Penang Island, Malaysia, and subsequently was also found in three other geographical locations in Malaysia including Cameron Highlands, Hulu Langat, and Gerik (Cunningham 1965; Tan et al. 2010). In recent years, it can occasionally be collected from the remote areas of Pahang and Perak in Malaysia.

This fungus is commonly known in the Malay language as "cendawan susu rimau (harimau)" or "kulat susu rimau (harimau)," meaning the Tiger Milk mushroom. According to folklore, this mushroom was sprouted from the very spot where the milk of a roaming tigress has dropped on the ground – hence known as Tiger Milk mushroom.

Tan et al. (2013) described two new species based on collections from the tropical forests of Pahang, Malaysia. These were named as *Lignosus tigris and Lignosus cameronensis*. Morphologically, these species looked almost similar but upon detailed inspection were found to have differences in their pore and basidiospore sizes (Table 14.2, Fig. 14.3).

Prior to the discovery of *Lignosus tigris* and *Lignosus cameronensis*, these two species were often used interchangeably as *Lignosus rhinocerus*. Even now the wild types of these three species are being used as *Lignosus rhinocerus*, since the sclerotia are often collected without intact stipe and cap, which make it hard to differentiate them. The discovery of these two new species has sparked interest in domesticating them and investigating their medicinal properties.

Preliminary studies on successful small-scale cultivation of *Lignosus tigris* (Cultivated strain *Lignosus tigris K*) and *Lignosus cameronensis* (Cultivated strain *Lignosus cameronensis* S4) have enabled preclinical investigation into their sub-acute toxicity (Kong et al. 2016a; Lee et al. 2017). Nutritional properties on *Lignosus tigris* (Kong et al. 2016b) and *Lignosus cameronensis* (unpublished data) have also been investigated. They are found to be dissimilar yet comparable to *Lignosus rhinocerus*. Both these new species have been found to be safe for consumption with no adverse effect on blood biochemistry and primary organs. Preliminary investigation on the bioactivity of *Lignosus tigris* showed interesting comparisons to *Lignosus rhinocerus*, especially for breast cancer (unpublished results). At present, active studies are ongoing to elucidate the exclusivities of these two newly recognized species.

14.5 The Cultivation

The increasingly high demand and availability of modern medications have caused a decline in the usage of Tiger Milk mushroom. One important reason for its scarce use is that the mushroom can only be located by "luck," since the sporophore will only sprout from the underground sclerotium when the environment and conditions are optimum. The most sclerotium can remain underground for months, years, or even decades, before they are found and harvested. The nutrient in the sclerotium is used to sprout the sporophore causing the mass of harvested sclerotium to vary when harvested. Myth has it that only one mushroom would grow within a radius of 5 km, compounding the fact that locating this precious fungus in the wild forest is indeed a difficult task. From our observation, the rationale for this is that the Tiger Milk mushroom is nontoxic in nature (refer to toxicity studies conducted, Sect. 14.7) and its large network underground (known as the mycelium) would be degraded by insects (as feed) or other degradation factors; hence, it does not grow in groups like other mushrooms.

This popular mushroom is also under threat of depletion due to deforestation as a result of modern development and pollution (Vikineswary and Chang 2013).

Efforts to cultivate the mushroom have been initiated since 1999. Huang cultivated *P. rhinocerus* (taxonomically synonymous with *Lignosus rhinocerus*) using substrate bags [sawdust (80%), wheat bran (18%), sugar cane (1%), CaCO3 (1%), and water (1:1–1.4)] inoculated with spawn (young *P. rhinocerus* mycelia grown on sawdust-wheat bran medium) (Huang 1999b). Sclerotia were reported to form after 1 and a half years.

Rahman and colleagues noted the growth of *Lignosus rhinocerus* mycelium in mushroom complete medium (MCM) by employing the submerged culture technique (Rahman et al. 2012). Lai and colleagues further optimized the submerged culture conditions for the production of mycelial biomass and exopolysaccharides from *Lignosus rhinocerus* by controlling the culture conditions and modifying the medium's composition (Lai et al. 2014). Prior to that, they have also reported the optimal solid-state culture condition for *Lignosus rhinocerus* mycelial growth (Lai et al. 2011).

Abdullah and colleagues (Abdullah et al. 2013) showed that substrate formulation consisting of sawdust, paddy straw, and spent yeast at 7.9:1:1 in bags supported the best mycelial growth of this fungus. Sclerotia were formed 3–4 weeks after burial. They also reported the production of mycelium from submerged fermentation in flasks under shaking and static conditions (Lau et al. 2011, 2013a, b). Chen and colleagues (2013) described the preparation of *Lignosus rhinocerus* mycelium in bioreactors, also via submerged fermentation.

LiGNO™ Biotech Sdn. Bhd., a Malaysian company, developed an in-house proprietary method for the quick cultivation and mass production of *Lignosus rhinocerus* in 2009 (which was named TM02) using specially formulated culture medium consisting of rice, water, and other food-based materials. Upon inoculation, plastic storage containers were incubated in environmentally controlled culture room for up to 6 months to cultivate the sclerotia before harvesting. This product of cultivation (TM02) was used for most, if not all, of the scientific validation studies and research, ranging from safety studies (pre-commercialization) which led to the registration of *Lignosus rhinocerus* as a dietary supplement (registration number: MAL 13025025TC) to various bioactivity validation studies.

LiGNO™ Biotech Sdn. Bhd. is now recognized as the world's first commercial producer of *Lignosus rhinocerus*, and their efforts have enabled *Lignosus rhinocerus* to be commercially available. As of September 2010, the Tiger Milk mushroom has since been listed under the traditional medicine active ingredient list in the Malaysian National Pharmaceutical Control Bureau and was approved as the first registered commercial Tiger Milk mushroom product in April 2012.

14.6 Stages of Growth of the Three *Lignosus* Species

The different stages of growth for *Lignosus rhinocerus* are rather unusual. The fruiting body (also known as sporophore) has central stipitate pilei that grow from a subterranean sclerotium in a humid environment. The developmental stages of *Lignosus rhinocerus* are shown in Fig. 14.4a–e.

Under microscopic examination of the mycelial growth, the expansion of the mycelium due to the repeated branching of the germ tube (short, initial hypha) is seen to develop into a circular form which looks like a pair of "tiger eyes" (Fig. 14.4a). The colony color is usually white to beige or light yellow upon maturation and appears to look fluffy or velvety. Radiating hyphae cross-link to facilitate nutrient uptake and also for expansion to greater areas. After 1–2 months upon mycelium inoculation on solid medium, the mycelia will fully colonize the substrate, and sclerotia will begin to form. Vigorous mycelial growth promotes the development of mature sclerotia that can be harvested after 4–6 months. At this time, one can see shrinkage of the substrate block as the sclerotia mature. The growth rate of these three species differs: *Lignosus tigris* > *Lignosus rhinocerus* > *Lignosus cameronensis*.

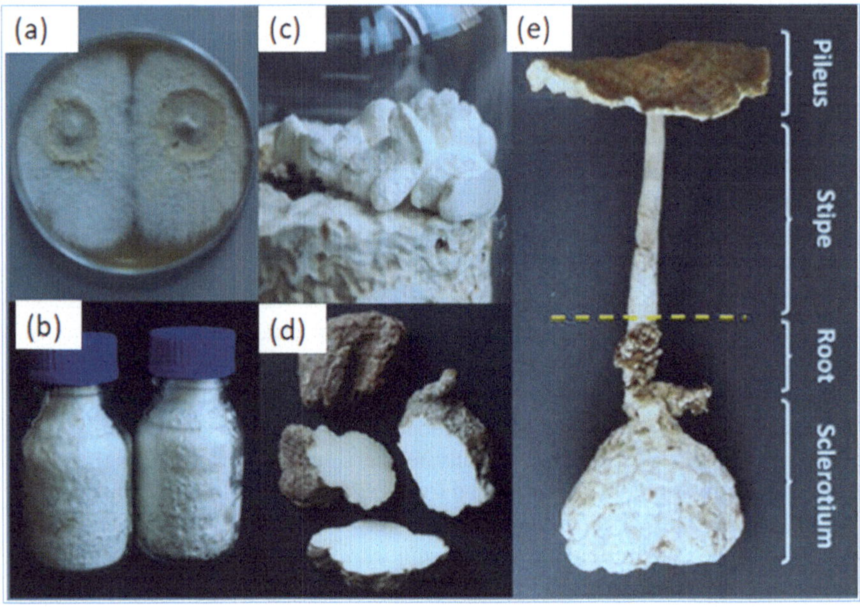

Fig. 14.4 Various stages of *Lignosus rhinocerus* development. (**a**) The culture of *Lignosus rhinocerus* mycelium on nutrified agar, also known as the "tiger eyes" (2 week culture). (**b**) Mycelial cultures of *Lignosus rhinocerus* on solid medium (1–2 month cultures). (**c**) Mature sclerotia on the surface of culture medium (4–6 month culture). (**d**) *Lignosus rhinocerus* sclerotia, the part with medicinal value. (**e**) The fruiting body of *Lignosus rhinocerus*. The yellow dotted line represents the division of aboveground and underground parts

Three-month-old mycelial culture can be buried in soil, and a stipe would sprout from the sclerotium after approximately 12 months, preceded by the formation of pileus (Fig. 14.4e).

The sclerotium of *Lignosus rhinocerus* comes in different shapes and sizes (spherical to oval or irregular with a diameter of 4–5 cm). The pale to the grayish brown outer skin (rind) appears rough and wrinkly to keep the internal compacted hyphal mass from drying out (Fig. 14.4d). From our observation, it is interesting to note that the sclerotium is made up of loose cells (Fig. 14.5). While observing its growth on an agar plate, each of these single, loose cells germinates on its own to form mycelia.

14.7 The Verification of *Lignosus rhinocerus*'s Traditional Uses: Scientific Evidences for Its Medicinal Usage

The *Lignosus rhinocerus* is traditionally used for ailments such as cough, asthma, fever, sinusitis, nausea, allergic reaction (including rashes), vomiting, indigestion, breast cancer, chronic hepatitis, gastric ulcer, and food poisoning, for wound healing, and as a general tonic to strengthen the body (Chang and Lee 2006). Prior to

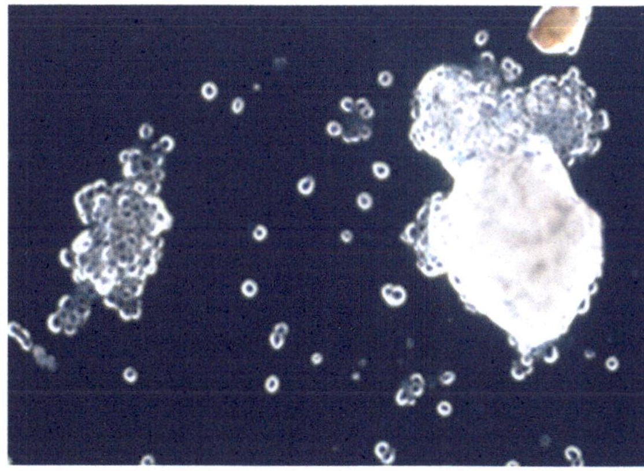

Fig. 14.5 Microscopic observation of sclerotium loose cells

mass cultivation of this species, no scientific studies were possible to validate the use of this fungus for these ailments. Small-scale studies were done using wild type, but the results were always inconsistent, which is not surprising as the nutritional content in the collected sclerotia always varies as the nutrient in the sclerotium is used to sprout the sporophore.

Scientific investigations to validate the traditional claims recorded by ethnomycological surveys have focused on the investigation of the composition and functional properties of *Lignosus rhinocerus* sclerotium.

The Medicinal Mushroom Research Group (MMRG) took the initiative to verify the safety of this mushroom for consumption. A thorough study which encompassed chronic, subacute, and acute toxicity, genotoxicity, antifertility, and teratogenic effects was conducted using the TM02, a cultivated strain of this mushroom. The work done was in compliance with the Organization for Economic Cooperation and Development (OECD) (Anonymous 2009). It was concluded that TM02 did not pose any threats of toxicity at dosages up to 1000 mg/kg, a much higher dosage than the recommended daily dosage of 0.5 g per day (average body weight of 50 kg). Cancer patients who often take much higher dosages (up to 100 mg/kg daily) also fall within the safe dosage range.

There were also no significant changes in body weight, urinalysis, hematological examination, and clinical biochemistry. No alterations were seen in the microscopic examinations of the organs. The mushroom also did not halt pregnancy nor alter the number of offspring nor produce any congenital malformation in offspring.

14.7.1 Anticancer

Its anticancer activities are well explored (Lai et al. 2008; Lau et al. 2013b; Lee et al. 2012; Suziana Zaila et al. 2013; Wong and Cheung 2009; Wu et al. 2013). Antiproliferative activities of cultivated and wild-type sclerotium of *Lignosus rhinocerus* against human leukemia, liver cancer, lung cancer, colon cancer, nasopharyngeal cancer, skin cancer, breast cancer, and prostate cancer cell line(s) have been explored (Lai et al. 2008; Lee et al. 2012; Yap et al. 2013; Lau et al. 2013b; Suziana Zaila et al. 2013). Among all tested cancer cell lines, human breast adenocarcinoma cell line (MCF-7) is the most susceptible cytotoxic target of cold-water extract of the sclerotium of *Lignosus rhinocerus* in which no significant cytotoxicity was found against the corresponding non-tumorigenic cell line (184B5) (Lee et al. 2012; Lau et al. 2013b; Yap et al. 2013). These findings provide scientific evidence for the traditional use of the sclerotium in breast cancer treatment. However, the identity of the bioactive component(s) responsible for the antiproliferative activity has yet to be established. The cold-water extract of *Lignosus rhinocerus* sclerotium contains 75% carbohydrate and 1.2% proteins (Lee et al. 2012). Lee and colleagues demonstrated that the antiproliferative properties were due to the proteins or the protein-carbohydrate complexes found in the high molecular weight fraction of the cold-water extract. This method of extraction prevents excessive degradation of thermolabile constituents, such as proteins and peptides. This was further verified when the cytotoxicity of the cold-water extracts was diminished when subjected to heat treatment from 60° to 100 °C (Lau et al. 2013a).

14.7.2 Anti-inflammatory

Many of the traditional ailments which led to the use of *Lignosus rhinocerus* sclerotium are linked to anti-inflammatory activity. The in vitro and in vivo anti-inflammatory activities of TM02 were investigated. The cold-water extract exhibited anti-acute inflammatory activity by reducing paw edema induced by carrageenan. All three phases of edema development were affected, and it was comparable to the effect of a nonsteroidal anti-inflammatory drug, indomethacin. In fact, the cold-water extract showed greater paw edema inhibition (~88%) than the standard drug (Lee et al. 2014). This anti-inflammatory effect was shown to be due to high molecular weight fractions of the cold-water extract, not withholding synergistic effect with lower molecular weight fractions. This observation is linked to an in vitro study done with cold-water extract, along with high and medium molecular weight fractions that were found to exhibit an inhibitory effect on TNF-alpha production. From the cotton pellet-induced granuloma test, it was shown that cold-water extract did not inhibit transudative and proliferative phases of chronic inflammation (Lee et al. 2014). The active principle of this bioactivity remains unelucidated but is hypothesized to be protein-polysaccharide complexes. The link between anti-inflammatory activity and its possible application to respiratory ailments are yet to be established, although a product named "BreathEzi" with sclerotium from Tiger Milk mushroom as its main active ingredient (Abdullah et al. 2014) has been developed to reduce common asthma symptoms and promote better respiratory health.

14.7.3 Antioxidative Activity and Inhibition of Protein Glycation

A positive correlation between inhibition of protein glycation and antioxidative potency has been shown to protect against free radicals derived from glycation and may have therapeutic potential (Elosta et al. 2012; Ramkissoon et al. 2013). The fractions from the cold-water extract of TM02 showed lower glycation inhibitory activity than the cold-water extract itself. Lower molecular weight fractions were observed to have better glycation inhibitory activities than those of higher molecular weight, and this is complementary to the secondary metabolites content and antioxidant potential of the fractions. Further in vitro analyses indicated that the SODs present in the mushroom (mainly found in the medium molecular weight fraction) might be responsible for the potent inhibitory effects against CML, Pent, and other related AGE structure formation (Yap et al. 2015, 2018).

Pathway annotation of *Lignosus rhinocerus* genome-transcriptome identified several interesting transcripts that are related to the biosynthesis of several types of terpenes and terpenoids (secondary metabolites) including carotenoid and abscisic acid, as well as genes that code for glyoxalase I, the key enzyme in the anti-glycation, in addition to CAT-peroxidases and SODs that are able to reduce the oxidative damage in cells through the superoxide radical degradation pathway, suggesting a potential role of TM02 in diabetes management (Yap et al. 2018).

Some other functional properties of *Lignosus rhinocerus* sclerotium such as immunomodulatory, neurite outgrowth-stimulating, antinociceptive, antiviral, and antimicrobial properties have also been described (Abdul Razak 2009; Eik et al. 2012; Phan et al. 2013; Mohanarji et al. 2012; Wong et al. 2009, 2011; Kavithambigai et al. 2013). Nondigestible carbohydrates (NDCs) from wild-type sclerotia of *Lignosus rhinocerus* have also been the subject of research. Gao and colleagues showed that the NDCs significantly stimulated the growth of beneficial bacteria and *Bifidobacterium longum* and *Lactobacillus brevis* showed significantly strong inhibition against *Clostridium celatum*, an important pathogen in human colon. On that basis, the sclerotium of *Lignosus rhinocerus* might be developed into a novel prebiotic for gastrointestinal health (Gao et al. 2009). The study of physicochemical and functional properties and biopharmacological values of novel sclerotial dietary fiber (DF) preparation from *P. rhinocerus* (cultivated by Sanming Mycological Institute, Fujian, China) by a scale-up-modified AOAC procedure using industrial enzymes has also been the interest of Cheung and colleagues (Wong and Cheung 2005a, b; Wong et al. 2005, 2006).

14.8 Surveillance Study

A small-scale surveillance study has been performed on 101 subjects who were keen to volunteer their feedback after consumption of Tiger Milk mushroom (TM02). Each of these individuals has particular pre-existing health-related concerns that have prompted them to try consuming this medicinal mushroom. The concerns include asthma-related, persistent cough, allergic rhinitis, sinusitis, low immunity (susceptible to getting sick often), and joint pains (Fig. 14.6a). The results of the surveillance study showed that 75% of consumers find that TM02 was effective in alleviating their health concerns (Fig. 14.6b).

Subjects who took TM02 to alleviate the symptoms of cough have found it to be helpful in cough relief and had lesser episodes of a recurrent cough. Some subjects have reported phlegm discharge. Phlegm usually indicates a developing infection as a result of the respiratory tract becoming inflamed, which also leads to coughing. Subjects who have low immunity (constantly falling ill) have acknowledged that TM02 has made them feel more energetic and less lethargic and fall ill less frequently. Asthmatic subjects have reported that their breathing has improved and have been less dependent on inhalers. Consumption of TM02 has also helped in relieving symptoms of itchiness, allergies, and fewer episodes of allergic rhinitis

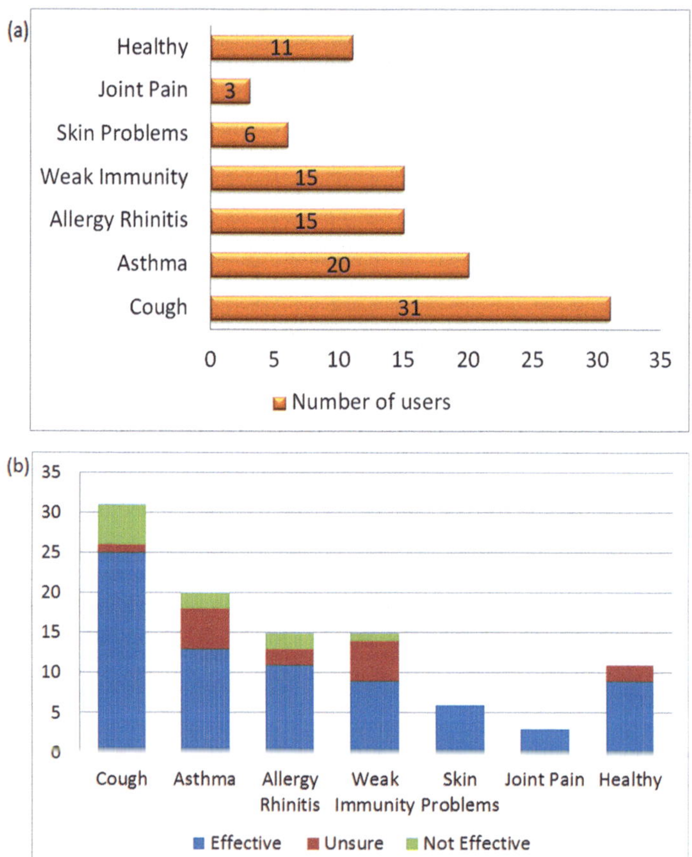

Fig. 14.6 (**a**) Pre-existing health-related concerns of subjects prior to consuming Tiger Milk mushroom. (**b**) Efficacy of Tiger Milk mushroom based on subject's own perception. Number of subjects: 101 (age range, 3 months–88 years old from both gender); surveillance study conducted via email and phone conversation

(especially upon waking up in the morning, less sneezing) and sinusitis. It is also interesting to note that a few subjects took TM02 to help relieve their joint pains.

This small-scale surveillance study has given useful insights into the efficacy of TM02 and provides the groundwork for future research.

14.9 The Prospects of Tiger Milk Mushroom as a Nutraceutical

The advent of genomic, transcriptomic, and proteomic data for *Lignosus rhinocerus* TM02 has opened doors for downstream work in developing new functional nutraceuticals. The de novo draft genome sequence of cultivated sclerotia of *Lignosus rhinocerus* (TM02) has been reported (Yap et al. 2014). A total of 1686 genes in *Lignosus rhinocerus* genome were predicted to encode for hypothetical proteins such as cytochrome P450s, lectins, fungal immunomodulatory proteins, laccases, as well as a repertoire of enzymes responsible for carbohydrate and glycoconjugate metabolism (Yap et al. 2014). The genome also contains genes encoding key enzymes involved in secondary metabolite biosynthesis from a nonribosomal peptide, polyketide, and triterpenoid pathways, in particular sesquiterpenoid biosynthesis genes in which up to 12 sesquiterpene cyclase genes were found. From the proteomic study, a total of 16 nonredundant, major proteins in sclerotium of *Lignosus rhinocerus* cultivar TM02 was detected by using two-dimensional gel electrophoresis coupled with mass spectrometry analysis. Putative lectins, immunomodulatory proteins, superoxide dismutase (0.91%), serine proteases, and aegerolysin (0.37%) in the sclerotium may have pharmaceutical potential (Yap et al. 2015).

The discovery of these proteins, along with the investigations into the bioactivity of *Lignosus rhinocerus* cultivar TM02, may lead to the potential development of nutraceuticals in the near future.

Active research is currently ongoing in our laboratories on the mechanism of action for the bioactivities of this medicinal mushroom. Substantial amounts of data have been generated since the inception of its research in 2009, and more are expected not only on *Lignosus rhinocerus* but also on *Lignosus tigris and Lignosus cameronensis*. We hypothesize that *Lignosus rhinocerus*'s role in respiratory and lung ailments may largely involve anti-inflammatory activities leading to a healthier lung (improvement in lung capacity) which will enable improved breathing, hence increasing/optimizing the amount of oxygen in the blood. Sufficient oxygen supply from a set of healthy lungs is the key factor to optimal health and dismissal of a variety of illness. Insufficient oxygen causes insufficient biological energy that result in anything from mild fatigue to life-threatening diseases.

It is also well-known that cancer cells thrive in low oxygen levels and are partial anaerobes. Hence, our hypothesis is that *Lignosus rhinocerus* has the ability to increase oxygen levels in the blood. Therefore, it can suppress the proliferation of cancer cells and simultaneously assist normal cells in optimal growth. It is possible that compounds in *Lignosus rhinocerus* play ancillary roles in nourishing cell growth. For cancer patients undergoing chemo-/radiotherapy (side effects include having a negative impact in healthy cells), the consumption of *Lignosus rhinocerus* may improve the growth of the healthy, normal cells. It is also interesting to note that pollutants and toxic components in our water and food can greatly reduce the oxygen level available to cells in our body. This anaerobic environment will upset the metabolism of the cell and, subsequently, will no longer participate in the healthy functioning of the body. It may also cause the cell to generate improper chemicals, forming a chain reaction in other surrounding cells, and weaken the immune system.

From another viewpoint, *Lignosus rhinocerus* may play a detoxification role in our biological system. The permissiveness of oxygen into the body in itself will lead to a powerful detoxifying mechanism. We would like to suggest that cancer (notwithstanding other causes) is due to a carcinogenic substance that could perhaps be an inducer for cells to proliferate. The detoxifying action may neutralize/form intermediary nontoxic components of the said carcinogenic substance. It would also include providing detoxifying reactions with substrates and metabolites for their reactions. Most, if not all, detoxifying processes require energy (in the form of ATP). We hypothesize that *Lignosus rhinocerus* plays a role in increasing/optimizing the amount of oxygen in the blood; this will then lead to increased energy production for the process of detoxification.

14.10 Conclusions

Three species of genus *Lignosus* have been identified in Malaysia based on morphological differences and the genetic DNA sequences of ITS regions: *Lignosus rhinocerus*, *Lignosus tigris*, and *Lignosus cameronensis*. Scientific investigations to validate the traditional claims recorded by ethnomycological surveys have focused on the investigation of the composition and functional properties of *Lignosus rhinocerus* sclerotium. Chronic, subacute and acute toxicity, genotoxicity, antifertility and teratogenic, nutritive, anticancer, anti-inflammatory, antioxidative activity and inhibition of protein glycation effects have been reported using the cultivated species (TM02) of this mushroom. The advent of genomic, transcriptomic, and proteomic data for *Lignosus rhinocerus* TM02 has opened doors for downstream work in developing new functional nutraceuticals.

References

Abdul Razak MS (2009) Antinociceptive of *Lignosus rhinocerus* Spp (Research report, Universiti Industri Selangor). Retrieved from http://www.scribd.com/doc/73538759/Antinoceceptive-of-Lignosus-Rhinoserus-Spp

Abdullah N, Haimi MZD, Lau BF, Annuar MSM (2013) Domestication of a wild medicinal sclerotial mushroom, Lignosus rhinocerotis (Cooke) Ryvarden. Ind Crops Prod 47:256–261. https://doi.org/10.1016/j.indcrop.2013.03.012

Abdullah NA, Gan SH, Wan Ezumi MF, Faezahtul AH, Wan Majdiah WM, Malagobadan JA/L (2014) BreathEzi (Tiger Milk mushroom): a natural remedy for the management of asthma. 25th International Invention, Innovation & Technology Exhibition, 8–10 May, Kuala Lumpur, Malaysia

Anonymous (2009) Chronic toxicity studies. OECD guideline for the testing of chemicals. Section4: health effects, no.452. Adopted by the Council on 7th September 2009, Paris, Cedex 16, France

Burkill IH, Haniff M (1930) Malay village medicine. Gard Bull Straits Settlements 6:165–321

Burkill IH, Birtwistle W, Foxworthy FW, Scrivenor JB, Watson JG (1966) A dictionary of the economic products of the Malay peninsula, Volume 1. Governments of Malaysia and Singapore by the Ministry of Agriculture and Co-operatives, Kuala Lumpur

Chang YS, Lee SS (2004) Utilisation of macrofungi species in Malaysia. Fungal Diver 15:15–22

Chang YS, Lee SS (2006) Food for thought – Fungi focus. FRIM Focus March 2006:5–6

Chen TI, Zhuang HW, Chiao YC, Chen CC (2013) Mutagenicity and genotoxicity effects of *Lignosus rhinocerotis* mushroom mycelium. J Ethnopharmacol 149(1):70–74. https://doi.org/10.1016/j.jep.2013.06.001

Cooke MC (1879) XV. Enumeration of Polyporus. Trans Bot Soc Edinburgh 13(1–4):131–159. https://doi.org/10.1080/03746607909468768

Corner EJH (1989) Ad Polyporaceas VI. The genus Trametes. Beihefte zur Nova Hedewigia 97:1–197

Cui BK, Tang LP, Dai YC (2011) Morphological and molecular evidences for a new species of Lignosus (Polyporales, Basidiomycota) from tropical China. Mycol Prog 10(3):267–271

Cunningham GH (1965) Polyporaceae of New Zealand. N Z Dep Sci Ind Res Bull 164:1–304

Douanla-Meli C, Langer E (2003) A new species of Lignosus (Polyporaceae) from Cameroon. Mycotaxon Ithaca Ny 86:389–394

Eik LF, Naidu M, David P, Wong KH, Tan YS, Sabaratnam V (2012) *Lignosus rhinocerus* (Cooke) Ryvarden: a medicinal mushroom that stimulates neurite outgrowth in PC- 12 cells. Evid Based Complement Alternat Med. https://doi.org/10.1155/2012/320308

Elosta A, Ghous T, Ahmed N (2012) Natural products as anti-glycation agents: possible therapeutic potential for diabetic complications. Curr Diabetes Rev 8(2):92–108

Evelyn J (1664) The diary of John Evelyn, vol 1. Nabu Press, Charleston

Fries EM, Nyman CM (1837) Polyporus sacer Afzel. ex Fr. In: Adami Afzelii fungi Guineenses: quos ad schedulas et specimina inventoris, descriptos 1–8

Gao S, Lai CKM, Cheung PCK (2009) Nondigestible carbohydrates isolated from medicinal mushroom sclerotia as novel prebiotics. Int J Med Mushrooms 11(1):1–8. https://doi.org/10.1615/IntJMedMushr.v11.i1.10

Haji Taha A (2006) Orang Asli: the hidden treasure. Department of Museums, Malaysia

Huang NL (1999a) Identification of the scientific name of hurulingzhi. Acta Edulis Fungi 6:32–34

Huang NL (1999b) A new species of medicinal mushroom – Cultivation of Hurulingzhi. Edible Fungi China 18:8–9

Imazeki R (1952a) *Rigidoporus lignosus* (Klotzsch) Imazeki. Bull Gov Forest Exp Stn Tokyo 57:113

Imazeki R (1952b) Microporus sacer (Afzel. ex Fr.). Bull Gov Forest Expl Station Meguro 57:113

Index Fungorum (2014) *Lignosus rhinocerotis* (Cooke) Ryvarden. Retrieved 12, March, 2018, from http://www.indexfungorum.org/names/NamesRecord.asp?RecordID=534962

Kavithambigai E, Vikineswary S, Thayan R (2013) Antiviral activity and mode of action of mushroom extracts against dengue virus type-2. In: Abstracts of the 3rd international conference on dengue and dengue haemorrhagic fever, Bangkok, Thailand, 21–23 October 2013

Kong BH, Fung SY, Tan NH, Pailoor J (2016a) Sub-acute toxicity study of Tiger Milk mushroom *Lignosus tigris* Chon S. Tan cultivar E sclerotium in Sprague Dawley rats. Front Pharmacol 7:246. https://doi.org/10.3389/fphar.2016.00246

Kong BH, Tan NH, Fung SY, Pailoor J, Tan CS, Ng ST (2016b) Nutritional composition, antioxidant properties and toxicology evaluation of the sclerotium of Tiger Milk mushroom *Lignosus tigris* cultivar E. Nutr Res 36(2):174–183. https://doi.org/10.1016/j.nutres.2015.10.004

Lai CKM, Wong KH, Cheung PCK (2008) Antiproliferative effects of sclerotial polysaccharides from *Polyporus rhinocerus* Cooke (Aphyllophoromycetideae) on different kinds of leukemic cells. Int J Med Mushrooms 10(3):255–264. https://doi.org/10.1615/IntJMedMushr.v10.i3.60

Lai WH, Siti Murni MJ, Fauzi D, Abas Mazni O, Saleh NM (2011) Optimal culture conditions for mycelial growth of *Lignosus rhinocerus*. Mycobiology 39(2):92–95. https://doi.org/10.4489/MYCO.2011.39.2.092

Lai WH, Loo SS, Rahmat N, Shaharuddin S, Daud F, Zamri Z, Saleh NM (2013) Molecular phylogenetic analysis of wild tiger's milk mushroom (*Lignosus rhinocerus*) collected from Pahang, Malaysia and its nutritional value and toxic metal content. Int Food Res J 20(5):2301–2307

Lai WH, Salleh SM, Daud F, Zainal Z, Othman AM, Saleh NM (2014) Optimization of submerged culture conditions for the production of mycelial biomass and exopolysaccharides from *Lignosus rhinocerus*. Sains Malaysiana 43(1):73–80

Lau BF, Aminudin N, Abdullah N (2011) Comparative SELDI-TOF-MS profiling of low-molecular-mass proteins from Lignosus rhinocerus (Cooke) Ryvarden grown under stirred and static conditions of liquid fermentation. J Microbiol Methods 87(1):56–63. https://doi.org/10.1016/j.mimet.2011.07.005

Lau BF, Abdullah N, Aminudin N (2013a) Chemical composition of the Tiger's Milk mushroom, *Lignosus rhinocerotis* (Cooke) Ryvarden, from different developmental stages. J Agric Food Chem 61(20):4890–4897. https://doi.org/10.1021/jf4002507

Lau BF, Abdullah N, Aminudin N, Lee HB (2013b) Chemical composition and cellular toxicity of ethnobotanical-based hot and cold aqueous preparations of the tiger's milk mushroom (Lignosus rhinocerotis). J Ethnopharmacol 150(1):252–262. https://doi.org/10.1016/j.jep.2013.08.034

Lee ML, Tan NH, Fung SY, Tan CS, Ng ST (2012) The antiproliferative activity of sclerotia of *Lignosus rhinocerus* (Tiger Milk mushroom). Evid Based Complement Alternat Med. https://doi.org/10.1155/2012/697603

Lee SS, Tan NH, Fung SY, Sim SM, Tan CS, Ng ST (2014) Anti-inflammatory effect of the sclerotium of Lignosus rhinocerotis (Cooke) Ryvarden, the tiger milk mushroom. BMC Complement Alternat Med 14:359. https://doi.org/10.1186/1472-6882-14-359

Lee SS, Fung SY, Tan NH, Pailoor J (2017) Safety evaluation of sclerotium from a medicinal mushroom, *Lignosus cameronensis* (cultivar): preclinical toxicology studies. Front Pharmacol. https://doi.org/10.3389/fphar.2017.00594

Leveille JH (1846) Descriptions des champignons de l'herbier du Museum de Paris. Ann Sci Nat Bot 5:111–167

Lloyd CG (1909–1912) Index of the Mycological Notes of C.G. Lloyd, Volume VII. Cincinnati, Ohio. https://archive.org/details/indexofmycologic07lloy/page/n4

Mohanarji S, Dharmalingam S, Kalusalingam A (2012) Screening of *Lignosus rhinocerus* extracts as antimicrobial agents against selected human pathogens. J Pharm Biomed Sci 18(11):1–4

Núñez M, Ryvarden L (2001) East Asian polypores 2. Polyporaceae s. lato. Synopsis Fungorum 14:170–522

Patouillard N (1900) Leucoporus sacer (Afzel. ex Fr.) Pat.In: Essai taxonomique sur les familles et les genres des Hyménomycètes, 82. France, Lons-Le-Saunier, https://doi.org/10.5962/bhl.title.40287

Phan CW, David P, Naidu M, Wong KH, Sabaratnam V (2013) Neurite outgrowth stimulatory effects of culinary-medicinal mushrooms and their toxicity assessment using differentiating Neuro-2a and embryonic fibroblast BALB/3T3. BMC Complement Alternat Med 13(1):261. https://doi.org/10.1186/1472-6882-13-261

Rahman NA, Daud F, Kalil MS, Ahmad S (2012) Tiger milk mushroom cultivation by using submerged culture technique. WSEAS Trans Biol Biomed 9(3):83–92

Ramkissoon JS, Mahomoodally MF, Ahmed N, Subratty AH (2013) Antioxidant and anti-glycation activities correlates with phenolic composition of tropical medicinal herbs. Asian Pac J Trop Med 6(7):561–569. https://doi.org/10.1016/S1995-7645(13)60097-8

Ridley HN (1890) On the so-called tiger's milk "Susu Rimau" of the Malays. J Straits Branch Royal Asiat Soc 22:341–344

Ryvarden L (1972) Lignosus goetzei (Henn.) Ryvarden. Norwegian J Bot 19:232

Ryvarden L (1975) Studies in the Aphyllophorales of Africa 3. Three new polypores from Zaïre. Bulletin du Jardin Botanique National de Belgique 45:197–203

Suziana Zaila CF, Farida Zuraina MY, Norfazlina MN, Lek Mun L, Nurshahirah N, Florinsiah L, Rajab NF (2013) Antiproliferative effect of Lignosus rhinocerotis, the tiger milk mushroom on HCT 116 human colorectal cancer cells. Open Conf Proc J 4:65–70. (Bentham Open)

Tan CS, Ng ST, Vikineswary S, Lo FP, Tee CS (2010) Genetic markers for identification of a Malaysian medicinal mushroom, *Lignosus rhinocerus* (Cendawan susu rimau). Acta Horticulturae (ISHS) 859:161–167

Tan CS, Ng ST, Tan J (2013) Two new species of *Lignosus* (Polyporaceae) from Malaysia – *L. tigris* and *L. cameronensis*. Mycotaxon 123:193–204. https://doi.org/10.5248/123.193

Vikineswary S, Chang ST (2013) Edible and medicinal mushrooms for sub-health intervention and prevention of lifestyle diseases. Tech Monitor 3:33–43

Wong KH, Cheung PCK (2005a) Dietary fibers from mushroom sclerotia: preparation and physicochemical and functional properties. J Agric Food Chem 53(24):9395–9400. https://doi.org/10.1021/jf0510788

Wong KH, Cheung PCK (2005b) Dietary fibers from mushroom sclerotia: 2. In vitro mineral binding capacity under sequential simulated physiological conditions of the human gastrointestinal tract. J Agric Food Chem 53(24):9401–9406. https://doi.org/10.1021/jf0510790

Wong KH, Cheung PCK (2009) Sclerotia: emerging functional food derived from mushrooms. In: Cheung PCK (ed) Mushrooms as functional foods. John Wiley & Sons, Inc, Hoboken, pp 111–146

Wong KH, Wong KY, Kwan HS, Cheung PCK (2005) Dietary fibers from mushroom sclerotia: 3. In vitro fermentability using human fecal microflora. J Agric Food Chem 53(24):9407–9412. https://doi.org/10.1021/jf051080z

Wong KH, Katsumata S, Masuyama R, Uehara M, Suzuki K, Cheung PCK (2006) Dietary fibers from mushroom sclerotia. 4. In vivo mineral absorption using ovariectomized rat model. J Agric Food Chem 54(5):1921–1927. https://doi.org/10.1021/jf052619w

Wong KH, Lai CKM, Cheung PCK (2009) Stimulation of human innate immune cells by medicinal mushroom sclerotial polysaccharides. Int J Med Mushrooms 11(3):215–223. https://doi.org/10.1615/IntJMedMushr.v11.i3.10

Wong KH, Lai CKM, Cheung PCK (2011) Immunomodulatory activities of mushroom sclerotial polysaccharides. Food Hydrocoll 25(2):150–158. https://doi.org/10.1016/j.foodhyd.2010.04.008

Wu H, Zhu H, Li X, Liu Z, Zheng W, Chen TF, Yu B, Wong KH (2013) Induction of apoptosis and cell cycle arrest in A549 human lung adenocarcinoma cells by surface-capping selenium nanoparticles: an effect enhanced by polysaccharide-protein complexes from *Polyporus rhinocerus*. J Agric Food Chem 61(41):9859–9866. https://doi.org/10.1021/jf403564s

Yap YHY, Tan NH, Fung SY, Aziz AA, Tan CS, Ng ST (2013) Nutrient composition, antioxidant properties, and anti-proliferative activity of *Lignosus rhinocerus* Cooke sclerotium. J Sci Food Agric 93(12):2945–2952. https://doi.org/10.1002/jsfa.6121

Yap HYY, Fung SY, Tan NH, Ng ST, Tan CS (2014) DNA barcode markers for two species of tiger milk mushroom: Lignosus tigris and L. cameronensis. Int J Agric Biol 16:841–844

Yap HYY, Fung SY, Ng ST, Tan CS, Tan NH (2015) Genome-based proteomic analysis of Lignosus rhinocerotis (Cooke) Ryvarden sclerotium. Int J Med Sci 12(1):23–31. https://doi.org/10.7150/ijms.10019

Yap HYY, Tan NH, Ng ST, Tan CS, Fung SY (2018) Inhibition of protein glycation by Tiger Milk mushroom [*Lignosus rhinocerus* (Cooke) Ryvarden] and search for potential anti-diabetic activity-related metabolic pathways by genomic and transcriptomic data mining. Front Pharmacol 9:103. https://doi.org/10.3389/fphar.2018.00103

Yokota A (2011) Tropical forests and people's livelihood: stalls of traditional medicine vendors in Kota Kinabalu. JIRCAS Newslett 62:9–10

Diversity and Medicinal Value of Mushrooms from the Himalayan Region, India

15

Sanjana Kaul, Malvi Choudhary, Suruchi Gupta, Dinesh Chandra Agrawal, and Manoj K. Dhar

Contents

Abstract

Mushrooms being considered as 'elixir of life' are valued for their culinary and therapeutic purpose throughout the world. They are widely appreciated as human food for centuries and represent an untapped reservoir of bioactive metabolites. This chapter gives a comprehensive overview on the diversity and indigenous knowledge of wild and cultivated mushrooms in Himalayan region of India. The diverse topographical features and altitudinal variation in Himalayan region of India favours luxuriant biodiversity, assemblage and distribution of macrofungi. The Himalayan belt of India encompasses north western region including the states Jammu and Kashmir, Himachal Pradesh and Uttarakhand and north eastern region including Sikkim, Arunachal Pradesh, West Bengal and Assam. The accumulated knowledge summarised in this chapter also focuses on the therapeutic benefits of mushroom related to their biological activity. The present study will serve as a foundation for further research on the exploration and utilisation of mushrooms in India.

S. Kaul (✉) · M. Choudhary · S. Gupta · M. K. Dhar
School of Biotechnology, University of Jammu, Jammu, Jammu and Kashmir, India

D. C. Agrawal
Department of Applied Chemistry, Chaoyang University of Technology, Taichung, Taiwan

D. C. Agrawal, M. Dhanasekaran (eds.), *Medicinal Mushrooms*,
https://doi.org/10.1007/978-981-13-6382-5_15

Keywords
Mushrooms · Diversity · Medicinal value · Himalayan region · Bioactive metabolites

Abbreviations

DNA Deoxyribonucleic acid
PUFA Polyunsaturated fatty acids

15.1 Introduction

Ongoing interdisciplinary scenario has unveiled mushrooms as unexplored resource in terms of metabolic characterisation as well as bioactivity profiling. Mushrooms are defined as cosmopolitan heterotrophic macrofungi which are directly or indirectly involved in maintaining ecological balance and stability (Choudhary et al. 2015; Sheikh et al. 2015). Out of the 1.5 million fungi estimated in the world, approximately 14,000 species produce fruiting bodies that are large enough to be judged as mushrooms (Chang 2006).

Archaeological evidence reveals that mycophily was customary in the human society from the prehistoric times and people belonging to ancient China, India and Iran used mushrooms for culinary purposes as well as for ritualistic performances. They have been considered as 'the gift from God Osiris' by early Egyptians, 'the food of the Gods' by ancient Romans and 'the elixir of life' by Chinese (Muhammad and Suleiman 2015). Human relationship with mushrooms is always treasured and appreciated since they have been used as nutritional functional food with possible medical applications (Khan and Tania 2012). Edible, medicinal and wild are the three main categories of the mushroom industry which has expanded dramatically in the last few years due to their nutritional and pharmacological effects (Royse et al. 2016).

Substantial compilation shows that the fruit body of mushrooms accumulates biologically active secondary metabolites including phenolic compounds, polyketides, terpenes and steroids (Boonsong et al. 2016). These begifted nutriceuticals and nutraceuticals are purported with many therapeutic properties such as anticancer, antihypercholesterolemic, antihypertensive, antidiabetic, antiobesity, hepatoprotective, antiaging, antimicrobial, antiallergic and antioxidant activities (Vaz et al. 2012; Qin et al. 2015). Recent reports have revealed more than 2000 species of mushrooms all over the world with 283 edible mushroom species from India only. India is endowed with high ecological diversity, ranging from alpine to tropical ecosystems covering a total area of more than 3 million km^2 (Vattakaven et al. 2016). India is ranked as eighth mega-diverse hotspots in the world, with a high diversity of habitat containing high level of endemicity (Mittermeier et al. 2011). It is spawning with diverse agroclimatic conditions. Abundance of wide range of microclimatic conditions is nourishing the growth of a large number of mushrooms (Das 2010). The medicinal importance of mushrooms can be better understood through the study of biodiversity and will also promote an interest towards understanding the fungal genome for metabolic potential (Chambergo and Valencia 2016). Also, mushrooms represent as underutilised and untapped resource for large-scale recycling of agro-wastes in an agricultural country like India (Tuli et al. 2013). The Himalayan range of India comprising of Northeastern and Northwestern regions have extensive areas under forest cover and are known to harbour diverse array of mushrooms. These regions occupy lucrative position for exploring the mushroom diversity for its bioactive chemical components and medicinal properties. This book chapter summarises mushroom diversity in Himalayan region of India which can serve as a repository for future studies.

15.2 Nutritional Composition of Mushrooms

Since the ancient times, mushrooms are not only valued for taste but also possess therapeutic attributes (Sharma and Gautam 2017). They are appreciated all over the world for splendid flavour and nutritional value (Malik et al. 2017). Nowadays, the term 'culinary-medical' mushrooms is being used for edible mushrooms as they serve considerable health benefits to consumers in addition to nourishment (Deb et al. 2018.) The main nutritional constituents present in mushrooms are polysaccharides, oligosaccharides, proteins, amino acids, vitamins, iron, zinc, selenium, sodium, chitin, fibres and minerals (Sharma and Gautam 2017). Approximately, the fruit body of a mushroom contains about 56.8% carbohydrate, 25.0% protein, 5.7% fat and 12.5% ash on a dry weight basis (Acharya et al. 2017) (Fig. 15.1). Mineral nutrients in mushrooms are considered as key factors for normal functioning of the body (Dutta and Acharya 2014). The protein content of mushrooms is even higher than most of the vegetables and is, therefore, considered as poor people's protein. Mushrooms possess digestible proteins and contain all the essential amino acids, particularly, lysine and leucine; however, sulphur containing amino acids cysteine

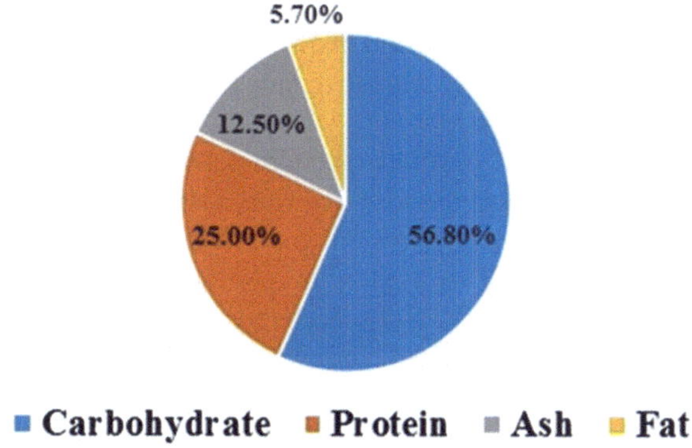

Fig. 15.1 Nutritional composition of mushrooms

and methionine are limiting in them (Chaudhary and Tripathy 2016; Vishwakarma et al. 2011). Health value of mushroom is attributed to the fact that fat content of mushroom is low and proportion of polyunsaturated fatty acids is more, relative to the total fatty acids (Malik et al. 2017). They are also known to be rich in antioxidant compounds predominantly polyphenols and flavonoids. These compounds can scavenge free radicals and are thus regarded as excellent antioxidants (Acharya et al. 2017). Mushrooms also hold a great importance in our diet due to its nutraceutical attributes that provide therapeutic benefits to the consumers. Polyunsaturated fatty acids (PUFA) especially the n-3-fatty acid family in mushrooms are claimed to exert a protective effect against the development of cardiovascular and inflammatory diseases (Sheikh et al. 2015). Consumption of mushrooms has been reported to improve immune function in cancer patients and thus prolong survival time in many types of cancers (Sharma and Gautam 2017). Besides, mushrooms are the source of wide variety of bioactive molecules like triterpenoids, steroids, phenols, etc. These compounds in addition to other nutrients serve its value as a source of medicine. Considering the valued advantages of mushrooms as tasty, nutritious, flavourful food in addition to its health benefits, continuing research towards unraveling the diversity of mushrooms seems imperative for exploring the potential of mushrooms as a source of bioactive compounds.

15.3 Mushroom Diversity in the Himalayan Region

The Himalayan ensemble is endowed with rich biological diversity of Mycota. The region is surrounded by vast stretches of mountains and sloping terrain. It has the distinction of wide range of climatic conditions, altitudinal variation, topographical features and soil characteristics which synergistically promote the luxurious growth of mushrooms. The Himalayan region of India is further divided into Northwestern and Northeastern Himalayas. The north western region includes the states of Jammu and Kashmir, Uttarakhand and Himachal Pradesh and north eastern region comprises of Sikkim, West Bengal, Assam and Arunachal Pradesh (Figs. 15.2 and 15.3). These states constitute different ecological zones that serve as critical centres for endemism and ectomycorrhizal diversity in Himalayas.

Northwestern Himalayan region of India consists of three states including Jammu and Kashmir, Uttarakhand and Himachal Pradesh. The unique geography and subtropical to temperate climate regime in these regions favour habitation of large number of mushrooms.

Uttarakhand constitutes western part of the central Himalayas and is one of the hotspots of global biodiversity. Garhwal region of Uttarakhand is very suitable for the growth of mushrooms due to its vast climatic conditions, plant distribution and field features. This region is glorified by the rich diversity of mushrooms (Vishwakarma et al. 2011). There are reports about the exploration of different species of mushrooms from Northwestern Himalayan region of India. For instance,

Sharma and Gautam (2017) during their frequent surveys of Northwestern Himalayan regions have reported six species of coral mushrooms belonging to

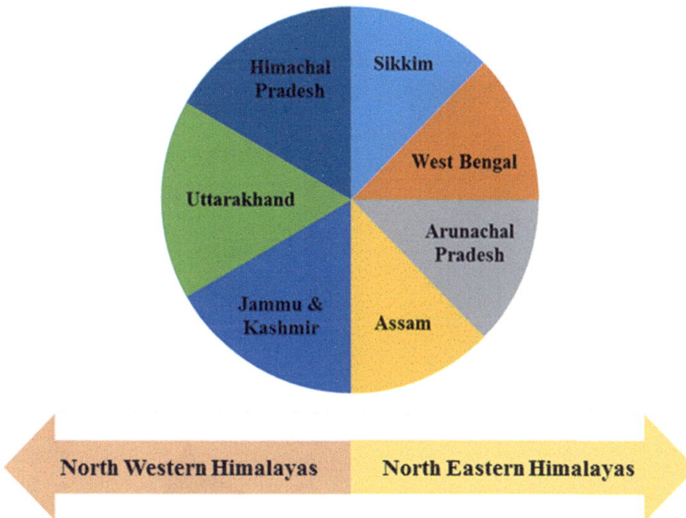

Fig. 15.2 Himalayan regions of India

genus *Ramaria*, viz. *R. botrytis*, *R. rubripermanens*, *R. flava*, *R. flavescens*, *R. aurea* and *R. stricta*, and six species belonging to genus *Clavaria*, viz. *C. fragilis*, *C. coralloides*, *C. purpurea*, *C. vermicularis*, *C. amoena* and *C. rosea*. Information regarding the culinary status of collected mushrooms gathered from local inhabitants have also been documented by them. Besides, they have also reported the detailed biochemical profiling and biological activities of the mushrooms from this region for the first time.

Another study carried out in Garhwal Himalayas reported 21 species of wild mushrooms belonging to 15 genera and 13 families. Of all the recorded wild edible macrofungi, the most significant seasonal food species were *Agaricus augustus*, *Hericium coralloides*, *H. erinaceus*, *Laetiporus sulphureus*, *Macrolepiota procera*, *Chlorophyllum rachodes*, *Pleurotus ostreatus* and *Ramaria sanguinea*. The rest of the species belonged to families *Agaricaceae*, *Russulaceae*, *Tremellaceae*, *Gomphaceae* and *Hericiaceae* (Singh et al. 2017a, b). In addition to this, Bhatt et al. (2016) have also collected 219 specimens of macrofungi from Garhwal Himalayas and categorised them into 15 species, 12 genera and 08 families. *Morchella esculenta*, *Cantharellus cibarius*, *Cantharellus minor* and *Grifola frondosa* were few among the collected varieties that were consumed by the local people.

Similarly, medicinal importance of *Ganoderma lucidum*, *Agaricus campestris*, *Hydnum repandum*, *Coprinus comatus*, *Morchella esculenta* and *Cantharellus cibarius* have also been documented from Garhwal region (Vishwakarma et al. 2011).

Fifteen species of wild medicinal mushrooms belonging to 15 genera and 14 families have been reported from Uttarakhand. Among them, *Cordyceps militaris*, *Ophiocordyceps sinensis* and *Morchella esculenta* are the members of *Ascomycota*, and the rest belonged to *Basidiomycota* (Bhatt et al. 2018).

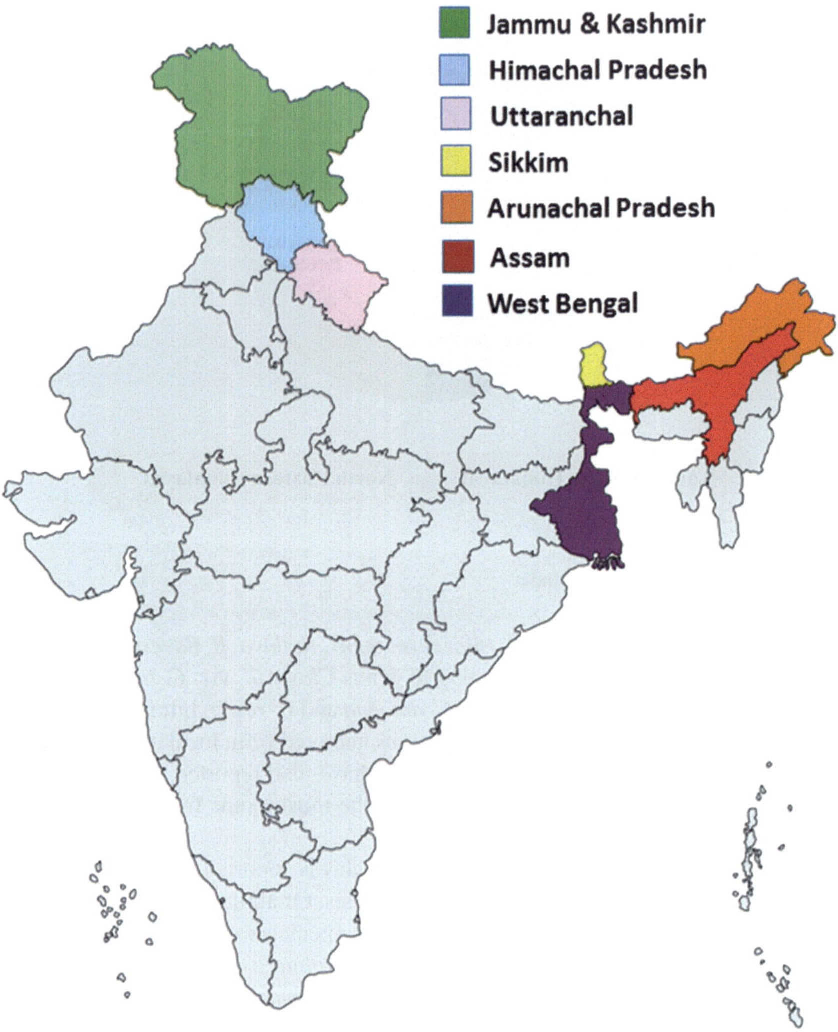

Fig. 15.3 Map of India showing the Himalayan region

During the period 2000–2013, a study was conducted in two states of Northwestern Himalayas, viz. Uttarakhand and Himachal Pradesh. The information about edible mushrooms consumed by the local people as well as for trade purpose was gathered. Accordingly, it was found that *Cordyceps sinensis* and several species of *Morchella* were collected specifically for trade purposes in the spring season at high altitude regions of Himalayas. Species of *Amanita*, *Agaricus*, *Astraeus*, *Hericium*, *Macrolepiota*, *Morchella*, *Pleurotus* and *Termitomyces* have been reported to be consumed very commonly by the local people, whereas genera that have been less commonly collected and consumed belong to include *Auricularia*, *Cantharellus*, *Sparassis*, *Lactarius*, *Ramaria* and *Russula* (Semwal et al. 2014).

Among all the edible mushrooms, *Morchella* is one of the most highly priced mushroom found in the world (Nitha et al. 2007). It is sold in the market for the price ranging from Rs. 2000 (29.22 $) to Rs. 2500 (36.53 $) per kg (Negi 2006). *Morchella esculenta*, commonly known as Guchhi, morels, common morel, true morel, yellow morel, sponge morel, is one of the most expensive fungi of *Morchellaceae* family (Ajmal et al. 2015). Due to its unique flavour, taste and texture, it is considered as a one of the delicacies among vegetarians and is used all over the world in different recipes. Morels are widely distributed in India and are common in temperate forests of Northwestern states including Himachal Pradesh, Punjab, Jammu and Kashmir and Uttarakhand. Local people of these areas believe that sprouting of morels has a strong correlation with thunder and lightning (Negi 2006). Apart from this, Morels have been traditionally used as a therapeutic agent for curing ailments such as pneumonia, fever, cough, stomach pain and in pregnant and lactating mothers. It is also known to cure all the respiratory disorders. Moreover, the morels also possess significant pharmacological activities like antimicrobial, anti-inflammatory, immunostimulatory, etc. (Bala et al. 2017).

Considerable diversity exists among different species of Morels. Although morphological studies and classical taxonomy are helpful in identification of morels, however, for better understanding, molecular techniques including DNA sequencing could be of great help in sorting out the taxonomic confusion concerning many species of *Morchella*. In this regard, Kanwal et al. (2011) have collected *Morchella* from Western Himalayas of India. Based on sequence analysis, yellow and black morels have been found to be prominent along with *Verpa* spp. Moreover, a clear distinction was revealed between yellow and black morels based on phylogenetic analysis by maximum parsimony, maximum likelihood and Bayesian inference.

A study on Morels has also been carried out in Darma valley in Pithoragarh district, Kumaun Himalaya of Uttaranchal. The study aimed to shed light about further exploration of these macrofungi. Characteristic feature of common edible species of genus *Morchella* like *M. semilibera*, *M. angusticeps*, *M. deliciosa*, *M. crassipes*, *M. esculenta* and *M. conica* has also been described in this study (Negi 2006).

Jammu and Kashmir State of India which also lies in the Northwest Himalayas is bordered on the north and east by main Himalayan ranges and to the south by Punjab plains. Due to its varied climatic and topographic conditions, this state provides a congenial environment for the lavish growth of mushrooms (Kumar and Sharma 2011; Wani et al. 2013). The macrofungal species richness of the state is directly related to its expansive forest communities and diverse weather patterns. Most of the regions of this state have not been explored extensively for mushroom diversity (Pala et al. 2012). Therefore, various workers have carried out a systematic survey of different regions of the state frequently. This has led to the discovery of different species of mushrooms. Some of the mushroom species have been recorded for the first time from the state of Jammu and Kashmir, and some of them are even first reports from India.

In a study, conducted by Kumar and Sharma (2011), 150 samples of wild mushrooms were collected and studied for their macro–micro morphological and ethnomycological features. Accordingly, these were categorised into 66 taxa of wild mushrooms belonging to 33 genera spread over 22 families, 10 orders and 3 classes. The identified mushrooms belonged to following genera, viz. *Agaricus*, *Astraeus*, *Amanita*, *Auricularia*, *Boletus*, *Bovista*, *Cantharellus*, *Chalciporus*, *Clavaria*, *Clavulina*, *Coprinus*, *Flammulina*, *Geopora*, *Gyromitra*, *Helvella*, *Lactarius*, *Lentinus*, *Leucopaxillus*, *Lycoperdon*, *Macrolepiota*, *Morchella*, *Otidea*, *Peziza*, *Pleurotus*, *Ramaria*, *Rhizopogon*, *Russula*, *Schizophyllum*, *Scleroderma*, *Sepultaria*, *Sparassis Termitomyces* and *Verpa*. In order to have better understanding of these fungi and their impact on economy, ethnomycological information was gathered from tribal men, women, village heads and Ayurvedic hakims. It was found that some of the varieties of mushrooms are consumed as fresh vegetables including *Agaricus arvensis*, *Boletus* spp., *Coprinus comatus*, *Peziza badia*, *Clavaria vermicularis*, *Clavulina* spp., *Geopora arenicola*, *Gyromitra* spp., *Helvella* spp., *Macrolepiota procera*, *Morchella* spp., *Otidea leporina*, *Pleurotus* spp., *Ramaria* spp., *Sparassis* spp., *Sepultaria sumneriana*, *Russula* sp. and *Termitomyces* spp. While certain varieties like *Morchella* spp., *Geopora arenicola*, *Sepultaria sumneriana*, *Sparassis* spp. *Pleurotus* spp. and *Verpa conica* are specifically consumed during winters by inhabitants of hilly area when vegetables are not available in plenty, and due to harsh weather conditions, there is restriction in the movement of local people. Apart from above-mentioned species of mushrooms from Jammu and Kashmir, Kumar and Sharma (2011) have also explored the diversity of boletoid macrofungi in Jammu and Kashmir. Their study described the habit, habitat and macro and micro features of six boletoid taxa including *Austroboletus malaccensis*, *Boletus edulis*, *Boletus formosus*, *Boletus granulatus*, *Boletus luridus* and *Suillus cavipes*. In continuing their work on exploring different varieties of mushrooms of Jammu and Kashmir state, Kumar and Sharma (2009) have illustrated with photographs, habit, habitat and macro- and microscopic details of mushroom belonging to order *Pezizales* with nine taxa spread over four families. These included *Helvella atra*, *Helvella crispa*, *Helvella elastica*, *Geopora arenicola*, *Geopyxis catinus*, *Morchella esculenta*, *Otidea leporina*, *Peziza badia* and *Sepultaria arenosa*. Further, other reports from Northwestern region have been tabulated (Table 15.1).

15.4 Northeastern Himalayan Region

Northeastern Himalayan region of India consists of four states including Sikkim, West Bengal, Assam and Arunachal Pradesh. The regional ethnomycological knowledge along with rich biodiversity supports the growth and availability of mushrooms in these states.

Sikkim, a small Himalayan state in India, serves as a treasure trove for fleshy wild mushrooms. Altitudinal variation along with wide range of microclimatic conditions favours and nourishes the growth of a large number of mushrooms in the

Table 15.1 Statewise distribution/diversity of mushrooms in the Northwestern Himalayan region of India

S. no.	State	Region/area	Mushrooms	References
Northwestern Himalayan region				
1.	Himachal Pradesh	Kangra, Kullu, Shimla, Solan and Lahaul Spiti	*Lactarius*, *Laetiporus*, *Laccaria*, *Stropharia*, *Marasmius*, *Cortinarius*, *Ramaria*, *Russula* and *Strobilomyces*	Chaudhary and Tripathy (2016)
2.	Jammu and Kashmir	Jammu province	*Cantharellus cibarius*, *Coprinus comatus*, *Geopora arenicola*, *Ramaria Formosa*, *Ramaria flavo-brunnescens*, *Sparassis crispa* and *Termitomycetes striatus*	Kumar and Sharma (2009)
			Phallus macrosporus, *Phallus rubicundus* and *Phallus hadriani*	Yangdol et al. (2016b)
		Kashmir Himalayas	*Amanita ceciliae*, *Amanita flavoconia*, *Amanita muscaria*, *Amanita pantherina*, *Amanita phalloides*, *Amanita vaginata*, *Amanita virosa*, *Russula aeruginea*, *Russula atropurpurea*, *Russula aurea*, *Russula cyanoxantha*, *Russula delica*, *Russula emetica* and *Russula nobilis*	Pala et al. (2012)
			Gyromitra sphaerospora (Peck) Sacc. and *Mutinus caninus*	Wani et al. (2013)
			Cortinarius flexipes, *Cortinarius fulvoconicus* and *Cortinarius infractus*	Itoo et al. (2015)
		North Kashmir	*Collybia chrysoropha* and *Russula albida*	Kaur and Rather (2016)
		Rajouri district	*Verpa bohemica*	Anand and Chowdhry (2013)
		Yusmarg, Gulmarg, Mammer, Kellar and Pahalgam	*Morchella esculenta*, *Coprinus comatus*, *Fomes fomentarius*, *Ganoderma lucidum*, *Neolentinus* sp., *Suillus sibiricus*, *Suillus granulates*, *Lactarius deliciosus*, *Russula atropurpurea*, *Russula aurea*, *Calvatia* sp., *Lycoperdon* sp., *Agaricus bisporus* and *Cantharellus cibarius*	Farooq et al. (2017)
		Ladakh Region	*Cyathus olla*	Dorjey et al. (2013)
			Laetiporus sulphureus	Yangdol et al. (2014)
			Peziza ammophila, *Peziza ampliata*, *Peziza badia*, *Peziza succosa* and *Peziza vesiculosa*	Dorjey et al. (2016)
			Inocybe curvipes and *Inocybe sororia*	Yangdol et al. (2016a)

(continued)

Table 15.1 (continued)

S. no.	State	Region/area	Mushrooms	References
			Geopora arenicola, *G. sepulta*, *Pulvinula convexella*, *P. Miltina*, *Anthracobia macrocystis* and *Geopyxis majalis*	Yangdol et al. (2017)
		Poonch district	*Strobilomyces echinocephalus* and *Strobilomyces mollis*	Kour et al. (2013)
			Clitocybe dilatata, *Clitocybe hydrogramma* and *Clitocybe nebularis*	Kour et al. (2015)
3.	Uttarakhand	Garhwal	*Ganoderma lucidum*, *Agaricus campestris*, *Hydnum repandum*, *Coprinus comatus*, *Morchella esculenta* and *Cantharellus cibarius*	Vishwakarma et al. (2011)
		Garhwal Himalaya	*Agaricus augustus*, *Hericium coralloides*, *H. erinaceus*, *Laetiporus sulphureus*, *Macrolepiota procera*, *Chlorophyllum rachodes*, *Pleurotus ostreatus*, *Ramaria sanguinea*, *Coprinus comatus*, *Macrolepiota procera*, *Cantharellus lateritius*, *Ramaria botrytis*, *Ramaria sanguinea*, *Helvella crispa*, *Laetiporus sulphureus*, *Psathyrella candolleana*, *Aleuria aurantia*, *Lactifluus volemus*, *Lactifluus corrugis*, *Russula cyanoxantha*, *Stropharia rugosoannulata*, *Tremella foliacea*, *T. mesenterica* and *T. fuciformis*	Singh et al. (2017a, b)

state. Many researchers have compiled the ethnomycological knowledge and mushroom diversity within this state.

Das (2010) reported 126 wild mushrooms with their distribution, growing period and status of edibility from Barsey Rhododendron Sanctuary of the state Sikkim. Out of 126 identified species, 19 belong to the class *Ascomycota* and 107 belong to *Basidiomycota*. He has also enlisted about 46 medicinally important mushrooms with dominant genera being *Cordyceps*, *Xylaria*, *Coprinus*, *Ganoderma*, *Hygrocybe*, *Lycoperdon*, *Russula*, *Trametes* and *Xylaria*. Furthermore, during his macrofungal survey of Sikkim from 2014 to 2015, he rediscovered a species of *Strobilomyces*, viz. *Strobilomyces polypyramis*, from North district of the state after a gap of 164 years (Das et al. 2014). Additionally, he also reported two species belonging to the family *Cortinariaceae*, viz. *Cortinarius varicolor* (subg. *Phlegmacium*) and *C. salor* (subg. *Myxacium*), for the first time from India coupled with their detailed macro- and micromorphological characterisation (Das and Chakraborty 2015).

Cordyceps sinensis, commonly known as Yarsa gumba/Keera jhar, is a reputed longevity promoting herb as reported by Panda and Swain (2011). Its pharmacological potential is contributed in modulating hepatic, renal, cardiovascular and endocrine functions and in erythropoiesis and immunomodulation besides antitumor function. Apart from above-mentioned species of mushrooms, Chakraborty et al. (2017) have also described for the first time two species of boletoid mushrooms, namely, *Aureoboletus nephrosporus* and *Strobilomyces mirandus*, with their detailed morphological characterisation. In another study regarding the discovery of new species, from Sikkim a marasmioid fungus, *Marasmius indopurpureostriatus* was reported by Dutta et al. (2015) during their ethnomycological survey of forest area of East district. These are some of the reports on the compilation of mushroom diversity within this state revealing the status of mushroom richness and productivity.

Assam, situated in Northeast India, is blessed with diverse hills and vales. It experiences varied geo-climatic conditions and altitudinal variations which harbour and sustain diversity of ecological habitats such as forests, grasslands and wetlands supporting the luxuriant growth of wild mushrooms. Besides, it is a home to reserve forests, viz. Eastern Himalayan upper Bhabar Sal forest, Eastern Himalayan Lower Bhabar Sal forest, Eastern Terai Sal forest, Eastern heavy alluvial Plain Sal forest, Eastern Hill Sal forest, Northern Secondary moist deciduous forest, Evergreen forest, Lower alluvial Savanah, Woodland, Eastern west alluvial grass land, Riparian Fringe Forest and Khoir Sissoo forest. Most of these forests are inhabited by a large number of ethnic groups holding an invaluable information on the rich mushroom diversity and domestication. Therefore, an attempt has been made to benchmark the diversity of macrofungi with respect to their morphological distribution, habitat and edibility within the state.

In a study conducted by Sarma et al. (2010), the diversity and evaluation of few edible mushrooms used by some ethnic tribes of Western Assam have been documented. He identified 26 different species of mushrooms belonging to 14 genera and 13 families. The highest distribution frequency was that of *Ganoderma lucidum* followed by *Cantharellus tubaeformis* and *Agaricus bisporus* (83.33%), and least found were species of *Agaricus*, *Boletus*, *Lenzites*, *Lycoperdon* and *Termitomyces* (16.66%). He has also summed up the edibility status of 25 mushrooms belonging to genera *Auricularia*, *Agaricus*, *Boletus*, *Calvatia*, *Cantherallus*, *Ganoderma*, *Lentinus*, *Laetiporus*, *Lycoperdon*, *Morchella*, *Schizophyllum*, *Termitomyces* and *Tricholoma*.

Furthermore, the importance of taxonomy and diversity of mushrooms was studied by Parveen et al. (2017). During his macrofungal foray, 44 samples were collected from different locations and analysed for their edibility, pharmacological properties and industrial applications. Few of the dominant genera reported were *Lentinus, Ganoderma, Pleurotus, Agaricus, Lycoperdon* and *Macrolepiota*. In addition to this, another report shedding light on the relationship between ethnobotanical knowledge and traditional folk practices has been highlighted in a study carried out by Nath and Sarma (2018) who identified 14 species of edible mushrooms along with their preferred substrata. *Auricularia auricula, Ganoderma lucidum* and *Trametes pubescens* are few of the species used by the tribes to overcome food and nutraceutical deficiencies. The consumption of wild edible mushrooms by many ethnic tribes as reported by many workers sheds light on the enrichment of the socio-economic status of the tribal people of the state.

West Bengal, constituent unit of Eastern Himalayas, is geographically nested in one among the top 25 global biodiversity hotspots in the world. It serves as an 'oriental region' providing a profusion of phytogeographical conditions. It is topographically extended from Himalayas in the north to the Bay of Bengal in the south. It experiences wide range of climatic pattern and hydrological regimes suitable for promoting tropical forest ecosystem for the gregarious growth of mushrooms. Much of the contribution in evaluating mushroom diversity in the state of West Bengal is owed to Acharya and his co-workers. An extensive survey to explore the knowledge about mushroom utilisation for medicinal and culinary purposes was described by Dutta and Acharya (2014). A 5-year macrofungal foray was conducted among eight districts of West Bengal. The study explored 34 macrofungi with some species like *Amanita, Astraeus, Russula, Termitomyces, Armillaria, Auricularia, Fistulina, Grifola, Hericeum, Coprinus, Pholiota, Meripilus, Pleurotus, Calocybe, Lentinus, Tricholoma* and *Volvariella* being relished as food and species of *Cordyceps, Ganoderma, Schizophyllum, Termitomyces, Daldinia* and *Pisolithus* being used for curing human ailments.

Three different wild edible mushrooms, viz. *Lentinus squarrosulus*, *Russula albonigra* and *Tricholoma giganteum*, have been assessed for their nutriceutical potential by Giri et al. (2013). Another study conducted by Khatua et al. (2015) on *Russula senecis* not only revealed the therapeutic potential but also taxonomically classified the species based on its macro- and micromorphological features.

An extensive survey of wild mushrooms in the Gurguripal Ecoforest of West Bengal by Singha et al. (2017a, b) provided a detailed analysis of nine species of mushrooms (*Termitomyces heimii*, *Astraeus hygrometricus*, *Leucopaxilus* sp., *Amanita vaginata*, *Agaricus campestris*, *Russula delica*, *Schizophyllum commune*, *Pleurotus ostreatus* and *Cantharellus* sp.). They have described their importance which can be explored for their potential in the field of human health, nutrition and disease prevention. They also worked on the eco-diversity, productivity and distribution frequency of mushrooms; a total of 71 species have been enlisted from Paschim Medinipur region of the state. The species reported belonged to 41 genera spread over 24 families with 32 edible, 39 inedible and altogether 19 having pharmacological active substances. Other reports on diversity of mushrooms from the north eastern region have been tabulated (Table 15.2).

15.5 Medicinal Importance of Mushrooms

Mushrooms are not only valued for their splendid, tasteful flavour but are also known to have significant medicinal possibilities such as antioxidant, antimicrobial, cardioprotective, hypoglycemic and hepatoprotective (Rai et al. 2013). Mushroom research has focused on discovery of various bioactive metabolites that could serve as an important component of today's pharmaceutical compendium. Although much work is focused on the identification or diversity of mushrooms in the Northwestern Himalayan region, however, potential of mushrooms as therapeutic agents in these regions is bleakly explored. Few of the reports have been enlisted in Table 15.3 to benchmark the bioactive potential of various mushroom species reported from the Himalayan region.

Table 15.2 Statewise distribution/diversity of mushrooms in the Northeastern Himalayan region of India

S. no	State	Region/area	Variety of mushrooms	Reference
Northeastern Himalayan region				
1.	Sikkim	Himalayan region	*Agaricus* sp., *Cortinarius* sp., *Clitocybe* sp., *Collybia* sp., *Gilera* sp., *Hygrophorus* sp., *Hypholoma* sp., *Flammula* sp., *Lactarius* sp., *Lentinus* sp., *Mycena* sp., *Pleurotus* sp., *Pluteus* sp. and *Russula* sp.	Acharya et al. (2010a, b)
2.	West Bengal	Midnapur district	*Lysurus* sp.	Acharya et al. (2010a, b)
		Howrah and South 24 Parganas	*Tulostoma chudaei*	Chakraborty et al. (2013)
		Lateritic	*Amanita hemibapha*, *A. vaginata*, *A. vaginata* var. *alba*, *Astraeus hygrometricus*, *Russula albonigra*, *R. brevipes*, *R. cyanoxantha*, *Russula* sp. *R. senecis*, *R. lepida*, *Termitomyces clypeatus*, *T. heimii* and *T. microcarpus*	Dutta and Acharya (2014)
		Himalayan	*Armillaria mellea*, *Auricularia auricula*, *Fistulina hepatica*, *Grifola frondosa*, *Hericeum* sp., *Coprinus comatus*, *Pholiota squarrosa*, *Meripilus giganteus* and *Pleurotus* sp.	
		Coastal	*Calocybe indica*, *Lentinus squarrosulus*, *Pleurotus ostreatus*, *M. gigantea*, *Macrocybe lobayensis* and *Volvariella volvacea*	
		Darjeeling hills	*Russula* sp.	Paloi et al. (2015)

15.6 Conclusions

Nature has bestowed the Himalayan region with rich biodiversity, varied geographical settings, special climatic conditions and forest cover, perfectly suited for sustaining bewildering species of mushrooms. The present chapter has enlisted the various reports regarding the existence, diversity and bioactive potential of mushrooms in Western and Eastern Himalayan region of India. Also, the culinary credentials have revealed that mushroom species are nutritionally rich with high protein and carbohydrate content and low levels of fat. In addition, the ethnomycological aspects of macrofungi in these regions have promoted mushroom as a functional ingredient, important in maintaining good health and promoting life longevity. The documented literature described here could preferably attract the researchers and mycologists to indulge their interest in mushrooms and their production. Such information will not only help in preserving and protecting the fast eroding traditional knowledge but also pave a way for the exploration of some new species of

Table 15.3 Bioactive potential of mushrooms from Northwestern and Northeastern Himalayas (2012–2017)

S. no.	Mushroom species	Location	Extract/ compound	Medicinal properties	Reference
1.	*Agaricus bisporus*, *Pleurotus ostreatus*, and *Coprinus atramentarius*	Jammu and Kashmir	Methanolic extract	Antioxidant	Khan et al. (2016)
2.	*Verpa bohemica* and *Morchella esculenta*	Kashmir Himalaya	Ethanolic and methanol extract	Antioxidant	Shameem et al. (2015)
3.	*Calocybe indica*		Aqueous extract	Antidiabetic	Singh et al. (2017a, b)
4.	*Cordyceps (C. gracilis, C. cicadae, C. sinclairii)* and *Metacordyceps (M. dhauladharensis)*,	Indian Himalayas	–	Antioxidant and antibacterial activity	Sharma and Gautam (2017)
5.	*Ramaria subalpina*	Sikkim Himalayas	Methanolic extract	Antioxidant	Acharya et al. (2017)
6.	*Psathyrella spadicea*	Cold desert of Ladakh	Methanolic extract	Antioxidant	Yangdol et al. (2016c)
7.	*Agaricus bisporus*, *Coprinus atramentarius*, *Ganoderma lucidum*, *Morchella esculenta*, *Pleurotus ostreatus* and *Verpa bohemica*	Western Himalayas	Methanolic, ethanolic and distilled water extract	Antioxidant	Sheikh et al. (2015)
8.	*Meripilus giganteus*, *Entoloma lividoalbum*, *Ramaria aurea and Pleurotus flabellatus*	Darjeeling Hills, West Bengal	Ethanolic and ethyl acetate extract	Antibacterial and anticandidal activity	Rai et al. (2013)

(continued)

Table 15.3 (continued)

S. no.	Mushroom species	Location	Extract/ compound	Medicinal properties	Reference
9.	*Agaricus campestris*, *Amanita vaginata*, *Armillaria mellea*, *Astraeus hygrometricus*, *Auricularia auricula*, *Auricularia* sp., *Calocybe indica*, *Fistulina hepatica*, *Hygrophorus miniatus*, *Lepiota procera*, *Lepiota* sp., *Pleurotus djamor*, *Pleurotus ostreatus*, *Pleurotus sajor-caju*, *Lentinus squarrosulus*, *Polyporus grammocephalus*, *Ramaria botrytis*, *Russula albonigra*, *Russula delica*, *Russula laurocerasi*, *Russula lepida*, *Schizophyllum commune*, *Termitomyces clypeatus*, *Termitomyces eurhizus*, *Termitomyces microcarpus*, *Tricholoma giganteum*, *Tricholoma lobayense*, *Tricholoma* sp., *Tricholoma crassum* and *Volvariella volvacea*	West Bengal	Methanolic extract	Antimicrobial	Giri et al. (2012)
10.	*Termitomyces heimii*, *Astraeus hygrometricus*, *Leucopaxilus* sp., *Amanita vaginata*, *Agaricus campestris*, *Russula delica*, *Schizophyllum commune*, *Pleurotus ostreatus* and *Cantharellus* sp.	Gurguripal Ecoforest, West Bengal	Acetone extract	Antibacterial	Singha et al. (2017a, b)

mushrooms. In view of increasing urbanisation, changing climatic patterns and swelling population, more studies on the ethnomycology of mushrooms in the Himalayan region are called for. Further, these findings will open up a new horizon regarding harvesting strategies and management plans of these socially and economically important species of mushrooms.

References

Acharya K, Pradhan P, Chakraborty N, Dutta AK, Saha S, Sarkar S, Giri S (2010a) Two species of *Lysurus* Fr.: additions to the macrofungi of West Bengal. J Bot Soc Bengal 64(2):175–178

Acharya K, Rai M, Pradhan P (2010b) Agaricales of Sikkim Himalaya: a review. Researcher 2(5):29–38. https://doi.org/10.5530/pj.2017.1.6

Acharya K, Das K, Paloi S, Dutta AK, Hembrom ME, Khatua S, Parihar A (2017) Exploring a novel edible mushroom *Ramaria subalpina*: chemical characterization and antioxidant activity. Pharmacogn J 9(1):30–34

Ajmal M, Akram A, Ara A, Akhund S, Nayyar BG (2015) *Morchella esculenta*: an edible and health beneficial mushroom. Pak J Food Sci 25(2):71–78

Anand N, Chowdhry PN (2013) Taxonomic and molecular identification of Verpa bohemica: a newly explored fungi from Rajouri (J&K), India. Rec Res Sci Technol 5(1):09–12

Bala P, Gupta D, Sharma YP (2017) Mycotoxin research and mycoflora in some dried edible morels marketed in Jammu and Kashmir, India. J Plant Dev Sci 9(8):771–778

Bhatt RP, Singh U, Stephenson SL (2016) Wild edible mushrooms from high elevations in the Garhwal Himalaya-I. Curr Res Environ Appl Mycol 6(2):118–131. https://doi.org/10.5943/cream/6/2/6

Bhatt RP, Singh U, Uniyal P (2018) Healing mushrooms of Uttarakhand Himalaya, India. Curr Res Environ Appl Mycol 8(1):1–23

Boonsong S, Klaypradit W, Wilaipun P (2016) Antioxidant activities of extracts from five edible mushrooms using different extractants. Agric Nat Res 50(2):89–97. https://doi.org/10.1016/j.anres.2015.07.002

Chakraborty N, Dutta AK, Pradhan P, Acharya K (2013) *Tulostoma chudaei* Pat an addition to macrofungal flora of India. J Mycopathol Res 51(1):185–187

Chakraborty D, Semwal KC, Adhikari S, Mukherjee SK, Das K (2017) Morphology and phylogeny reveal two new records of boletoid mushrooms for the Indian mycobiota. Trop Plant Res 4(1):62–70. https://doi.org/10.22271/tpr.2017.v4.i1.009

Chambergo FS, Valencia EY (2016) Fungal biodiversity to biotechnology. Appl Microbiol Biotechnol 100(6):2567–2577. https://doi.org/10.1007/s00253-016-7305-2

Chang ST (2006) The world mushroom industry: trends and technological development. Int J Med Mushrooms 8(4):297–314. https://doi.org/10.1615/IntJMedMushr.v8.i4.10

Chaudhary R, Tripathy A (2016) Diversity of wild mushroom in Himachal Pradesh (India). Int J Innov Res Sci Eng Technol 5(6):10859–10886

Choudhary M, Devi R, Datta A, Kumar A, Jat HS (2015) Diversity of wild edible mushrooms in Indian subcontinent and its neighboring countries. Rec Adv Biol Med 1:69–76. https://doi.org/10.18639/RABM.2015.01.200317

Das K (2010) Diversity and conservation of wild mushrooms in Sikkim with special reference to Barsey Rhododendron Sanctuary. NeBIO 1(2):1–13

Das K, Chakraborty D (2015) Two new records of *Cortinarius* from Sikkim (India). J New Biol Rep 4(1):1–6

Das K, Hembrom ME, Parihar A, Mishra D, Sharma JR (2014) *Strobilomyces polypyramis*–rediscovery of a wild mushroom from Sikkim, India. Ind J Plant Sci 3(2):13–18

Deb U, Jagannath A, Anilakumar KR, Mallesha CA (2018) Nutritional studies and antioxidant profile of pickled oyster mushrooms of north east India. Def Life Sci J 3(1):64–70

Dorjey K, Kumar S, Sharma YP (2013) *Cyathus olla* from the cold desert of Ladakh. Mycosphere 4:256–259. https://doi.org/10.5943/mycosphere/4/2/8

Dorjey K, Kumar S, Sharma YP (2016) Studies on Genus *Peziza* from Ladakh (Jammu & Kashmir), India. Kavaka 46:18–22

Dutta AK, Acharya K (2014) Traditional and ethno-medicinal knowledge of mushrooms in West Bengal, India. Asian J Pharm Clin Res 7(4):36–41

Dutta AK, Das K, Acharya K (2015) A new species of *Marasmius* sect. Globulares from Indian Himalaya with tall basidiomata. Mycosphere 6(5):560–567. https://doi.org/10.5943/mycosphere/6/5/5

Farooq R, Shah MA, Reshi ZA (2017) Morphological and molecular characterization of some mushrooms in Kashmir Himalayan forests. International Conference on Recent Innovations in Science, Agriculture, Engineering and Management ISBN 978-93-86171-80-1335-342.

Giri S, Biswas G, Pradhan P, Mandal SC, Acharya K (2012) Antimicrobial activities of basidiocarps of wild edible mushrooms of West Bengal, India. Int J PharmTech Res 4(4):1554–1560

Giri S, Mandal SC, Acharya K (2013) Proximate analysis of three edible mushrooms of West Bengal, India. Int J Pharm Tech Res 5(2):365–369

Itoo ZA, Reshi ZA, Basharat Q, Majeed ST, Andrabi KI (2015) Identification and characterization of ectomycorrhizal *Cortinarius* Species (Agaricales, Basidiomycetes) from temperate Kashmir Himalaya, India, by ITS Barcoding. Adv Mol Biol 2015:1–9. https://doi.org/10.1155/2015/507684

Kaur M, Rather HA (2016) Records of agarics: new to Jammu and Kashmir. J New Biol Rep 5(3):165–169

Kanwal HK, Acharya K, Ramesh G, Reddy MS (2011) Molecular characterization of Morchella species from the Western Himalayan region of India. Curr Microbiol 62(4):1245–1252. https://doi.org/10.1007/s00284-010-9849-1

Khan MA, Tania M (2012) Nutritional and medicinal importance of *Pleurotus* mushrooms: an overview. Food Rev Int 28(3):313–329. https://doi.org/10.1080/87559129.2011.637267

Khan AA, Gani A, Ahmad M, Masoodi FA, Amin F, Kousar S (2016) Mushroom varieties found in the Himalayan regions of India: antioxidant, antimicrobial, and antiproliferative activities. Food Sci Biotechnol 25(4):1095–1100

Khatua S, Dutta AK, Acharya K (2015) Prospecting *Russula senecis*: a delicacy among the tribes of West Bengal. Peer J. https://doi.org/10.7717/peerj.810

Kour H, Kumar S, Sharma YP (2013) Two species of *Strobilomyces* from Jammu and Kashmir, India. Mycosphere 4:1006–1013. https://doi.org/10.5943/mycosphere/4/5/14

Kour S, Kour H, Kumar S, Sharma YP (2015) New Records of *Clitocybe* Species from Jammu and Kashmir, India. Ind J For 38(1):43–46

Kumar S, Sharma YP (2009) Some potential wild edible macrofungi of Jammu Province (Jammu and Kashmir), India. Ind J For 32(1):113–118

Kumar S, Sharma YP (2011) Additions to Boletes from Jammu and Kashmir. J Mycol Plant Pathol 41(4):579–583

Malik AR, Wani AH, Bhat MY, Parveen S (2017) Ethnomycological knowledge of some wild mushrooms of northern districts of Jammu and Kashmir, India. Asian J Pharm Clin Res 10(9):399–405

Mittermeier RA, Turner WR, Larsen FW, Brooks TM, Gascon C (2011) Global biodiversity conservation: the critical role of hotspots. In: Zachos F, Habel J (eds) Biodiversity hotspots. Springer, Berlin, Heidelberg, pp 3–22. https://doi.org/10.1007/978-3-642-20992-5_1

Muhammad BL, Suleiman B (2015) Global development of mushroom biotechnology. Int J Emerg Trends Sci Technol 2(06):2660–2669

Nath RK, Sarma TC (2018) Edible macrofungi of Kaliabar sub-division of Nagaon district, Assam, India. Annals Plant Sci 7(3):2161–2165. https://doi.org/10.21746/aps.2018.7.3.12

Negi CS (2006) Morels (*Morchella* spp.) in Kumaun Himalayas. Nat Prod Rad 5(4):306–310

Nitha B, Meera CR, Janardhanan KK (2007) Anti-inflammatory and antitumour activities of cultured mycelium of morel mushroom, *Morchella esculenta*. Curr Sci 92(2):235–239

Pala SA, Wani AH, Bodha RH, Mir RA (2012) Three hitherto unreported macro-fungi from Kashmir Himalayas. Pak J Bot 44(6):2111–2115

Paloi S, Dutta AK, Acharya K (2015) A new species of *Russula* (Russulales) from Eastern Himalaya, India. Phytotaxa 234(3):255–262. https://doi.org/10.11646/phytotaxa.00.0.0

Panda AK, Swain KC (2011) Traditional uses and medicinal potential of *Cordyceps sinensis* of Sikkim. J Ayurveda Integr Med 2(1):9–13. https://doi.org/10.4103/0975-9476-78183

Parveen A, Khataniar L, Goswami G, Hazarika DJ, Das P, Gautom T, Barooah M, Boro RC (2017) A study on the diversity and habitat specificity of macrofungi of Assam, India. Int J Curr Microbiol Appl Sci 6(12):275–297. https://doi.org/10.20546/ijcmas.2017.612.034

Qin DW, Gu Z, Guo JY (2015) Medicinal mushroom for prevention of disease of modern civilization. Evid Based Complement Alternat Med 2015: Article ID 812725, 2 pages, https://doi.org/10.1155/2015/812725

Rai M, Sen S, Acharya K (2013) Antimicrobial activity of four wild edible mushrooms from Darjeeling hills, West Bengal, India. Int J PharmTech Res 5(3):949–956

Royse DJ, Baars J, Tan Q (2016) Current overview of mushroom production in the world. In: Deigo CZ (ed) Edible and medicinal mushrooms: technology and applications. Wiley Online Library, pp 05–13. https://doi.org/10.1002/9781119149446.ch2

Sarma TC, Sarma I, Patiri BN (2010) Wild edible mushrooms used by some ethnic tribes of Western Assam. Bioscan 3:613–625

Semwal KC, Stephenson SL, Bhatt VK, Bhatt RP (2014) Edible mushrooms of the Northwestern Himalaya, India: a study of indigenous knowledge, distribution and diversity. Mycosphere 5(3):440–461. https://doi.org/10.5943/mycosphere/5/3/7

Shameem N, Kamili AN, Ahmad M, Masoodi FA, Parray JA (2015) Radical scavenging potential and DNA damage protection of wild edible mushrooms of Kashmir Himalaya. J Saudi Soc Agric Sci 16(4):314–321. https://doi.org/10.1016/j.jssas.2015.10.005

Sharma SK, Gautam N (2017) Chemical and bioactive profiling and biological activities of coral fungi from Northwestern Himalayas. Sci Rep 7:46570. https://doi.org/10.1038/srep46570

Sheikh PA, Dar GH, Dar WA, Shah S, Bhat KA, Kousar S (2015) Chemical composition and anti-oxidant activities of some edible mushrooms of Western Himalayas of India. Vegetos- Int J Plant Res 28(2):124–133. https://doi.org/10.5958/2229-4473.2015.00046.4

Singh V, Bedi GK, Shri R (2017a) In vitro and in vivo antidiabetic evaluation of selected culinary-medicinal mushrooms (Agaricomycetes). Int J Med Mushrooms 19(1):17–25. https://doi.org/10.1615/IntJMedMushrooms.v19.i1.20

Singh U, Bhatt RP, Stephenson SL, Uniyal P, Mehmood T (2017b) Wild edible mushrooms from high elevations in the Garhwal Himalaya—II. Curr Res Envi Appl Mycol 7(3):208–226. https://doi.org/10.5943/cream/6/2/6

Singha K, Banerjee A, Pati BR, Das PK (2017a) Eco-diversity, productivity and distribution frequency of mushrooms in Gurguripal Eco-forest, Paschim Medinipur, West Bengal, India. Curr Res Environ Appl Mycol 7(1):8–18

Singha K, Pati BR, Mondal KC, Mohapatra PKD (2017b) Study of nutritional and antibacterial potential of some wild edible mushrooms from Gurguripal Ecoforest, West Bengal, India. Indian J Biotechnol 16:222–227

Tuli HS, Sharma AK, Sandhu SS, Kashyap D (2013) Cordycepin: a bioactive metabolite with therapeutic potential. Life Sci 93(23):863–869. https://doi.org/10.1016/j.lfs.2013.09.030. Epub 2013 Oct 10.

Vattakaven T, George RM, Balasubramanian D, Rejou-Mechain M, Muthusankar G, Ramesh BR, Prabhakar R (2016) India Biodiversity Portal: an integrated, interactive and participatory biodiversity informatics platform. Biodivers Data J 4:1–15. https://doi.org/10.3897/BDJ.4.e10279

Vaz JA, Ferreira IC, Tavares C, Almeida GM, Martins A, Vasconcelos MH (2012) Suillus collinitus methanolic extract increases p53 expression and causes cell cycle arrest and apoptosis in a breast cancer cell line. Food Chem 135(2):596–602. https://doi.org/10.1016/j.foodchem.2012.04.127

Vishwakarma MP, Bhatt RP, Gairola S (2011) Some medicinal mushrooms of Garhwal Himalaya, Uttarakhand, India. Int J Med Arom Plants 1(1):33–40

Wani BA, Pala SA, Wani AH, Bodha RH et al (2013) Two hitherto unreported macro-fungi from Kashmir Himalayas. Bioresour Bull 2(2):1–05

Yangdol R, Kumar S, Sharma YP (2014) A new edible variety of *Laetiporus sulphureus* from the cold desert of Ladakh. J Mycol Plant Pathol 44(4):463–465

Yangdol R, Kumar S, Lalotra P, Sharma YP (2016a) Two species of Inocybe from Trans- Himalayan Ladakh (J&K), India. Curr Res Environ Appl Mycol 6(4):305–311

Yangdol R, Kumar S, Sharma YP (2016b) Three species of Phallus (Basidiomycota: Agaricomycetes: Phallaceae) from Jammu & Kashmir, India. J Threat Taxa 8(1):8403–8409

Yangdol R, Sharma YP, Bhattacherjee S, Acharya K (2016c) Molecular, physical and biochemical characterization of an edible Mushroom, Psathyrella spadicea (P. Kumm.) Singer, from cold desert of Ladakh, India. Curr Res Environ Appl Mycol 6(4):334–343. https://doi.org/10.5943/cream/6/4/12

Yangdol R, Kumar S, Sharma YP (2017) Some pyronemataceous macrofungi from Ladakh (J&K), India. Curr Res Environ Appl Mycol 7(4):293–300

L-Ergothioneine: A Potential Bioactive Compound from Edible Mushrooms

16

Saraswathy Nachimuthu, Ruckmani Kandasamy, Ramalingam Ponnusamy, Jack Deruiter, Muralikrishnan Dhanasekaran, and Sivasudha Thilagar

Contents

S. Nachimuthu (✉) · R. Ponnusamy
Department of Biotechnology, Kumaraguru College of Technology, Coimbatore, Tamil Nadu, India
e-mail: saraswathy.n.bt@kct.ac.in

R. Kandasamy
Department of Pharmaceutical Technology, University College of Engineering, Bharathidasan Institute of Technology Campus, Anna University, Tiruchirappalli, Tamil Nadu, India

J. Deruiter · M. Dhanasekaran
Department of Drug Discovery and Development, Harrison School of Pharmacy, Auburn University, Auburn, AL, USA

S. Thilagar
Department of Environmental Biotechnology, School of Environmental Sciences, Bharathidasan University, Tiruchirappalli, Tamil Nadu, India

D. C. Agrawal, M. Dhanasekaran (eds.), *Medicinal Mushrooms*,
https://doi.org/10.1007/978-981-13-6382-5_16

Abstract

Edible mushrooms possess health-promoting bioactive compounds and are used throughout the world for their extensive medicinal properties. One of the unique bioactive compounds found in mushrooms is the water-soluble thiol-containing antioxidant amino acid, L-ergothioneine (LE). It is a stable antioxidant molecule and does not degrade at high temperature and high pH. L-Ergothioneine is not synthesized by plants and animals, including human. Therefore, LE is obtained through dietary sources, mainly from edible mushrooms such as *Agaricus bisporus*, *Lentinula edodes*, *Pleurotus ostreatus* and *Grifola frondosa.* These mushroom species have been reported to have a high LE content in them. LE is considered as a molecule for longevity and referred to as longevity vitamin. In this chapter, we have described the sources, the cytoprotective mechanisms, and therapeutic potential of LE.

Keywords

Anti-inflammatory · Antioxidant · Cytoprotectant · Health supplement · L-Ergothioneine

Abbreviations

ASF	Amphiphilic solute facilitator
ARE	Antioxidant response elements
ESSE	Ergothioneine disulfide
EFSA	European Food Safety Authority
8-BrdG	8-Bromo-2′-dG
ETT	Ergothioneine transporter
FDA	Food and Drug Administration
GRAS	Generally recognized as safe
GSH	Glutathione
HA	Hyaluronic acid
HPLC	High-performance liquid chromatography
HILIC	Hydrophilic interaction chromatography
HOBr	Hypobromous acid
HOCL	Hypochlorous
IL-6	Interleukin-6
LC-MS	Liquid chromatography-mass spectrometry
LE	L-Ergothioneine

MMP1	Matrix metalloproteinase 1
MPO	Myeloperoxidase
ND	Neurodegenerative diseases
OCTN1	Organic cation transporter
PLP	Pyridoxal 5-phosphate
ROS	Reactive oxygen species
RNS	Reactive nitrogen species
RT-PCR	Reverse transcription polymerase chain reaction
SAM	*S*-Adenosylmethionine
TNFα	Tumor necrosis factor alpha
UV	Ultraviolet

16.1 Introduction

Mushroom belongs to a unique group in kingdom *Fungi* which produces attractive fruiting bodies and is placed under a division called *Eumycota* "the true fungi." Selected nonpoisonous mushrooms have been used as food in many countries for centuries. Attempts have been made to optimize cultivation techniques for these edible mushrooms for commercial consumption (Dubost et al. 2007). The edible mushrooms are important due to their high digestible carbohydrate, protein, and low-fat content (Chang and Wasser 2012). Apart from their nutritional value, many edible mushrooms possess health beneficial bioactive compounds which belong mainly to the phenolic group. Aqueous and organic solvent extracts or purified bioactive metabolites from mushrooms are used as therapeutic agents. The major therapeutic activities of edible mushrooms have been proved as antitumor, anticarcinogenic, antimicrobial, and antioxidant.

Living organisms are constantly under threat of oxidative stress. Redox reactions in the living cells take care of detoxification of reactive oxygen species (ROS) and reactive nitrogen species (RNS) generated under oxidative stress. To counteract the oxidative stress, various antioxidant molecules are synthesized by the cells or obtained exogenously. Glutathione (GSH) is one of the primary antioxidant molecules, attributed with antioxidant properties in prokaryotes and eukaryotes, whereas in certain prokaryotes alternate thiol-containing low molecular weight compounds such as bacillithiol, trypanothione, mycothiol, ovothiol, and ergothioneine are synthesized.

16.2 Edible Mushrooms as Sources of L-Ergothioneine (LE)

Edible mushrooms produce an array of phenolic compounds, and one such potential secondary metabolite is L-ergothioneine (LE). LE was first isolated from fungus (*Claviceps purpurea*) in infected grains of rye called ergot in 1909. Later in 1926, LE was discovered in human blood which stimulated the interest in the compound.

Edible mushrooms are the richest source of LE. It has been reported that mushrooms have 45 times higher LE than other food items. It is one of the important thiol-imidazole amino acids which maintain the redox state of the cell. LE is one of the special types of thiol-containing molecules with high redox potential (E^0 = −0.06 V), compared to glutathione (GSH) (E^0 = −0.24 V) (Cheah and Halliwell 2012).

LE is not synthesized by higher eukaryotes like plants and animals (Melville et al. 1955). But plants absorb LE from the soil, where it is synthesized by soil microbes. Among many sources analyzed for LE, mushrooms showed the highest LE content, but its concentration greatly vary among mushroom species. It was also found in some of the economically important mushrooms like *Ganoderma neojaponicum*, *G. applanatum*, and *Paecilomyces tenuipes* (Dubost et al. 2006, Ito et al. 2011, Kalaras et al. 2017). LE is accumulated in mycelia as well as in fruiting bodies. Studies conducted to enhance the LE production in mushrooms by adding additives showed that methionine, aspartic acid, and lysine showed an increased LE biosynthesis (Lee et al. 2009, Pramvadee et al. 2012). LE production increased with the increased pH and was higher under alkaline condition (Lin et al. 2015). This has led the production of LE from edible mushrooms in more economical way and its subsequent commercialization in the market (Table 16.1).

16.3 Biosynthesis of LE

Microbes synthesize LE to protect themselves from the harmful oxidative stress. Mushrooms, non-yeast fungi, mycobacterium, and actinomycetes are capable of synthesizing LE. However, solid-state fermentation of mycelia of edible mushrooms requires a long time to obtain LE; attempts have been made to culture mycelia under submerged condition. It has now been established that LE is synthesized by the sequential action of five enzymes, encoded by the genes *egtA*, *egtB*, *egtC*, *egtD*, and *egtE* as shown in Fig. 16.1. Histidine is methylated by an *S*-adenosylmethionine (SAM)-dependent methyltransferase EgtD, to give the trimethyl ammonium betaine and hercynine. Hercynine is then converted to γ-glutamylcysteinylhercynine sulfoxide by iron (II)-dependent oxidase (EgtB) which requires oxygen and γ-glutamylcysteine. The γ-glutamylcysteine is formed by EgtA, a γ-glutamylcysteine ligase. Subsequently, glutamine amidotransamidase homolog EgtC mediates the hydrolysis of the N-terminus glutamic acid providing *S*-(β-amino-β-carboxyethyl) ergothioneine sulfoxide. Finally, EgtE, a pyridoxal 5-phosphate (PLP)-dependent β-lyase, forms LE.

In a report by Dubost et al. (2006), it was found that the concentration of LE increased with the harvest cycle. They recorded 0.6 mg/g, 1.0 mg/g, and 1.32 mg/g dry weight of LE in the first, second, and third harvest cycle, respectively. Since LE is a secondary metabolite, its synthesis may increase with stress level (Lin et al. 2015). Optimization of culture condition for edible mushrooms has to be carried out to enhance the LE. Cultivation of mycelia of edible mushrooms under submerged conditions has advantages of ease of LE production, high yield, and low cost of the

Table 16.1 LE contents in mushrooms

Mushroom species	Quantity of LE reported mg/g dw (Mean ± SD)	References
Sparassis crispa	2.37 ± 0.42	Lee et al. (2009)
Tremella foliacea	0.61 ± 0.03	Lee et al. (2009)
Lepista nuda	5.54 ± 0.26	Lee et al. (2009)
Suillus luteus	2.27 ± 0.24	Lee et al. (2009)
Ramaria botrytis	0.29 ± 0.03	Lee et al. (2009)
Tricholomopsis rutilans	2.50 ± 0.30	Lee et al. (2009)
Tricholoma matsutake	0.74 ± 0.08	Lee et al. (2009)
Suillus granulates	0.09 ± 0.03	Lee et al. (2009)
Russula virescens	0.68 ± 0.04	Lee et al. (2009)
Sarcodon aspratus	1.79 ± 0.02	Lee et al. (2009)
Hydnum repandum	0.78 ± 0.02	Lee et al. (2009)
Suillus bovines	1.09 ± 0.07	Lee et al. (2009)
Lampteromyces japonicus	0.43 ± 0.16	Lee et al. (2009)
Lactarius torminosus	0.82 ± 0.15	Lee et al. (2009)
Hericium erinaceus	0.96 ± 0.07	Lee et al. (2009)
Armillaria mellea	1.94 ± 0.01	Lee et al. (2009)
Neolentinus lepideus	2.41 ± 0.09	Lee et al. (2009)
Hygrophorus russula	4.98 ± 0.31	Lee et al. (2009)
Cantharellus cibarius	4.09 ± 0.20	Lee et al. (2009)
Polyozellus multiplex	0.51 ± 0.01	Lee et al. (2009)
Boletus auripes	2.40 ± 0.05	Lee et al. (2009)
Pleurotus ostreatus	2.20 ± 0.13	Lee et al. (2009)
Agaricus bisporus	1.21 ± 0.82	Lee et al. (2009)
Lentinula edodes	1.86 ± 0.73	Lee et al. (2009)
Ganoderma applanatum	0.06 ± 0.02	Lee et al. (2009)
Fomitopsis pinicola	0.07 ± 0.01	Lee et al. (2009)
Ganoderma lucidum	0.08 ± 0.02	Lee et al. (2009)
Ganoderma neo-japonicum	0.07 ± 0.00	Lee et al. (2009)
Grifola gargal	2.04 ± 0.20	Ito et al. (2011)
Grifola frondosa	0.67 ± 0.06	Ito et al. (2011)
Pleurotus ostreatus	1.98 ± 0.04	Ito et al. (2011)
Pleurotus eryngii	1.41 ± 0.12	Ito et al. (2011)
Lentinula edodes	0.40 ± 0.03	Ito et al. (2011)
Pholiota nameko	0.46 ± 0.11	Ito et al. (2011)
Agrocybe cylindracea	1.29 ± 0.07	Ito et al. (2011)
Leucopaxillus giganteus	1.70 ± 0.14	Ito et al. (2011)
Agaricus blazei	0.00 ± 0.00	Ito et al. (2011)
Tricholoma sp.	0.91 ± 0.14	Ito et al. (2011)
Cyttaria espinosae	0.00 ± 0.00	Ito et al. (2011)

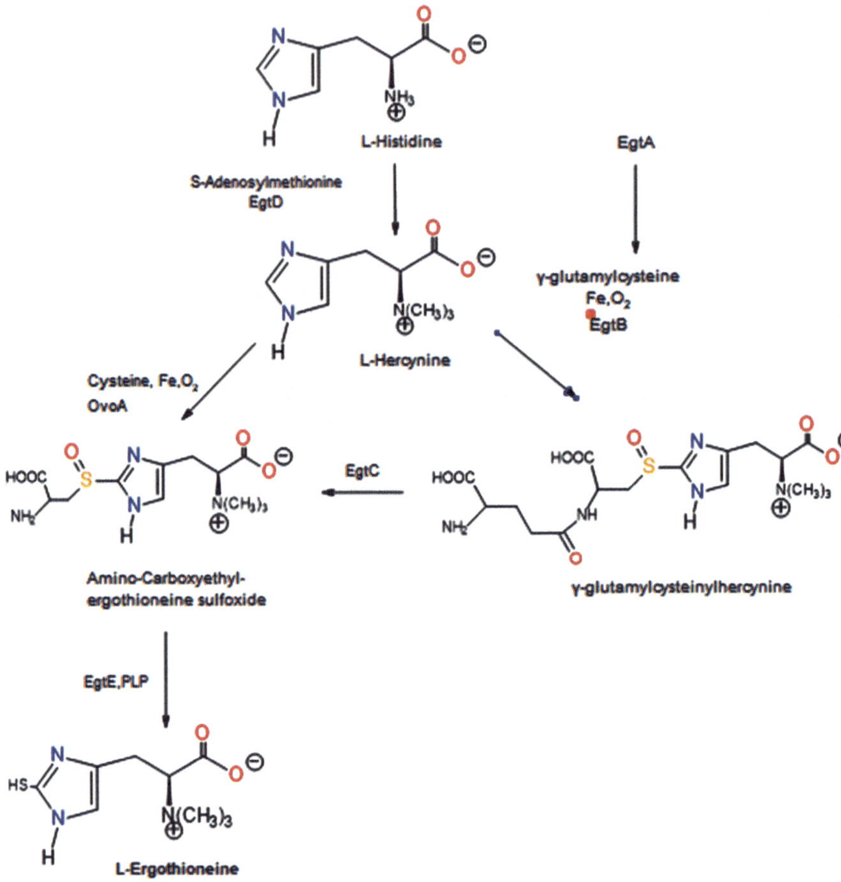

Fig. 16.1 Biosynthetic pathway of L-ergothioneine

product. Under the submerged conditions, there was an increase in synthesis of LE by 3.5-fold with highest concentration of 1.89 mg/g of dry weight compared to 0.6 mg/g dry weight. Also, under fermentation conditions, highest level of LE production has been reported to be 200 mg/L (Jiang et al. 2014). Considering the market demand, synthesis of LE was attempted with pure chemical reactions, and a method was patented by Oxis International Inc., USA.

Production of secondary metabolites from fruiting bodies of mushrooms is a time-consuming process, and hence studies are driven to enhance the production of bioactive compounds from mycelial cultures. In mycobacterium, LE is synthesized in five steps catalyzed by five genes, viz., egtA, egtB, egtC, egtD, and egtE. Five genes (egtA, B, C, D, E) from mycobacterium were transferred to *E.coli*, and LE could be produced using recombinant DNA technology (Osawa et al. 2018).

16.4 Metabolism of LE

LE is accumulated in cells and is resistant to autoxidation. The half-life of LE in mammalian cells is approximately 1 month. Cycling of sulfur group of LE takes place through nonenzymatic reaction. LE exists in two tautomeric forms such as thiol and thione (Fig. 16.2). At physiological pH, it predominantly exists in thione form. This makes it stable antioxidant as it does not undergo autoxidation. When the SH group is oxidized, it immediately gets reduced through redox cycle.

Degrading metabolites of LE are excreted in urine, and the level of excretion is not directly correlated with LE concentration in plasma. It is shown that LE is reabsorbed by the kidney and circulated. LE is accumulated in the cells and slowly released as evidenced by continuous excretion of LE after stopping oral administration. It is documented that among all tissues in the human body, LE is preferentially accumulated in the liver and whole blood (Tang et al. 2018). L-Ergothioneine sulfonate, L-hercynine, and S-methyl-ergothioneine are the detectable metabolites of LE (Fig. 16.3) (Cheah et al. 2017). These metabolites are produced upon oxidation of LE with ROS such as hypochlorite, iron peroxide, peroxynitrite, and hydrogen peroxide. S-Methyl-ergothioneine in urine is much higher than plasma and is cleared much faster. LE is oxidized to disulfide or mixed sulfide, and it is regenerated by reduction of disulfide.

16.5 Quantification Methods of LE

LE, an intracellular metabolite, is isolated from the mycelia and fruiting bodies of edible mushrooms by treating the samples with hot water (Ito et al. 2011). The hot water extract is freeze-dried and stored at −20 °C for further use. The quantity of LE is reported as mg/g of dry weight. Several methods are available to quantify LE. As per the method described by Hunter (1928), the reaction of LE in deproteinized solution is reacted with diazotized sulfanilic acid and subsequently allowed to react in alkaline condition. The bluish-red-colored product is measured by spectrophotometric analysis. The method has two drawbacks such as interference by compounds present in the blood such as purines and L-cysteine. Other products such as tyrosine and histidine also react with the substrate and form colored products. To over interference by colored compounds, a modified method involves separation of LE by chromatography using alumina column. But then it involves a lengthy procedure (Melville and Lubschez 1953).

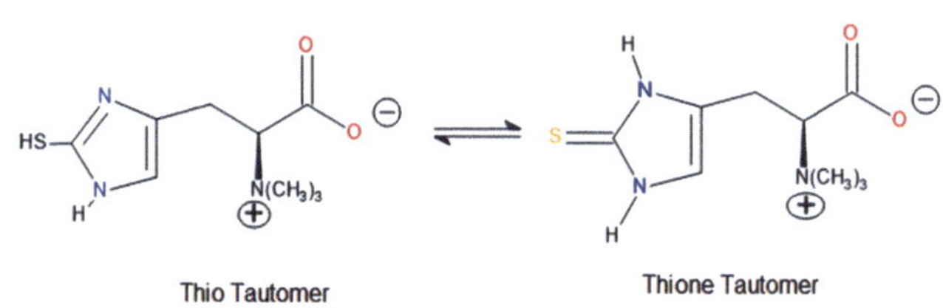

Fig. 16.2 L-Ergothioneine tautomers

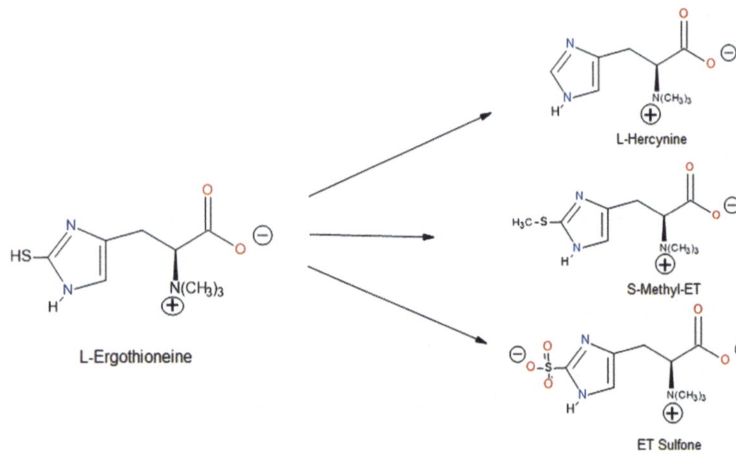

Fig. 16.3 Metabolism of L-ergothioneine

Spectrophotometric method is specifically suited for analysis of blood samples. After deproteinization and heat precipitation, the thiols are oxidized using CU^{2+} under alkaline condition, but LE does not oxidize. Titration of the sample against 2,2′-dipyridyl disulfide at 1.3 pH gives instantaneous color which is measured spectrophotometrically at 440 nm (Carlsson 1974). HPLC-UV and LC-MS are sensitive methods to quantify LE in edible mushroom and other samples (Dubost et al. 2006; Zhou et al. 2010). Quantification of fermentation broth requires a method which could separate the impurities in the fermentation broth.

HILIC (hydrophilic interaction chromatography) is used to separate highly polar compounds. The functional group containing acylamino supplies large number of hydrogen bonds to imidazole N of LE. Fermentation broth with mixture of compounds is highly polar to nonpolar. LE exists as thione form at pH 7.4 which is maintained in mobile phase and can be separated using hydrophilic stationary phase (Liu et al. 2016).

16.6 Properties of LE

LE is an odorless, colorless solid with 229.3 g/mol molecular weight and is highly soluble in water (0.9 M at 25 °C). LE exists in two tautomeric forms such as thiol and thione (Fig. 16.2). Thione form is the dominant form in aqueous solution or under physiological conditions compared to thiol form. LE does not undergo autoxidation in physiological condition; hence, it remains a stable antioxidant compared to other water-soluble antioxidants such as glutathione which gets oxidized quickly. LE makes complexes with metals ions like $CU^{2+}$and Fe^{2+} in the ratio of two molecules of LE with one metal ion (Zhu et al. 2011) and thus prevents their role in redox cycle.

16.7 Antioxidant Mechanism of LE

Sulfhydryl group of LE is responsible for its antioxidant property. As LE exists as thione in aqueous solutions, it is capable of scavenging strong free radicals, hydroxyl radicals, hypochlorite, peroxynitrite, and also chelates redox active cations like Cu^{2+} (Akanmu et al. 1991). LE antioxidant pathway is mediated by Nrf2/ARE signaling cascade. Nrf2 (nuclear factor 2) is a transcription factor which triggers the expression of many cytoprotective antioxidant pathway genes, and hence it is an oxidative stress marker. This gene product is translocated into the nucleus under stress condition. It recruits sMaf (small musculoaponeurotic fibrosarcoma) protein and forms Nrf2/sMaf complex. Subsequently, Nrf2/sMaf complex binds to "antioxidant response elements" (ARE) and triggers the expression of many antioxidant genes (Hseu et al. 2015). LE disulfide is unstable under physiological condition but stable in strong acidic condition. Ergothioneine disulfide (ESSE) formed under physiological condition is maintained in its reduced state by other thiols in the redox cycle such as glutathione (GSH) (Cumming et al. 2018).

16.8 Transport of LE

The higher plants and animals take up the LE from other sources. In vertebrates, there is a transmembrane protein called "cation transporter," ETT (ergothioneine transporter). LE therefore is distributed unevenly in the cells of the vertebrates. ETT is a biomarker to measure the LE activity in the cell. Mutation in ETT makes the cells more vulnerable to oxidative stress. It is also observed that a mutation in ETT locus is associated with autoimmune diseases like Crohn's disease and rheumatoid arthritis. LE is transported into the cells by special protein called "organic cation transporter" (OCTN1) belonging to "solute carrier family 22 member 4" (SLC22A4) gene. There are two types of transmembrane proteins, OCTN1 and OCTN2. OCTN1 is specific to LE and less specific to carnitine, while OCTN2 is specific to carnitine. Plasma membrane is impermeable to hydrophilic compounds. This transporter is highly conserved across species and performs important role. Blocking or inactivating ETT leads to decrease in LE uptake. OCTN1 gene knockout in mouse showed no accumulation of LE, indicating the importance of transporter for LE (Kato et al. 2010). It was observed that deletion of OCTN1 transporter in *Caenorhabditis elegans* resulted in increased oxidative damage (Cheah et al. 2013). The intracellular LE concentration directly correlates with the expression of the ETT. It has been confirmed through RT-PCR quantification of mRNA of ETT and quantification of LE using LC-MS/MS. Analysis of polymorphism in the SLC22A4 gene in human showed a strong correlation between mutation in SLC22A4 and susceptibility to chronic inflammatory diseases. Transport of LE by ETT is driven by sodium ion concentration. Several studies have shown that OCTN1 is primarily specific to LE and has less affinity toward cation inhibitors like verapamil and methimazole (Tamai et al. 2004; Gründemann et al. 2005; Grigat et al. 2007; Nakamura et al. 2008).

Cell membrane is impermeable to LE (Gründemann 2012); hence there was a continuous search for transport mechanism. Cellular transport of LE is known to be dependent on temperature and the concentration of sodium ions. Silencing of OCTN1 gene resulted in the inhibition of uptake of LE. Knockout mouse (octn1−/−) showed a complete lack of LE accumulation in many tissues of mouse (Kato et al. 2010). Erythrocytes accumulated the highest amount of LE which is in correlation with the highest ETT expression in OCTN1. Transport of LE is low at lower pH and lower sodium concentration (Tang et al. 2018). Absence of LE uptake in OCTN1 double knockout mouse showed no alternate transport mechanism for LE. It was also confirmed in the Zebra fish knockout (Pfeiffer et al. 2015). ETT is an amphiphilic solute facilitator (ASF) family of integral transport protein. The gene coding for OCTN1 was cloned and studied for its in vitro specificity and was found to have broad specificity. It can also transport cations like carnitine and tetraethylammonium. OCTN1 has high affinity (Km = 21 μM) toward LE than methimazole and hercynine. OCTN1 expression is the highest in the bone marrow, small intestine, fetal liver, kidney, cerebellum, and spinal cord. It is also reported that basal level of LE in human varies and attributed to polymorphism in SLC44A2 ETT efficiency in transporting LE (Grundemann et al. 2005; Grundemann 2012).

16.9 LE and Human Diseases

16.9.1 Antidepressant Activity of LE

Depression is one of the neural disorders, and it is reported that approximately 20% of the world population is affected by this causing a huge loss to the individuals and society (Browne and Lucki 2013). Antidepressant drugs prescribed are known to have significant side effects. There is a demand for bioactive compounds with no or minimal side effects. In the depressed patients, oxidative stress increases and level of antioxidants decreases. This is an indication that there is a need for suitable intervention to improve the physiological condition using potential antioxidant compounds. Interestingly LE showed neuronal differentiation in addition to antioxidant activity. Studies using LE from golden oyster mushroom extract was carried out with mice to evaluate the antidepressant activity. It was found that LE exerted neuronal differentiation and antidepressant activity (Yang et al. 2012; Ishimoto et al. 2014; Nakamichi et al. 2016).

16.9.2 Neurodegenerative Diseases and LE

Pathogenesis of neurodegenerative diseases (ND) like Alzheimer, Parkinson, and Huntington's are attributed to increased production of free radicals. It is understood that the free radical scavenging is the mechanism of clearing the stress in the neural cells. Anti-neurogenetic action of edible mushroom extracts in in vitro cell line studies demonstrated that the bioactive compounds are responsible for neuroprotective activity (Phan et al. 2017). In animal models, oral ingestion of edible mushroom containing LE has been reported to improve the learning abilities in Alzheimer-affected animals. Nature-based nutraceuticals like LE from medicinal mushrooms are suggested as better therapeutic agent for neurodegenerative diseases. Edible mushroom extracts promoted neural progenitor cell differentiation into neurons (Song et al. 2010; Yang et al. 2012; Song et al. 2014).

16.9.3 Preeclampsia and LE

This is one of the late pregnancy-associated disorders due to the formation of defective placentation. Every year approximately 18% of the pregnant women die due to preeclampsia worldwide (Khan et al. 2006). The etiology of the disorder is associated with poor immunomodulatory reactions. This leads to placenta ischemia and subsequently oxidative stress. Plasma mediators are released into plasma. These mediators enter into maternal circulatory system and induce anti-angiogenic factors which ultimately lead to endothelial dysfunction. It is reported that in preeclamptic women, due to lysis of red blood cells, iron concentration increases and it results in the production of ROS in the placenta. LE is shown to chelate iron and reduces the injury due to ROS (Kerley et al. 2018).

16.9.4 Chronic Inflammatory Disease and LE

Under chronic inflammatory disease condition, myeloperoxidase in neutrophils reacts with H_2O_2 and chloride ions and forms strong oxidant hypochlorous (HOCL), a potent cytotoxin (Burner et al. 2000). HOCL released by the neutrophils reacts with lysine residues of protein forming lysine chloramines, which subsequently reacts with tyrosine residues and forms 3-chloro tyrosine. It is reported that extract from an edible mushroom *Grifola gargal* containing LE effectively inhibited the protein chlorination by HOCL (Ito et al. 2011). Under in vitro cell culture of 3T3-L1, pretreatment of LE leads to inhibition of IL-6 in a dose-dependent manner. In another study, free fatty acid-induced lipotoxicity could be reduced by treating with LE (Laurenza et al. 2008) (Table 16.2).

Table 16.2 Commercially pure LE and LE supplement in the market

Sl. no.	Name of the commercial product	Mode of synthesis	Name of the company
1	ErgoActive	Fermentation method	Blue California, USA
2	ERGOLD(TM)	Chemical method	Oxis International Inc., California, USA
3	Ergo4Health/kidney	Proprietary high-intensity UV light technology	Entia Biosciences, Inc., Oregon, USA

16.10 Application of LE in Healthcare

16.10.1 Role of LE in Human Body

Preferential accumulation of LE in certain tissue types kindled the scientific curiosity in this unusual amino acid (Grundemann et al. 2005). It is obtained from dietary sources like edible mushrooms. LE is detected in various types of human tissues which include the brain, red blood cells, liver, kidney, ocular tissues, and seminal fluids. Physiological role of LE was established after the discovery of specific transporter protein transcribed by SLC22A4 (solute carrier family 22, member 4). Expression profiling of "ergothioneine transporter" (ETT) in cells showed a strong correlation between level of expression of ETT and ET uptake. LE modulates immune system under toll-like receptor stimulation. Upon TLR stimulation by LE, cytokines such as IL-6 and IL-12p40 are upregulated. Since LE is accumulated in erythrocytes, and its role is identified as a protection agent in the cells against oxidative stress, it has been recommended for the treatment of red blood cell disorders.

16.10.2 LE in Dietary Supplement

LE has been given a "generally recognized as safe (GRAS)" status by FDA, USA. It is sold in the market as a food supplement, nutraceuticals, or a functional food (Ey et al. 2007, Bao et al. 2010). LE processed from mushroom waste was formulated as a dietary supplement for yellow fish and cattle. A significant accumulation of LE was shown in yellow fish. It was observed that addition of mushroom extract reduced the accumulation of hydroperoxide in fish muscle and significantly reduced the astaxanthin degradation. It has been noted that exposure to UVA rays results in generation of reactive oxygen species, DNA damage, and apoptosis in cells. Mitochodria loses its membrane potential upon exposure to UVA rays (Cheah et al. 2013). LE pretreatment to cells before UVA exposure prevented the loss of membrane potential and increased glutathione content in cells. Increased LE content in cells results in overexpression of antioxidant pathway genes such as HO-1, NQO-1, and GCLC.

16.10.3 LE in Cosmetic Products

Cosmetic industry is growing rapidly over the decades, and nowadays many of the ingredients in the cosmetic products are derived from natural sources due to adverse side effects by chemicals. There is a booming market for natural ingredient-based cosmetic products. Among the natural sources for preparation of skincare products, bioactive molecules from edible mushrooms are viewed as great potential (Hyde et al. 2010). Exposure to intense sunlight for long duration results in ultraviolet ray-induced oxidative damage of biomolecules in the skin cells subsequently causing sunburn, photoaging, and skin cancer. Approximately 4% of the sunlight consists of UV-B which activates the production of reactive oxygen species (ROS) upon exposure. Excessive release of ROS results in generation of 8-hydroxy 2'doxyguanosine (8-OHdG) which results in the pathogenesis of skin disorders. Asahi et al. (2016) studied the UV-B protection property of *Coprinus comatus* mushroom extract which has a high amount of LE. In vitro studies showed that the extract could inhibit myeloperoxidase (MPO). MPO-mediated synthesis of halogenated molecules such as HOCL and hypobromous acid (HOBr) is responsible for production of ROS molecules and inflammation in the skin. Edible mushroom extract strongly inhibited formation of 8-bromo-2′-dG (8-BrdG) (Asahi et al. 2016). The level of inhibition was comparatively higher than GSH, one of the prominent physiological antioxidants. Disulfide formation occurs when hypohalous acids react with thiols. LE contains thiol group which is responsible for formation of the hypohalous acids.

LE is used as an important ingredient in cosmetic products, especially in sunscreen lotions as it has UVA protection property (Bazela et al. 2014). It is readily absorbed when applied on the skin. The *Coprinus comatus* extract is used as ingredient in many skincare products. Photoaging is the cleavage of collagen and destruction of vital molecules. *C. comatus* extract and LE showed effective inhibition of the collagen cleavage and inflammation by inhibiting the MMP1 and TNFα (Obayashi et al. 2005). Skin whitening property of the mushroom extract is attributed to inhibition of tyrosinase, thereby preventing synthesis of melanin. LE showed a dose-dependent inhibition of tyrosinase. Skin integrity is maintained by matrix binding biomolecules like hyaluronic acid (HA), and mushroom extract with LE is shown to inhibit HA-degrading enzyme. Therefore edible mushroom extract containing high LE is widely used in cosmeceuticals (skincare products).

16.11 Safety of LE

LE has been accepted as one of the important food supplements with significant health benefits. The pharmacokinetic of LE in human subjects showed no known adverse effect up to 25 mg/day for 1 week. In animal models, LE administration up to 1600 mg per kg of body weight did not show any adverse effects behavioral, histopathological changes (Schauss et al. 2011; Forster et al. 2015). Also, no mutagenic activity was observed in a bacterial reverse mutagenesis assay of the LE with a concentration of 5000 $\mu g\ ml^{-1}$. The Oxis International Inc., CA 90210, USA, holds a patent on chemical synthesis of LE and markets it as Ergoneine®. A clear scientific evidence of ergothioneine uptake in human being through regulated transporter by kidney is available (Shinozaki et al. 2017). According to the study report submitted to the European Commission, no adverse effect was detected at 800 mg/kgbw (body weight) per day. No adverse effects could be shown by administration of pure LE in human beings as evident in the results of liver function tests and lipid profile (Cheah et al. 2017). European Food Safety Authority (EFSA) based on scientific studies revealed that there has been no specific relationship between consumption of LE supplement and development of diabetes mellitus and chronic inflammatory disease. It has been reported that up to 30 mg/day for adults and 20 mg per day for children no genotoxicity was observed (EFSA panel report 2016).

16.12 Future Perspectives and Conclusion

Oxidative stress is one of the major metabolic loads exerted by the physiological reactions in the living cells. The ROS and RNS produced in the redox reactions are taken care by the cellular antioxidant molecules. Significant quantities of the antioxidants are synthesized by the cells, but a few biological antioxidants like LE are obtained from proper diet or dietary supplement. It is confirmed by scientific experiments that LE is a stable antioxidant under physiological condition. In animals, LE is taken up by the cells through regulated gateway of a specific transporter. LE is a special type of antioxidant as it is preferentially accumulated in cells which are exposed to oxidative stress. Edible mushrooms are known to possess many bioactive compounds having health beneficial effects. LE in edible mushroom is higher than any other dietary sources. LE is known to be involved in many physiological functions like anti-inflammatory, antioxidant, and protection of DNA and protein from oxidative damage. To meet the commercial demand of the LE in the market, it is produced through chemical method by certain companies. In order to get the natural LE, fermentation conditions need to be optimized to enhance its yield. Safety of the LE has been tested in animal models and human volunteers, and no adverse effect has been reported so far. Therefore, LE can be considered as an important health beneficial bioactive compound.

References

Akanmu D, Cecchini R, Aruoma OI et al (1991) The antioxidant action of ergothioneine. Arch biochem Biophys 288(1):10–16

Asahi T, Wu X, Shimoda H et al (2016) A mushroom-derived amino acid, ergothioneine, is a potential inhibitor of inflammation-related DNA halogenations. Biosci Biotech Bioch 80(2):313–317. https://doi.org/10.1080/09168451.2015.1083396

Bao HN, Osako K, Ohshima T (2010) Value-added use of mushroom ergothioneine as a colour stabilizer in processed fish meats. J Sci Food Agr 19(15):1634–1641. https://doi.org/10.1002/jsfa.3992

Bazela K, Solyga-Zurek A, Debowska R et al (2014) L-Ergothioneine protects skin cells against UV-induced damage-a preliminary study. Cosmetics 1:51–60. https://doi.org/10.3390/cosmetics1010051

Browne CA, Lucki I (2013) Antidepressant effects of ketamine: mechanisms underlying fast-acting novel antidepressants. Front Pharmacol 4:161. https://doi.org/10.3389/fphar.2013.00161

Burner U, Furtmüller PG, Kettle AJ, Koppenol WH, Obinger C (2000) Mechanism of reaction of myeloperoxidase with nitrite. J Biol Chem 275(27):20597–20601

Carlsson J, Kierstan MPJ, Brocklehurst K (1974) A convenient spectrophotometric assay for the determination of l-ergothioneine in blood. Biochem J 139(1):237–242

Chang ST, Wasser SP (2012) The role of culinary-medicinalmushrooms on human welfare with pyramid model for human health. Int J Med Mushrooms 14(2):95–134

Cheah IK, Halliwell B (2012) Ergothioneine; antioxidant potential, physiological function and role in disease. Biochim Biophys Acta 1822:784–793. https://doi.org/10.1016/j.bbadis.2011.09.017

Cheah IK, Ong RL, Gruber J et al (2013) Knockout of a putative Ergothioneine transporter in *Caenorhabditis elegans* decreases lifespan and increases susceptibility to oxidative damage. Free Radic Res 47:1036–1045. https://doi.org/10.3109/10715762.2013.848354

Cheah IK, Richard MY, Terry SZ et al (2017) Administration of pure ergothioneine to healthy human subjects: uptake, metabolism, and effects on biomarkers of oxidative damage and inflammation. Antioxid Redox Signal 26(5):193–206. https://doi.org/10.1089/ars.2016.6778

Cumming BM, Chinta KC, Reddy VP et al (2018) Role of ergothioneine in microbial physiology and pathogenesis. Antioxid Redox Signal 28(6):431–444. https://doi.org/10.1089/ars.2017.7300

Dubost NJ, Beelman RB, Peterson D et al (2006) Identification and quantification of ergothioneine in cultivated mushrooms by liquid chromatography-mass spectroscopy. Int J Med Mushrooms 8:215–222. https://doi.org/10.1615/IntJMedMushr.v8.i3.30

Dubost NJ, Beelman RB, Royse DJ (2007) Influence of selected cultural factors and postharvest storage on ergothioneine content of common button mushroom Agaricusbisporus (J. Lge) Imbach (Agaricomycetideae). Int J Med Mushrooms 9:163–176

EFSA Journal (2016) Scientific opinion on the safety of synthetic L-ergothioneine (Ergoneine®) as a novel food pursuant to Regulation (EC) No 258/97. Parma, Italy: European Food Safety Authority. EFSA J No. 4629 14(11). https://doi.org/10.2903/j.efsa.2016.4629

Ey J, Schomig E, Dirk T (2007) Dietary sources and antioxidant effects of ergothioneine. J Arg Food Chem 55(16):6466–6474. https://doi.org/10.1021/jf071328f

Forster R, Spezia F, Papineau D et al (2015) Reproductive safety evaluation of l-ergothioneine. Food ChemToxicol 80:85–91. https://doi.org/10.1016/j.fct.2015.02.019

Grigat S, Harlfinger S, Pal S et al (2007) Probing the substrate specificity of the ergothioneine transporter with methimazole, hercynine, and organic cations. Biochem Pharmacol 74(2):309–316. https://doi.org/10.1016/j.bcp.2007.04.015

Gründemann D (2012) The ergothioneine transporter controls and indicates ergothioneine activity--a review. Prev Med 54(Suppl):S17–S14. https://doi.org/10.1016/j.ypmed.2011.12.001

Grundemann D, Harlfinger S, Golz S et al (2005) Discovery of the ergothioneine transporter. Proc Natl Acad Sci USA 102:5256–5261. https://doi.org/10.1073/pnas.0408624102

Hseu YC, Lo HW, Korivi M (2015) Dermato-protective properties of ergothioneine through induction of Nrf2/ARE-mediated antioxidant genes in UVA-irradiated human keratinocytes. Free Radic Bio Med 86:102–117. https://doi.org/10.1016/j.freeradbiomed.2015.05.026

Hunter G (1928) The estimation of carnosine in muscle extract-A critical study. Biochem J 16(5):640–654

Hyde KD, Bahkali AH, Moslem MA (2010) Fungi-an unusual source for cosmetic. Fungal Diversity 43:109

Ishimoto T, Nakamichi N, Hosotani H et al (2014) Organic cation transporter-mediated ergothioneine uptake in mouse neural progenitor cells suppresses proliferation and promotes differentiation into neurons. PLoS One 9(2):e89434. https://doi.org/10.1371/journal.pone.0089434

Ito T, Kato M, Tsuchida H, Harada E et al (2011) Ergothioneine as an anti-oxidative/anti-inflammatory component in several edible mushrooms. Food Sci Technol Res 17(2):103–110

Jiang WX, Liu Q, Mei B L et al (2014) The preparation of ergothioneine. China Patent CN 201410243400.5 (2014)

Kalaras MD, Richie JP, Calcagnotto A, Beelman RB (2017) Mushrooms: a rich source of the antioxidants ergothioneine and glutathione. Food Chem 233:429–433. https://doi.org/10.1016/j.foodchem.2017.04.109

Kato Y, Kubo Y, Iwata D et al (2010) Knockout and metabolome analysis of carnitine/organic cation transporter OCTN. Pharm Res 27:832–840. https://doi.org/10.1007/s11095-010-0076-z

Kerley RN, McCarthy C, Kell DB, Kenny LC (2018) The potential therapeutic effects of ergothioneine in pre-eclampsia. Free Radic Biol Med 117:145–157. https://doi.org/10.1016/j.freeradbiomed.2017.12.030

Khan KS, Wojdyla D, Say L et al (2006) WHO analysis of causes of maternal death: a systematic review. Lancet 367(9516):1066–1074. https://doi.org/10.1016/S0140-6736(06)68397-9

Laurenza I, Colognato R, Migliore L et al (2008) Modulation of palmitic acid-induced cell death by ergothioneine: evidence of an anti-inflammatory action. BioFactors 33:237–247

Lee WY, Park EJ, Ahn JK (2009) Supplementation of methionine enhanced the ergothioneine accumulation in the Ganoderma neo-japonicum mycelia. Appl Biochem Biotech 158:213–221. https://doi.org/10.1007/s12010-008-8322-0

Lin SL, Chien SC, Wang SY et al (2015) Submerged cultivation of mycelium with high ergothioneine content from the culinary-medicinal golden oyster mushroom, *Pleurotuscitrinopileatus* (higher basidiomycetes). Int J Med Mushrooms 17(8):749–761. https://doi.org/10.1615/IntJMedMushrooms.v17.i8.50

Liu QI, Zhang W, Wang H et al (2016) Validation of a HILIC method for the analysis of ergothioneine in fermentation broth. J Chromatogr Sci 54(6):934–938. https://doi.org/10.1093/chromsci/bmw023

Melville DB, Lubschez R (1953) A method for the determination of ergothioneine in blood. J Biol Chem 1:275–285

Melville DB, Horner WH, Otken CC et al (1955) Studies on the origin of ergothioneine in animals. J Biol Chem 213:61–68

Nakamichi N, Nakayama K, Ishimoto T et al (2016) Food-derived hydrophilic antioxidant ergothioneine is distributed to the brain and exerts antidepressant effect in mice. Brain Behav 6(6):e00477. https://doi.org/10.1002/brb3.477

Nakamura T, Yoshida K, Yabuuchi H et al (2008) Functional characterization of ergothioneine transport by rat organic cation/carnitine transporter Octn1 (slc22a4). Biol Pharm Bull 31(8):1580–1584

Obayashi K, Kurihara K, Okano Y et al (2005) L-Ergothioneine scavenges superoxide and singlet oxygen and suppresses TNF-α and MMP-1 expression in UV-irradiated human dermal fibroblasts. J Cosmet Sci 56:17–27

Osawa R, Kamide T, Satoh Y et al (2018) Heterologous and high production of ergothioneine in *Escherichia coli*. J Argric Food Chem 66(5):1191–1196. https://doi.org/10.1021/acs.jafc.7b04924

Pfeiffer C, Bach M, Bauer T et al (2015) Knockout of the ergothioneine transporter ETT in zebrafish results in increased 8-oxoguanine levels. Free Radic Biol Med 83:178–185. https://doi.org/10.1016/j.freeradbiomed.2015.02.026

Phan CW, David P, Sabaratnam V (2017) Edible and medicinal mushrooms: emerging brain food for the mitigation of neurodegenerative disease. J Med Food 20(1):1–10. https://doi.org/10.1089/jmf.2016.3740

Pramvadee T, Anupam G, Fumito S et al (2012) Mycobial enhancement of ergothioneine by submerged cultivation of edible mushroom mycelia and its application as an antioxidative compound. Food Chem 131:247–258

Schauss AG, Beres E, Vertesi A et al (2011) The effect of ergothioneine on clastogenic potential and mutagenic activity: genotoxicity evaluation. Int J Toxicol 30:405–409

Shinozaki Y, Furuichi K, Kitajima S (2017) Impairment of the carnitine/organic cation transporter 1-ergothioneine axis is mediated by intestinal transporter dysfunction in chronic kidney disease. Kidney Int 92(6):1356–1369. https://doi.org/10.1016/j.kint.2017.04.032

Song TY, Chen CL, Liao JW et al (2010) Ergothioneine protects against neuronal injury induced by cisplatin both *in vitro* and *in vivo*. Food Chem Toxicol 48(12):3492–3499. https://doi.org/10.1016/j.fct.2010.09.030

Song TY, Lin HC, Chen CL et al (2014) Ergothioneine and melatonin attenuate oxidative stress and protect against learning and memory deficits in C57BL/6J mice treated with D-galactose. Free Rad Res 48:1049–1060. https://doi.org/10.3109/10715762.2014.920954

Tamai I, Nakanishi T, Kobayashi D et al (2004) Involvement of OCTN1 (*SLC22A4*) in pH-dependent transport of organic cations. Mol Pharm 1(1):57–66

Tang RM, Cheah IK, Yew TS et al (2018) Distribution and accumulation of dietary ergothioneine and its metabolites in mouse tissues. Sci Rep 8:1601. https://doi.org/10.1038/s41598-018-20021-z

Yang NC, Lin HC, Wu JH et al (2012) Ergothioneine protects against neuronal injury induced by β-amyloid in mice. Food Chem Toxicol 50(11):3902–3911. https://doi.org/10.1016/j.fct.2012.08.021

Zhou NB, Zhu YQ, Yin QH (2010) Determination of ergothioneine in *Agaricusbisporus* by HPLC. Food Sci Technol 35(10):271–272. https://doi.org/10.1093/chromsci/bmw023

Zhu BZ, Mao L, Fan RM et al (2011) Ergothioneine prevents copper-induced oxidative damage to DNA and protein by forming a redox-inactive ergothioneine-copper complex. Chem Res Toxicol 24:30. https://doi.org/10.1021/tx100214t

Mycotherapy of *Antrodia salmonea*: A Taiwanese Medicinal Mushroom

17

Palaniyandi Karuppaiya and Abdul Khader Akbar

Contents

Abstract

Generally, mushrooms possess all four functionalities of food, i.e., nutritional value, tastiness, physiological effects, and cultural aspects. For the physiological effects, mushrooms have become a valuable health food due to their several bioactive substances. Hence, they are considered as a vast and yet largely untapped source of powerful new pharmaceutical products. In particular, and most importantly for modern medicine, they represent an unlimited source of polysaccharides with antitumor and immunostimulating properties. In the present review,

P. Karuppaiya (✉)
Institute of Nutrition, College of Biopharmaceutical and Food Sciences, China Medical University, Taichung, Taiwan

A. K. Akbar
Department of Botany, C. Abdul Hakeem College, Vellore, Tamil Nadu, India

D. C. Agrawal, M. Dhanasekaran (eds.), *Medicinal Mushrooms*,
https://doi.org/10.1007/978-981-13-6382-5_17

we have described various properties of *Antrodia salmonea* (syn. *Taiwanofungus salmoneus*), a newly identified medicinal fungal species.

Keywords
Antrodia salmonea · Anticancer · Antioxidant · Medicinal mushroom · Functional food · Triterpenoid

Abbreviations

AS	*Antrodia salmonea*
UPLC	Ultra performance liquid chromatography
MS	Mass spectroscopy
TNF	Tumor necrosis factor
γ-GCLC	γ-Glutamylcysteine synthetase
iNOS	Inducible nitric oxide synthase
SOD	Superoxide dismutase
COX-2	Cyclooxygenase-2
CAT	Catalase
GPx	Glutathione peroxidase
AVOs	Acidic vesicular organelles
ROS	Reactive oxygen species

17.1 Introduction

Medicinal mushrooms have a long history of use in traditional medicine system as curatives for various health problems. In traditional medicine system, dried mushrooms and mushroom extracts have been used for human diseases (Money 2016). Mushrooms comprise a vast and yet largely unexploited powerful miniature pharmaceutical factory. The contemporary research identified a number of bioactive compounds including anticancer substances in many medicinal mushroom species. In particular polysaccharides are best known bioactive substances derived from medicinal mushrooms with potent antitumor and immunomodulating activities (Borchers et al. 1999; Lorenzen and Anke 1998; Mizuno et al. 1999; Wasser and Weis 1999; Mizuno et al. 1996; Mizuno 1999a, b, 2002; Ooi and Liu 1999; Tzianabos 2000; Reshetnikov et al. 2001). Reports show that different strains of one *Basidiomycetes* species can produce polysaccharides with different properties.

Antrodia salmonea (Taiwanofungus salmoneus T. T. Chang et W. N. Chou), a therapeutic medicinal mushroom species, was identified in 2004 in Taiwan

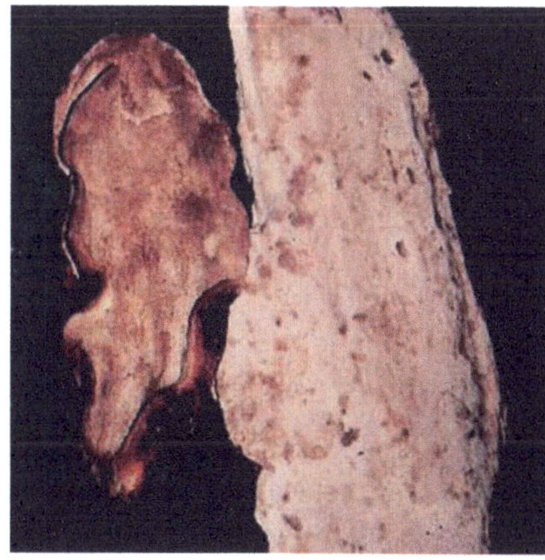

Fig. 17.1 Basidiomata of *Antrodia salmonea*. (Source: Chang and Chou 2004)

(Fig. 17.1). *A. salmonea* grows on the empty rotten trunk of *Cunninghamia konishii* which is endemic to Taiwan, thus indicating *A. salmonea* is endemic as well (Chen et al. 2016). The morphology of *A. salmonea* is similar to that of *A. camphorata* (*T.T. Chang et W.N. Chou*) (known as Chang-Chih in Chinese), which causes brown heart rot of endemic evergreen *Cinnamomum kanehirai* Hayata (Niu-Chang-Chih in Chinese) in Taiwan. The fruiting bodies of both *A. salmonea* and *A. camphorata* taste bitter but are different in color. Like *A. camphorata*, fruiting bodies of *A. salmonea* have been used in the treatment of diarrhea, hypertension, abdominal pain, itchy skin, and liver cancer and also used as a detoxicant in Taiwanese folk medicine (Shen et al. 2008).

According to an estimate by Wang and coworkers, there is a US $100 million market annually for fresh fruiting bodies *A. cinnamomea* in Taiwan, while dry fruiting bodies cost more than US $15000/Kg (Wang et al. 2013). High market value and production constraint of *A. cinnamomea* have opened a new venue for *A. salmonea* as a substitute. It contains many physiologically active components such as polysaccharides, triterpenoids, and adenosine. Chemical investigation has revealed that *A. camphorata* contain triterpenes, steroids, biphenyl compounds, and a sesquiterpene (Chen et al. 1995; Cherng et al. 1995; Cherng et al. 1996; Chiang et al. 1995; Huang et al. 2001; Shen et al. 2003; Shen et al. 1997; Wu and Chiang 1995; Yang et al. 1996). Pharmacological studies of *A. cinnamomea* fungus have shown cytotoxicity against P-388 murine leukemia cells (Chang and Chou 2004) and anti-inflammatory (Hsieh et al. 2007; Wang et al. 2003) and antiviral (Huang et al. 2003) properties. The active compounds isolated from *A. cinnamomea* by Shen and coworkers were found having anti-oxidative effect in human leukocytes (Shen et al. 2006, 2007). The safety level and nontoxic characteristics of the fermented culture broth of *A. salmonea* were evaluated using acute toxicity studies in mice (Shen et al. 2006). In another study, Hseu and coworkers reported the antioxidant property of fermented culture broth, filtrate, and mycelia of *A. salmonea* (Hseu et al. 2014).

17.2 Taxonomical Position of *Antrodia salmonea*

A fungus was collected from *Cunninghamia konishii* tree and named as Shiang-Shan-Chih. It was found similar to another native fungus known as Niu-Chang-Chih (*Antrodia camphorata*). Later based on two morphological features (wood rot and hyphal system) as reported by Dai and Niemelä (2002) and Ryvarden (1991), Shiang-Shan-Chih was renamed as *Antrodia salmonea* (Fig. 17.1) (Chang and Chou 2004; Wu et al. 2004).

Taxonomy

Kingdom	*Fungi*
Division	*Basidiomycota*
Class	*Agaricomycetes*
Order	*Polyporales*
Family	*Fomitopsidaceae*
Genus	*Antrodia*
Species	*Salmonea*

17.3 Metabolic Profiling of *A. salmonea*

A. salmonea is used as an alternative source in place of *A. cinnamomea*, despite the qualitative and quantitative difference of bioactive compounds of the two species. Only a few reports are available on metabolic profiling of *A. salmonea*.

Shen and coworkers identified four compounds, namely, lanosta-8,24-diene-3β,15α,21-triol, 24-methylenelanost-8-ene-3 β,15 α,21-triol, 2,3-dimethoxy-5-(2′,5′-dimethoxy-3′4′-methylenedioxyphenyl)-7-methyl-[1,4]-naphthoquinone, and 2,3-dimethoxy-6(2′,5′-dimethoxy-3′,4′-methylenedioxyphenyl)-7-methyl-[1,4]-naphthoquinone, which proved to possess anti-oxidative activity in human leukocytes (Shen et al. 2006). Later by spectroscopic analysis, three new anti-oxidative ergostanes, (i) methyl anticinate L, (ii) antcin M, and (iii) methyl antcinate K, were isolated and identified as (i) methyl 3α,7α,12α-trihydroxy-4α-methylergosta-8,24(29)-dien-11-on-26-oate, (ii) 3α,12α-dihydroxy-4α-methylergosta-8,24(29)-dien-11-on-26-oic acid, and (iii) methyl 3α,4β,7β-triphydroxy-4-methylergosta-8,24(29)-dien-11-on-26-oate, respectively (Fig. 17.2). These three compounds found exhibited anti-inflammatory activities in neutrophils and microglial cells (Shen et al. 2007).

In another study, Shen et al. (2008) isolated three new compounds, (2,4-dimethoxy-6-methylbenzene-1,3-diol, salmoquinone, and 3-(4-hydroxyphenyl)-4-isobutyl-1H-pyrrole-2,5-dione), along with six known compounds (2-methoxy-6-methyl-p-benzoquinone, 2,3-dimethoxy-5-methyl-p-benzoquinone, 2-hydroxy-5-methoxy-3-methyl-p-benzoquinone, eburcoic acid, fomefficinic acid C, and a pyrrolidone) (Fig. 17.3) from the mycelia of *A. salmonea* using $CHCl_3$ extraction. Further studies on these isolated compounds showed that 2-methoxy-6-methyl-p-benzoquinone and 2,3-dimethoxy-5-methyl-p-benzoquinone exhibited potent anticancer activity against KB, HepG2, and H2058 cell lines in vitro (Shen et al. 2008).

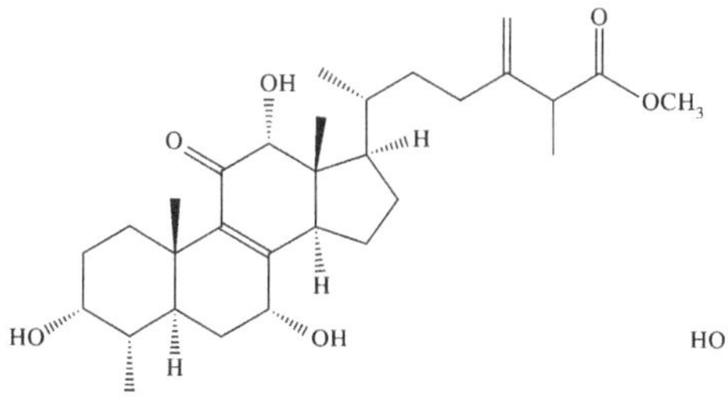

Methyl antcinate L

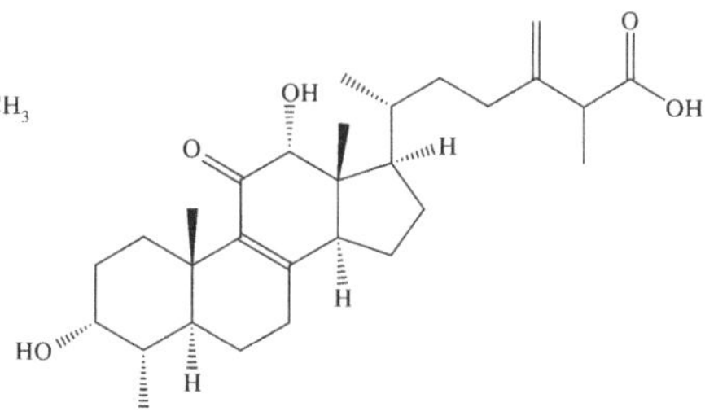

Antcin M

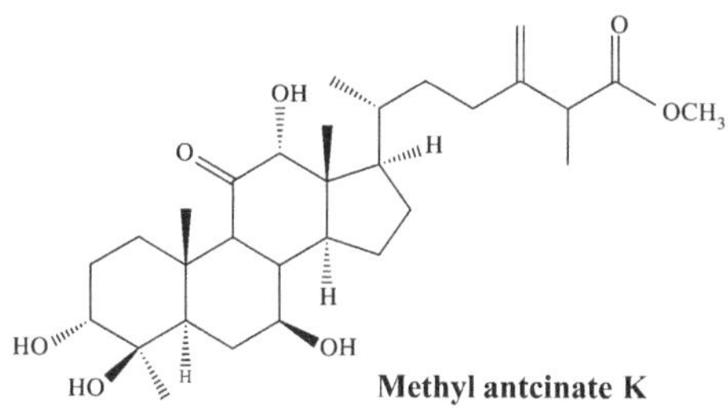

Methyl antcinate K

Fig. 17.2 Ergostanes isolated from *A. salmonea*

17.4 Biologically Active Compounds

Unlike *A. cinnamomea*, there are only a few studies on *A. salmonea*. Chen and coworkers (Chen et al. 2016) successfully isolated, purified, and confirmed the structures of the major triterpenoids, including five ergostane types and two lanostane types (Fig. 17.4). Based on these compounds, they prepared an index for identification and quantification of triterpenoids in *A. cinnamomea* and *A. salmonea*. Also they reported different types of volatile compounds in the fruiting bodies of *A. salmonea* (Chen et al. 2016).

17.5 Nutritional Composition of *A. salmonea*

A. salmonea fruiting bodies and mycelia are used as food additives with high nutritional value. The proximate composition analysis revealed the nutritional values of *A. salmonea* (Table 17.1) (Ulziijargal and Mau 2011).

17.6 Use of *A. salmonea* as a Supplement in Bread Making

Due to the high nutritional value of *A. salmonea*, its mycelia were used in bread making. It was found that when wheat flour (7%) obtained from grain fermented with *A. salmonea* mycelium was used for bread making, it substantially increased its nutritional value with regard to adenosine (0.92–1.96 μg/g), ergosterol

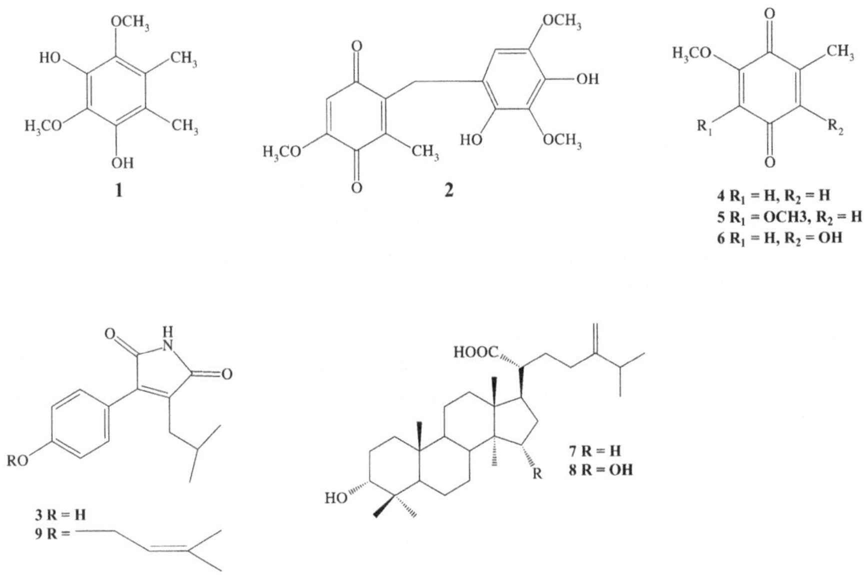

Fig. 17.3 Compounds isolated from the mycelia of *A. salmonea* using $CHCl_3$ extraction
(1) 2,4-Dimethoxy-6-methylbenzene-1,3-diol
(2) Salmoquinone
(3) 3-(4-Hydroxyphenyl)-4-isobutyl-1H-pyrrole-2,5-dione
(4) 2-Methoxy-6-methyl-p-benzoquinone
(5) 2,3-Dimethoxy-5-methyl-p-benzoquinone
(6) 2-Hydroxy-5-methoxy-3-methyl-p-benzoquinone
(7) Eburcoic acid
(8) Fomefficinic acid C
(9) 3-Isobutyl-4-{4-[(3-methyl-2-butenyl)oxy]phenyl}-1H-pyrrole-2,5-dione

(24.53–30.12 μg/g), ergothioneine (2.16–3.18 μg/g), and γ-aminobutyric acid (2.20–2.45 μg/g). Also, it contained lovastatin (0.43 μg/g) (Chien et al. 2016). Thus, fermentation of wheat grain with AS has potential application in the bread making for enhancing the nutritional value. Further research can be carried out for its application in other bakery and food products.

17.7 Pharmacological Activities of *A. salmonea*

Different pharmacological activities of fermented broth, filtrate, mycelia, and isolated bioactive compounds of *A. salmonea* have been depicted in Fig. 17.5.

17.7.1 Antioxidant Activity

Hseu et al. (2014) used different models to examine the antioxidant activity of whole fermented broth, filtrate, and mycelia of *A. salmonea* (AS). Also, the effect of

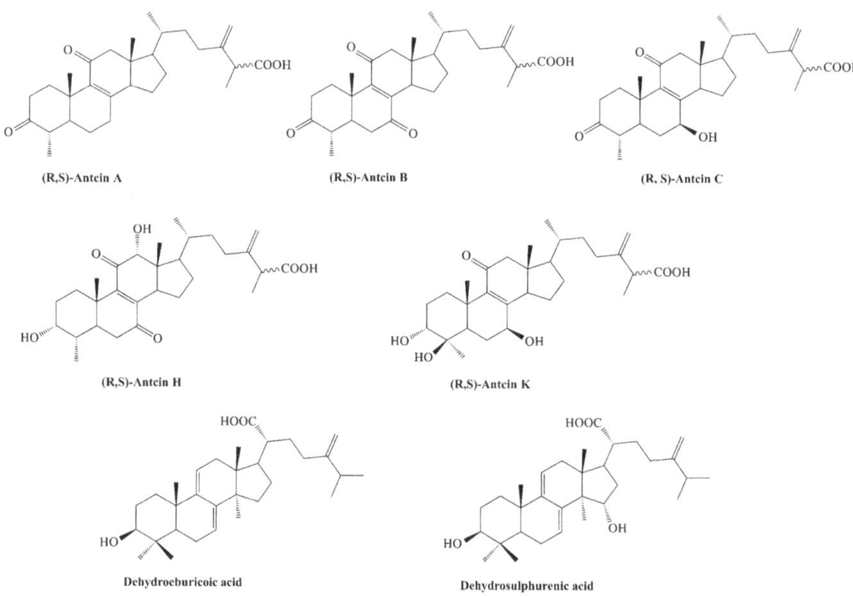

Fig. 17.4 Triterpenoids isolated from the fruiting bodies of *A. salmonea*

Table 17.1 Nutritional composition of dry mycelia of *A. salmonea*

Items	The content of dry mycelia
Ash (%)	5.57
Carbohydrates	
Reducing sugar (%)	37.97
Soluble polysaccharides (%)	4.74
Fiber (%)	3.28
Dietary fiber (%)	8.02
Total (%)	45.99
Protein (%)	45.21
Energy (kcal/100 g)	361.79

the AS on AAPH-induced oxidative hemolysis of human erythrocytes and $CuSO_4$-induced oxidative modification of human low-density lipoproteins was investigated. The fungus showed significant antioxidant activity against different oxidative systems and prevented AAPH-induced oxidative hemolysis in erythrocytes. Among the three substrates, whole fermented broth showed the maximum antioxidant activity (Hseu et al. 2014).

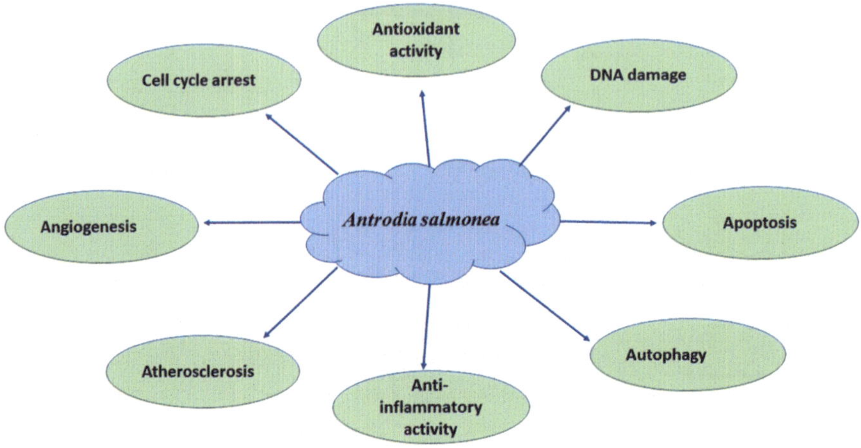

Fig. 17.5 Schematic representation of pharmacological activities of *Antrodia salmonea*

17.7.2 Anti-inflammatory Activity

Huang and coworkers studied the anti-inflammatory effect of ethanolic extract of *A. salmonea* (EAS) and its molecular mechanism in the lipopolysaccharide (LPS)-stimulated RAW246.7 macrophages and the carrageenan (Carr)-induced edema paw model. They reported that the anti-inflammatory mechanism of ethanolic extract was correlated with the downregulation of inflammatory cytokine and upregulation of antioxidant enzymes activities, which resulted in a reduction of iNOS, COX-2, and MDA and subsequent inflammatory response (Huang et al. 2012).

In a separate study, Chiang and coworkers investigated the anti-inflammatory response of ethanolic extracts of mycelia of this fungus on lipopolysaccharide-induced nitric oxide and tumor necrosis factor-α production in RAW 264.7 cells. They found a positive effect on the inhibition of nitric oxide production and tumor necrosis factor-α (Chiang et al. 2013).

Yang and coworkers (Yang et al. 2015) reported the effect of fermented culture broth of *A. salmonea* on the activation of Nrf2-mediated antioxidant genes in RAW264.7 macrophages and resulting protection against lipopolysaccharide (LPS)-stimulated inflammation and also elucidated the molecular mechanism underlying the protection.

17.7.3 Angiogenesis and Atherosclerosis

Yang and coworkers examined the anti-angiogenic and anti-atherosclerotic effects of the fermented culture broth of *A. salmonea* against tumor necrosis factor-α (TNF-α)-stimulated human endothelial (EA.hy 926) cells. Significant inhibition of TNF-α-induced migration/invasion and capillary-like tube formation in EA.hy 926 cells was observed. They further studied the molecular mechanism associated with the effect and demonstrated the usefulness of this fungus in the prevention of angiogenesis and atherosclerosis (Yang et al. 2014).

17.7.4 Antitumor Activity

Chang and coworkers in their in vivo study revealed that *A. salmonea* treatment was effective in delaying the tumor incidence and in reducing the tumor growth in MDA-MB-231-xenografted nude mice. It was found that *A. salmonea* significantly modulated the xenografted tumor progression as demonstrated by the induction of apoptosis, autophagy, and cell-cycle arrest (Chang et al. 2017a). Also, *A. salmonea* played an active role not only in execution/propagation of autophagic or apoptotic death of MDA-MB-231 cells but also in controlling the tumor growth in xenografted nude mice (Chang et al. 2017b).

In a separate study, an ethanolic extract from mycelia of *Taiwanofungus salmoneus* (synonym *Antrodia salmonea*) (TsE) alone or in combination with cisplatin was investigated using SK-Hep-1 cells to examine the apoptotic effect and possible mechanism involved in it. Significant inhibition of proliferation of SK-Hep-1 cells was observed at various concentrations of TsE alone or in combination with cisplatin. The cells apoptosis was attributed to the elevated sub-G1 phase, DNA damage, and activation of caspases 3, 8, and 9 activities (Chien et al. 2015).

17.8 Conclusion

A. salmonea used as a substitute of *A. cinnamomea* has been reported to possess several pharmacological properties. The new bioactive compounds isolated from fruiting bodies and mycelia of *A. salmonea* have shown anti-oxidative and anti-inflammatory activities in activated macrophage cells. However, unlike *A. cinnamomea* the species is meagerly explored and needs further research efforts for harnessing advantages of its full potential.

References

Borchers AT, Stern JS, Hackman RM, Keen CL, Gershwin ME (1999) Mushrooms, tumors, and immunity. Proc Soc Exp Biol Med 221:281–293. https://doi.org/10.1046/j.1525-1373.1999.d01-86.x

Chang TT, Chou WN (2004) *Antrodia cinnamomea* reconsidered and *A. salmonea* sp. nov. on *Cunninghamia konishii* in Taiwan. Bot Bull Acad Sin 45:347–352

Chang CT, Hseu YC, Thiyagarajan V et al (2017a) *Antrodia salmonea* induces G2 cell-cycle arrest in human triple-negative breast cancer (MDA-MB-231) cells and suppresses tumor growth in athymic nude mice. J Ethnopharmacol 196:9–19. https://doi.org/10.1016/j.jep.2016.12.018

Chang CT, Korivi M, Huang HC et al (2017b) Inhibition of ROS production, autophagy or apoptosis signaling reversed the anticancer properties of *Antrodia salmonea* in triple-negative breast cancer (MDA-MB-231) cells. Food Chem Toxicol 103:1–17. https://doi.org/10.1016/j.fct.2017.02.019

Chen CH, Yang SW, Shen YC (1995) New steroid acids from *Antrodia cinnamomea*, a fungal parasite of *Cinnamomum micranthum*. J Nat Prod 58:1655–1661

Chen CY, Chien SC, Tsao NW et al (2016) Metabolite profiling and comparison of bioactivity in *Antrodia cinnamomea* and *Antrodia salmonea* fruiting bodies. Planta Med 82:244–249. https://doi.org/10.1055/s-0035-1558141

Cherng IH, Chiang HC, Cheng MC et al (1995) Three new triterpenoids from *Antrodia cinnamomea*. J Nat Prod 58:365–371

Cherng IH, Wu DP, Chiang HC (1996) Triterpenoids from *Antrodia cinnamomea*. Phytochemistry 41:263–267

Chiang HC, Wu DP, Cherng IW Ueng CH (1995) A sesquiterpene lactone, phenyl and biphenyl compounds from *Antrodia cinnamomea*. Phytochemistry 39:613–616

Chiang SS, Wang LT, Chen SY, Maul JL (2013) Antibacterial and anti-inflammatory activities of mycelia of a medicinal mushroom from Taiwan, *Taiwanofungus salmoneus* (higher Basidiomycetes). Int J Med Mushrooms 15(1):39–47. https://doi.org/10.1615/IntJMedMushr.v15.i1.50

Chien RC, Hsieh YJ, Mau JL (2015) Apoptotic effect of extract from medicinal mushroom from Taiwan *Taiwanofungus salmoneus* (higher Basidiomycetes) mycelium combined with or without cisplatin on hepatocellular carcinoma cells. Int J Med Mushrooms 17(6):567–577. https://doi.org/10.1615/IntJMedMushrooms.v17.i6.70

Chien RC, Ulziijargal E, Mau JL (2016) Quality of bread supplemented with *Antrodia salmonea*-fermented grains. Food Technol Biotechnol 54:180–188. https://doi.org/10.17113/ftb.54.02.16.4336

Dai YC, Niemelä T (2002) Changbai wood-rotting fungi 13. *Antrodia* sensu lato. Ann Bot Fenn JSTOR 39:257–265

Hseu YC, Lee CC, Chen YC et al (2014) *Antrodia salmonea* in submerged culture exhibits antioxidant activities *in vitro* and protects human erythrocytes and low-density lipoproteins from oxidative modification. Food Chem Toxicol 66:150–157. https://doi.org/10.1016/j.fct.2014.01.035

Hsieh YH, Kuo PM, Chien SC (2007) Effects of Chamaecyparis formosensis Matasumura extractives on the lipopolysaccharide-induced release of nitric oxide. Phytomedicine 14:675–680. https://doi.org/10.1016/j.phymed.2006.11.029

Huang KF, Huang WM, Chiang HC (2001) Phenyl compounds from *Antrodia cinnamomea*. Chin Pharm J 53:327–331

Huang RL, Huang Q, Chen CF (2003) Anti-viral effects of active compounds from *Antrodia camphorata* on wild-type and lamivudine-resistant mutant HBV. Chin Pharm J 55:371–379

Huang GJ, Pan CH, Liu FC, Wu TS, Wu CH (2012) Anti-inflammatory effects of ethanolic extract of *Antrodia salmonea* in the lipopolysaccharide-stimulated RAW246. 7 macrophages and the λ-carrageenan-induced paw edema model. Food Chem Toxicol 50:1485–1493. https://doi.org/10.1016/j.fct.2012.01.041

Lorenzen K, Anke T (1998) Basidiomycetes as a source for new bioactive natural products. Curr Org Chem 2:329–364

Mizuno T (1999a) The extraction and development of antitumor-active polysaccharides from medicinal mushrooms in Japan. Int J Med Mushrooms 1:9–29. https://doi.org/10.1615/IntJMedMushrooms.v1.i1.20

Mizuno T (1999b) Bioactive substances in *Hericium erinaceus* (Bull.: Fr.) Pers. (Yamabushitake), and its medicinal utilization. Int J Med Mushrooms 1:105–119. https://doi.org/10.1615/IntJMedMushrooms.v1.i2.10

Mizuno T (2002) Medicinal properties and clinical effects on *Agaricus blazei* Murr. Int J Med Mushrooms 4:4. https://doi.org/10.1615/IntJMedMushr.v4.i4.30

Mizuno T, Yeohlui P, Kinoshita T, Zhuang C, Ito H, Mayuzumi Y (1996) Antitumor activity and chemical modification of polysaccharides from Niohshimeji mushroom, *Tricholoma giganteum*. Biosci Biotechnol Biochem 60:30–33. https://doi.org/10.1271/bbb.60.30

Mizuno T, Zhuang C, Abe K et al (1999) Antitumor and hypoglycemic activities of polysaccharides from the sclerotia and mycelia of *Inonotus obliquus* (Pers.: Fr.) Pil.(Aphyllophoromycetideae). Int J Med Mushrooms 1:301–316. https://doi.org/10.1615/IntJMedMushr.v1.i4.20
Money NP (2016) Are mushrooms medicinal? Fungal Biol 120:449–453. https://doi.org/10.1016/j.funbio.2016.01.006
Ooi VEC, Liu F (1999) A review of pharmacological activities of mushroom polysaccharides. Int J Med Mushrooms 1:195–206. https://doi.org/10.1615/IntJMedMushrooms.v1.i3.10
Reshetnikov SV, Wasser SP, Tan KK (2001) Higher Basidiomycota as source of antitumor and immunostimulating polysaccharides. Int J Med Mushrooms 3:361–394. https://doi.org/10.1615/IntJMedMushr.v3.i4.80
Ryvarden L (1991) Genera of polypores: nomenclature and taxonomy, Synopsis fungorum Series 5. Fungiflora AS, Oslo
Shen YC, Yang SW, Lin CS et al (1997) Zhankuic acid F: a new metabolite from a Formosan fungus *Antrodia cinnamomea*. Planta Med 63:86–88. https://doi.org/10.1055/s-2006-957614
Shen CC, Kuo YC, Huang RL et al (2003) New ergostane and lanostane from *Antrodia camphorata*. J Chin Med 14:247–258
Shen CC, Shen YC, Wang YH et al (2006) New lanostanes and naphthoquinones isolated from *Antrodia salmonea* and their antioxidative burst activity in human leukocytes. Planta Med 72:199–203. https://doi.org/10.1055/s-2005-916175
Shen CC, Wang YW, Chang TS et al (2007) Anti-inflammatory ergostanes from the basidiomata of *Antrodia salmonea*. Planta Med 73:1208–1213. https://doi.org/10.1055/s-2007-981591
Shen CC, Lin CF, Huang YL et al (2008) Bioactive components from the mycelium of *Antrodia salmonea*. J Chin Chem Soc 55:854–857. https://doi.org/10.1002/jccs.200800127
Tzianabos AO (2000) Polysaccharide immunomodulators as therapeutic agents: structural aspects and biological function. Clin Microbiol Rev 13:523–533
Ulziijargal E, Mau JL (2011) Nutrient compositions of culinary-medicinal mushroom fruiting bodies and mycelia. Int J Med Mushrooms 13(4):343–349. https://doi.org/10.1615/IntJMedMushr.v13.i4.40
Wang SY, Chang HN, Lin KT et al (2003) Antioxidant properties and phytochemical characteristics of extracts from *Lactuca indica*. J Agric Food Chem 51:1506–1512. https://doi.org/10.1021/jf0259415
Wang HC, Chu FH, Chien SC et al (2013) Establishment of the metabolite profile for an *Antrodia cinnamomea* health food product and investigation of its chemoprevention activity. J Agric Food Chem 61:8556–8564. https://doi.org/10.1021/jf402849b
Wasser SP, Weis AL (1999) Medicinal properties of substances occurring in higher basidiomycetes mushrooms: current perspectives. Int J Med Mushrooms 1:31–62. https://doi.org/10.1615/IntJMedMushrooms.v1.i1.30
Wu DP, Chiang HC (1995) Constituents of *Antrodia cinnamomea*. J Chin Chem Soc 42:797–800
Wu SH, Yu ZH, Dai YC et al (2004) *Taiwanofungus*, a polypore new genus. Fungal Sci 19:109–116
Yang SW, Shen YC, Chen CH (1996) Steroids and triterpenoids of *Antrodia cinnamomea* – a fungus parasitic on *Cinnamomum micranthum*. Phytochemistry 41:1389–1392. https://doi.org/10.1016/0031-9422(95)00767-9
Yang HL, Chang HC, Lin SW et al (2014) Antrodia salmonea inhibits TNF-α-induced angiogenesis and atherogenesis in human endothelial cells through the down-regulation of NF-κB and up-regulation of Nrf2 signaling pathways. J Ethnopharmacol 151:394–406. https://doi.org/10.1016/j.jep.2013.10.052
Yang HL, Lin SW, Lee CC et al (2015) Induction of Nrf2-mediated genes by *Antrodia salmonea* inhibits ROS generation and inflammatory effects in lipopolysaccharide-stimulated RAW264.7 macrophages. Food Funct 6:229–240. https://doi.org/10.1039/c4fo00869c